D0164245

Ecological Studies

Analysis and Synthesis

Edited by

J. Jacobs, München · O. L. Lange, Würzburg

J. S. Olson, Oak Ridge · W. Wieser, Innsbruck

Volume 5

Arid Zone Irrigation

Edited by

B. Yaron, E. Danfors and Y. Vaadia

With 181 Figures

Springer-Verlag Berlin · Heidelberg · New York 1973

ISBN 3-540-06206-8 Springer-Verlag Berlin · Heidelberg · New York
ISBN 0-387-06206-8 Springer-Verlag New York · Heidelberg · Berlin
ISBN 0-412-12570-6 Chapman & Hall Limited London

Typesetting, printing and bookbinding: Georg Appl, Wemding.

Preface

A book previously published within the framework of the Ecological Studies Series, entitled "Physical Aspects of Soil Water and Salts in Ecosystems" included a wide spectrum of research papers devoted to new findings in the field of soil-plant-water relationships. *"Arid Zone Irrigation"* has been written specifically as a textbook for agronomists, soil scientists, agrometeorologists, water engineers and plant physiologists who want a clear presentation of irrigation fundamentals in arid and semi-arid zones. It was our intention to provide an understanding of the basic principles governing irrigation technology and to help overcome the problem of water shortage in arid zone agriculture.

This book, written by a large number of specialists and covering a broad spectrum of different disciplines, is based on general up-to-date information, as well as on the results of the authors' own research.

The idea of preparing such a textbook was conceived during a series of international advanced courses on irrigation held annually at the Institute of Soils and Water, Agricultural Research Organization, Volcani Center, Bet Dagan, Israel. The final organization of the material has been influenced by discussions with colleagues from Sweden and Holland and the participants in our summer courses.

Grateful acknowledgements are due to Professor CALVIN C. ROSE, CSIRO, Canberra, Australia, Professor DALE SWARTZENDRUBER, Purdue University, Lafayetta, U.S.A., and Dr. SHLOMO P. NEUMAN, Agricultural Research Organization, Bet Dagan, Israel, for their many helpful suggestions during critical reading of the manuscript. We thank also Mrs. ETTA SHUR for her dedicated work in typing the manuscript, and Mrs. NIHA GESTETNER for her meticulous drawing of the figures.

Bet Dagan, October 1973

BRUNO YARON
ERIK DANFORS
YOASH VAADIA

Contents

I. Arid Zone Environment

II. Water Resources

III. Water Transport in Soil-Plant-Atmosphere Continuum

IV. Chemistry of Irrigated Soils – Theory and Application

V. Measurements for Irrigation Design and Control

VI. Salinity and Irrigation

VII. Irrigation Technology

VIII. Crop Water Requirement

Contributors

ASSAF, RAFAEL, Institute of Horticulture, Agricultural Research Organization, Newe Ya'ar Experimental Station, Israel

BIELORAI, HANOCH, Institute of Soils and Water, Agricultural Research Organization, Bet Dagan, Israel

BRESLER, ESHEL, Institute of Soils and Water, Agricultural Research Organization, Bet Dagan, Israel

BURAS, NATHAN, Faculty of Agricultural Engineering, Israel Institute of Technology, The Technion, Haifa, Israel

DAN, JOEL, Institute of Soils and Water, Agricultural Research Organization, Bet Dagan, Israel

DANFORS, ERIK, Department of Land Improvement and Drainage, Royal Institute of Technology, Stockholm Sweden

FUCHS, MARCEL, Institute of Soils and Water, Agricultural Research Organization, Bet Dagan, Israel

GAIRON, SCHABTAI, National Council for Research and Development, Jerusalem, Israel

HADAS, AMOS, Institute of Soils and Water, Agricultural Research Organization, Bet Dagan, Israel

HELLER, JOSEPH, Extension Service, Ministry of Agriculture, Tel Aviv, Israel

KAFKAFI, UZI, Institute of Soils and Water, Agricultural Research Organization, Bet Dagan, Israel

LEVIN, ISRAEL, Institute of Soils and Water, Agricultural Research Organization, Bet Dagan, Israel

LEVY, RACHEL, Institute of Soils and Water, Agric. Res. Organization, Bet Dagan, Israel

MANDEL, SHMUEL, Institute of Groundwater, Hebrew University, Jerusalem, Israel

MANTELL, ALVIN, Institute of Soils and Water, Agricultural Research Organization, Bet Dagan, Israel

MEIRI, AVRAHAM, Institute of Soils and Water, Agricultural Research Organization, Bet Dagan, Israel

MORESHET, SHMUEL, Institute of Soils and Water, Agricultural Research Organization, Bet Dagan, Israel

PLAUT, ZVI, Institute of Soils and Water, Agricultural Research Organization, Bet Dagan, Israel

RAWITZ, ERNEST, Faculty of Agriculture, Hebrew University, Rehovot, Israel

SHALHEVET, JOSEPH, Institute of Soils and Water, Agricultural Research Organization, Bet Dagan, Israel

SHAINBERG, ISAAC, Institute of Soils and Water, Agricultural Research Organization, Bet Dagan, Israel

SHMUELI, ELIEZER, Institute of Soils and Water, Agricultural Research Organization, Bet Dagan, Israel

SHIMSHI, DANIEL, Institute of Soils and Water, Agricultural Research Organization, Gilat Experimental Station, Israel

VAADIA, YOASH, Agricultural Research Organization, Bet Dagan, Israel

VINK, ANTHONY, P. A. Laboratory of Physical Geography and Soil Science, University of Amsterdam, Amsterdam, Holland

YARON, BRUNO, Institute of Soils and Water, Agricultural Research Organization, Bet Dagan, Israel

I. Arid Zone Environment

Climate and soil are the main environmental factors to be considered when dealing with irrigation development under arid and semiarid conditions.

The reader will readily understand why dry areas need irrigation: they completely lack precipitation, or have an unequal distribution of precipitation during the year. The main characteristics of arid and semiarid areas are long, hot, nonrainy summers, with precipitation varying between 20 and 1000 mm and concentrated in a short, rainy period during the winter season.

An understanding of the soil-forming processes under arid conditions as a function of environmental factors is necessary in order to handle irrigation design and management successfully. Thus, rather detailed information on these processes are presented before proceeding to a discussion of the soil distribution in arid-zone landscapes. Here, the soils of a broad area are considered beginning with the typical desert soils of the extremely arid zone, through the soils of subhumid areas, which are characterized by alternating wet and dry seasons. In the soil classifications, both the American (seventh approximation) and FAO definitions are used.

From this chapter the student who is unfamiliar with arid- and semiarid-zone environments should given some understanding of the conditions that provide a natural setting for irrigated agriculture.

1

Climate and Irrigation

M. FUCHS

General Considerations

Irrigation is an agricultural practice designed to supplement a deficiency of climate: the imbalance between the water supplied by precipitation and the evaporative demands of the atmosphere. During a period of rainfall the soil stores a certain quantity of the supplied water. When the rain ceases, evapotranspiration reduces the stored water. If the rate of depletion by evapotranspiration is such that the water status of the soil reaches a critical level for plant growth, the need for irrigation arises. In this context it is obvious that the impact of climate on the water balance of crops, and thus on irrigation, depends upon the physical properties of the soil and the water physiology of the crop. Nevertheless, it is possible to define, in broad terms, the dependence of irrigation requirements upon the type of climate.

Precipitation and evaporation are antagonistic processes. Abundant rainfall is associated with high content of water vapor in the air, cloudy skies, which reduce solar radiation, and low temperatures. The evaporation rate is enhanced by the opposite conditions, that is, by low air humidity, high solar radiation, and high air temperature.

Various methods have been devised for combining precipitation and evaporation data into integrated indices that can characterize different types of climates.

Indices as proposed by THORNTHWAITE (1948) and KÖPPEN (1936) have been used to classify and map world climates in great detail, according to their degree of aridity. These indices divide the climates into a large number of precisely defined categories. However, most of the boundaries are set rather arbitrarily, and consequently the information provided is qualitative only and of little help to the irrigationist.

Based on a consideration precipitation versus evaporation, we can divide the climates of the world into four broad categories that have a direct bearing on the management of irrigation: humid, subhumid, semiarid, and arid. Within each of these categories there is, of course, a wide spectrum of different climatic patterns. However, we shall limit our description to those features directly related to irrigation.

Whereas precipitation patterns are well documented and are mapped on a worldwide scale, data on evaporation are scarce. A crude world map by BUDYKO (1963) is the only extensive study of evaporation that has been made. This lack of information is attributable to the fact that the evaporation rate depends upon a large number of meteorological factors, many of which are difficult to measure. Since solar radiation is a basic determining factor of evaporation, we shall use this parameter to characterize the evaporativity of the climates described.

The Arid Climates

When there is no season during which crops can be raised without irrigation, the climate is defined as arid. The limit of 250 mm mean annual rainfall is generally used to characterize this type of climate. However, the rains are infrequent and their occurrence completely unpredictable. In many desert areas several years may elapse between rainfalls. When rains do occur, they are often of high intensity and cause devastating floods.

An arid climate is generally favorable to very high evaporation rates, as solar radiation values are very high and the humidity of the air is low. A notable exception is the desert of Rjasthan, India, where the relative humidity is high and where the high dust content of the atmosphere reduces the intensity of solar radiation.

Table 1. Distribution of average monthly precipitation and air temperature (after TREWARTHA, 1954), and yearly solar radiation (after BUDYKO, 1958) of arid stations

Month	Aswan, Egypt		Astrakhan, U.S.S.R.		Alice Springs, Australia	
	T (° C)	P (mm)	T (° C)	P (mm)	T (° C)	P (mm)
Jan.	15	0	–7	13	29	46
Feb.	17	0	–6	8	28	43
March	21	0	0	10	25	33
April	26	0	9	13	20	23
May	29	0	18	18	16	15
June	32	0	23	18	12	15
July	33	0	25	13	11	10
Aug.	32	0	23	13	14	10
Sept.	31	0	17	13	19	10
Oct.	28	0	10	10	23	18
Nov.	22	0	3	10	27	23
Dez.	17	0	–3	13	28	33
Average	25		9.3		21	
Total		0		152		279
Solar radiation Kcal cm^{-2} yr^{-1}	220		130		190	

Table 1 summarizes temperature, rainfall, and solar radiation data for three examples of arid climates. The Aswan climate is typical of a tropical desert. There is practically no rainfall. Temperatures throughout the year are high, but summers are torrid. Average solar radiation values are among the highest recorded in the world. The station at Astrakhan, in the delta of the Volga near the Caspian Sea, characterizes continental, mid-latitude, arid steppes. The slightly higher rainfall of Alice Springs, another arid steppe, in the center of Australia, is offset by the strong evaporative demand of the atmosphere, as indicated by the intensity of solar radiation and the high air temperatures.

The introduction or irrigation permits the development of agriculture. However, because of the very low rainfall, the local groundwater resources suffice only for the irrigation needs of small areas. Where groundwater is readily available, oases are created. Otherwise the water economy of irrigated areas depends upon remote catchment areas and involves the transport of water over distances that can be considerable; for example, the National Water Carrier system in Israel.

The Semiarid Climates

Semiarid regions have characteristics even more variable than those of arid regions, and thus it is difficult to define a single climatological group encompassing all such regions. We shall label as semiarid all those regions in which the rainfall deficiency calls for crop irrigation during part of the growing season. Here, dry farming productivity is low and is restricted to a few drought-resistant species.

Several climatological entities can be included under this pragmatic definition: the tropical savanna climate, the subtropical Mediterranean climate, and the steppe climate. These three climates show considerable differences in total precipitation as well as in annual rainfall distribution. As indicated by Fig. 1, the relation between the annual trends of air temperature and precipitation are also different.

a) The *Savanna climate,* which prevails in the largest part of the intertropical zone, is warm throughout the year, with the highest temperature occurring during the rainy season. During the dry season a recession of vegetative growth occurs. Perennial species shed their leaves and annuals disappear. However, the temperature conditions can accommodate a year-round growing season for most tropical and mesothermal crops. The benefits of irrigation here are the lengthening of the potential growing season and the possibility of diversifying crop rotation.

b) By contrast, the *Mediterranean climate* has a mild but well – defined winter during which maximum rainfall occurs. The summer is typically hot and rainless.

The Mediterranean climate does not cover very large areas, but is scattered around the world at the fringes between the tropical and the temperate zones. It is found along the shores of the Mediterranean Sea, in South Africa, in southern Australia, in California, and in Chile.

The natural vegetation is chiefly composed of small evergreen trees and shrubs, which are well adapted to the dry environment. Nonirrigated agriculture is based on winter crops such as wheat, which is sown after the first rains and harvested in the late spring. Unirrigated summer crops are limited to a few species that have very short growing seasons, or to plants provided with powerful root systems that can use moisture stored in the deep layers of the soil and in the bedrock crevasses.

The amount of rainfall varies from year to year, and its occurrence is unpredictable. Consequently, the risk of crop failure and unsatisfactory yields is high. The introduction of irrigation not only eliminates most of this risk, but makes it possible to grow a large number of crops during the summer months. It also permits the establishment of perennial crops, such as fruit trees, which otherwise would not produce. These irrigated crops are very productive because the high solar radiation of the summer months brings about favorable conditions for photosynthesis, provided a suitable water balance is maintained.

c) The *steppe climate* occupies larger land areas than any other climatic category. It is found in the central and western part of North America, in Central Asia, in India, in the Middle East, and in North Africa. The largest steppe areas of the Southern Hemisphere are located in Australia, South Africa, and Argentina. Whereas the savanna and the Mediterranean climates are homogeneous entities, the steppe climate presents many variations. It forms the transition climate between the desert climate and either the savanna, the Mediterranean, or the temperate subhumid climate. In each case the transition climate is a drier version of one of these three climatic types.

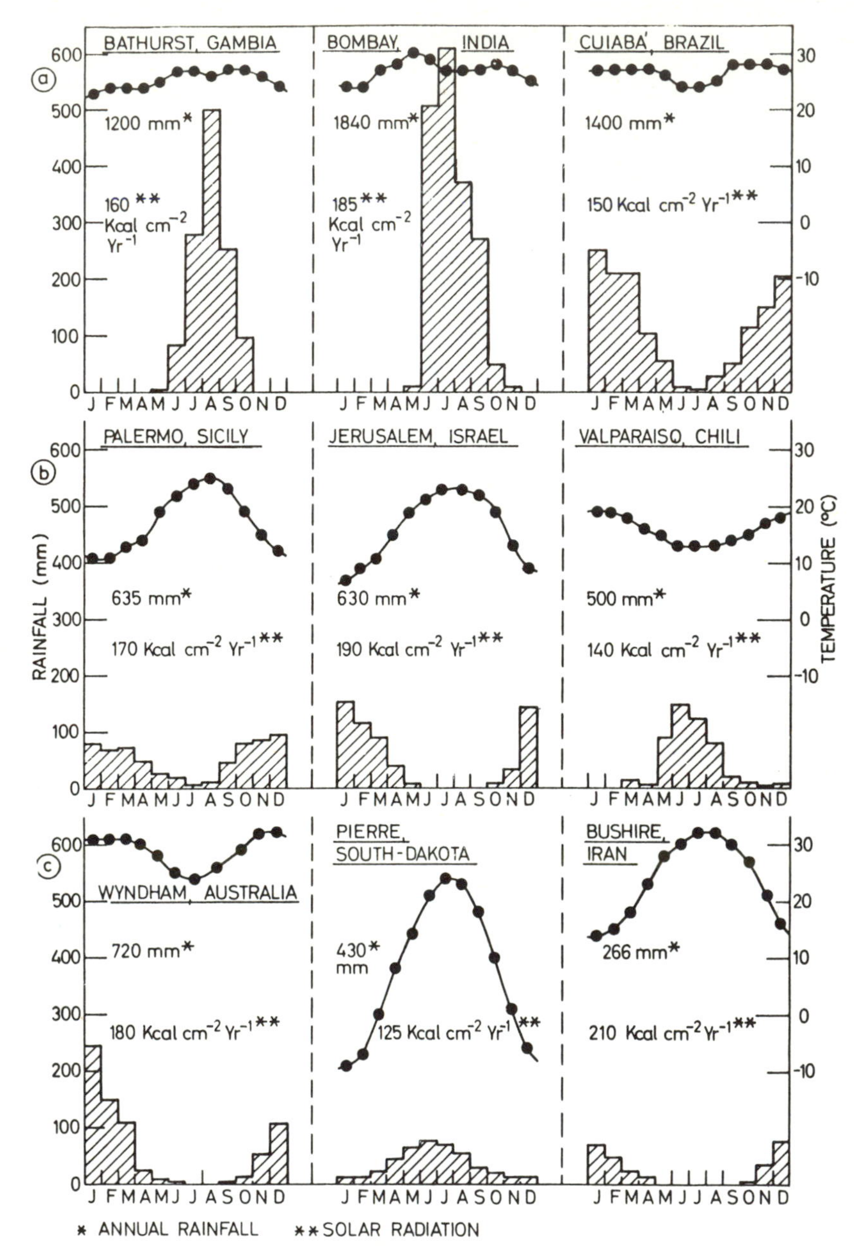

Fig. 1. Monthyl ombrothermic diagram for various types of semiarid climates (after TREWARTHA, 1954) and mean annual solar radiation data. (After BUDYKO, 1958)

The common feature of all steppe climates is a rainfall deficiency associated with a high evaporative demand. As a result, intensive agriculture without irrigation is generally not possible. However, the steppe climate is well suited for most grain crops. The wettest variants of the steppe climate are in fact the world's largest wheat producers. Yields per unit area decrease and crop failures are more frequent as aridity incrases. Open grassland is the typical landscape and free-range cattle raising the most common agricultural land use. As with the Mediterranean climate, the high solar radiation is favorable to high productivity, and therefore large irrigation schemes transform the steppe into very rich agricultural land.

Subhumid Climates

Subhumid climates are widespread throughout both hemispheres. They are found at the fringes of the equatorial rain forest. The monsoon climates of Central America, Southeast Asia, and India are in this group, as are large areas of the North American Midwest, Central Europe, the U.S.S.R., and China.

Table 2. Distribution of average monthly precipitation (after TREWARTHA, 1954) and yearly solar radiation (after BUDYKO, 1958) of various subhumid climates

Month	Precipitation (mm)						
	Manaos, Brazil	Saigon, Vietnam	Kiev, Ukraine	Omaha, Neb.	Mukden, Manchuria	Uppsala, Sweden	Bucharest, Rumania
Jan.	211	23	28	18	5	33	30
Feb.	203	4	20	23	8	28	28
March	206	8	38	33	22	30	43
April	213	43	43	71	28	30	51
May	167	205	43	104	56	43	63
June	99	320	61	119	86	51	84
July	46	281	76	102	160	69	71
Aug.	33	279	61	81	152	71	48
Sept.	35	328	43	76	84	51	38
Oct.	117	281	43	58	46	53	38
Nov.	114	94	38	28	25	43	48
Dec.	208	79	38	23	5	41	43
Yearly total	1.652	1.955	532	736	677	543	585
Yearly incoming solar radiation Kcal cm^{-2} yr^{-1}	150	140	110	110	110	80	130

The precipitation patterns of the subhumid climates differ little from those of the humid climates. However, as shown in Table 2, total rainfall is slightly lower and the annual distribution is more variable. Solar radiation values are generally slightly higher, indicating that evaporation may be considerable.

In many cases there is a short, but marked, dry period with lower rainfall. In other cases, although the long-range average rainfall is high throughout the year, there are frequent drought periods. During these dry periods, the evaporative demand of the atmosphere is generally high. Consequently, the water balance of the crops can be upset. If this occurs during critical stages of crop development, irrigation becomes a necessity. On the other hand, because of the large year-to-year variations of the rainfall patterns, it is often difficult to optimize the design and management of irrigation systems. Large-scale irrigation schemes in these areas are usually avoided by planting drought-resistant species and varieties. Irrigation is generally restricted to high-value crops and to those associated with processing plants, such as vegetable crops for the canning industry.

Humid Climates

Humid climates are those for which precipitation equals or exceeds the evaporative demand of the atmosphere for most periods of the year. In principle, irrigation is unnecessary here. However, sandy soils with very low water-holding capacity may by nonproductive without its help. Irrigation also reduces the risk of failure of sensitive garden crops and plant nurseries.

On the world map humid climates have a wide zonal distribution. Warm, humid climates are found in the 10° to 20° wide latitudinal strip centered on the Equator. The rain-forest areas of central Africa and northern Brazil are typical examples. Colder humid climates are widespread in the areas under oceanic influence in North and South America, Europe, and Asia.

The characteristic these climates have in common is heavy precipitation, mainly rain in the equatorial zone, but also snow and fog in the oceanic areas at high latitudes. Rainfall is evenly distributed throughout the year, and the year-to-year variations are usually small. TREWARTHA (1954) indicates that droughts may occasionally occur, but they rarely endanger the crops.

The large percentage of cloud cover reduces incident solar radiation and decreases evaporation, thus maintaining a favorable water balance for the vegetation, even if the total annual rainfall is low. However, the resulting low temperatures and low light levels may adversely affect crop yields.

Conclusion

Irrigation practices have to satisfy a set of conditions and requirements imposed by the climate. The brief review of the world's climates presented here suggests that a broad-scale assessment of the relationship between climate and irrigation will provide guidelines for the planning of agricultural policy. For example, on the basis of the description of the Mediterranean climate, the crop rotation and the design of the irrigation system should take into account the fact that the climate requires the supply of irrigation water to peak during the summer months. The irrigation system should be able to meet the maximum requirement, but the crop rotation should be planned to alleviate a concentrated demand on the water supply.

However, in actual field applications, the irrigationist also needs precise quantitative information on irrigation dosage and scheduling. The detailed investigation of the atmospheric and soil environment of the crops provides some of the required data. The reader will be introduced to these methods in Chaps. III and V.

Literature

BUDYKO, M. I.: The Heat Balance of the Earth's Surface, 259 pp. Washington, D. C.: Office of Technical Services. U.S. Dept. of Commerce 1958.

BUDYKO, M. I.: Evaporation under Natural Conditions. Washington, D.C.: Office of Technical Services. U.S. Dept. of Commerce 1963.

KÖPPEN, W.: Das geographische System der Klimate. In: KÖPPEN, W., GEIGER, R. (Eds.): Handbuch der Klimatologie, vol. I, part C. Berlin: Gebrüder Borntraeger 1936.

THORNTHWAITE, C. W.: An approach toward a rational classification of climate. Geo. Rev. 38, 55–94 (1948).

TREWARTHA, G. T.: An Introduction to Climate, 402 pp. New York–Toronto–London: McGraw-Hill 1954.

2

Arid-Zone Soils

J. DAN

The soil (pedosphere) is the active transition zone between the lithosphere and the atmosphere where the various processes responsible for the transformation of matter and energy occur. The same may be said for plant and animal life (biosphere), so that sometimes both may actually comprise one active unit (the ecosystem, according to JENNY, 1961).

The Soil Profile

Soil horizons develop as a result of weathering. Pedogenesis refers to the original geologic deposit as the parent material, which underlies the soil profile and is referred to as the C-horizon. A very young soil, especially in arid regions, may contain only a physically disintegrated horizon above the parent material (C-horizon). Later, some accumulation of organic matter in the topsoil initiates the development of an upper A-horizon. This horizon may acquire a good granular or subangular blocky structure due to a considerable concentration of saturated organic material. In these cases it is designated as a melanic* or mollic** A-horizon. In dry areas, however, there may be only a limited formation or organic material, thus forming a poorly structured pallid* or ochric** A-horizon. As a result of chemical weathering (a relatively slow process), clay is formed and the B-horizon between the A- and C-horizons develops. The B-horizon is often colored by iron compounds. It is designated as cambic B if it differs from the upper horizon mainly in structure and usually also in color. However, clay movement may occur from the A- to the B-horizon, whereby an argillic** or argilluvic* B-Horizon is established. The physical disintegration of the rock continues and slowly attacks the parent material deeper in the soil, so that the C-horizon is found at greater depths. In many soils, especially in arid regions, the soluble salts, gypsum or lime, reprecipitate at some depth in the C- or B-horizons. These horizons are designated as ca (for the lime-enriched horizons), cs (for gypsum-enriched horizons), and sa (for salt-enriched horizons). These may be indurated to form a petrocalcic** horizon (in the case of indurated ca horizons) or hard gypsum or even saline pans. There are also other secondary horizons in the soil, such as a g (gley) horizon, etc. A schematic soil profile for arid regions is presented in Fig. 1.

* Terms herein followed by one asterisk refer to the FAO definitions (DUDAL, 1968).

** Terms followed by two asterisks refer to the American definition (Soil Survey Staff, 1960, 1967).

Soil-forming Factors

Soil formation is a result of various processes that are induced by soil-forming factors. Five soil-forming factors—parent material, climate, living material, time, and relief—have been distinguished (JENNY, 1941). The parent material may include different hard and soft rocks, or even various unconsolidated sediments. These rocks or sediments comprise the raw material for soil formation. The parent material is acted upon by the

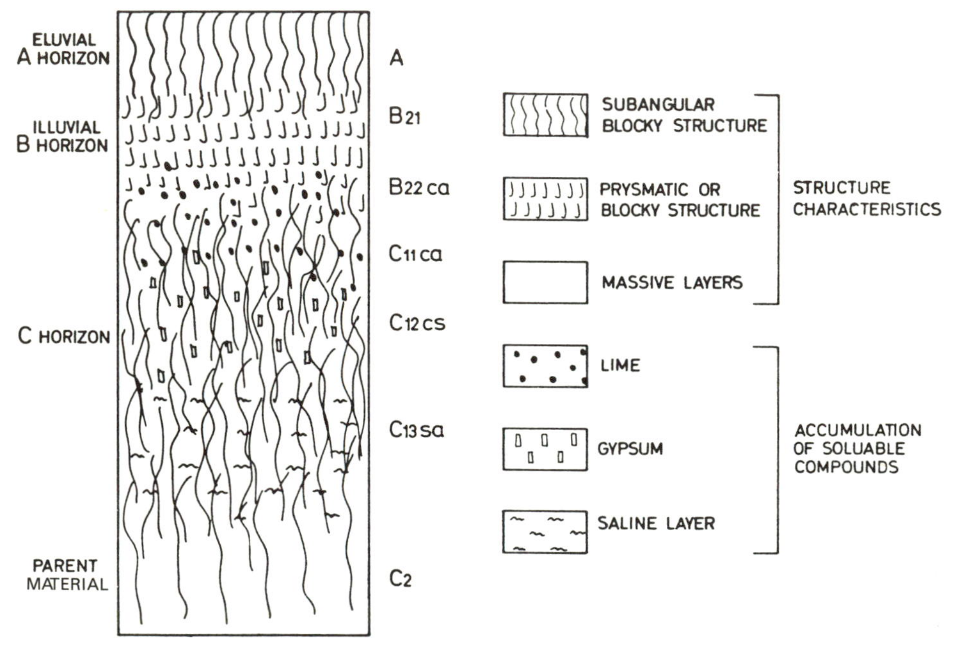

Fig. 1. Schematic soil profile in arid zones

climate and vegetation, which change its original constituents according to the intensity of these factors. Time is also an important soil-forming factor, since it determines the extent of activity of the other factors and therefore the degree of maturity of the profile development. Finally, relief is important because it modifies the effects of climate (produces different microclimates) and affects such processes as erosion, sedimentation, and drainage.

The Soil-forming Processes

The soil-forming processes are induced by a certain combination of the five soil-forming factors. These processes, like the aforementioned factors, are interdependent in their features, and sometimes it is difficult to differentiate between them. They may be physical, chemical, or biological in nature, but some, like the disintegration of rocks, can be all three. We may differentiate between quick and slow processes. The former need only a few hundred years in which to leave their mark on the soils, so that even the young soils are affected. The slow processes require several thousands of years in order to affect the soils and to bring them to a mature stage. The chemical processes are much quicker in hot and humid climates, and the physical processes are more pronounced in cold and arid climates.

Physical Disintegration of Rock

This process is relatively quick in arid regions, especially when the parent material consists of soft or heterogeneous rock. It is markedly affected by the climate. As a result of differences in nighttime and daytime temperatures, various mineral components of such heterogeneous rocks as granite expand and contract to various degrees, leading to the breakdown of the rock. The outer cover of the rock expands and contracts differently relative to the interior (as the temperature differences mainly affect the exterior cover of the rock), so that the outer cover may be peeled off; this process is called exfoliation. Variations in humidity affect hydration and dehydration, and, as a result, the volume of the minerals changes. Due to the above processes, the rock slowly crumbles and disintegrates.

Rock disintegration is also affected by various chemical processes. As a result of processes such as oxidation, the rock mass gets larger and stresses exist within it that help to break it down. Most of the processes also involve water.

Vegetation—when it exists—also greatly affects the disintegration of rock. Plant roots enter the cracks in the rocks and slowly widen them through osmotic pressure. The plants and microorganisms liberate acids that attack the rock chemically, hastening its disintegration.

The degree of rock disintegration is greatly affected by the features of the parent material. Here we may differentiate between hard rock (whether homogeneous or heterogeneous), soft rock, and unconsolidated material. Hard rocks, especially homogeneous ones, are sometimes very resistant to physical weathering, and it may happen, especially in warm and humid areas, that the chemical weathering is quicker than the physical breakdown of rock (as in the terra rossa on limestone). Soft rocks, like chalk, marl, and shales, break down very easily, and plant roots also may enter even the undisintegrated rock. Unconsolidated sediments are already physically weathered to various degrees, so that this process is restricted to the stones and gravel that appear in these sediments.

Movement of Solutes in the Soil

This group of processes affects the leaching, removal, and recrystallization of the soluble salts, gypsum, lime, and, to some extent, SiO_2 as well. Salt movement and leaching are quick processes. Gypsum, being less soluble, is leached more slowly. Lime and magnesium carbonate are very insoluable in water, but are slowly dissolved in the presence of CO_2. Therefore, in the long run, even these salts are leached.

Usually an equilibrium exists, at least to some degree, between the content and appearance of lime and soluble salts in mature upland soils and the climate of the region. Even if the parent material is free from such compounds, some of these salts always appear in the rainwater and dust sediments. This equilibrium and the efficiency of leaching depend mainly on the actual evapotranspiration in the various climatic zones. In arid regions, salts and lime will be leached to the depth of the rainfall water penetration. At this depth they will be redeposited, and ca, cs, and sa horizons will appear. Soluble salts and gypsum will be leached deeper than lime, which is less soluble (see Fig. 2). In these areas, the soil is dried out completely by the plant roots by the next rainy season, at which time wetting of the soil will again begin at the top.

Fig. 2 shows the relation between the leaching stage, the climate, soil texture, and relief. Part (a) characterizes medium-textured soils in somewhat undulating upland regions of Israel. The soils are really affected mainly by the climate, classified according

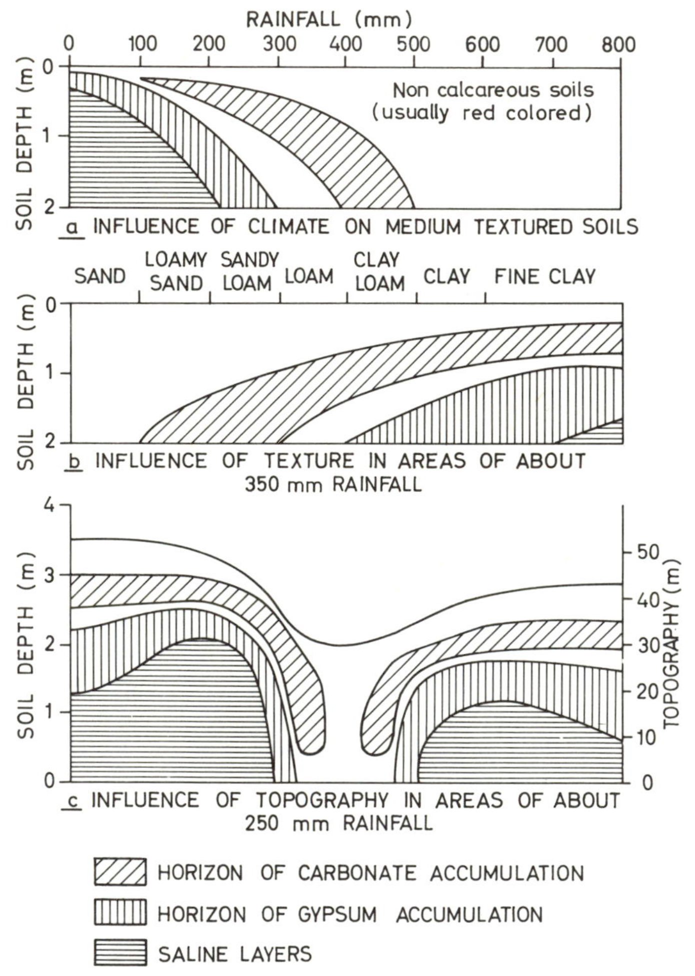

Fig. 2. The effect of rainfall, soil texture, and topography on soil leaching in arid regions. a) In their leaching stage, well-drained, medium-textured upland soils reflect the influence of climate. In humid areas chloride and salt are leached throughout. When aridity incrases, leaching is restricted, so that lime, gypsum, and finally even soluble salts are redeposited in the deeper soil layers. b) Although 350 mm is sufficient to leach all the salt and carbonates from sandy soils, it is not sufficient to leach medium- and fine-textured soils. This lime is redeposited in the deeper layers of medium-textured soils; in fine-textured soils gypsum and salts also are redeposited at depth. c) In arid areas, 250 mm or less rainfall is insufficient to leach lime, gypsum, and soluble salts that are reprecipitated in the deeper soil layers. In well-drained depressions, however, the runoff water accumulates and leaches the soluble compounds

to the effective rainfall (i.e., rainfall minus evapotranspiration) in each area. Therefore, in cold regions, where evapotranspiration rates are low, 100 mm of rainfall is classified as a semiarid zone, whereas in temperate regions 300–400 mm are needed for the same climatic classification; in subtropical areas 400–600 mm, and in tropical areas even higher amounts, are required. Also, there is a big difference between summer and winter rains. Since evapotranspiration rates are low during the winter season, the rainfall may accumulate, and consequently be more efficient than summer rains for leaching.

The leaching processes are affected by the parent material as well. Here we are concerned mainly with the effects of soil texture, the amounts of carbonates, and the porosity of the underlying material-rock or sediment. Basic parent materials that liberate much clay during weathering will retard leaching to some extent, due to the higher water-holding capacity; these soils will thus resemble those of more arid areas in properties related to leaching. As a result, ca, cs, and sa horizons will be found at shallower depths, and the type of clay minerals will resemble those of more arid areas. The contrary may be true for soils on sandy or acid parent materials, where the soils resemble those of more humid areas (Fig. 2b). Calcareous rocks, especially soft ones, also retard leaching. Underlying permeable rocks, like the hard calcareous rocks in the mountains of Israel, accelerate the leaching process, whereas impermeable rocks, like calk or marl, retard this process.

The relief factor also effects the degree of leaching. On steep slopes there is much runoff, and consequently the soils are kept relatively dry and the leaching process will be similar to that in drier areas. The contrary may be true for naturally drained depressions (which occur mainly in arid areas). Here the soils are usually more leached than on the uplands (Fig. 2c). In the arid areas of the Israel Negev, for instance, one finds nonsaline silty or fine sandy soils in depressions among the saline loessial sierozems.

In undrained depressions and impeded drainage areas, leaching is restricted, and the capillary movement of water may bring soluble salts to the surface. As a result, we get saline soils (solonchaks) in areas where saline groundwater occurs at shallow depths. This phenomenon characterizes the center of swampy areas in arid regions as well as some coastal swamps. Solonchaks with somewhat deeper groundwater occur at the periphery of the undrained depressions in extremely arid regions as the restricted rainfall cannot leach the soluble salts. In less arid areas, the soils with somewhat deeper groundwater (like those found at the periphery of swamps) are leached to some extent in their upper horizons. The Cl^- anion is leached, the Na^+ cation remains in the exchange complex, and the soil becomes highly alkaline (solonetz soils). These soils are characterized by a pronounced B-horizon (natric B), since clay saturated with Na^+ is very mobile. In still more humid areas, as leaching continues, the Na^+ cation may be exchanged by H^+ in the upper soil layers (especially if the soil is noncalcareous), and the soil becomes acid. These soils are called solodi. In the deeper layers we might still find alkaline, or even saline horizons, and toward the center of the swamp alkaline and saline soils may still characterize areas of shallower groundwater. In calcareous regions, solonetz and solodi soils are less widespread.

In semiarid to subhumid, and sometimes even humid areas, the groundwater may enrich the soils with carbonates, and as a result most swampy soils of these regions are calcareous. When groundwater is somewhat deeper, as at the periphery of swamps, leaching of lime may be normal, and acid noncalcareous soils such as planosols may occupy these areas.

Illuviation and Churning

The illuviation process refers to the movement of clay from the upper eluvial A-horizon to the lower illuvial B-horizon. This process depends on the amount of available clay, on the type of exchangeable cations, and on the rate of natural churning (mixing) of the various soil layers. The process is quite rapid, but usually does not occur in young soils where no clay minerals are available. Exchangeable Ca^{++} retards this process to some extent, whereas exchangeable H^+, and especially Na^+, hastens the movement of clay into the deeper soil horizons. In very arid regions this process is quite slow, or even nonexistent, as little clay is usually available and the scanty rain is not sufficient to move it into the deeper soil horizons. In less arid and semiarid regions, this process may be recognizable, but it is often somewhat retarded as Ca^{++} or Mg^{++} cations usually saturate the exchange complex. Only in solonetz or solonetzic soils, where exchangeable Na^+ is abundant, is this process quick, and as a result there is a good differentiation of an illuvial B-horizon. In subhumid areas where there is generally some exchangeable H^+ present, illuviation is quite pronounced.

The illuviation process does not exist in sandy soils, and is rapid or very pronounced in coarse- to medium-textured soils (which characterizes soils on acid to intermediate rocks like granite). It is somewhat retarded, or sometimes even nonexistent, in fine-textured soils that characterize basic rock, shales, and clays (Fig. 3).

The natural churning of soils counteracts the illuviation process. As a result, soils with rapid churning do not show the result of this process, even if it is actually present. The churning is caused by various animals-such as rodents (mice or moles), ants, termites (mainly in the tropics), earthworms (in soils rich in saturated organic material such as chernozems and rendzinas), etc.

In fine-textured montmorillonitic soils such as vertisols, mechanical churning does occur. This churning is caused by the shrinkage of these soils during the dry season (due to the expanding lattice feature of montmorillonite), and the slipping of soil particles from the upper layer into the cracks. During the rainy season, the soil expands, but as the cracks were already partly filled, the soil mass stresses upwards and downwards, and consequently slickenslides are developed in the deeper soil layers (Fig. 3 upper part). In some medium to fine-textured soils with predominantly montmorillonitic clay, both churning and illuviation are recognizable. An eluvial A horizon characterizes these soils, but in addition we find slicksides in the deeper soil horizons, and the top of the B horizon has a rounded top surface because it stresses upwards in the area between the cracks, while A horizon materials falls in between (Fig. 3 middle part).

Hydration, Dehydration, Oxidation and Reduction

When a soil is moistened, hydration of several minerals, like iron oxides, occurs; but when the soil dries, these minerals dehydrate and loose the hydration water. In permanently wet soil layers, on the other hand, hydration proceeds until all the free iron oxides and some other minerals are hydrated:

$$Fe_2O_3 + nH_2O = Fe_2O_3\,nH_2O$$

The typical dehydrated iron oxide is the reddish haematite. The minerals change by hydration, and we get yellow and brown minerals of the limonite group.

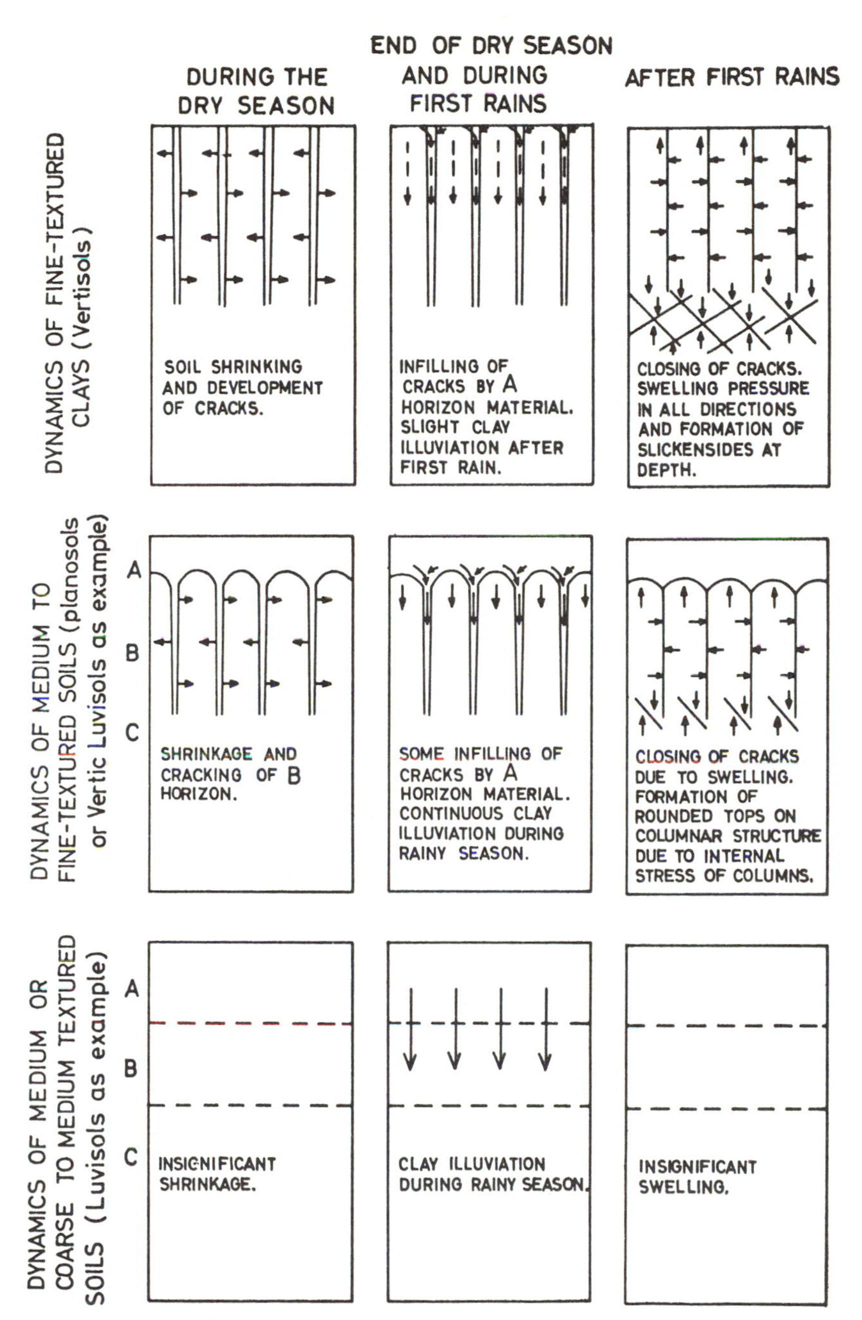

Fig. 3. The dynamics of illuviations and churning processes in soils of various textures in arid regions that are characterized by expanding clay lattice.

Fine-textured montmorillonitic clays crack during the summer. Afterward, materials from the top layer fall into the cracks. This soils swells again after the first rains. The same process also occurs in the B-horizon of medium- to fine-textured soils but with less intensity. Clay illuviation may be quite significant and an illuvial A-horizon is formed. In medium-textured soils no pressure cracks occur and only the illuviation process is well marked

Most of the aerated, well-drained soils are usually oxidized. In poorly-drained areas, below the water-table, or in very deep soil layers, chemical reduction may occur because of the lack of free oxygen. This process is activated mainly by reducing soil bacteria. Various minerals may be reduced, especially iron oxides in humid regions, and various sulfates, like gypsum, in brackish swamps.

$$2\,Fe_2O_3 = 2\,FeO + O_2$$
$$CaSO_4 + H_2O + CO_2 = H_2S + CaCO_3 + 2\,O_2$$

Gypsum reduction is widespread in brackish swamps, and as a result we detect the typical odor of H_2S. Brackish swamps are widespread in poorlydrained areas in arid regions.

In subhumid regions some free iron may occur, and as a result the planosol (for example, the nazzaz soil in Israel), which occupies the same position in the catena as the laterites, may have some iron-oxide concretions. Soil color and mottling are much affected by the aforementioned processes. The red colors are the result of dehydrated iron oxide (haematite), and the yellows are caused by hydrated iron. Layers affected by reduction and movement of iron are either white or grey.

Organic Material Accumulation and Decomposition

Soil organic material, or humus, is produced as a result of the decomposition of animal and plant residue by various bacteria. Its production depends on the rate of plant growth, and hence it is found mainly in semiarid to humid areas where plant growth is fast. The decomposition of organic material is affected by the activity of various bacteria. Since bacteria are much more active in warm climates than in cold ones, the accumulation of organic material is more pronounced in temperate and cold climates. In poorly drained areas, and especially in swampy places where the water table is close to the soil surface, breakdown of organic material is incomplete, and it accumulates. In extreme cases, peat or organic soils will be formed.

The type of organic material is affected by the presence of lime and soil reaction. In calcareous soils the organic material is saturated, and it is reworked mainly by bacteria and earthworms and other animals. These earthworms and animals may even mix it with the deeper layers, as in the chernozems or rendzina soils.

As a rule, accumulation and decomposition of organic material are fast processes, and may even affect young soils such as lithosols, regosols, and alluvial soils.

Erosional and Depositional Processes

These processes, in contrast to those previously mentioned, are not really soil-forming, but are geologic processes that oppose and destroy profile development. Water erosion occurs in upland areas, and the eroded material is deposited on the footslopes and floodplains. The coarser material is deposited on the footslopes and river levees, and the finer particles accumulate on the toeslope and backswamp areas. Wind erosion may affect any type of relief. This process is usually prevalent in very dry areas, where vegetation is scanty or even nonexistent. Accumulation of wind-eroded material may occur on any type of landscape; it is found mainly in desert fringe regions like southern Israel, where the dust content in the air may be very great.

Erosional and depositional processes are thus affected by relief, climatic, and biotic factors. On steep slopes erosion is usually very fast, and as a result young soils like lithosols and regosols characterize these areas (DAN and YAALON, 1968). Fast accumulation of eroded material occurs below these slopes, and thus these areas are characterized by coarse-textured young alluvial soils at the footslopes, and finer alluvial soils further down in the toeslope areas. On moderate slopes erosion is more restricted, and mature soils may occur. Consequently, deposition at the footslope and toeslope is limited, and soil-forming processes acting on the sediments also produce mature soils here (Fig. 4). These mature soils lack a typical C-horizon, as the parent material is deposited on the original topsoil. The deeper layers are characterized by old, buried B-horizon material. These soils may be said to be "accumulative" or cumulic, to differentiate between them and other mature zonal soils. Mature soils on aeolian parent material in desert fringe areas also are usually accumulative or cumulic like the loess and aeolian loess-derived soils in southern Israel.

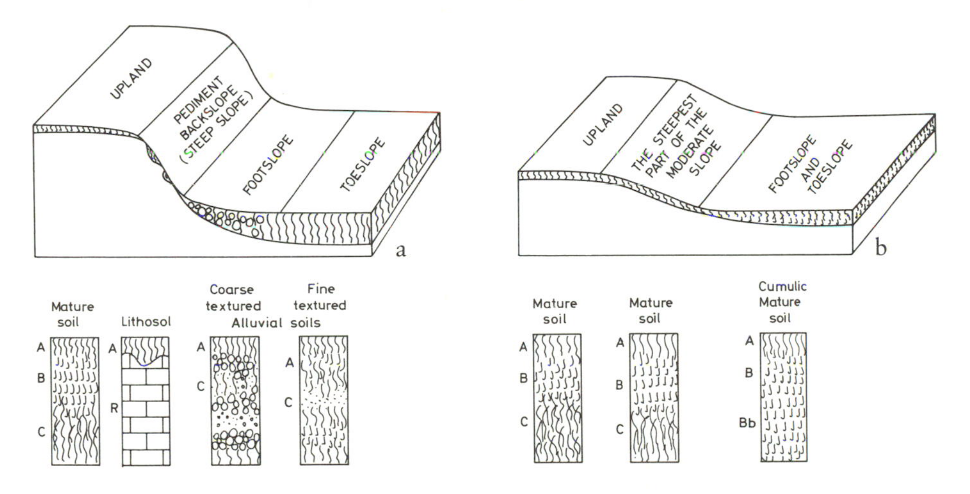

Fig. 4. The relation between the slope elements and soil profile features in well-drained subhumid, semiarid, or mildly arid regions. a) Typical soil distribution along a steep slope. b) Typical soil distribution along a moderate slope

Climate and vegetation have very important effects on erosion processes. In arid areas, vegetative cover is sparse, and the soil is easily eroded by runoff water. In humid areas the natural vegetation protects the soil from accelerated erosion, and only after clearing of the vegetation does the erosion hazard in upland areas become severe. The same may be true in connection with wind erosion, and therefore it is usually restricted to extremely arid areas, where the vegetative cover is scanty, or even completely absent.

Erosion is also affected by the soil's infiltration capacity and rock permeability. Permeable soils and rocks like terra rossa on limestone prevent the accumulation of runoff water, and therefore retard erosion. The contrary is true for impermeable rocks like shale.

Soil Distribution in Arid-Zone Landscapes

On a regional basis soil-forming processes are found to change in a pattern that is closely related to the geomorphology of the earth surface; that is, to the type of landscape that prevails. As a result, quite a good correlation exists between the landscapes in the various climatic zones and the soil distribution.

This type of distribution in arid regions, together with the main properties of the dominant soil, will be described in the following paragraphs.

Soils of the Extremely Arid-Zone (Typical Desert Soils)

The vegetation of this zone is very sparse; what exists is usually restricted to dry riverbeds, rocky pockets, and various other depressions. Some vegetation may also be found in dune areas. As a result of the poor vegetative cover, the zone is characterized by severe water and wind erosion.

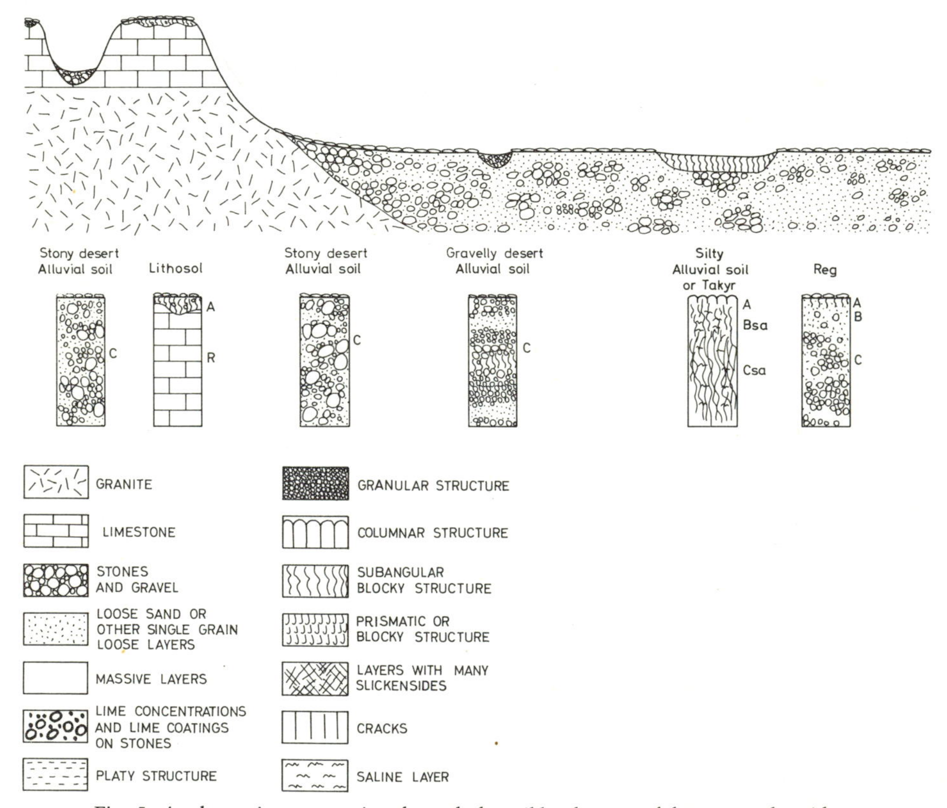

Fig. 5. A schematic cross section through the soil landscapes of the extremely arid zone

Several types of soil landscapes can be distinguished in this zone. *Gravelly desert plains* are most widespread. Various reg soils characterize most of this landscape. These soils are covered by a dense desert pavement of polished gravel. The upper horizon is usually a vesicular loam or sandy loam, which may cover a dusty, or occasionally somewhat prismatic, B-horizon. The deeper layers may include various materials, from silt and clay to gravel or even hard rock. Most of the regs are saline, and gypsum or salt hardpans are widespread. Regs with good mechanical composition can be reclaimed and prepared for irrigated agriculture.

Numerous dry riverbeds cross the reg plains. The deposits of these riverbeds range from coarse desert alluvium in the more active and torrential riverbeds to silty alluvium in the less active streams. Similar silty deposits also cover the various undrained desert depressions. The gravelly deposits and some of the silty sediments do not reveal any profile differentiation. However, many of the silty deposits, especially those in the depressions, have been affected by soil-forming processes. Saline soils (solonchaks) form in poorly drained despressions due to the capillary rise of saline groundwater. These soils are very saline, especially in their upper layers. In other depressions, takyrs and takyr-like soils are formed. These soils are characterized by a very hard alkaline surface crust, which usually breaks into polygons upon drying. The deeper layers are usually saline. These soils are widespread in noncalcareous regions, as in the vicinity of granitic mountains.

Extensive desert areas are also occupied by sand dunes of various types, which usually have no agricultural value. Shallow aeolian wind deposits may also cover other soils and landscapes in the vicinity of the dunes, especially at the leeward side of the large sandy areas. In these cases, the sand usually improves the ecological conditions and some of these soils could be cultivated and irrigated – especially where the sand is mixed with fine sandy or silty desert alluvium.

Many desert areas are mountainous. Steep, bare, rocky slopes characterize most of this landscape. Young, undeveloped, stony, and gravelly desert alluvium is found in the footslope areas, alluvial fans, and dry riverbeds. Some typical regs may be found in larger valleys. A complex landscape type including both mountains and large valleys that may be termed a "range and basin landscape" is also quite widespread. The valleys in this case are filled mainly by various regs. However, the depressions in these valleys are usually rather large and, as a result, solonchaks, takyrs, and other similar soils also cover wide areas.

Soils of the Arid Zone (Desert and Semidesert Soils)

The natural vegetation of this zone is also sparse and does not cover the entire soil surface. However, plants, especially small shrubs, are found throughout the area, but this vegetation is not dense enough to inhibit severe erosion. Ecological conditions are better in the depressions as a result of the concentration of runoff water. In these places, the vegetation may cover the whole surface and thus reduce erosion.

The soil landscapes of this zone are rather similar to those of the typical deserts. The soils, however, are somewhat different and usually exhibit deeper and more pronounced development.

Various grey, greyish-brown, and red desert soils cover the plains pediments, footslopes, and inactive alluvial fans. These soils exhibit a typical ABcaC- or ABCca-horizon sequence. The A-horizon is, as a rule, light-colored with a low chroma[1], and is

1 The "chroma" indicates the strength of the color. High chromas characterize strong colors and low chromas characterize dull colors.

usually also coarse-textured. The B-horizon may be finer textured and brown or reddish-brown in color. A secondary carbonate ca horizon is found at a quite shallow depth. This layer may be indurated to form a petrocalcic horizon[2]. The deeper soil layers are frequently gypsipherous or saline, and secondary cs or sa horizons might be found at this depth.

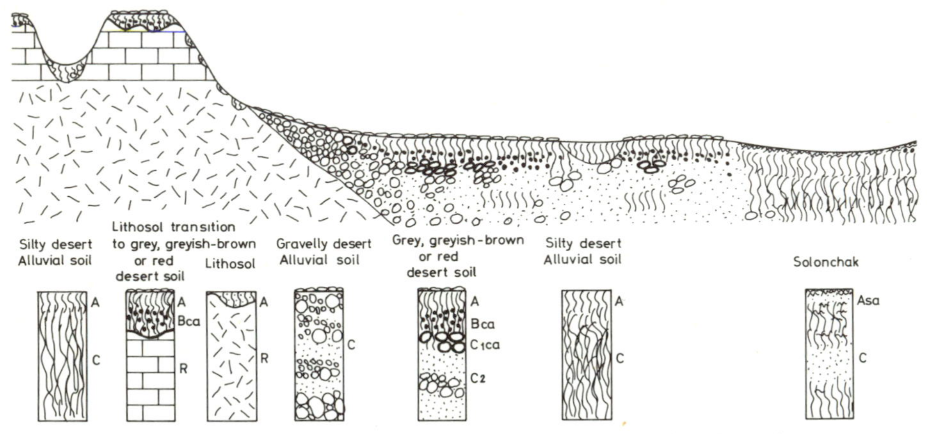

Fig. 6. A schematic cross section through the soil landscapes of the dry arid zone

Fine silty, and occasionally also clayey, deposits are found in small, dry riverbeds and depressions. The soils formed resemble those of the same landscape in true deserts. However, the silty deposits and young alluvial silty soils may cover larger areas, including the small, well-drained valleys that may occur in hilly and mountainous regions such as the central Negev in southern Israel.

Desert dunes are also quite common. However, many of these are stabilized due to the denser vegetative cover. Some soil formation may proceed in these stabilized dunes, and may include mainly a redistribution of lime and occasionally also some clay formation.

Lithosols are found on mountain slopes; they include various shallow, physically weathered, rocky material, which may be mixed, eventually, with some aeolian deposits. The mountain plateaus may be partly covered by similar lithosols and partly by shallow grey, greyish-brown, and red desert soils (Fig. 6). Stony grey, greyish-brown, and red desert soils usually cover footslope areas, and silty desert alluvial soils may be found in the centers of the small valleys, along the streambeds. Range-and-basin landforms are also widespread. The basins are mainly covered by grey, greyish-brown, and red desert soils, whereas takyrs, solonchaks, and eventually fine silty and clayey desert alluvial soils are found in the depressions that occupy the center of these valleys.

Soils of the Arid Desert Fringe Areas

These areas include, among others, the great loess belts. The ecological conditions enable the development of a continuous grass carpet that reduces the natural erosion to a minimum. This vegetation also traps the fine aeolian dust brought from the desert

2 Hard indurated layer due to secondary carbonate deposition.

regions. As a result, most of the soils in this zone have developed from fine aeolian sediments or from redeposition of these sediments.

The soil landscapes of this zone include hilly and mountainous regions, loessial and clay plains and similar undulating areas, and old, stabilized dunes. Most of the soils developed from aeolian deposits. These sediments may weather to form typically

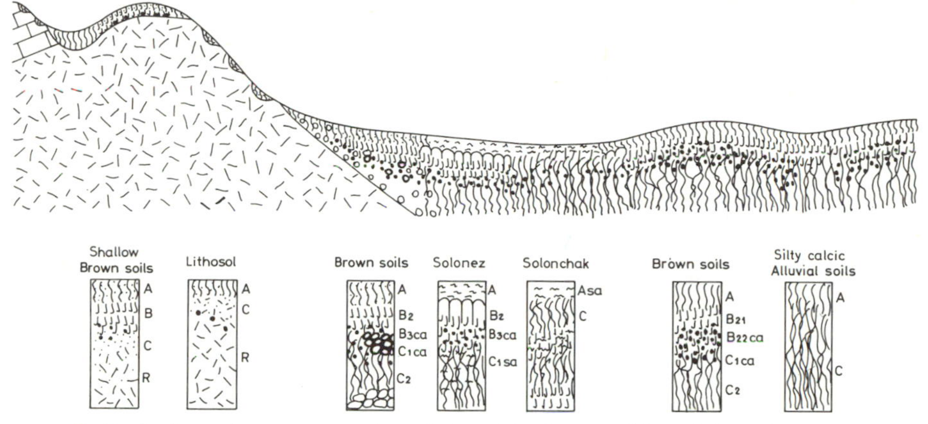

Fig. 7. A schematic cross section through the soil landscapes of the arid desert fringe areas

coarse-, medium-, or medium- to fine-textured ABCca, ABcaC, or ACca brown, reddish-brown, or chestnut soils, in the case of typical loess, or mixed loessial and sandy deposits. The finer-textured parna and similar deposits usually weather to form vertisols.

Many reddish-brown and brown soils are found in subtropical and tropical regions. The soils in these areas are characterized by a well-developed pallid A-horizon and a cambic or argilluvic (textural) B-horizon. A ca-horizon is usually found at quite shallow depths. The saturated-paste conductivity of the deeper soil layers still exceeds 2 mmho/cm, but as a rule these are not really saline. The brown soils of cooler regions, as well as the chestnut soils, are similar, with a higher organic matter content in the upper layers; as a result, they are characterized by a mollic epipedon or a melanic A-horizon. The B-horizon is usually less developed, and frequently is even absent.

The vertisols are fine-textured soils that crack badly during the dry season. These cracks are filled during the rainy season by soil material from the upper layer and due to the swelling of the montmorillonitic clay. As a result, the various soil layers mix, mechanical stresses develop, and slickensides are formed in the deeper layers (see p. 17).

Alluvial soils, especially silty calcareous alluvial soil, mainly from water-redeposited loessial material, cover large, well-drained flood plains and lower river terraces. Solonchaks and solonetz are found in poorly drained depressions. The solonetz are highly alkaline soils with a coarse-textured upper horizon and a very dense, argilluvic B-horizon. Various young soils, the character of which is intermediate between the alluvial soils and typical brown soils, may be found on higher terraces.

The mountain slopes in this zone are usually covered by various lithosols. The moderately stepp slopes, however, may be covered partly by shallow, silty brown and reddish-brown soils. This phenomenon is usually concentrated in the wetter slope exposures, such as the north-facing slopes in the Northern Hemisphere.

Soils of the Semiarid Zone

The ecological conditions of this zone are quite good. The natural vegetation usually consists of a well-developed grass cover or, in the tropics, of a quite dense savanna vegetation. This zone differs from the former mainly in degree rather than in general properties. The soils and vegetation are somewhat better developed. In reality, these two belts merge without a definite boundary.

Passing from the drier areas, the soils get deeper and somewhat more leached. The A-horizon is usually better developed, the B-horizon usually has a typical argilluvic feature, and the ca horizon is found quite deep. Some soils are even noncalcareous throughout, especially in granitic or sandy areas and among coarse-textured soils. However, the soils are usually finer textured than in the more arid zone, and vertisols or soils with vertic features are widespread. The saturated-paste conductivity of the deeper layers also decreases as we pass from the brown soils to red and brown Mediterranean soils. The latter are usually noncalcareous soils with a typical argilluvic B-horizon. In the cooler temperate zone chernozems with a deep mollic or melanic A-horizon characterize most of the areas, although chestnut soils are also found, especially in the drier part of this zone. Vertisols cover large areas, particularly on the periphery of the large deserts or on basalts or other rocks rich in ferromagnesium minerals.

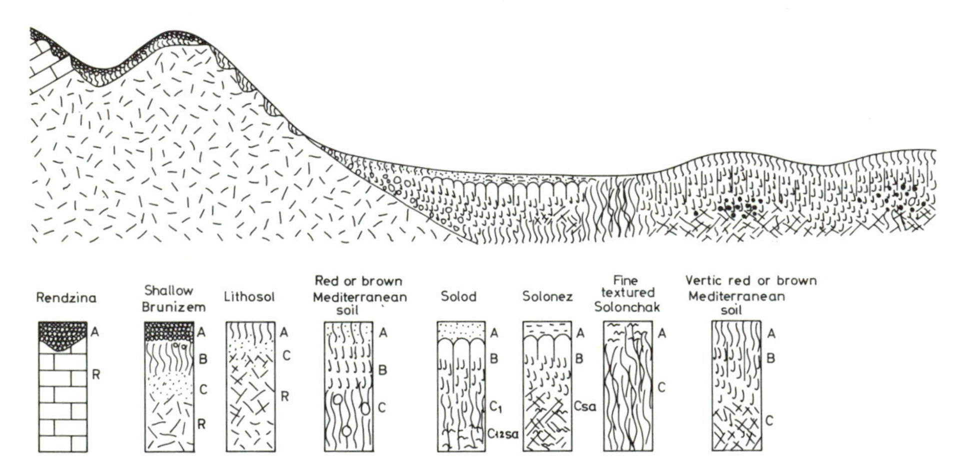

Fig. 8. A schematic cross section through the soil landscapes of the semiarid zone

The soils of flood plains and depressions also generally show a more advanced stage of development. Alluvial soils are usually restricted to active flood plains. The poorly drained depressions are often characterized by solonetz soils, which may be somewhat leached in their upper layers and thus form solodized solonetz or even solods. The solonchaks are usually restricted to the centers of these depressions.

The mountainous areas are covered either by lithosols or, especially on moderate slopes, by somewhat deeper and more developed soils. These may include various soils with melanic A-horizons and without a B-horizon like rendzinas or with a cambic or argilluvic B like shallow chestnut or shallow brunizems and brown forest soils. Many soils do not exhibit a melanic A; these include various soils with a cambic B, like eutric or calcic cambisols, and others with an argilluvic B, like shallow red and brown Mediterranean soils. Some shallow fine-textured cracking clays are also found; these are classified as shallow vertic soils or, if they are somewhat deeper, as vertisols.

Soils of the Subhumid Areas

Most of these areas are characterized by an alternating wet and dry season. Whether the dry season is in summer or in winter may affect soil development quite differently. The vegetation consists mainly of xerophytic scrub or forest in the winter-rainfall areas. High savanna or other types of quite xerophytic vegetation are found in the tropical summer-rainfall areas, whereas a high-grass prairie usually characterizes the temperate regions.

Most of the soils in this zone are leached to various degrees. Many of them are also polygenetic and may have developed during a somewhat moister period, especially in the summer-rainfall areas. Typical ABC red and brown Mediterranean soils or similar soils with a well-developed argilluvic B-horizon are widespread. The clay minerals of the soils in the tropics may contain a high percentage of kaolinite whereas montmorillonite is more widespread in soils of winter-rainfall areas. Brunizems with well-developed melanic A-horizons are more widespread in cooler areas. Various vertisols also cover large

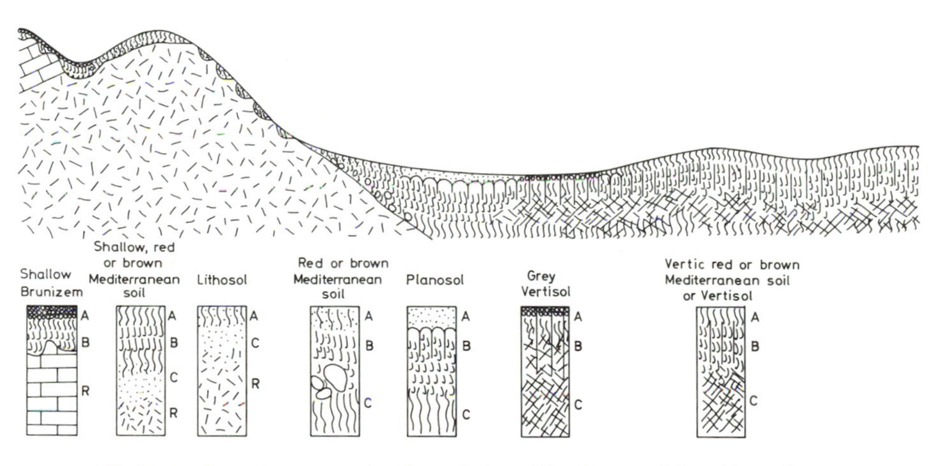

Fig. 9. A schematic cross section through the soil landscapes of the subhumid zone

areas, especially in the subtropical and tropical zones. These include brown types on parent material rich in ferromagnesium compounds on plateaus, and undulating relief and grey to black hydromorphic types in undrained depressions. Various planosols with a coarse-textured A-horizon and a fine-textured, impervious and gleyed B-horizon also cover large flat areas and depressions.

Various shallow red and brown Mediterranean soils, eutric cambisols and brown forest soils and brunizems, together with similar leached lithosols, are found in mountainous areas. They usually resemble the soils on similar landscapes in the semiarid zone, but are generally more leached. Moreover, the somewhat deeper types are more widespread in the subhumid climatic zone.

Soils of the Large River Flood Plains

The main agricultural areas in the deserts were concentrated, in the past, along the large river flood plains that crossed these regions. The Nile in Egypt, the Euphrates and Tigris in Mesopotamia, and the Indus in India are typical examples. The waters of these rivers originate in moister areas. The same may hold true, at least partly, for the alluvial material that has been deposited in these plains and valleys. This material is usually quite finely-textured and has a silty or clayey nature. Many of these sediments have been subsequently salinized or alkalinized because of impeded drainage conditions or poor irrigation methods.

The soils of these flood plains include various alluvial soils, solonchaks, solonetz soils, and other hydromorphic soils and vertisols, many of which are also saline or alkaline. In all these flood plains, especially in their lower parts, a well-planned drainage system is usually required in order to retain the fertility level of these lands.

Appendix: Comparison of Arid Soil Designations

Commonly used soil designations were presented in the article. These are mainly based on the former U.S. classification (BALDWIN, KELLOG, and THORP, 1938) and the Russian classification (IVANOVA, 1956). The soils of these classifications were not always well defined, especially where soil intergrades are concerned. The soils are much better and more accurately defined in the new U.S. classification (Soil Survey Staff, 1960, 1967) and in the FAO-UNESCO definitions of soil units for the soil map of the world (DUDAL, 1968, 1969). The soils mentioned in the article will thus be compared with these new classifications. It should be remembered, however, that the criteria and definitions for these new classifications are somewhat different from those of the former ones and the comparison may sometimes be approximate. This comparison does not always cover the entire range of the soils dealt with. Also, in many cases the new soils are not equivalent to the older concepts. Thus, the same soil group may be correlated with several older groups or with parts of these groups, and vice versa.

Common Terms	FAO Classification	U. S. Classification
Coarse desert alluvium or stony and gravelly desert alluvium	Eutric or calcaric fluvisols	Torripsamments or torrifluvents
Sand and sandy soils		Psamments, mainly xeropsamments, torripsamments, quartzipsamments, and also ustipsamments
Silty alluvium or silty desert alluvial soils	Calcaric fluvisols	Torrifluvents
Alluvial soils	Fluvisols	Fluvents (mainly xerofluvents and ustifluvents; also torrifluvents)
Lithosols	Lithosols	Lithic torriorthents, lithic xerorthents; also lithic ustorthents
Reg	Yermosols (mainly gypsic yermosols) or orthic solonchaks	Camborthids or calciorthids; also some haplargids

Common Terms	FAO Classification	U.S. Classification
Grey, greyish brown, and red desert soils	Yermosols; also some xerosols	Argids and orthids (among them also paleargids[a] and paleorthids[a]
Saline soils, solonchaks	Solonchaks	Salorthids
Takyrs	Takyric solonchaks	Solarthids, natrargids, calciorthids, camborthids
Solonetz	Solonetz	Natrargids, natrixeralfs, natrustalfs, natraqualfs; also some natraquolls, natrustolls, and natrixerolls
Solodized solonetz		Natrixeralfs, natrustalfs, natraqualfs
Solods	Solodic planosols	Natrixeralfs, natrustalfs, natraqualfs
Eutric or calcic cambisols	Eutric or calcic cambisols	Ustochrepts, xerochrepts, ustropepts
Brown and reddish-brown soils	Xerosols; also poorly developed castanozems	Argids or orthids; also poorly developed ustolls or xerolls
Chestnut soils	Castanozems	Ustolls, mainly calciustolls; sometimes xerolls (calcixerolls)
Chernozems	Chernozems	Ustolls, or sometimes also xerolls and udolls
Brumizems	Phaeozems	Udolls, mainly argiudolls; sometimes also xerolls
Rendzinas	Rendzinas	Rendolls
Brown forest soils	Haplic Phaeozems, haplic castanozems, haplic chernozems	Haplustolls, calciustolls, haploxerolls, calcixerolls; in humid regions also other soils
Red and brown Mediterranean soils	Chromic Luvisols; also brunic luvisols	Ustalfs or xeralfs
Planosols	Planosols	Albaqualfs; also glossaqualts and other similar soils
Vertisols	Vertisols	Vertisols
Brown vertisols	Chromic vertisols	Chromusterts or chromoxererts
Grey and black hydromorphic vertisols	Pellic vertisols	Pellusterts or pelloxererts

[a] Soils containing a petrocalcic horizon.

Literature

BALDWIN, M., KELLOGG, Ch. E., THORP, J.: Soil classification. In Soils and man. USDA Yearbook, pp. 979–1001 (1938).
BUCKMAN, M. O., BRADY, N. C.: The Nature and Properties of Soils (7th ed.). New York: Macmillan 1969.
DAN, J., YAALON, D. H.: Trends of soil development with time in the Mediterranean environments of Israel. Trans. Conf. Mediterranean Soils, Madrid, Spain, pp. 139–145 (1966).
DAN, J., YAALON, D. H.: Pedomorphic forms and pedomorphic surfaces. Trans. 9th Internat. Cong. Soil Sci., vol. IV, pp. 577–584 (1968).
DURAL, R.: Definitions of soil units for the soil map of the world. FAO, Rome (1968).
IVANOVA, E. N.: Essai de classification générale de sols. 6th Int. Cong. Soil Sci. E, pp. 387–394 (1956).
JENNY, H. F.: Factors of Soil Formation. New York: McGraw-Hill 1941.
JENNY, H. F.: Derivation of state factor equations of soils and ecosystems. Proc. Soil Sci. Soc. Amer. **25/5**, 385–388 (1961).
RUHE, R. V.: Elements of soil landscape. Trans. 7th Congr. Soil Sci. **4**, 165–170 (1955).
Soil Survey Staff: Soil classification. The 7th approximation. USDA, Washington, D. C. (1960).
Soil Survey Staff: Supplement to soil classification. USDA, Washington, D. C. (1967).

II. Water Resources

Only a small part of the water that exists on this planet can be developed and utilized beneficially for irrigation. Therefore, the proper handling of water resources requires a thorugh understanding of the natural hydrologic processes. In the present discussion, emphasis will be placed on arid-zone hydrology and the development and management of water resources in general.

Correct evaluation of the available water supplies is extremely important in selecting the appropriate water resource to be used for irrigation.

In this chapter the reader will find a general discussion of the hydrologic fundamentals relevant to arid zones. Once understood, these fundamental principles are applied to specific problems of arid-zone hydrology. The third section of this chapter is devoted to problems of water development and management. Emphasis is placed on water storage, and the integration of surface water and groundwater in a single water-resource system.

Irrigation waters always contain dissolved matter and guite often carry solid materials in suspension. The quality of such waters is determined by their effect on plant growth, soil properties, environmental equilibrium, and irrigation technology. The fourth section of this chapter deals with the criteria that determine the quality of water for irrigation and provides a critical review of the current trends in quality evaluation.

1

Hydrologic Fundamentals

N. Buras

Introduction

Hydrology is the science that describes the occurrence of water on earth, its areal and temporal distribution, and its quantitative and qualitative aspects. It is estimated that approximately 97% of all the water in the world -1.2×10^{18} m^3–is in the oceans. The remaining 3%, or about 4×10^{16} m^3, is fresh water that may be usable for irrigating lands. More than $^3/_4$ of the fresh water cannot be developed and utilized with the current technology: 3×10^{16} m^3 is frozen into polar ice caps and mountain glaciers, and more than 4×10^{15} m^3 is estimated to lie in aquifers at depths exceeding 800 m below the ground surface, leaving only about 4.4×10^{15} m^3 that could be used each year for the benefit of mankind. The available amount is divided between surface water (rivers, streams, fresh-water lakes), groundwater (to a depth of 800 m below the ground surface), and atmospheric water (atmospheric moisture, clouds, etc.) as follows:

Surface water	0.13×10^{15} m^3
Groundwater	4.27×10^{15} m^3
Atmospheric water	0.01×10^{15} m^3
Total	4.41×10^{15} m^3

The huge amount of more than 4 million billion m^3 of water is quite impressive. However, two points are worth emphasizing from the outset:

a) The fresh water available today in the world is a limited source. It may well happen that further advancement of the human race will hinge on the development and management of the worlds water resources.

b) The amount of 4.41×10^{15} m^3 represents a long-term annual average. Not only does this quantity fluctuate from year to year and from season to season, but its distribution over the land area is extremely unequal, ranging from very large amounts in the water-drenched tropical forests to very small amounts in the desiccated rocky deserts.

These observations alone should be sufficient to motivate an interest in hydrology – a science whose aim is to obtain basic information for the purpose of developing and managing water resources. The discipline of water resources engineering is concerned primarily with reconciling the discrepancy that exists between the *availability* of water (in space, time, and quality) and its *desirable properties* (amounts required, their distribution, and their quality). Some fundamental concepts in hydrology will be briefly outlined below, leaving the applicative aspects related to the development and utilization of water resources (see p. 5).

Surface Hydrology

Precipitation and Runoff

Water in nature is in a continuous state of motion and transformation. It moves from the oceans to the atmosphere, and from the atmosphere, directly or by devious paths, back to the oceans. This circulation of the water is termed the *hydrologic cycle.*

Only parts of the phases of the hydrologic cycle are of interest to the water resources engineer. Of these, the relationship between rainfall and runoff seems to be an important indicator of the development potential of the water resources in a given region. To be sure, this relationship is quite complex. It is influenced by a large number of variable factors, and is beclouded by uncertainty regarding the magnitude of the natural phenomena involved. Many detailed descriptions and studies of natural hydrologic phenomena have been published in the literature.

Because of the complexity of the hydrologic phenomena and our unsufficient understanding of the participating factors, many of the formulas used to describe the relationship between precipitation and runoff have been developed empirically. These formulas are limited in their application to those special conditions under which they were developed, and caution must be exercised in using them when the conditions are different. One particular formula whose use has become widespread is the so-called "rational formula," although there is hardly anything rational about it:

$$Q = C\,I\,A \tag{1}$$

where Q is the peak discharge, in m^3/sec, from a watershed whose area is A km^2; C is the runoff coefficient, which varies with the characteristic of the watershed; and I is the rainfall intensity, mm/hr. Some typical values for C are given in Table 1.

Table 1. Values of runoff coefficient C (from CHOW, 1962)

Soil type	Vegetative cover		
	Cultivated crops	Pasture	Woodland
Light-textured: sandy or gravelly	0.20	0.15	0.10
Medium-textured: on clay pans	0.40	0.35	0.30
Fine-textured and shallow soils	0.50	0.45	0.40

As can be suspected, the rational formula is too general and far from satisfactory for many practical cases. Another formula that expresses the precipitation-runoff relationship in a more refined manner was developed by CHOW (1962).

$$Q = 0.285\,A\,X\,Y\,Z' \tag{2}$$

Here Q is the discharge, m^3/sec; A is the catchment (watershed) area, km^2; X is the runoff coefficient, mm/hr; Y is a dimensionless climatic factor; and Z' is a peak reduction factor, also dimensionless.

Eq. (2) attempts to describe a very complex natural phenomenon – the transformation of precipitation into runoff – in a simplified manner. At the same time, the quantities that enter into this equation should be both relevant and measurable.

It is quite clear that the watershed, where rainfall produces runoff, is an important factor in the rainfall-runoff relationship. And it is reasonable to assume that the amount of runoff, measured in terms of the discharge Q at the natural outlet of the watershed, is directly proportional to the area A of the catchment usptream from this point.

It is also recognized that not all the atmospheric water that reaches a given catchment area will eventually appear as runoff at the point of measurement. The fact that only a portion of the total precipitation is transformed into streamflow is represented by the runoff coefficient X, where

$$X = \frac{Re}{t} \tag{3}$$

Here, Re is the "rainfall excess," in mm, which is generated during the time period t (measured in hours).

The magnitude of the "rainfall excess" depends on the actual precipitation measured during time t, on the soil type of the watershed, on the vegetative cover, and on the cropping pattern. CHOW (1962) derived an empirical relationship for Re, which, when adapted to conditions prevailing on the coastal plain of Israel (MILEVSKY, 1969), takes the form

$$Re = \frac{\left(R - \frac{5,000}{N'} + 50\right)^2}{R + \frac{20,000}{N'} - 200} \tag{4}$$

In this equation R is the actual precipitation measured in mm during time t, and N' is a runoff number that depends on soil type, cropping pattern, and vegetative cover. For the coastal plain of Israel, N' varies between 65 and 90 (MILEVSKY, 1969).

Returning to Chow's formula (2), one observes that the measurement of precipitation may present some problems. As the amount of precipitation varies over the area of the watershed, one finds that the larger the watershed is, the greater is the variability in the precipitation. This problem may be partially overcome by installing an appropriate number or rain gauges over the entire catchment area. This, indeed, is often done in large watersheds. In the case of small watersheds, where there are few rain gauges, a climatic factor Y must also be introduced into the rainfall-runoff equation. The magnitude of this coefficient is usually in the neighborhood of 1.0. Its purpose is to relate the rainfall in a particular watershed to rainfall data measured at a meteorological station that is taken to represent a larger area.

Eq. (2) also takes into account the time factor that is a part of any hydrologic process. As a result of this factor, the peak discharge at a given point on the stream is attenuated by the retention properties of the watershed above it. This attenuation is represented by the factor Z', which is a function of the retention properties of the watershed. Specifically, it is customary to associate with every watershed a characteristic "detention time," t_p, which is given by the expression

$$t_p = 0.00505 \left(\frac{L}{S}\right)^{0.64} \tag{5}$$

Here, L is the length of the watercourse in meters, measured from its highest point in the watershed to its outlet, and S is the slope of the watercourse in percent. Z' is a function of the ratio t/t_p and has a maximum value of 1.0 (no reduction in peak discharge). The relationship Z' versus t/t_p must be derived empirically for each specific physiographic and climatic region. For the coastal plain of Israel (MILEVSKY, 1969) this relationship is shown graphically in Fig. 1.

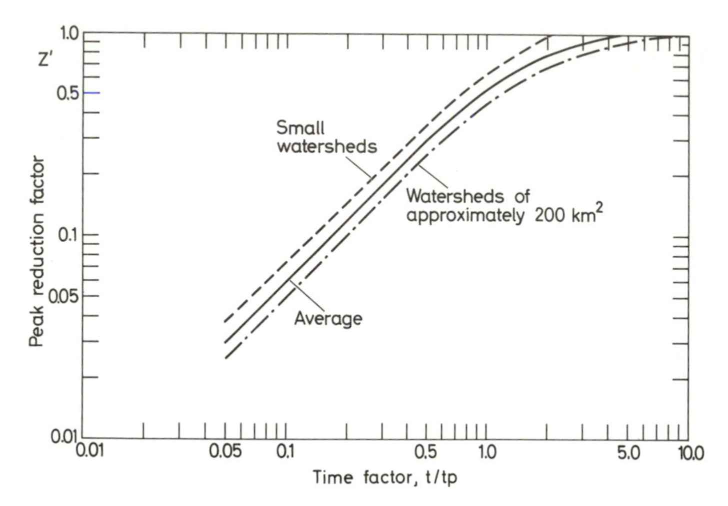

Fig. 1. Peak reduction factor in the coastal plain of Israel

Eq. (2) is very useful in the design of hydraulic structures (channels, culverts, etc.) in small watersheds. The most important design criterion is the magnitude of the peak discharge. To estimate peak discharges using Chow's formula, values of X and Z' are calculated for different values of t and the results are introduced into Eq. (2) to yield Q. The largest value of Q corresponds to the estimated peak discharge.

When rainfall intensity–duration–frequency curves are available, they can be used to advantage in conjunction with Chow's formula when estimating R. This estimate is given by

$$R = It \qquad (6)$$

where I is the intensity in mm/hr that corresponds to a given frequency. In this way, the frequency of the various peak discharges can also be estimated.

The Unit Hydrograph

From the previous discussion it is clear that runoff varies not only in space (from watershed to watershed) but also in time. A graphical representation of the variation of water flow with time at a given point in space is known as a *hydrograph*. Since we are dealing with natural hydrologic phenomena where it is assumed that precipitation and runoff are associated in a cause-and-effect relationship, it is desirable to introduce this relationship into time-distribution curves of runoff. The hydrograph of direct runoff resulting from one unit (say, one inch) of effective rainfall generated uniformly over a drainage area (catchment) at a uniform rate during a specified period of time is called a *unit hydrograph*.

The unit hydrograph can be used to derive the hydrograph of runoff that results from an arbitrary amount of effective rainfall. The definition of a unit hydrograph embodies a number of basic assumptions (CHOW, 1964):

a) The effective rainfall is uniform throughout its duration or throughout the specified period of time.

b) The effective rainfall is uniformly distributed throughout the entire area of the watershed.

c) The base (time duration) of the hydrograph of direct runoff, due to an effective rainfall of unit duration, is constant.

d) The ordinates of the direct-runoff hydrographs of a common time base are directly proportional to the total amount of direct runoff represented by each hydrograph.

e) For a given drainage basin, the hydrograph of runoff due of a given period of rainfall reflects all the characteristics of the watershed.

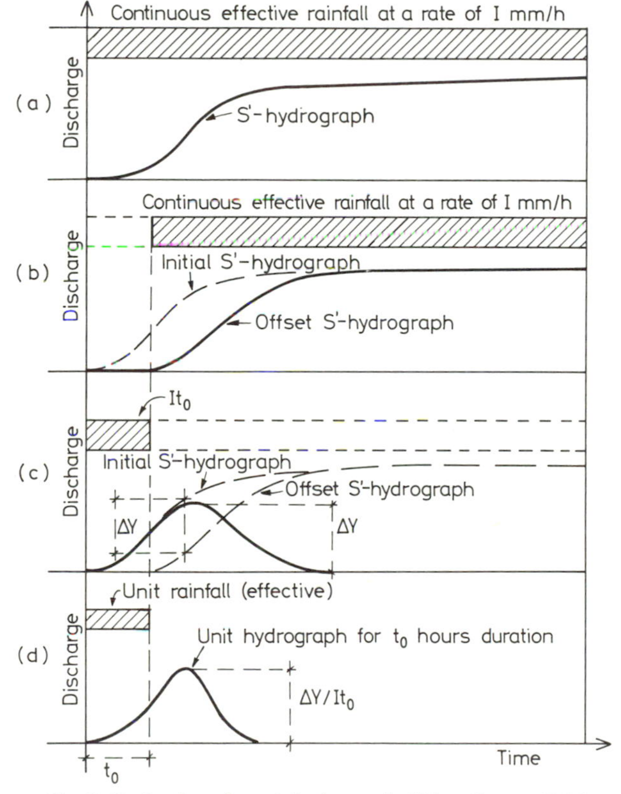

Fig. 2. Derivation of a unit hydrograph. (After CHOW, 1964)

These assumptions, of course, are never satisfied exactly under natural conditions of rainfall and physiography. However, this does not detract from the usefulness of the unit hydrograph in estimating peak discharges and total volumes of runoff (streamflow) that are generated by a rainstorm of a given magnitude. For example, violation of assumptions (a) and (b) may lead to overestimation of the actual peak flows, while changes in the basin characteristics as a result of man-made structures or natural fluctuations of the vegetative cover may lead to underestimation of the actual streamflows. Such departures from the true values of runoff merely reflect our incomplete knowledge of the hydrological phenomena, at the same time, the unit hydrograph remains a useful tool in planning the development and utilization of water resources.

To derive a unit hydrograph, one starts from a theoretical *S-hydrograph,* which represents the runoff produced by a continuous effective rainfall at a constant rate of I mm/hr for an indefinite period (Fig. 2a). Ultimately, the ordinate of the S-hydrograph approaches the rate of the effective (constant) rainfall asymptotically or at equilibrium. Of course, the rate I of effective rainfall is selected so that it is meaningful for the specific climatic zone within which the watershed in question is located.

If the specified period of time for which the unit hydrograph is to be derived is t_o, the S-hydrograph is offset by this amount (Fig. 2b). Again, the desired duration t_o (usually measured in hours) should reflect the length of rainstorms usually experienced in the given region.

The difference between the ordinates of the initial S-hydrograph and the offset S-hydrograph, Δy, plots as a bell-shaped curve (Fig. 2c). If one divides the increments Δy by the product It_o, one obtains the unit hydrograph for t_o hours' duration.

From this derivation it is clear that the unit hydrograph is characterized by the following three parameters: the maximum (peak) discharge, the time lag from the beginning of the rain to the peak flow, and the total base time. In order to determine these parameters in the field, a certain amount of instrumentation is required. Furthermore, the field measurements must represent a period of several years in order to be statistically meaningful. This may become an obstacle to the development of ungauged river basins and watersheds. In order to overcome this difficulty, *synthetic unit hydrographs* have been developed (LINSLEY *et al.,* 1958). By performing a correlation analysis between the three unit hydrograph parameters and a number of watershed characteristics that are invariant and easily measurable (such as area of the river basin, slope of basin area-elevation curve, number of major streams in the basin), it is possible to estimate the peak flow, the time lag between the beginning of the precipitation and the peak discharge, and the total base time for an ungauged area.

The concept of a unit hydrograph is essential for the design of hydraulic structures connected with the development and use of surface-water resources. The application of this concept to the design process belongs to the discipline of hydraulic engineering (LINSLEY and FRANZINI, 1964).

Groundwater

Groundwater Movement

In many parts of the world, groundwater is a major source of water for agriculture as well as for domestic and industrial purposes. On a worldwide basis groundwater forms nearly 97% of all the available water resources. In Israel, more than 2/3 of all the water used is pumped from aquifers.

In addition to acting as a source of water, aquifers may also perform other functions within a water resource system:

1) *Water Storage.* Water development and utilization projects usually require the availability of large storage reservoirs. Quite often, suitable sites for surface storage

reservoirs are scarce, but it may be possible to store large amounts of water in subsurface aquifers.

2) *Conveyance of Water.* An aquifer may serve as a conduit, by transmitting water from recharge areas to collection points where it is withdrawn for beneficial use. Water-table elevations can be controlled through recharge and pumping operations, thus affecting flow rates in the aquifers.

3) *Affecting Water Quality.* Due to their granular structure, some aquifers can serve as effective natural filters. In addition, it is possible to change the mineral quality of groundwaters by mixing them with waters introduced into the aquifer through artificial recharge.

4) *Regulation of Base Flow.* Spring discharge and stream base flow depend on the rate at which groundwater seeps out for an aquifer. They may be regulated to some extent by controlling the piezometric surface in the aquifer through pumping.

These considerations indicate that it is necessary to have at least some understanding of the basic principles that govern the movement of groundwater in aquifers.

The movement of water through soils, as well as through porous media in general, is best described by Darcy's law:

$$v = \frac{Q}{A} = -K \frac{dh}{dL} \tag{7}$$

where v is the velocity of flow (TODD, 1959), Q is the flow rate (discharge), A is the cross-sectional area of the medium through which flow takes place, K is the hydraulic conductivity of the medium, and dh/dL is the hydraulic gradient. Of course, the velocity v varies from point to point (both in magnitude and in direction) in most naturally occurring porous media. In addition, aquifer materials such as sand and gravel show different degress of heterogeneity in their physical and mechanical properties. Nevertheless, for all practical purposes, the velocity is usually given an average value in the direction of flow. Darcy's law is applicable whenever the flow is laminar, as is usually the case in nature.

Multiplying both sides of Eq. (7) by the thickness of the aquifer b, one obtains

$$vb = -Kb \frac{dh}{dL} \tag{8}$$

The quantity vb, whose units are L^2/T, is often referred to as the *specific discharge* of the aquifer and is designated by the symbol q. It represents the rate of flow per unit width of the aquifer. The product Kb is called *transmissivity* (with symbol T) and is very convenient in dealing with horizontal groundwater flow.

The velocity v defined in Eq. (7) has three direction components:

$$v_x = -K \frac{\partial h}{\partial x} \tag{9}$$

$$v_y = -K \frac{\partial h}{\partial y} \tag{10}$$

$$v_z = -K \frac{\partial h}{\partial z} \tag{11}$$

Continuity of flow, together with the assumption that the water and the porous medium are incompressible, lead to the follow equation:

$$\frac{\partial v_x}{\partial x} + \frac{\partial v_v}{\partial y} + \frac{\partial v_z}{\partial z} = 0 \tag{12}$$

Substituting the directional components of velocity (Eq. (9), (10), (11) into Eq. (12) leads to the general expression for steady flow of water in a homogeneous and isotropic porous medium:

$$\frac{\partial^2 h}{\partial x^2} + \frac{\partial^2 h}{\partial y^2} + \frac{\partial^2 h}{\partial z^2} = 0 \tag{13}$$

When flow takes place only along one direction, Eq. (13) simplifies to

$$\frac{\partial^2 h}{\partial x^2} = 0 \tag{14}$$

It should be emphasized that this expression is valid only for confined aquifers, and no exact analytical solutions are available for the phreatic case. A common approach in the case of unconfined aquifers is to adopt the so-called Dupuit assumptions and consider the specific discharge:

$$q = -K\,h\,\frac{dh}{dx} \tag{15}$$

Integrating this between the limits h_o (elevation of the water table above a fixed datum at point x_o) and h (Fig. 3), one obtains

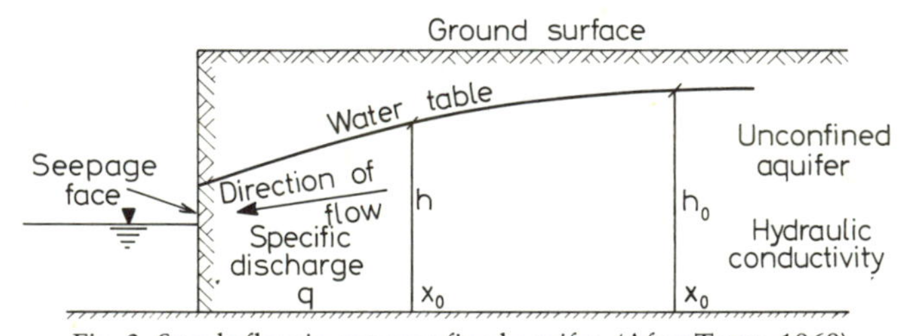

Fig. 3. Steady flow in an unconfined aquifer. (After TODD, 1969)

$$q = \frac{K}{2x}\,(h^2 - h_o^{\,2}) \tag{16}$$

In this equation, the quantities x, h and h_o are easily measured in the field. Methods for estimating the parameter K are discussed in Chap. V.

For unsteady flow, one must also consider changes in groundwater storage within the aquifer. Hence, a coefficient of storativity, S, is included in the mathematical equation (TODD, 1959):

$$\frac{\partial^2 h}{\partial x^2} + \frac{\partial^2 h}{\partial y^2} + \frac{\partial^2 h}{\partial z^2} = \frac{S}{T}\,\frac{\partial h}{\partial t} \tag{17}$$

Eq. (17) is valid only for confined flow. The mathematical expressions representing nonsteady flow of groundwater in unconfined aquifers are nonlinear in nature and therefore difficult to solve by exact analytical methods.

Groundwater Fluctuations

Distinction should be made between confined and phreatic (unconfined) aquifers. In the former, the water-bearing formation is sandwiched between layers having considerably lower hydraulic conductivity. In the latter, there is no upper confining layer. In the confined aquifer, the flow of water is somewhat similar to that in a pressure pipe, and if a borehole is drilled into the saturated zone, the water in it will rise above the upper boundary of the aquifer. Thus, one cannot speak of a water table under confined conditions, but of a *piezometric surface*. The term *water table* is reserved for phreatic aquifers.

When the sides of a borehole that penetrates a confined aquifer are insulated (e.g., by a steel pipe), the water level there is indicative of the pressure at the bottom of the borehole. Such a device is called a *piezometer*.

Changes in groundwater levels, including water tables and piezometric surfaces, are caused by changes in groundwater storage, whether due to natural or to artificial causes. There are cases in which groundwater fluctuations are induced by external loads.

Groundwater levels exhibit long-term and seasonal fluctuations that are commonly attributed to the rainfall pattern. Alternating series of dry and wet years will be reflected by corresponding drops and rises in groundwater levels. Similar fluctuations can also be observed within any one year, in regions where there are distinct wet and dry seasons. The seasonal fluctuation of the groundwater levels may lag behind the rainfall pattern, because of the slow movement of water within the porous aquifer. Thus, the aquifer has a damping effect on the annual variation of the available water resources in a given region.

Unconfined aquifers that are in direct contact with a stream channel may show groundwater fluctuations as a result of water-level changes in the stream. At certain times, the stream acts as a source, supplying water to the aquifer; at other times it acts as a sink. Similarly, in aquifers that are adjacent to the seacoast, groundwater levels often fluctuate in response to oceanic tides.

Diurnal fluctuations of the water tables can often be observed in shallow aquifers, as a result of evaporation and transpiration by vegetation. Piezometric surfaces may vary diurnally due to changes in the barometric pressure.

Summary

The major portion of all the waters on earth is concentrated in the oceans, with most of the remaining fresh water immobilized in polar ice caps and mountain glaciers. Thus, only a small part of the water that exists on this planet can be developed and utilized beneficially with present-day technology. As the world population increases and as the overall standard of living rises, water becomes more and more scarce. To use the available water more efficiently, a better understanding of the natural hydrologic processes is necessary.

The hydrologic phenomena that are most relevant to water resources engineering include the rainfall – runoff relationship and the be havior of moisture in soils and rocks.

The rainfall–runoff relationship represents the generation of a certain runoff discharge in response to a given rainfall intensity and pattern. It is complicated by the physiographic, vegetative, and cultural properties of the catchment (watershed) in question. Although these relationships are known to be almost always nonlinear (AMOROCHO and ORLOB, 1961), linearity assumptions are usually introduced, either implicitly or explicitly, in most hydrologic analyses. On the basis of Chow's (1962) runoff formula, a graph was developed for the coastal plain of Israel (Fig. 1) that describes the interdependence between the various physiographical factors, time, and runoff in the area.

A useful concept in estimating streamflow from rainfall data is the unit hydrograph (SHERMAN, 1932). This is a graphical representation of runoff discharge versus time that results from one unit of effective rainfall, falling uniformly over the area of a given catchment, at uniform rate during a given period of time. The usefulness of the unit hydrograph is linked to the linearity assumption that underlies this concept, i.e., 2 inches of effective rainfall would generate a peak discharge twice as large as that due to 1 inch of rainfall, if both fall during the same time. Although the assumption of linearity is not exactly correct, the unit hydrograph has found wide application in hydraulic engineering practice.

The investigation of groundwater flow phenomena gained great impetus during the last decade when the use of aquifers in water resources development became widespread in many parts of the world. The flow of groundwater in aquifers generally obeys Darcy's law. Although this law is a simplified expression for a complex natural phenomenon, it is very useful in calculations connected with the development and utilization of groundwater resources.

Groundwater tables in phreatic aquifers, as well as piezometric surfaces in confined aquifers, fluctuate with time due to the hydrologic regime prevalent in the region. Such fluctuations are analogous in nature to the variations in streamflow exhibited by surface hydrologic system. However, in the subsurface the fluctuations have lower amplitudes and longer wavelengths than is the case with surface systems. This is very helpful in planning the development of aquifer systems.

Literature

AMOROCHO, J., ORLOB, G. T.: Nonlinear analysis of hydrologic systems, Contribution 40. Water Resources Center. Los Angeles: Univ. of California 1961.

CHOW, V. T.: Hydrologic determination of waterway areas for the design of drainage structures in small basins, Bull. 462. Urbana: Univ. of Illinois 1962.

CHOW, V. T. (Ed.): Handbook of Applied Hydrology. New York: McGraw-Hill 1964.

LINSLEY, R. K., Jr., KOHLER, M. A., PAULHUS, J. L. H.: Hydrology for Engineers. New York: McGraw-Hill 1958.

LINSLEY, R. K., Jr., FRANZINI, J. B.: Water-Resources Engineering. New York: McGraw-Hill 1964.

MILEVSKY, A.: Master plan for the drainage of Nahal Poleg basin (in Hebrew). M. Sc. Thesis, Technion, Haifa 1969.

SHERMAN, L. K.: Stream flow from rainfall by the unit graph method. Eng. News-Record, **108**, 501–505 (1932).

TODD, D. K.: Ground Water Hydrology. New York: Wiley 1959.

2

Hydrology of Arid Zones

S. MANDEL

Hydrological Principles

The term *water crop* denotes the resources of liquid water (surface water as well as groundwater) that are generated by precipitation at more or less regular intervals. The part of the hydrologic cycle that generates the water-crop is not essentially different under conditions of aridity than under any other climatic conditions but is modified to a lesser or greater degree. Before discussing these modifications, it is useful to review briefly the relevant part of the hydrologic cycle and to define a few terms that will be used in the subsequent discussion.

For the sake of simplicity we shall assume that precipitation always falls in the form of rain. A certain part of the rainfall is intercepted by vegetation or accumulates in shallow puddles and is quickly lost by evaporation.

Effective rainfall, that ist, total rainfall minus the evaporation losses, *infiltrates* into the soil, forms the *soil moisture,* and generates the water crop. *Infiltration rates* vary within a wide range between approximately 25 mm/hr in light soils and 0.25 mm/hr in heavy clayey soils. When a soil is wetted, the infiltration rate decreases gradually to a given limit. (Infiltration rates represent the amount of rainfall that the soil is able to soak up within a given period of time. These rates are, of course, considerably smaller than the actual velocities at which the particles of water travel downward through the interstices of the soil.)

Field capacity is defined as the maximum amount of water that a given soil is able to hold against the pull of gravity and is usually expressed in mm. The soil moisture is subsequently depleted by evaporation and by transpiration through plants, a process known as *evapotranspiration.* The evaporative losses from a free water surface under given climatic conditions are called *potential evaporation* (mm/yr, mm/day, etc.).

Consider an area covered with soil and underlain by pervious rocks such as sand, gravel, fissured rocks, etc. After the soil has been wetted to field capacity, the rest of the water *percolates* downward through the pervious strata and eventually *replenishes* the groundwater in an aquifer. The area across which this percolation occurs constitutes the *replenishment area* of the *aquifer.* In the aquifer, groundwater flows slowly and laterally to some outlet, such as a spring, which often may be located some distance from the original replenishment area. We shall return to a more detailed discussion of groundwater systems on p. 46.

Since percolation takes place only when the soil moisture is at least at field capacity, groundwater replenishment is strongly dependent on rainfall duration. Surface runoff occurs whenever the rainfall intensity exceeds the rate of infiltration of the soil. Thus, it depends strongly on rainfall intensity, and is influenced by rainfall duration only to the extent that the longer it rains, the less pervious the soil becomes.

The above picture is of course highly schematic; in nature conditions are usually much more complex. An aquifer may drain into a river, or a river may replenish the groundwater in an aquifer. Evapotranspiration may take a considerable toll from shallow groundwater bodies and from the banks of flowing streams. It is difficult to estimate the "moisture storage" of areas covered by patchy soils, partially weathered rock debris, etc.

If all the paramters, such as potential evaporation, field capacity, infiltration rate, and geology of the subsoil are known, the hydrologic cycle can be simulated using a *bookkeeping procedure,* and surface runoff as well as groundwater replenishment can be calculated. In actual practice, most of the required parameters must be deduced from records of past observations, or at least must be calibrated against such records. Therefore, the book keeping procedure is less synthetic and more empirical than it may appear. Since complete hydrologic records are seldom available, simpler methods based on the vaguely defined concept of a "yearly average" are still widely used, especially in groundwater hydrology.

The Hydrologic Regime of Arid Zones

The Water Crop of Arid Zones

Arid zones can be characterized by the following conditions:

a) Annual potential evaporation is significantly greater than annual rainfall.

b) Precipitation occurs only during well-defined seasons and tends to vary strongly from year to year. In extremely arid zones, precipitation is a rare event and has no regular characteristics.

c) Plant cover is limited primarily by the availability of soil moisture. (Cold arid zones, as well as areas with excessively saline soils, form an exception to this rule.)

In arid zones, the values of potential evaporation can be expressed as a rather stable function of the year's season and precipitation remains the only stochastic parameter of interest. It is therefore important to ask in what way the available record of rainfall relates to the corresponding water crop record.

Since effective rainfall is always considerably smaller than measured rainfall, one must write off all those days during which the daily rainfall remains at or below the daily potential evaporation. In addition, a certain minimum amount of soil moisture is always abstracted by the native vegetation (agriculture tends to increase this amount to a considerable extent).

It follows that small changes in the annual rainfall can have a very significant effect on the annual water crop. In particular, years with very little or no water crop tend to be much more frequent than "drought years" as defined by climatic criteria.

Furthermore, due to seasonal changes in potential evaporation and influence of the soil moisture, years with similar amounts of rainfall may yield different water crops. In particular, a good agricultural year that is characterized by well-spaced, low-intensity rainfall is apt to be a poor year as far as the water crop is concerned, and vice versa.

Empirical rules are sometimes employed in estimating the annual water crop from rainfall data. One particularly useful rule has the form

$$R = a \times (P - b)$$

where R and P are the water crop and the amount of rainfall, in mm/yr, respectively, and a and b are empirical constants characteristic for a given region.

Implicit in this formula is the assumption that rainfall occurs according to a fixed seasonal pattern. The formula does not take into account any deviations from this assumed fixed pattern (say, rainfall in April instead of January). It was found that in the central and northern part of Israel, $a = 0.9$ and $b = 360$ mm/yr.

In addition to its simplicity, a formula of this kind has several attractive features: It is dimensionally correct it also clearly defines a threshold value at which the water crop becomes zero (e.g., as above, 360 mm/yr), in conformity with the most salient characteristic of arid zones moreover, only two parameters have to be deduced from empirical data.

In view of the great complexity of the phenomena involved, it seems doubtful whether a reasonably accurate "global" formula can be developed in order to estimate the water crop under all possible conditions.

Characteristics of Surface Runoff

Almost all the surface runoff in a typical arid zone occurs at irregular intervals, in the form of large floods that last only a few days or hours. During the rest of the time, the riverbed is dry or may carry just a small trickle of water. A seasonal regularity of the flow characteristics is observed only in rivers that are fed from snow melt or from less arid, mountainous regions.

The time lag between the beginning of precipitation and the onset of surface runoff depends on the properties of the surficial layers. Clayey soils often swell and become practically impervious once the uppermost layer has been wetted. On the other hand, the infiltration capacity of light soils diminishes gradually with time, reaching a minimum value when the entire soil section has been weeted. In the semidesertic northern Negev of Israel, in areas covered by loess (a friable aggregate composed mainly of clay particles), intense runoff usually starts after only about 10 mm of antecedent precipitation. In the coastal plain of Israel, which is covered mainly by light soils, up to 100 mm of antecedent precipitation are required before runoff can occur.

Since surface runoff depends strongly on rainfall intensity, the correlation between annual rainfall and riverflow is usually very poor. Large floods as well as large annual discharges may occur in rivers during climatically dry years, and vice versa.

High-intensity rainstorms tend to be of small areal extent and are seldom adequately recorced. Under these conditions, those methods of hydrologic analysis that rely on the assumption that rain is distributed over the entire catchment area must be applied with caution.

Characteristics of Groundwater Replenishment

Areas without soil cover where there are extensive outcrops of sand, gravel, fissured rock, etc., are the most favorable for groundwater replenishment. Even a patchy soil cover can cause surprisingly large losses of water by evapotranspiration. The seemingly bare, karstic western slopes of the Judean Hills lose about 400 mm/yr out of an average rainfall of 600 mm/yr due to this cause.

Deep percolation of water from the soil to an aquifer occurs only during rainstorms that are of exceptionally long duration of that follow each other at short intervals, so that the soil moisture constantly remains above or near field capacity. In shallow light soils this may happen more or less regularly each year, but in deep soils groundwater replenishment is a very rare phenomenon.

In very arid regions, even the small surficial moisture-holding capacity of porous rocks, fine sands, and partially weathered rocks is sufficient to dissipate practically all of the infiltrated rainwater through evapotranspiration. Under such conditions, groundwater is replenished only locally by seepage from infrequent river floods; this is why in desertic areas bodies of fresh and comparatively young groundwater are found in the vicinity of river channels, whereas regional aquifers usually contain old and brackishwater.

The Springs of Arid Zones

Two classes of springs of special interest are the very large perennial springs that frequently form the outlets of karstic aquifers, and saline springs or saline swamps in desertic regions.

Almost all of the large perennial springs in arid regions form the outlets of carbonate aquifers (limestone, dolomite, chalk, etc.). The genesis of these springs is associated with the formation of solution channels by groundwaters containing dissolved CO_2 and the progressive generation of flow channels in the aquifer that lead the water toward one common outlet. Measurements of dissolved CO_2 in waters from carbonate aquifers in Israel demonstrate that the solution process is active under the semiarid conditions currently prevailing at depths of 100–500 m. below the surface. In view of these findings, it is not necessary to invoke hypothetical pluvial periods in trying to account for the genesis of large springs in arid regions.

Although the reliable discharge of a perennial spring constitutes a very attractive source of water, a considerable part of this discharge is lost during the wet season because of lack of demand. If wells are drilled into an aquifer that feeds a spring, the total water supply (residual spring flow plus pumpage) can be regulated to minimize water losses. Although this regulation of the water supply is bought at the price of pumping, it can be achieved without the surface storage usually associated with high evaporative losses.

Saline swamps are indicative of discharge from aquifers in desertic regions where the water evaporates as soon as it appears at the surface. In the center of the swamp is a sterile region covered by salt incrustations surrounded by phreatophytic salt-resistant plants. Frequently, date palms are cultivated on the fringes of the swamp, thus forming an oasis. Better-quality groundwater can be obtained from boreholes drilled at some distance from the swamp where the salt content has not yet reached high levels.

Mineralization of Water Resources

Mineralization of water resources, and especially of groundwater, is the universal problem of arid regions. It is caused by the following mechanisms:

1) Evaporation from open water surfaces and from shallow groundwater. Extreme examples are the "playas" (seasonal inland lakes that evaporate and leave behind ex-

tensive salt incrustations), and the saline swamps or "sabkhas". The relatively high salinity of groundwater immediately below a shallow water table has been confirmed in many regional investigations. In such areas, better water can usually be obtained from deeper, suitably constructed boreholes.

2) Fossile brines from ancient lagoons and inland lakes. Vestiges of such brines often fill the deeper parts of aquifers and remain semistagnant due to their greater specific weight. Large reservoirs of fossile brines persist during quasi-geological periods, even in areas where groundwater circulation is relatively active. For example, the fossile brines in the Jordan rift valley seem to represent vestiges of a Miocene sea and are continuously being flushed out through saline springs, at least since the intensive faulting of the area at the end of the Pleistocene era. At present this process contributes Cl ions to the Jordan River at a rate of about 70,000 ton/yr.

3) Connate salts precipitated together with in fine-grained marine sediments. Hypothetical connate salts provide an explanation for the relatively high salinity of water derived from fine-grained rocks.

4) Airborne salts deposited by precipitation and in the form of dry fallout. This factor is of considerable magnitude in some areas, e.g. in the coastal plain of Israel where about 15 ton/km^2 of salt are deposited from the air annually.

5) The brackish springs that issue from karstic limestone formations in the Mediterranean region, neat the sea or offshore. These are probbly due to the continuous convection of seawater.

The investigator who tries to pin down the sources of salinity in a given area is faced with a number of plausible hypotheses. However, it is often difficult to adequately substantiate the explanation chosen.

The dominant chemical ions in natural waters are Cl, SO_4, HCO_3, Na, K, Ca, and Mg. Natural geochemical groups of water can be distinguished on the basis of the relative abundance of these elements, and sometimes also with the help of minor constituents such as F, Br, and S. The groups indicate (although not always in a completely unambiguous way) the natural environment through which the water has passed.

The following are the major characteristics of some of the more common groups (the symbols refer to equivalent units):

1. Water in contact with granitic rocks	$Na + K > Ca + Mg$
	$HCO_3 > SO_4 > Cl$
2. Water from carbonate rocks	$Ca \geqslant Mg > Na + K$
(limestones and dolomites)	$HCO_3 > SO_4 > Cl$
3. Waters from basaltic aquifers	$Na > Ca > Mg > K$
	$HCO_3 > SO_4 > Cl$
4. Seawater	$Na > Mg > Ca > K$
	$Cl > SO_4 > HCO_3$

If brackish water is encountered, geochemical investigations usually enable one to distinguish between one or two freshwater components and a certain admixture of the brine. However, the chemical composition of almost all brines converges toward that of seawater. When the brine enters a freshwater environment, ion-exchange and dilution processes obliviate the distinctive geochemical characteristics that the concentrated brine might have possessed. For this reason, the actual source of salinity can seldom be determined on the basis of geochemical studies alone.

The Development and Exploitation of Groundwater in Arid Zones

Techniques of Investigation

In principle, groundwater exploitation resembles that of a large surface-water reservoir. The problem consists of routing a stochastic input pulse (natural replenishment) through a deterministic system (the aquifer) and devising operations on the system (pumping) so that a certain objective (water supply) is met with due regard to certain constraints (probability of dangerously low water levels, etc.).

The physical properties and the basic equations that govern the behavior of a groundwater system are more complex than those that govern the behavior of a surface-water reservoir. However, the routine application of modern analog and digital simulation techniques make it possible to achieve the required aim, provided that the following data are available: (a) a map of the hydrological constants of the aquifer and its boundaries; (b) the discharge mechanism of the aquifer; (c) definition of "safe" water levels (here a time factor is normally involved; for example, although it may be safe to lower the water level drastically for a short period, mineralized water may be attracted if this low level is maintained for an extended period); (d) a long-term series of replenishment events; and (e) the location, discharge, and depth of boreholes.

Unfortunately, there is no straightforward method of obtaining these items of information; they must be put together from geological maps, subsurface geological data gathered from boreholes and geophysical measurements, records of water levels measured in precisely leveled boreholes, records of pumping, pumping tests, chemical analyses, climatic data, etc.

Most projects concerned with groundwater investigation pursue the twin objectives of simultaneously producing water and information on which the future rational exploitation of the source can be based. The best procedure is to develop the groundwater resource in stages and to "feed back" the information gained during each stage into the previous conceptual picture until a sufficiently detailed and accurate geohydrologic model emerges. In this way, some costly research items, such as special exploration boreholes, can be reduced to manageable levels and others, such as pumping tests, can be added to the utilitarian business of completing and testing exploitation boreholes. The entire research effort thus can be spread over a longer period. This procedure requires mature judgment, ingenuity, and a firm hand on the part of the investigator, a task that certainly cannot be entrusted to junior personnel.

New techniques, based on the use of environmental isotopes (mainly tritium, deuterium, and oxygen-18), may prove beneficial in reducing field work and furnishing information that cannot easily be obtained by other means.

Tritium dating enables the determination of the residence time of water in an aquifer. The ratios of deuterium/hydrogen and oxygen-18/oxygen-16 are indicative of the mechanism of replenishment. If the water stayed on the surface for some time prior to infiltration (e.g., if replenishment was due to percolation from a river), then the heavier isotopes are relatively enriched. Isotope methods are applicable only within certain theoretical limits.

The selection of sampling sites and the specification of sampling procedures require meticulous planning as well as some prior information on the hydrologic system under investigation. Meaningful results can be derived only from an adequate number of representative samples. Unfortunately, since isotope analyses are still time-consuming and costly, there is a tendency to be content with a small number of poorly defined samples that do not yield reliable information. As a result of such difficulties, these new techniques so far have gained only limited acceptance in hydrologic investigations.

It should be recognized that a complete description of an aquifer system is seldom achieved in practice. Very frequently guidelines for exploitation must be derived from incomplete hydrogeologic models that do not justify the use of highly sophisticated mathematical optimization procedures.

Exploitation of Groundwater Reserves

Since many aquifers contain large volumes of water in storage, groundwater is considered the most reliable water resource in arid regions. The aquifer functions as a seasonal storage reservoir during dry periods and also as a long-term storage reservoir whose effect is to smooth out large fluctuations in groundwater replenishment.

Usually a distinction is made between exploitation within the limits of the so-called "*safe yield*" and "*mining*" of a groundwater reserve. In safe-yield exploitation, the average annual groundwater withdrawal is kept smaller than the average annual groundwater replenishment. If withdrawal exceeds replenishment during an indefinite period of time, the groundwater resource is said to be "mined". Neither term is exact, but a more precise definition does not exist.

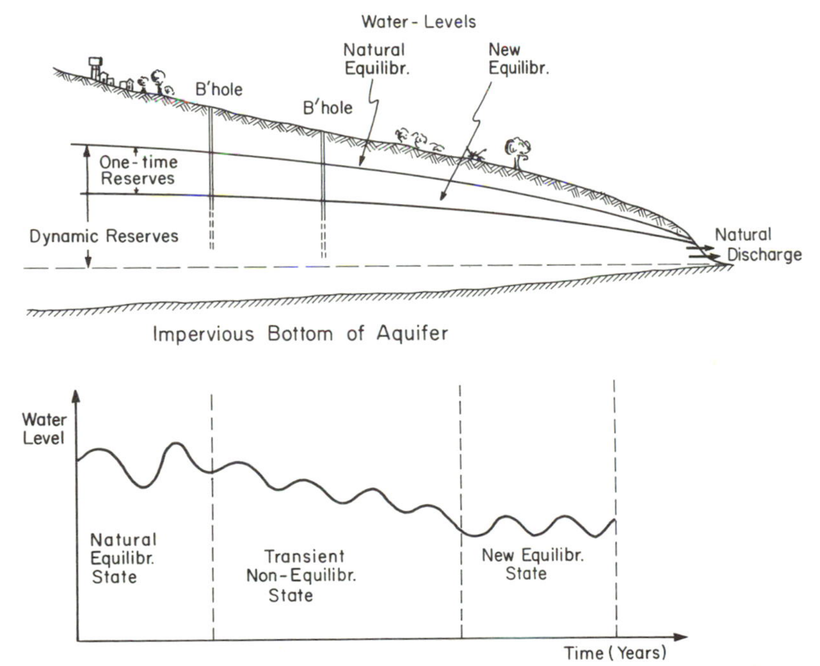

Fig. 1. The concept of groundwater storage

The concept of groundwater storage is schematically illustrated in Fig. 1. The volume of water that is held between the water table and the impervious substratum represents the total available reserve. The volume of water that lies below the water table and above the level of the natural discharge may be called the *dynamic reserve*. The entire volume of water flows slowly toward the natural outlet, and the energy for this movement is derived from the weight of the dynamic reserve. When natural replenishment takes place, the water table rises, the dynamic reserve increases, and the rate of natural discharge also increases. Between any two successive events of replenishment, both the dynamic reserve and the natural discharge decrease. If the dynamic reserve is large in comparison with the average annual replenishment, as is the case in most aquifers, these fluctuations are damped out until an equilibrium is reached between the average annual replenishment and the average annual natural discharge. When groundwater is withdrawn by pumpage, the water table falls, the dynamic reserve diminishes, and the natural discharge of the aquifer decreases until a new equilibrium is established. These hydraulic states of the aquifer can be succinctly represented by the following equations:

Natural equilibrium: average annual replenishment = average annual natural discharge

Transient conditions: average annual replenishment < annual natural discharge + pumpage

New equilibrium: average annual replenishment = diminished average annual natural discharge + pumpage

There can be no continuous groundwater exploitation without a simultaneous reduction in the dynamic reserve. If, for example, groundwater is withdrawn at a rate that constitutes 80% of the average annual replenishment, a new equilibrium will not be reached until about 80% of the dynamic reserve have been eliminated and the natural discharge has been diminished to about 20% of its original value. That part of the dynamic reserve that has dissipated is called a *one-time reserve*. In an arid zone, the reduced reserve must still be sufficiently large to enable the "safe" exploitation of an aquifer during a possible dry cycle.

The duration of the transient state depends on the ratio between the dynamic reserve and the average annual replenishment, and is typically 5 –50 years. During this long transition period, large volumes of water are often discharged from the aquifer without being utilized. These losses can be minimized and at least part of the "one-time reserve" can be recovered by temporarily stepping up the rate of groundwater withdrawal to above the safe-yield limit. Of course, the rate of withdrawal must be reduced to within the safe-yield limits as soon as an acceptable new equilibrium state is approached in the aquifer.

The exploitation of a "one-time reserve" has been successfully implemented in Israel. The purpose was to obtain urgently needed water supplies and at the same time to gain several extra years in which to complete the National Water Carrier from the Jordan-river.

Large-scale mining of groundwater currently takes place in such areas as the south western part of the United States, Mexico, and the northwestern part of Australia. In all of these cases mining has not been a deliberate policy but stemmed entirely from uncontrolled development of the groundwater resources. Any new plans for the efficient use of the remaining groundwater reserves must now contend with established trends that are difficult to control. One can only hope that the vast reserves of groundwater that have been recently discovered in Libya, the Sinai, the Negev desert in Israel, and elsewhere in the world will be mined according to well-defined and rational plans.

Literature

GAT, J. *et al.*: The stable isotope composition of mineral waters in the Jordan Rift-Valley of Israel. J. Hydro. **7**, 334 (1969).

GOLDSCHMIDT, M.: On the water-balance of several mountain undergroundwater catchments in Israel. Hydrol. paper **4**, Israel Hydrol. Service (1959).

HARPAZ, Y.: Hydrological investigations in the southern desert of Israel. IASH Pub. **56–57**, p. 124 (1961).

IASH: Isotopes in Hydrology. Vienna Symposium, Nov. 1966.

MANDEL, S.: A conceptual model of karstic erosion by groundwater. IASH Pub. **74**, p. 662 (1965).

MERO, F.: An Approach to Daily Hydro-Meteorological Water Balance. Computations for Surface and Groundwater-Basins. "Tahal" Tel Aviv, P. N. 1011, Aug. 1969.

MERO, F.: Application of groundwater-depletion curves in analyzing and forecasting spring-discharges influenced by well fields. IASH Pub. **63**, p. 107 (1964).

SCHOELLER, H.: Les Eaux Souterraines. Paris: Masson et Cie 1962.

SIMONS, M.: Deserts, the Problem of Water in Arid Lands. Oxford: Univ. Press 1967.

TAUSSIQ, K.: Natural Groups of Groundwater and Their Origin. Tel Aviv: Mekoroth Ltd 1961.

UNESCO: Paris Symposium on the Problem of Arid Zones, 1962.

UNESCO: Arid Zone Hydrology: Recent Developments (Ed. SCHOELLER, H.), 1959.

WALTON, K.: The Arid Zones. Chicago: Aldine 1969.

YAALON, H.: Salinzation and salinity, J. Chem. Edu. **44**, p. 59 (Oct. 1967).

3

Development and Management of Water Resources

N. Buras

Introduction

In the first part of this chapter, some basic hydrologic concepts were presented, with special emphasis on situations prevailing in arid zones. The hydrologic information collected in the field and processed in the form of hydrographs and groundwater maps, or in other adequate forms, is used as an input to the process of planning the development of water resources and of designing systems that will attain development objectives.

The general problem of development and management of water resources has recently received considerable attention (Maass *et al.,* 1962; Hall and Dracup, 1970; James and Lee, 1971; Buras, 1972). The problem has wide ramifications and is related to several disciplines. In dealing with arid zones, the planner must cope with the fact that there is a scarcity of water (especially surface-water) resources, and that there is a great variability in natural surface flows. For this reason, the present section will deal with the following two aspects of water resources engineering: the management of groundwater resources as a supplement to scarce surface-water supplies; and the storage of surface waters as a solution to the problem of variable streamflow. The latter aspect is related to the question of reservoir design and operation.

Application of Hydrologic Principles to Reservoir Design

The Rippl Method

Flow-Mass Curves. A major problem usually encountered in the development and utilization of surface waters is the discrepancy between demand and supply. The natural supply of water in a stream varies in accordance with the governing hydrologic phenomena, whereas the demand is a function of various climatic, technological, and economic factors. In order to overcome the imbalance between supply and demand, reservoirs are constructed in such a way that water flowing into them during periods of relative abundance can be used at times of low flow.

A reservoir can be looked upon as a queuing system (Langbein, 1958) in which the arriving quantities of water "queue up" behind the dam until released in accordance with a prescribed release rule. The inflows into a reservoir are considered to be stochastic variables that theoretically have no upper bound. It therefore follows that there is a finite probalility, albeit small, that a streamflow discharge that is higher than any on record will occur in the future, and there is a finite probability that a discharge that is lower (the lower limit being zero flow) than any on record will occur in the future. Considering this situation, one can postulate that given a natural hydrologic system, its maximum average yield is at most equal to the long-term average discharge; and in order to deliver indefinitely a yearly amount of water equal to the long-term average annual flow, it is necessary to build an infinite reservoir (HURST, 1951).

A rigorous proof of these postulates is beyond the scope of this book. However, suffice it to say that reservoirs provide us with a certain degree of streamflow regulation that is certainly less than 100%. The regulated flow that can be used for a variety of beneficial purposes is called the reservoir *yield.*

One method of estimating the yield of a reservoir utilizes a so-called flow-mass curve and is termed the *Rippl method* (RIPPL, 1883). An example of this is shown in Fig. 1, where the continuous solid line represents the cumulative flow at a site suitable for surface storage. The slope of the parallel broken lines, *AG* and *BF,* gives the outflow from the reservoir (the latter is also called reservoir *yield*). The reservoir yield in this case is 75×10^6 m³/yr.

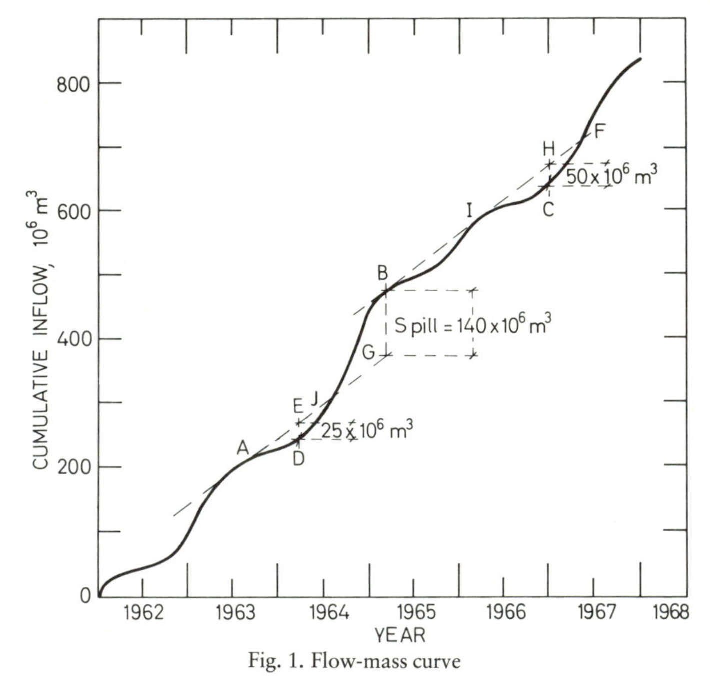

Fig. 1. Flow-mass curve

Consider the portion of the mass curve between points *A* and *F*. Assuming that the reservoir is full, at point *A*, the rate of outflow (broken line) is greater than the rate of inflow (solid line). The maximum distance between the outflow line *AG* and the mass curve is *ED,* which represents the shortage (or deficiency) that would have accrued if the demand of 75×10^6 m³/yr would have been satisfied by natural streamflow alone. The distance *ED* represents 25×10^6 m³, which is the magnitude of this deficiency. Hence, in order to insure a yield of 75×10^6 m³ yr, it is necessary to provide a storage facility with a capacity of 25×10^6 m³.

Beyond point *D*, the rate of inflow is greater than the rate of outflow, and the reservoir is filling again. In fact, at point *J* the reservoir is full and spilling. The full reservoir will continue to spill as long as the rate of inflow is greater than the rate of outflow (point *B*). The amount spilled is measured by the vertical distance *BG* (140×10^6 m³).

At point *B*, reservoir drawdown is resumed. The reservoir is again filled (but is not spilling) at point *I*, and then again drawn down. A new shortage of water may develop if there are no stored reserves, as shown by the portion of the inflow curve *ICF* that is below the straight outflow line *IF*. The maximum distance between the two curves is at *HC* (50×10^6 m^3), which also shows the size of the reservoir required for the maintenance of a specified yield of 75×10^6 m^3/yr.

This example shows how to estimate the volume of storage that must be provided to ensure a given rate of supply. One should emphasize that the longer the hydrologic record, the more accurate is the estimate.

The Rippl Mass Curve as a Production Function. Production is defined as the process of converting given resources (inputs) into other resources (outputs) whose form and location are more useful for further production and consumption. In the case of reservoirs, input resources such as land, steel, cement, machinery, and manpower are measured in terms of cost; they are then converted into irrigation water, industrial water, hydroelectric power, etc., which are expressed as reservoir yield. The relationship between the *input* and the *output* of the system is expressed by the *production function*. This function also indicates the maximum outputs that can be obtained from given inputs.

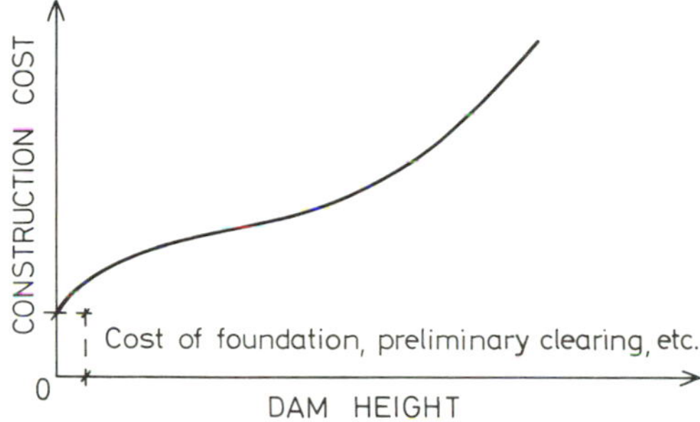

Fig. 2. Construction cost as a function of a dam height

An example of how the flow-mass curve is used to derive the dam height might be considered. Each dam involves a different construction cost, and one can plot a curve as shown in Fig. 2. The site can be characterized by a curve of reservoir capacity versus dam height, which is a function of topography. This curve is shown in Fig. 3.

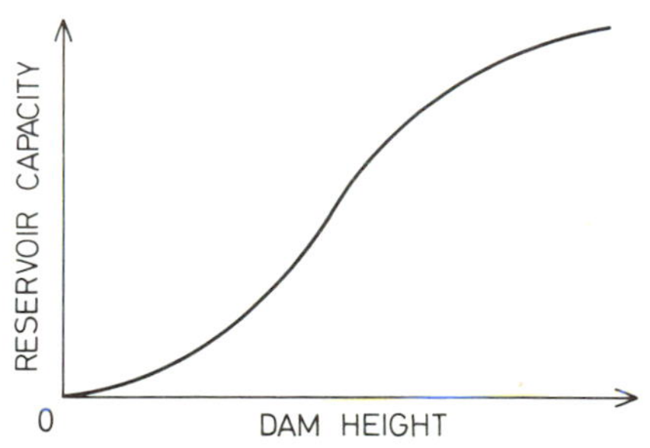

Fig. 3. Reservoir capacity function of dam height

The Rippl mass curve can now be used. For a specified reservoir yield, a certain reservoir capacity is necessary. Thus, one can plot the reservoir yield versus the reservoir capacity. The general shape of this curve is shown in Fig. 4.

Figs. 2, 3, and 4 represent the basic data for the problem. By combining the graphs in Figs. 2 and 3, one obtains the cost of construction as a function of reservoir capacity, as shown in Fig. 5. Finally, by combining Figs. 4 and 5, one obtains the reservoir yield

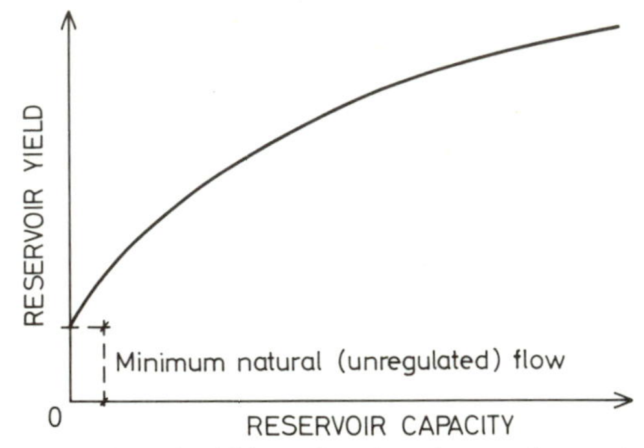

Fig. 4. Reservoir yield as a function of reservoir capacity

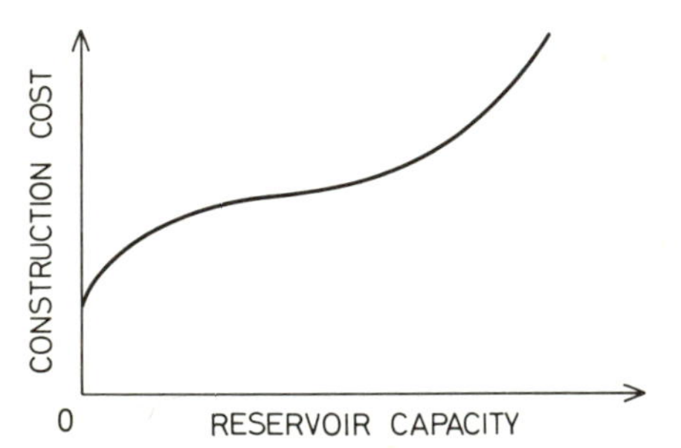

Fig. 5. Construction cost as a function of reservoir capacity

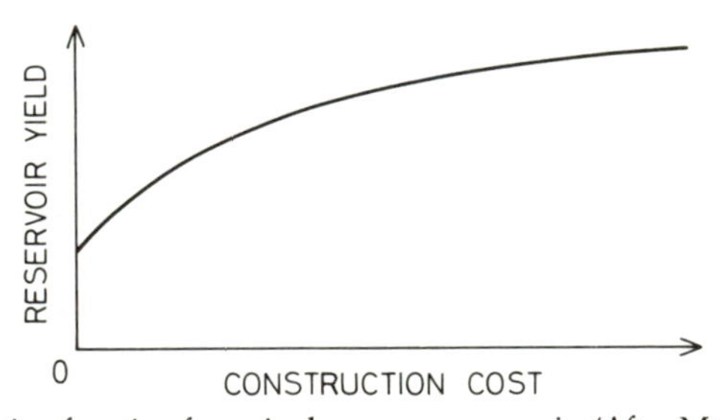

Fig. 6. Production function for a single-purpose reservoir. (After MASS *et al.*, 1962)

as a function of construction cost. This is the desired production function, inasmuch as the reservoir yield represents the output and the construction cost the input into the storage system. Fig. 6 shows the final production function.

Hydrologic Uncertainty

Introduction. Hydrologic phenomena such as streamflows and groundwater-table fluctuations exhibit a certain amount of variability, which is observed not only seasonally (summer flows are different from winter flows) but also annually. The range of this variability is much narrower in a temperate zone than in an arid region because in the former the precipitation is more evenly distributed throughout the year. In an arid zone, where the wet seasons are interspersed with long dry periods, some water courses may carry discharges as high as several thousand cfs or run completely dry. Furthermore, the yearly flows may vary form almost zero to many billions of cubic meters per year.

This situation poses serious problems in planning the development and utilization of water resources. First, the uncertainty involved in the hydrologic phenomena must be expressed quantitatively with the aid of probabilistic expressions. Second, the probability of occurrence of the hydrologic events has to be estimated by performing frequency analyses on data such as streamflow, discharge, water-table elevation, etc. Recently, several methods have been developed for the synthetic generation of streamflow data in order to study the effect of various sequences of high and low flows on the design and operation of a specific water resource system.

Hydrologic Frequency Studies. It is important to recognize that hydrologic uncertainty is not the only probabilistic aspect in the design of a water resource system. Uncertainties involving costs and future returns introduce considerable complications into the design process. At this stage, however, we shall address ourselves to the uncertainty connected with the natural hydrologic process from which water has to be developed for beneficial use.

Early hydrologic frequency studies attempted to fit a Gaussian probability curve to field data such as yearly discharges (HURST, 1951). However, the Gaussian (normal) probability law has a serious shortcoming (as far as hydrologic events are concerned) in that its range spans the entire domain of natural numbers from $-\infty$ to $+\infty$. This is inadmissible in hydrology because one cannot speak of negative flows. In answer to this criticism some hydrologists claimed that if the mean value of the distribution is shifted far enough to the right, the area remaining under the negative tail of the frequency distribution curve is so small that for all practical purposes it can be neglected.

Later it was discovered that when the logarithms of the streamflow data are used, a good agreement with the normal distribution law is obtained without necessarily resorting to the concept of negative flows.

When dealing with hydrologic uncertainty, it is desirable to estimate the following two interrelated quantities.

1) *Recurrence interval* is defined as the average interval, in years, between the occurrence of a hydrologic event of a specified magnitude and an equal or larger event. Such an event may be, for example, the peak flow or the largest daily discharge. If there exists a hydrologic record of n years and the events (e.g., peak flows) are ranked in decreasing order of magnitude, ($m = 1,2, \ldots, n$), a simplistic approach would be to estimate the recurrence interval, t_p, as n/m. This may appear perfectly reasonable, since for $m = 1$ (the highest peak flow on record) a good estimate of the average interval between this event and an equal or larger peak flow is n years. However, if we consider $m = n$) the lowest peak flow on record), then the recurrence interval is $n/n = 1$, i.e., every year we can expect a flow as low as the lowest recorded so far. What about peak flows *lower* than those on record? We have no way of estimating their recurrence interval. For this reason, the recurrence interval is given by the formula

$$t_p = \frac{n+1}{m} \tag{1}$$

2) *Probability of Occurrence.* If an event has a recurrence interval of $_{tp}$ years, then the probability that it will be equalled or exceeded in any one year is given by

$$p = \frac{1}{t_p} \tag{2}$$

In addition to the probability that a flow of a given magnitude will be equalled or exceeded in any one year, we are often interested in estimating the probability that an event of a given recurrence interval will be equalled or exceeded in a given period of time. For example, if one designs a storage system with a 25-year economic life, one may ask what the probability is that during these 25 years a 100-year flood will occur. To answer this question, let p denote the probability that a 100-year flood will happen in any one year. The probability that this flood will *not* occur in any one year is $1-p$. Assuming that the yearly flood events are statistically independent, the probability that a 100-year flood will not happen in two consecutive years is $(1-p)^2$. Extending this reasoning to the economic life of the storage system where $N = 25$ years, the probability that a 100-year flood *will not* occur during this period is $(1-p)^N$. Finally, the probability that this flood *will* happen during the economic life of the project is

$$P = 1-(1-p)^N \tag{3}$$

The probability of occurrence of a 100-year flood therefore is $p = 1/100 = 0.01$; $1-p = 0.99$; $(1-p)^N = 0.99^{25} = 0.78$; $P = 0.22$. Thus there is a chance of better than 1 in 5 that during the economic life of the project the storage system will have to cope with a 100-year flow, or even with a larger flow.

When studying frequencies of flows, one is often interested in estimating the frequency of rare events such as floods and droughts. A theoretical frequency distribution of extreme events has been derived by GUMBEL (1958). According to him, the probability π of occurrence of a flood whose magnitude is equal to or greater than some given value x is

$$\pi = 1-e^{-e^{-b}} \tag{4}$$

Because of its form, Eq. (4) is also called the double-exponential probability-distribution function. The parameter b is given by

$$b = \frac{x-\bar{x} + 0.450}{0.78\,\sigma} \tag{5}$$

where $\bar{x}$ is the magnitude of the average flood on record and σ is the standard deviation of the observed series of floods. Values of b corresponding to various return periods t_p and probabilities π are given in Table 1. Using Eqs. (4) and (5) to extend Table 1 in both directions, one can easily construct an appropriate probability chart for the graphical representation of flood (or drought) data.

Hydrologic frequency studies are made to obtain quantitative estimates of the uncertainty associated with hydrologic events and are an accepted practice in water resources engineering. The end product of these studies is a table (or a curve) that indicates the recurrence interval and the probability of occurrence of streamflows or other hydrologic events of various magnitudes. Of particular interest are extremely low and extremely high flows, which are usually extrapolated from existing data.

Generation of Synthetic Data. In analyzing streamflow data for design purposes (here "design" includes the preparation of operating policies), one must ask the following question: If the sequential pattern of the hydrologic events were different, how would this affect the design and operation of the water-resource system?

When so viewed, the hydrologic data exhibit another uncertainty in the possible sequences of dry years and wet years, in addition to the uncertainty connected with the magnitude of flows. It therefore appears that our set of hydrologic data represents a unit sample, taken from an infinite population of equally likely hydrologic records, all of which have the same mean and variance but differ from each other with regard to the internal sequence of events. To overcome the difficulties that one encounters with such a small sample, it has been proposed to generate additional synthetic data that would have the same statistical properties (mean and variance) as the original record.

Table 1. Values of b in Gumbel's double exponential function. (After LINSLEY and FRANZINI, 1964)

b	Return period t_p	Probability π
0.000	1.58	0.368
0.367	2.00	0.500
0.579	2.33	0.571
1.500	5.00	0.800
2.250	10.0	0.900
2.970	20.0	0.950
3.902	50.0	0.980
4.600	100.0	0.990
5.296	200.0	0.995
6.000	403.0	0.9975

To evolve such a synthetic series, one assumes that the streamflow observed during any given period (say, a month) is linearly related to the streamflow of the previous period. This is the same as saying that the outcome of each trial depends only on the outcome of the previous trial. This assumption forms the basis for what is known as stochastic *Markov processes.*

If one defines the symbols

Q_i–streamflow during the ith month, where i is counted from the beginning of the generated sequence

$\bar{Q}_j$–mean monthly flow in calendar month j, $j = 1,2,\ldots, 12$

B_j–regression coefficient for estimating flow in month $j + 1$ from the flow in month j

s_j–standard deviation of flows in month j

r_j–correlation efficiency between flows in calendar months j and $j + 1$

E_i–random variable normally distributed with zero mean and unit variance (obtained from a table of normally distributed random numbers with zero mean and unit variance)

A particular model for generating a synthetic sequence of streamflows than can be expressed as (HUFSCHMIDT and FIERING, 1966):

$$Q_{i+1} = \bar{Q}_{j+1} + B_j (Q_i - \bar{Q}_j) + s_{j+1}(1-r_j^2)^{1/2} E_i \tag{6}$$

Two remarks must be made in connection with this model: (1) Since E_i is normally distributed with zero mean, it is conceivable that negative values of E_i may be obtained, thus yielding negative values of Q_{i+1}. This, of course, is inadmissible. In practice, such negative values very seldom appear and are neglected. (2) The model described by Eq. (6) admits only a serial correlation of lag one; i.e., a correlation between flows in adjacent months. In some hydrologic systems, one may find that the streamflow in any given month may be significantly correlated with other flows occuring two or three months earlier or later. Models including serial correlations of lag two, three, or more are more complex and are beyond the scope of this chapter.

Operation of Reservoirs for Water Supply

Introduction

The foregoing discussion of reservoir design indicated some of the conceptual difficulties arising from the stochastic properties of hydrologic processes. In addition, the mode of operating the reservoir, i.e., the various amounts of water released (or abstracted) from it at different times, is also important in dermining its usefulness as a storage facility. Thus, the reservoir manager is particularly interested in two distinct situations: (1) The reservoir is full and spilling; that is, some of the water is not utilized and is lost to the system: (2) The reservoir is drawn down to its minimum level, which causes a shortage of water to develop. The remainder of this section is devoted to a brief discussion of these diametrically opposed situations.

Maximum Utilization of Stored Water

Consider the case of Lake Kinneret, the major surface storage facility in the Israel water network. It is necessary to derive a monthly operating rule for this reservoir.

The lake receives a certain net monthly inflow that is characterized by some probability

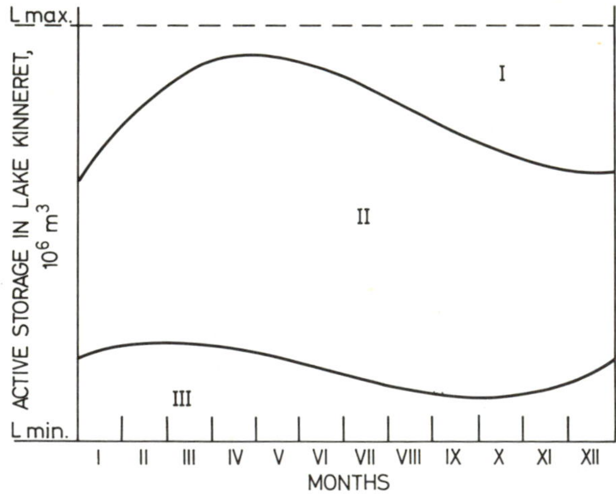

Fig. 7. Control curves for Lake Kinneret operations (schematic)

distribution. Every month, a certain amount of water has to be abstracted for irrigation, domestic supply, groundwater replenishment, and other purposes. During a yearly cycle, two states of the reservoir are of particular concern: (a) the reservoir is full and any additional inflow, will have to be spilled, thereby causing a reduction in the amount of water that can be utilized during the dry period, and (b) the reservoir is drawn down to such a low level that there is a danger of it not being able to supply the demands.

To overcome these difficulties, two control curves have been introduced to indicate the maximum and minimum levels of water in the lake, which ensure, with given probabilities, that there will be no spills and no shortages. These curves are shown schematically in Fig. 7.

The two control curves divide the volume-time continuum into three operational regimes:

I. Abstract water from reservoir at maximum capacity of existing facilities until lake level drops to the prescribed position on the curve for the month under consideration.

II. Release water from the reservoir to satisfy demand in the most economical manner.

III. Stop abstractions from the reservoir until the lake level rises to a prescribed elevation.

a) *Derivation of upper control function.* Define the following notation:

L – active storage in reservoir, in volume units

L_i – storage in reservoir at the beginning of month i, in volume units

$$0 \leqslant L_i \leqslant L \tag{7}$$

Q_i – inflow into the reservoir in month i, in volume units

x_i – cumulative inflow into the reservoir from month i to the end of the current hydrologic year, in volume units,

$$x_i = \sum_{j=1}^{12} Q_j \tag{8}$$

P – maximum abstraction of water from reservoir, in volume units per month

The following conditions must be satisfied simultaneously in order to ensure that no water will leave the reservoir through spillage:

$$\begin{aligned} Q_{12} + L_{12} - P &\leqslant L \\ Q_{11} + Q_{12} + L_{11} = 2P &\leqslant L \\ Q_{10} + Q_{11} + Q_{12} + L_{10} - SP &\leqslant L \\ \text{-----------------} \\ Q_1 + Q_2 -\ldots + Q_{12} + L_1 - 12P &< L \end{aligned} \tag{9}$$

Substituting Eq. (8) in Eq. (9), the latter becomes

$$\begin{aligned} x_{12} + L_{12} - P &\leqslant L \\ x_{11} + L_{11} - 2P &\leqslant L \\ \text{-----------} \\ x_1 + L_1 - 12P &\leqslant L \end{aligned} \tag{10}$$

and, in general,

$$x_i + L_i - (13-i)P \leqslant L, \quad i = 1,2,\ldots, 12 \tag{11}$$

Hence the maximum cumulative inflow x_i from the ith month until the end of the hydrologic year that will result in no spills is

$$x_i \leqslant L - L_i + (13-i)P \tag{12}$$

Let us define the quantity

$$A_i = L - L_i + (13-i)P \tag{13}$$

and assume that the cumulative inflows x_i are normally distributed (an assumption that must be carefully checked under a variety of natural conditions) with mean μ_i and standard deviation σ_i. No spills will occur when $x_i = \mu_i$. Conversely, spills will occur with a given probability, a, if

$$x_i = A_i = \mu_i + \alpha(a)\sigma_i \tag{14}$$

The values of the function $\alpha(a)$ can be found in all standard mathematical tables (c.f., BURRINGTON and MAY, 1953).

Thus, by accepting a given degree of uncertainty with regard to reservoir spills, the corresponding values of A_i can be established using Eq. (14). Having thus computed A_i, all that remains is to calculate L_i, the storage in the reservoir at the beginning of month i, from Eq. (13). Thus

$$L_i = L - A_i + (13-i)P \tag{15}$$

The upper control curve L_i is a function of the month i, the acceptable risk of spills as expressed by A_i, and the operational rule expressed by P.

This analysis can be extended to yield an estimate of the amount of water that will be lost through spills (DEAN and BURAS, 1963).

b) *Derivation of lower control function.* Since, in the particular case of Lake Kinneret, the reservoir is to be used for storage purposes in conjunction with the local groundwater aquifers, the lower control function takes both components into account.

The following notation is defined:

L_j- amount stored in the surface reservoir (Lake Kinneret) at the beginning of month j, $j = 1,2,\ldots,12$, in volume units

x_j- cumulative inflow into the surface reservoir from the beginning of month j through the end of hydrologic year, in volume units

$P-$ amount pumped from the aquifers, in volume units per month

P_j- cumulative withdrawal from the aquifers from the beginning of month j through the end of the hydrologic year, in volume units,

$$P_j = (13-j)P \tag{16}$$

d_j- cumulative demand for water from the beginning of month j through the end of the hydrologic year, in volume units

y_j- cumulative recharge of the aquifers from the beginning of month j through the end of the hydrologic year, in volume units

s_j- amount of water available for pumping in the aquifers at the beginning of month j, in volume units

In considering the problem of meeting a given demand, the following restrictions must be satisfied:

$$x_j + L_j + y_j + S_j \geq d_j \tag{17}$$

$$x_j + L_j + P_j \geq d_j \tag{18}$$

In addition, since storage cannot be negative, one must also require that

$$x_j + L_j \geq 0 \tag{19}$$

These three restrictions can be rewritten in a manner that emphasizes the cumulative inflows into the reservoir,

$$x_j \geqslant -L_j \tag{20}$$

$$x_j \geqslant -L_j + d_j - y_j - S_j \tag{21}$$

$$x_j \geqslant -L_j + d_j - P_j \tag{22}$$

Combining these three conditions, one obtains

$$x_j \geqslant -L_j + \max \begin{Bmatrix} 0 \\ d_j-(y_j + S_j) \\ d_j-P_j \end{Bmatrix} \tag{23}$$

Eq. (23) can be interpreted as the minimum cumulative inflow that would satisfy a demand d_j, when the surface reservoir contains an amount L_j, while an amount P_j is being withdrawn from the S_j volume units stored in the aquifers that receive a natural replenishment of y_j. Upon examining Eq. (23) more closely, one finds that the following relationship always holds.

$$s_j + y_j > P_j \tag{24}$$

In other words, the sum of the water available in the aquifers and the cumulative recharge of these aquifers is greater (in fact, much greater) than the pumping capacity of the wells. If this were not so, one would be faced with the dangerous possibility of exhausting (mining) the aquifers within one year, a situation that would render the development and utilization of such aquifers impractical. Consequently,

$$d_j-P_j > d_j-(y_j + S_j) \tag{25}$$

and one can therefore write

$$x_j \geqslant -L_j + \max \begin{Bmatrix} 0 \\ \\ d_j-P_j \end{Bmatrix} \tag{26}$$

Define

$$B_j = -L_j + \max \begin{Bmatrix} 0 \\ \\ d_j - P_j \end{Bmatrix} \tag{27}$$

We now make the assumption that x is normally distributed, just as we did with respect to x_i in Eq. (14) (for a detailed discussion of general probability distribution functions, see any good textbook on probability and statistics, such as PARZEN, 1960). Accordingly, the probability that $x_j \geqslant B_j$ is

$$P\ \{x_j > B_j\} = \frac{1}{\sqrt{2\pi}\sigma_j} \int_{B_j}^{\infty} \exp\ [-(x_j-\mu_j)^2/2\sigma_j^2]\ dx_j \tag{28}$$

where μ_j is the mean and σ_j^2 the variance or the cumulative inflow x_j. For any probability (or risk) of shortage a, the corresponding value of B_j is given by

$$B_j = \mu_j - \alpha(a)\sigma_j \tag{29}$$

where $\alpha(a)$ is obtained from a table of normal distribution.
Substituting the value of B_j in Eq. (27), one gets

$$L_j = \alpha\ (a\sigma_j - \mu_j + \max \begin{matrix} 0 \\ d_j - P_j \end{matrix} \tag{30}$$

This is the lower control function and it incorporates the following types of variables:

a) Control variables; i.e., values that can be determined by the system operators such as a, the probability (risk) of shortage, and P_j, the pumpage from aquifers.

b) A partially controlled variable, d_j, which is the demand that the system has to satisfy.

c) Hydrologc (natural) parameters, such as μ_j and σ_j.

The upper and the lower control functions constitute the guidelines for operating the reservoir in a manner that ensures optimum utilization of the stored water.

Reduction of Risk of Shortages

In the previous section, the question of shortages was touched upon from the point of view of operating a water resource system. In this section, a very simple example will be presented to show how the risk aspect can be introduced into the design of an irrigation project.

Consider an irrigation project based on a single surface reservoir. An economic analysis has indicated that the irrigation water that can be supplied on a firm (guranteed) basis is worth 15 monetary units per unit volume of water annually. For any unit volume of water that is not supplied, a loss of 30 monetary units is incurred. Water supplied in excess of the guaranteed (firm) amount has no value. Annual costs per unit storage capacity are 1.5 monetary units. The inflow into the reservoir is a random variable with the following probability distribution:

Table 2. Probability distribution of yearly inflows

Inflow, f_i, 10^6 volume units	1	2	3	4	5
Probability, p_i	0.1	0.1	0.2	0.4	0.2

The design problems are as follows: (1) How large a reservoir should be constructed? (2) What is the guaranteed (firm) amount of water that can be supplied annually?

Define

I = guaranteed annual amount of water

K = reservoir capacity

y_i = quantity of water supplied if inflow f_i occurs

If a shortage occurs, we shall define it as

$$u_i = I - y_i, \quad u_i \geqslant 0 \tag{31}$$

The problem can be formulated as a problem in linear programming. The objective function to be maximized is the expected value of annual net benefits,

$$B = 15\,I - 1.5\,K - 30 \sum_{i=1}^{5} p_i u_i \tag{32}$$

If S represents the initial storage in the reservoir, then

$$S \leqslant \sum_{i=1}^{5} p_i(S + f_i - y_i), \quad i = 1,2,\ldots,5 \tag{33}$$

By cancelling S on both sides, this becomes

$$\sum_{i=1}^{5} p_i y_i \leqslant \sum_{i=1}^{5} p_i f_i \quad I = 1,2,\ldots,5. \tag{34}$$

Note that $\sum_{i=1}^{5} p_i f_i$ is the average yearly inflow, which is equal to 3.5×10^6 volume units. Thus, Eq. (34) can be written explicitly as

$$0.1y_1 + 0.1y_2 + 0.2y_3 + 0.4y_4 + 0.2y_5 \leqslant 3.5 \quad (35)$$

Now the amount of water supplied, y_i, can never exceed the sum $S + f_i$, so that

$$y_i \leqslant S + f_i, \quad i = 1,2,..5, \quad (36)$$

Transferring the unknows to the left-hand side, one obtains

$$y_i - S \leqslant f_i, \quad i = 1,2,..,5 \quad (37)$$

In addition, the water balance in the reservoir does not admit spills, and therefore

$$S + f_i - y_i \leqslant K, \quad i + 1,2,..,5 \quad (38)$$

Again, transferring the unknown variables to the left-hand side, this becomes

$$S - y_i - K \leqslant -f_i, \quad i = 1,2,..,5 \quad (39)$$

Finally, one has

$$I - y_i - u_i \leqslant o, \quad i = 1,2,..,5 \quad (40)$$

Relations (35), (37), (39), and (40) are the constraints imposed on the system. Combined, thc previous expressions lead to the following linear programming formulation.

$$\max B = 15\, I - 1.5\, K - 30 \sum_{i=1}^{5} p_i u_i$$

subject to $0.1y_1 + 0.1y_2 + 0.2y_3 + 0.4y_4 + 0.2y_5 \leqslant 3.5$

$$\begin{aligned} &y_i - S \leqslant f_i \\ &S - y_i - K \leqslant f_i \\ &I - y_i - u_i \leqslant 0 \\ &I = 1,2,..,5 \end{aligned} \quad (41)$$

The linear programming formulation is presented in matrix form in Table 3.

The solution indicates that K (reservoir capacity) is optimal at 3.0×10^6 volume units, I (guaranteed annual irrigation supply) is 3.5×10^6 volume units of water (average yearly flow, S (initial storage in the reservoir) is 1.6×10^6 volume units of water, and B (expected annual benefits) is 46.5×10^6 monetary units. The reservoir thus regulates the flow and reduces the risk of shortages, although the flow into it has a certain probability distribution.

Management of Groundwater Resources

Introduction

It was mentioned on p. 36 that groundwater, on a global basis, forms most of the available water resources. In many regions of the world, groundwater provides the bulk of the water for irrigation, industry, and domestic use. For example, in Israel about 2/3 of the water used annually is extracted from aquifers.

The important property of groundwater resources is that they can be developed in a much more gradual fashion than surface water. Well fields can be expanded gradually as the demand for water risses, leaving relatively little pumping capacity idle for extended periods. On the other hand, surface waters almost always require large investments in order to take advantage of the economies of scale. As a result, surface-water-development projects are often constructed to meet projected demand and so are underutilized during the period of demand growth.

Table 3. Linear programming (simplex) matrix for problem (41)

Constraint No.	Eq. No.	K	S	y_1	y_2	y_3	y_4	y_5	I	u_1	u_2	u_3	u_4	y_5	$\leq$ constant
1	[35]			0.1	0.1	0.2	0.4	0.2							3.5
2	[37]		−1	1											1
3	[37]		−1		1										2
4	[37]		−1			1									3
5	[37]		−1				1								4
6	[37]		−1					1							5
7	[39]	−1	1	−1											−1
8	[39]	−1	1		−1										−2
9	[39]	−1	1			−1									−3
10	[39]	−1	1				−1								−4
11	[39]	−1	1					−1							−5
12	[40]			−1					1	−1					0
13	[40]				−1				1		−1				0
14	[40]					−1			1			−1			0
15	[40]						−1		1				−1		0
16	[40]							−1	1					−1	0
Objective f'cn		−1.5						15		−3	−3	−6	−8	−6	

For these and other reasons, the management of groundwater resources is a problem of considerable interest. This section touches upon two aspects of this problem: (1) integration of aquifers within larger regional projects, and (2) the exploitation and recharge of groundwaters.

Integration of Aquifers in Water Resource Systems

Some indication regarding the utilization of aquifers in conjunction with surface water was given in a previous section. An example of a more direct integration of aquifers within a large-scale water resource system is the Israel water plan (BURAS, 1963).

A schematic representation of the system under consideration is given in Fig. 8. The purpose of the integration was to operate the system in an optimal manner, which was achieved with the aid of dynamic programming (BELLMAN and DREYFUS, 1962).

The operation of the system is regarded as a multistage decision process whereby a decision has to be made each month regarding the amount of water to be abstracted from the lake (x), the pumpage from aquifer A (e), and the pumpage from aquifer B (b). Define f_n (Q_j, G_k, y_{j-1}) as the expected minimum cost of operating the system for the remaining n months, using an optimal policy (to be derived below) and starting with

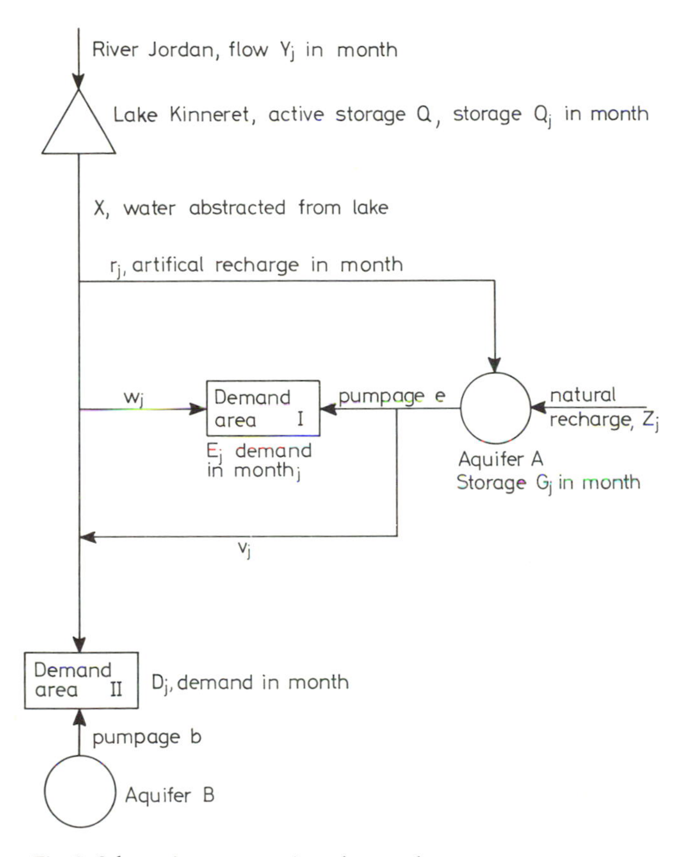

Fig. 8. Schematic representation of a complex water resources system

an amount Q_j of water in the surface reservoir, an amount G_j in aquifer A, and a record of the flow of the Jordan River into the lake during the previous month, y_{j-1}, where $j = 1,2,\ldots,12$. This definition implies that (1) there is a cost function that is related to the decision variables x, e, and b, and (2) there is a relationship between the river flows in any two successive months. This latter relationship is expressed by a conditional probability distribution function, $h(y_j \mid y_{j+1})$, which gives the distribution of the inflows during month j (y_j) in relation to the inflow during the preceding month (y_{j+1}). Assuming that the monthly flows are normally distributed, the explicit form of the conditional probability $h(y_j \mid y_{j-1})$ is given by PARZEN (1960).

$$h(y_j \mid y_{j-1}) = \frac{1}{\sqrt{2\pi}\,\sigma_j \sqrt{1-r_j^2}} \exp\left\{-\frac{1}{2\sigma_j^2(1-r_j^2)}\left[y_j-\mu_j-\frac{r_j\sigma_j}{\sigma_{j-1}}(y_{j-1}-\mu_{j-1})\right]^2\right\} \tag{42}$$

where μ_j is the mean and σ_j^2 the variance of the inflows in month j, and r_j is the serial correlation between the flows in months j and $j-1$.

The cost function has the form

$$\Phi = \Phi(x, e, b) \tag{43}$$

where the decision x takes into account the current inflow, y_j.

If only one monthly stage remains in the operation of the system, the minimum expected cost is

$$f_1(Q_1, G_1\, y_o) = \min_{\substack{x\\e\\b}} \left\{ \int_{-\infty}^{\infty} \Phi(x,e,b)\, h(y_1 \mid y_o)\, d_{y1} \right\} \tag{44}$$

When two monthly stages remain in the process, the decisions (x, e, b) are made and the state of the system is transformed from $(Q_2, G_2\, y_1)$ to (Q_1, G_1, y_o). The optimal operation of the system in state (Q_1, G_1, y_o) for one stage only is given by Eq. (44). Thus, the overall optimum for the last two stages is

$$f_2(Q_2, G_2, y_1) = \min_{\substack{x\\e\\b}} \left\{ \int_{-\infty}^{\infty} [\Phi(x,e,b) + f_1(Q_1, G_1, y_o)]\, h(y_2 \mid y_1) dy_2 \right\} \tag{45}$$

Using mathematical induction, one can derive a general expression for an n-stage decision process where n is any finite number:

$$f_n(Q_n, G_n, y_{n-1}) = \min_{\substack{x\\e\\b}} \left\{ \int_{-\infty}^{\infty} [\Phi(x,e,b) + f_{n-1}(Q_{n-1}, G_{n-1}, y_{n-2})]\, h(y_n \mid y_{n-1}) dy_n \right\} \tag{46}$$

The solution of Eqs. (44) and (46) yields the optimal policy for the operation of an integrated surface reservoir-aquifer system subject to constraints stemming from the pumping capacities of the wells, the conveyance capacities of the conduits, and the demands to be satisfied.

Exploitation and Recharge of Aquifers

a) *Groundwater Exploitation.* Groundwater is a renewable natural resource. Since its occurrence and availability in time are subject to the natural laws of hydrology (which include a significant stochastic aspect), the management of groundwater resources is directed toward matching the supply with the demand, and, at the same time, minimizing the possible degradation (quantitative and qualitative) of the resource itself. An important concept in groundwater management is that of *safe yield,* which may be defined as the amount of water that can be withdrawn annually from a groundwater basin without causing any undesired effects. The safe yield is usually derived from a mass balance equation that states that the net inflow into a basin must be equal to the net increase in storage.

The estimation of safe yield is based upon existing or some assumed conditions. When these conditions change, the safe yield has to be reevaluated. For example, the value of safe yield that is obtained when the aquifer is pumped at a constant rate with steady groundwater levels is different from the safe yield that is obtained while lowering the water table to a predetermined elevation and thereby "mining" part of the groundwater resource. Such mining operations have significant economic implications. By so doing, large amounts of water can be made available during a short time for the rapid development of irrigated agriculture, a fact that allows the large investments usually associated with surface-water supply systems to be deferred. The deferment of investment, of course, has a positive economic value.

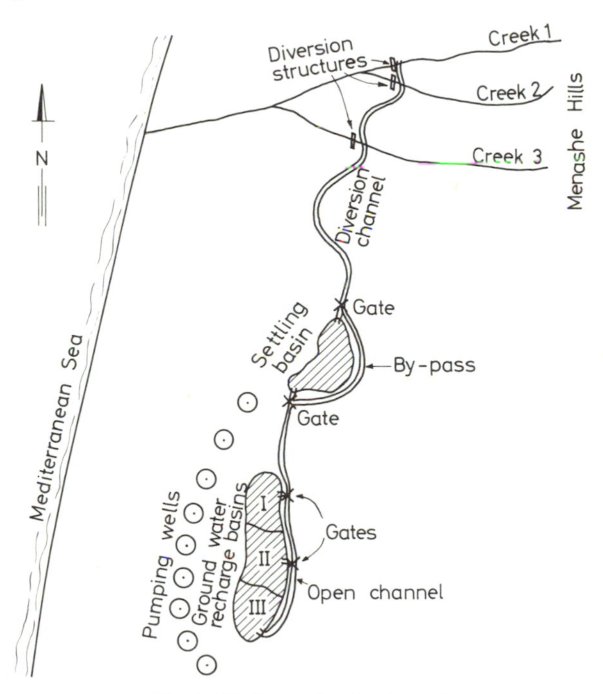

Fig. 9. The Menashe Creeks project

b) *Artificial Recharge.* One example of intensive groundwater management is the Menashe Creeks Project in Israel, shown schematically in Fig. 9. In this project, all winter flows of the Menashe Creeks, which are dry in summer, are diverted for groundwater replenishment. Most of the silt carried by the diverted flows is trapped by a settling basin. If the diverted water is relatively clear, it can flow through a bypass directly to the recharge basins.

A system of pumping wells provides water from the recharged aquifer according to the demand.

This technique of groundwater replenishment tbrough recharge basins is only one of the several currently used in artificial recharge. Two additional methods are descirbed briefly in the following.

1) *Recharge through Stream Beds.* Where conditions permit, advantage is taken of the high infiltration capacity of sandy stream beds. Check dams and dikes are often constructed to retard the flow and induce increased percolation into the underlying aquifer, which, of course, must be unconfined and in direct contact with the stream bed. This method of artificial recharge is practiced in many arid and semiarid zones, for example, in the southwestern part of the United States.

2) *Recharge through Wells.* This method of recharge, which is widely used in Israel, has two important advantages: (1) It is suitable for confined and unconfined aquifers, and (2) many of the wells can be used for pumping and recharge, so that a considerable saving can be effected in installation costs. In the artificial recharge of aquifers, one can use lake and river water, as well as treated sewage and wastewater.

It can be said that the utilization of aquifers as an element of a water resource system requires that they be judiciously exploited and recharged.

Summary

The regional development and management of water resources leads to problems that touch on several scientific disciplines. Such problems are not limited to hydraulic considerations of bridging the discrepancy between the demand for water and the availability of water in space and in time, but also include a very strong element of decision making: What facilities should be constructed? Where? How should they be operated? To answer these questions, one must examine a number of feasible alternatives before proceeding whit the design and the operative decisions. The decioions are usually made in the face of uncertainties in the hydrologic phenomena and the economic forecasts. The analysis and solution of design and operation problems that involve decision making form the domain of *water resources engineering* (BURAS, 1972).

A few important points connected with the use of water resources for irrigation were presented in this section. The problem of storage was emphasized because of its importance in arid zones where water is scarce and its occurrence exhibits a high degree of variability. The integration of surface and groundwater resources into one comprehensive system was also emphasized because of the specific hydro-economic properties of aquifers that make them well suited for gradual development.

The design of surface reservoirs was described as a process in which hydrologic principles interact with economic considerations and the stochastic aspects of the natural phenomena have a significant influence. The management of groundwater resources was approached from the point of view that aquifers are elements of more extensive water resource systems, and their exploitation and recharge has to be carried out in conjunction with surface-water systems.

Literature

BELLMAN, R. E., DREYFUS, S. E.: Applied Dynamic Programming. Princeton: Princeton Univ. 1962.

BURAS, N.: Conjunctive operation of a surface reservoir and a ground water aquifer. Symposium on Surface Waters. Pub. 63, Internat. Ass. Scientific Hydrology, Gentbrugge, Belgium, pp. 492–501 (1963).

BURAS, N.: Scientific Allocation of Water Resources. New York: Elsevier 1972.

BURRIGTON, R. S., MAY, D. C.: Handbook of Probability and Statistics. Handbooks Publishers, Sandusky, Ohio (1953).

DEAN, B. V., BURAS, N.: Effective Control and Economical Scheduling of Lake Kinneret Pumping Operations. Tahal, P. N. 282, Tel-Aviv (1963).

GUMBEL, E. J.: Statistics of Extreme Values. New York: Columbia Univ. Press 1958.

HALL, W. A., DRACUP, J. A.: Water Resources Systems Engineering. New York: McGraw-Hill 1970.

HUFSCHMIDT, M. M., FIERING, M. B.: Simulation Techniques for Design of Water-Resource Systems. Cambridge (Mass.): Harvard Univ. Press 1966.

HURST, H. E.: Long term storage capacity of reservoirs. Trans. Amer. Soc. Civil Eng. **116**, 770–799 (1951).

JAMES, L. D., LEE, R. R.: Economics of Water Resources Planning. New York: McGraw-Hill 1971.

LANGBEIN, W. B.: Queuing theory and water storage. Proc. Amer. Soc. Civil Eng. J. Hydraulics Div. **84**, 1–24 (1958).

LINSLEY, R. K., Jr., FRANZINI, J. B.: Water-Resources Engineering. New York: McGraw-Hill 1964.

MAASS, A., HUFSCHMIDT, M. M., DORFMAN, R., THOMAS, H. A., Jr., MARGLIN, S. A., FAIR, G. M.: Design of Water-Resource System. Cambridge (Mass.): Harvard Univ. Press 1962.

PARZEN, E.: Modern Probability Theory and Its Applications. New York: Wiley 1960.

RIPPL, W.: The capacity of storage reservoirs for water-supply. Proc. Inst. Civil Eng. **71**, 270–278 (1883).

4

Water Suitability for Irrigation

B. YARON

Introduction

Regardless of its source, irrigation water always contains impurities in the form of dissolved, or sometimes suspended, materials. The amount and nature of these materials under given environmental, climatic, soil, and plant conditions determine the usefulness and relative quality of the water.

Good indicators of irrigation water quality are the initial content of dissolved salts, amount of suspended solids, and the amount of pollutants from anthropogenic sources. The initial content of dissolved salts depends on the content of soluble salts in the surrounding rock or soil; the latter may dissolve to some extent during transport from the source to the irrigated field. The suspended material may come from various kinds of agricultural lands (irrigation, dry farming, grass, timber) and the composition of this material depends on the source of erosion. Pollutants reaching the irrigation water come from agrochemical residues such as inorganic fertilizers and biocides, and from industrial and domestic wastes.

The quality of irrigation water is defined with respect to its effect on plant growth, soil properties, soil biological equilibrium, and irrigation technology. Dissolved salts may affect plant growth and soil properties; agrochemical residues may destroy the biological equilibrium of the soil; and suspended material may have a negative effect on the technology of water supply and conveyance. Therefore, in considering the applicability of water resources for irrigation, it is important to know what effect water composition and quality will have on such factors as yield potential and environmental development.

In the past two decades, problems of irrigation water quality have been viewed primarily from the standpoint of salinity (U.S. Salinity Staff, 1954; GRILLOT, 1954; KOVDA, YARON, and SHALHEVET, 1967; WILCOX and DURUM, 1967; RHOADES and BERENSTEIN, 1971). The influence of agrochemicals carried by irrigation water on biological equilibrium has been widely discussed in a recent reports of the Committee on Water Quality Criteria (1968).

The present section deals with water resources in arid zones and considers problems of irrigation water quality in an environment that favors the accumulation of various constituents of the water in the upper layers of the soil during short periods of time.

The Constituents of Irrigation Water

The term "water quality" includes the individual and combined effects of the substances present in the water. The possibility of using a given water supply for irrigation purposes is determined to a large extent by its constituents.

Suspended Inorganic and Organic Materials

All river waters contain suspended materials. The fertility of river salts depends on the mineralogy and chemistry of the transported particles, thus, the quality of the suspended materials must be taken into consideration in irrigation design. Under given conditions, the fertility properties of the suspended matter insure a continuous high-level production for irrigated areas (e.g., the areas of the Nile, the Amu-Daria, and the Danube). In many cases, suspended solids can interfere with the flow of water through the distribution system, and thus their effect is negative.

Major Constituents

Rainwater contains the lowest salt concentration of all types of water used for irrigation. It includes dissolved gases (N_2, Ar, O_2, CO_2) and dissolved salts originating from terrestrial and marine sources. Generally, the amount of salts in rainwater (NH_3, Cl, Na) varies widely and is dependent on the distance from the sea. As an example, Table 1 illustrates the composition of rainwater collected close to Israel's coastline.

Table 1. The influence of distance from the sea and salt content of sea water on the chemical composition of rainwater. (After YAALON, 1961)

Location	Sea	Distance from sea (km)	HCO_3, mg/1	SO_4, mg/1	Cl, mg/1	Mg, mg/1	Ca, mg/1	Na, mg/1	K, mg/1	EC 10^6 at 25 °C
Jerusa-lem	Mediter-ranean	52.5	34.77	30.35	13.06	1.87	18.81	7.66	1.02	170
Haifa	Mediter-ranean	1.2	43.69	33.09	13.84	2.19	25.84	8.02	1.64	200
Eilat	Red	0.1	53.64	21.95	26.98	1.22	26.00	9.06	3.01	232
Sedom	Dead	0.1	85.79	118.70	13.05	9.95	49.20	18.62	8.13	467

Table 2. Mean composition of river waters of the world (ppm). (After LIVINGSTONE, 1961)

	HCO_3^-	$SO_4^=$	Cl^-	NO_3^-	Ca^{++}	Mg^{++}	Na^+	K^+	Sun
North America	68.0	20.0	8.0	1.0	21.0	5.0	9.0	1.4	142
South America	31.0	4.8	4.9	0.7	7.2	1.5	4.0	2.0	69
Europe	95.0	24.0	6.9	3.7	31.1	5.6	5.5	1.7	182
Asia	79.0	8.4	8.7	0.7	18.4	5.6	9.3		142
Africa	43.0	13.5	12.1	0.8	12.5	3.8	11.0	–	121
Australia	31.6	2.6	10.0	0.05	3.9	2.7	2.9	1.4	59
World	58.4	11.2	7.8	1.0	15.0	4.1	6.3	2.3	120

The Mg/Ca, K/Na, and $\frac{Cl-(Na+K)}{Cl}$ ratios in milliequivalent/liter for rainwater are similar to those for sea water in the vicinity of the sea, but differ as the distance from the sea increases (SCHOLLER, 1962). The salt concentration of rainwater in arid zones has a decided influence on the salt content of surface and groundwater.

The salt content of surface water is a function of the rocks prevalent at the water's source, of the climatic zone, and of the nature of the soil over which the water must flow.

Surface water can be classified into two groups: flowing water (rivers) and stagnant water (lakes). An analyses of the composition of river waters throughout the world (Table 2) shows that the predominant anions are HCO_3 and SO_4, and the main cations are Ca^{++} and Na^{+}.

Stagnant lakes in arid and semiarid regions are characterized by a much higher salt content than found in open lakes of the more humid zones. The salt composition of irrigation water is not static, but is in a state of continuous change. Thus, the evaluation of an irrigation water must be based on knowledge of seasonal variations in the salt content. The composition of flowing surface water changes under the influence of precipitation in the area. Fig. 1 shows the seasonal variation in salt content of the Tigris River at Baghdad as influenced by monthly rainfall.

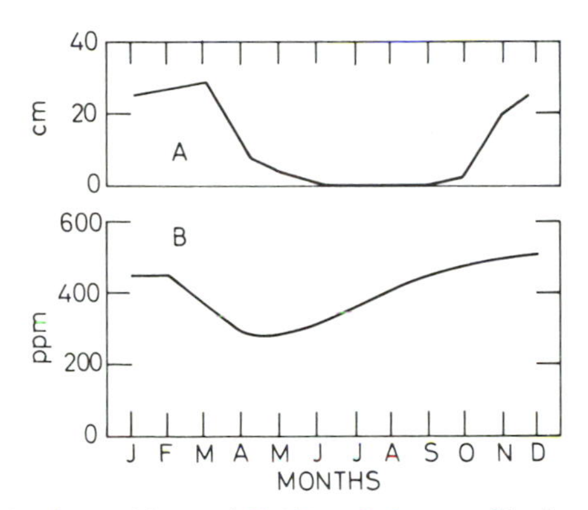

Fig. 1. The average depth of monthly rainfall (A) and the monthly fluctuations of average total dissolved salts of Tigris river at Baghdad. (DIELMAN *et al.*, 1963)

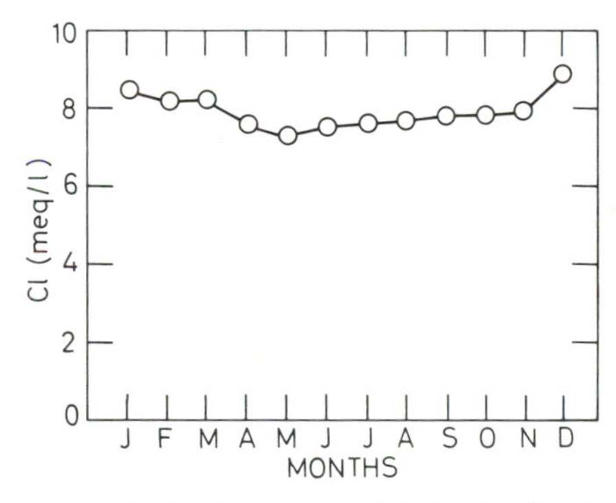

Fig. 2. The monthly fluctuations of chloride content of Lake Tiberias, Israel. (According to OREN, 1962)

BRYSINE (1961) noted that the chloride content of irrigation water at l'Oued-Oum in Morocco varies during the irrigation season from 200 to 1500 mg/1: The lack of rainfall and a high evaporation rate during the dry seasons contributes to an increased in the salt concentration of lakes. Whereas in large, open lakes such as Lake Tiberias rise in salt content as a result of such climatic conditions may be no greater than 20% small marshes may exhibit an increase in salt of as much as 100% (Figs. 2 and 3).

The salt content of groundwater depends on the source of the water and on the course over which it flows. Groundwater is mineralized according to the law of dissolution, based on the contact between the water and the rocks in its path. Changes in the salt content of groundwater during recharge result from reduction processes, base exchange, dispersion, etc.

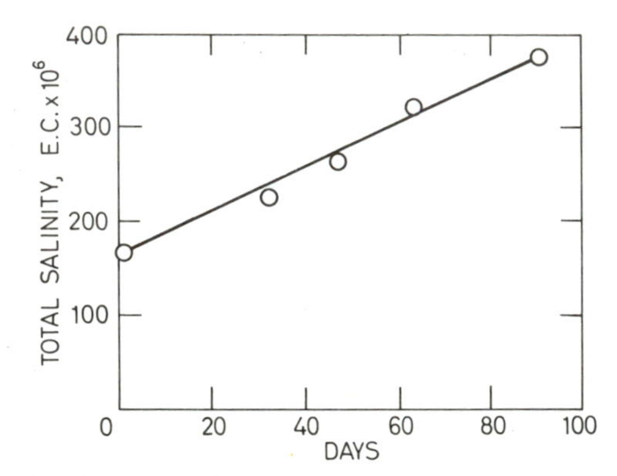

Fig. 3. The increase in chloride content of Netanya marshes, during Israeli summer. (According to YAALON, 1964)

Minor Constituents

Not all of the minor elements are found in any one source of irrigation water. They appear sporadically, singly or in groups, in different water sources. Bromine, fluorine, and iodine may be found when chloride is present. Most fresh water contains less than 1 ppm fluorine, 0.01 ppm bromine, and 0.2 ppm iodine. Irrigation water may also contain Li, Ru, Cs, Be, Sr, Ba, Ra, etc., but the quantities are so minute they have practically no influence on its quality. Other elements that may appear are the selenium

Table 3. Boron content of different lakes and rivers. (After LIVINGSTONE, 1961)

Location of water	Mean of samples	
River Tone, Japan	10	0.345
Watarase River, Japan	3	0.197
Agatsumo, Japan	4	1.970
Okuresawa, Japan	4	1.305
Great Salt Lake, Utah, U.S.A.		43.500
Florida streams	6	0.019
Greek River, Uganda		0.386
City water supplies, U.S.A.	24	1.000–0.010
River waters of U.S.S.R.		0.013

–arsenium–antimonium–bismuth group, and different metals such as Cu, Co, Ni, Zn, Ti, Zr, Va, Ch, and Mo. One particular microelement that is found in irrigation water and has a strong effect on plant growth is boron. Data on boron content of different lakes and rivers are presented in Table 3.

Man-Made Pollutants

The man-made pollutants that may be found in irrigation water come from municipal, industrial, and agricultural wastes. Among the nonionic constituents are various oily substances, phenols, synthetic detergents, dissolved gases, insecticides, and pesticides. Ionic constituents in irrigation water are mainly in the form of nitrates and soluble phosphorus, which are minor elements.

Generally, the composition is affected by the presence of man-made pollutants when irrigation return flow is used for irrigation.

Classification of Irrigation Water

The use of various types of water for irrigation has made it necessary to create a system of classification that is completely different from those used in geochemistry, the industry, and sanitary engineering. Such classifications are usually based on the total salt content of the water (the osmotic effect) and on the sodium content (since this is the ion responsible for the chemical and physical changes in soil and for the specific ion effect). Aside from these two main criteria, there are some additional factors that are freqently taken into consideration. Sodium carbonate is responsible for soil alkalinity, chloride and boron for toxic effects on plants, suspended material for adverse effects on irrigation technology, and pollutants for adverse effects on soil biological equilibrium.

Among the qualities to be considered when judging the suitability of water for irrigation are salinity, sodium, carbonate, chloride, boron, suspended material and biocide hazards. As these hazards do not always appear concurrently, in classifying water for irrigation, it is not necessary to consider them all. It must also be pointed out that it is not possible to develop a universal classification system that is suitable for all purposes. A comparative analysis of some of the classifications commonly used in different countries is presented in the following.

Salinity Hazard

The total salt content of irrigation water is one of the factors that indicates whether there is a danger that salt will accumulate in the soil. It can be determined by measuring the electric conductivity (*EC*) of the water.

Currently, the classification proposed by the U.S. Salinity Laboratory (1954) and modified by THORNE and PETERSON (1954) is the one most widely used. In this system, the limits between different classes of electric conductivities are (in micromho/cm): $<$ 250–low salinity; 250–750–moderate salinity; 750–2250–medium salinity; 2250–$>$ 4000–high, salinity; 4000–6000-very high salinity; and $>$ 6000–excessively high salinity.

In order to examine the validity of this classification for different areas, the classes of distribution of irrigation water in the western United States, Algeria, India, and Israel are presented in Table 4.

In Algeria, a relatively large percentage of the water falls above 2250 micromho/cm, and this water is extensively used for irrigation. In India (Rajahstan area), 40% of the water is classed as being highly and very highly saline, a definition that does not fit the particular conditions prevalent in India. Furthermore, in Israel 60% of the water is considered to be moderately saline according to the above classification, although this is not really so.

Thus, the salinity ranges chosen to characterize irrigation water in a given area must be modified according to the local environmental conditions. The total salt content of irrigation water will give only a general qualitative assessment of its quality.

Table 4. Percent of water supply that falls within indicated salinity ranges in four different countries, given as percentages. (From KOVDA, Yaron and SHALHEVET, 1967)

EC 10^6 at 25 °C	U.S.A. (West)	Algeria	India (Rajahstan area)	Israel
Samples	1018	79	576	1507
250– 750	53	5	13	36
750–2250	37	28	35	60
2250–4000		28	22	3
2250–5000	10			
4000–6000		18	18	1
5000–20000		21		

Sodium Hazard

Because of its effect on the soil and on the plant, sodium is considered one of the major factors governing water quality. Several methods have been proposed for expressing the sodium hazard. Previously, water quality was defined on the basis of its *sodium percentage (SP)* alone, the sodium percentage being the ratio of the total sodium content to the total cations held in solution, multiplied by 100. SCOTFIELD (1935) and MAGISTAD and CHRISTIANSEN (1944) considered water with an SP of 60% or more to be harmful. The sodium hazard, as determined by the SP of irrigation water, must be reflected in the exchangeable sodium percentage *(ESP)* of the soil. However, in research conducted in western Texas, no correlation has been found between the SP of the water and the ESP of the soil (LONGENECKER and LYERLY, 1958).

A value that has come into wide use in predicting the sodium hazard is the sodium adsorption ratio *(SAR)* proposed by the U.S. Salinity Laboratory (1954):

$$SAR = \frac{Na^+}{\sqrt{\frac{Ca^{++} + Mg^{++}}{2}}}$$

The classification of water according to SAR is also related to the total salt content of the water, and the range is divided into four groups-low, medium, high, and very high. For an electrical conductivity of 100 micromho/cm, the dividing points are at SAR values of 10, 18, and 26, and with an increase in salinity to 750 micromho/cm, the dividing points are at SAR values of 6, 10, and 18 (Fig. 4). This relationship represents the relative activity of the sodium ion in the cation exchange reaction with the soil and is derived from the classical Gapon equation. The validity of the sodium hazard prediction may be confirmed by examining the relationship between the SAR and the ESP of the soil. An empirical relationship between the SAR of the irrigation water and the ESP of the irrigated soil was established by the U.S. Salinity Laboratory (Fig. 5).

In general, a good correlation has been found between the ESP as calculated from the

SAR value and the ESP as determined experimentally. The following statistical correlation coefficients were found between calculated and measured ESP values for three different soil types in Israel: (1) grumusol, $r = 0.856$; (2) grumusolic dark brown, $r = 0.906$; (3) residual dark brown, $r = 0.925$ (Israel Salinity Survey, 1964). However, DURAND (1958) found large differences between calculated and determined values of ESP in Algeria. In 40% of the cases studied, these differences varied between 100 and 200 percent.

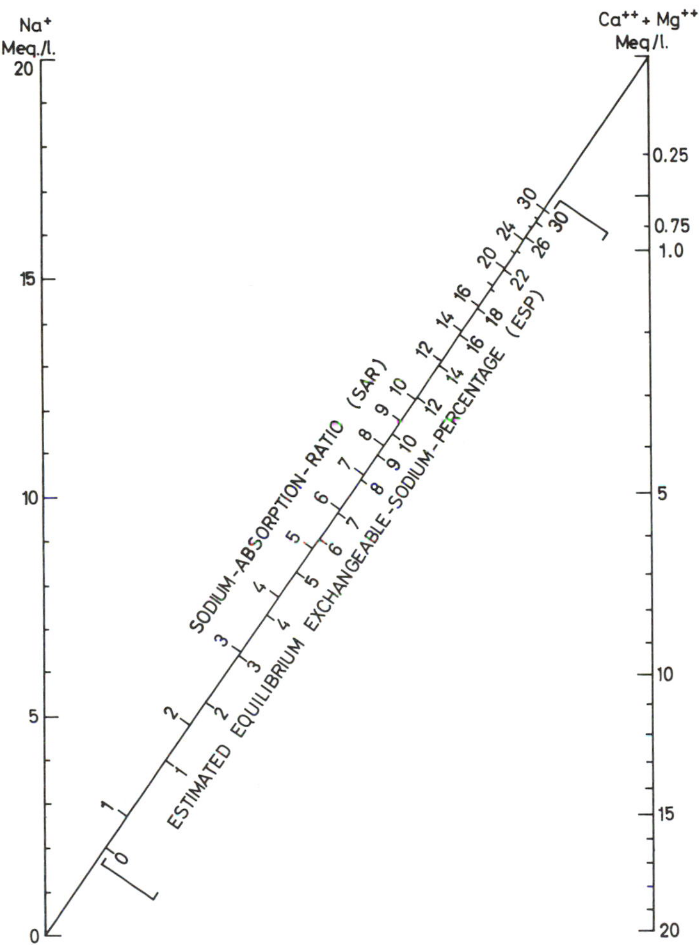

Fig. 4. Nomogram for determining the SAR value of irrigation water and for estimating the corresponding ESP value of a soil that is at equilibrium with the water. (From U.S. Salinity Lab. Staff, 1954)

Using a modified Langelier index together with SAR in evaluating the sodium hazard for water with high carbonate and without residual sodium bicarbonate, BOWER (1961, 1963) proposed the empirical equation

$$ESP = 2\,SAR + 2\,SAR\,(8.4 - pH_c)$$

The saturation index is defined by LANGELIERas the actual pH of the water (pH_a) minus the pH (obtained by calculation) that the water will have when it is in equilibrium with

$CaCO_3$ (pH_c). In the Bower equation, the term ($8.4-pH_c$) is analogous to Langelier's saturation index, except that 8.4, the approximate pH reading of a nonsodic soil in equilibrium with $CaCO_3$, is substituted for the actual pH value (pH_a) of the water. Applying this equation to a series of samples from wells in West Pakistan, a good correlation between calculated and measured values of ESP has been obtained.

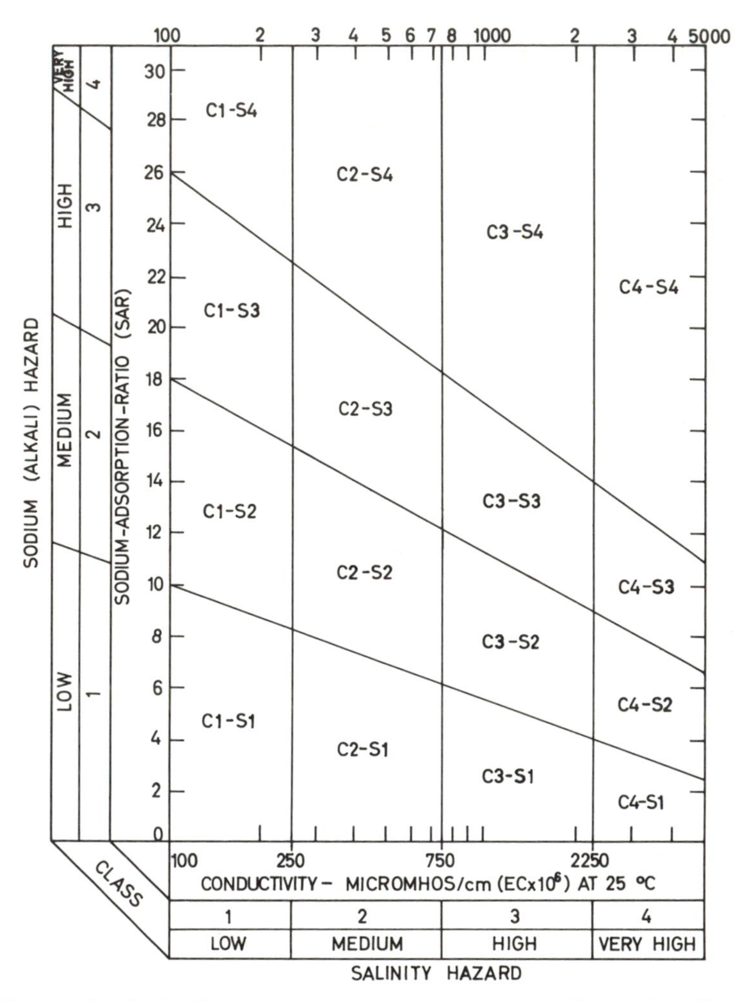

Fig. 5. Diagram for the classification of irrigation water. (From U.S. Salinity Lab. Staff., 1954)

In using the SAR value, HANDRA (1964) took into account the fact that water with a high sodium content is usually applied with gypsum. He therefore proposed that this gypsum should be included in the calculation of SAR, thereby reducing the SAR values of the irrigation water.

Bicarbonate Hazard

The bicarbonate anion is important in irrigation due to its tendency to precipitate calcium and magnesium from the soil in the form of calcium and magnesium carbonates. This brings about a change in the ratio between Na and the total amount of cations, thereby accentuating the effect of the sodium hazard of the irrigation water. EATON (1950) introduced the term "residual sodium carbonate" (RSC) as a means of characterizing the bicarbonate hazards, where

$$RSC = (CO_3^{--} + HCO_3) - (Ca^{++} + Mg^{++})$$

WILCOX concluded (1958) that water with more than 2.5 meq/l of residual sodium carbonate is not suitable for irrigation. Water containing 1.25–2.5 meq/l is considered marginal, and water with less than 1.25 is probably safe. ARANY (1956) showed that in evaluating the effect of residual sodium carbonate (referred to by the author as "soda equivalent"), one must cosider the soil type. Water with a given RSC may be dangerous for soil with an alkaline pH, but may have an ameliorating effect on soils with an acid pH.

Chloride Hazard

Since the chloride ion has no effect on the physical properties of the soil and is not absorbed in the soil complex, it has generally not been included in modern classification systems. It appears as a factor in some regional water classifications (GRILLOT, 1954). The term "chloride percentage" that is used in geochemistry was considered as "less directly agricultural" (SCOTFIELD, 1935). Nevertheless, the author included chloride in his water classification system. It consists of five groups of water (from excellent to unusable) with limits at 4, 7, 12, and 20 meq/l.

FIREMAN and KRAUS (1965) recommended that water be diveded into four groups according to chloride content, with limits at 2, 5, and 8 meq/l. DONNEEN (1963) introduced the term "potential salinity of irrigation water," which is equal to Cl + $^1/_2$ SO_4, in meq/l. For soils having good, medium, and low permeability, he recommended that the chloride limits be 5–20, 3–15, and 3–7 meq/l, respectively.

Table 5. The chloride hazard (Cl[a]) for citrus in the coastal area of Israel. (From KOVDA, YARON and SHALHEVET, 1967)

EC (10^6 at 25 °C)	Cl (meq/1)	Soil textures Sandy	Loamy	Clay
> 1200	6	Cl_1	Cl_1	Cl_1
1200–1500	6–7.5	Cl_1	Cl_1	Cl_2
1500–1750	7.5–9	Cl_1	Cl_1	Cl_3
1750–2250	9–15	Cl_1	Cl_2	Cl_4

[a] Cl_1 = no danger; Cl_2 = low risk; Cl_3 = medium risk; Cl_4 = dangerous

It must be realized that prediction of the chloride hazard is not general and can be made only for a specific area and for a specific crop. An example of a tentative regional classification designed for citrus culture in Israel is given in Table 5.

In this classification, the following criteria were used: (a) Correlation between the chloride (meq/l) and total salt (micromho/cm) content of water from 1466 wells, representing about 80% of the well water used for irrigation. As 98% of all the well water does not contain more than 15 meq/l chloride and 2250 micromho/cm of total salts, it was not considered necessary to include water of higher salinity in the classification. (b) Relation between the chloride in irrigation water and the accumulation of chloride in the soil solution. This relation is based on a study of 110 sites (Israel Salinity Survey, 1964), located on different soil types (16 in coarse-tectured soils, 25 in medium-textured soils, and 69 in fine-textured soils) in the coastal plain area, which is characterized by a mean annual rainfall of 400–600 mm. (c) The tolerance limit for sensitive citrus root stocks. This limit is 10 meq/l chloride in the saturated extract.

Boron Hazard

Boron is one of the microelements in irrigation water that harms plant growth. The boron content of irrigation water is classified on the basis of plant tolerance to this element (Table 6).

Table 6. Permissible limits of boron for several classes of irrigation water. (After SCHOFIELD, 1935)

Boron class	Sensitive crops (ppm)	Semitolerant crops (ppm)	Tolerant crops (ppm)
1	> 0.33	> 0.67	> 1.00
2	0.33–0.67	0.67–1.33	1.00–2.00
3	0.67–1.00	1.33–2.00	2.00–3.00
4	1.00–1.25	2.00–2.50	3.00–3.75
	< 1.25	< 2.50	< 3.75

In general, it is assumed that plant sensitivity to boron is influenced by the amount of boron in the soil solution and not by the boron adsorbed by the soil. When boron concentration in the soil solution is equal to its concentration in the irrigation water, it is possible to predict the boron hazard on the basis of the boron in the water (HATCHER *et al.*, 1959). The adsorption of boron in the soil varies with the soil texture (SHAH SINGH, 1964), and is higher in fine-textured soils. Therefore, in determining the equilibrium condition, the quantity of boron adsorbed must be deducted from the amount contained in the irrigation water.

Biocide Hazard

Insecticides, fungicides, rodenticides, and herbicides include both organic and inorganic compounds that can have a direct influence on the biological equilibrium of the soil. Synthetic insecticides, developed over the last 20 years, produce most of the hazard potential. They include various types, such as halogenated hydrocarbons, organo phosphates, carbamates, substituted ureas, and triazines. Many of the halogenated hydrocarbons seem to persist in the soil environment. Because herbicides, which are widely used in agriculture, are suspected of contaminating irrigation water, attention must be given to the permissible limits of biocides to be used in irrigated agriculture. Permissible levels established by the Federal Water Pollution Administration of USDI are shown in Table 7, together with information on treatment, rate of application, and estimated concentration in water in the field or in the crop (Water Quality Criteria, 1968).

Suspended Solid Hazard

The suspended solids are responsible for adverse effects on irrigation technology and soil permeability. When large quantities of suspended material are present in the irrigation water, the ability of sprinkler and trickle irrigation systems to carry and distribute water is reduced. In the sprinkler irrigation system the suspended material may accumulate on the leaves and cause biological disturbance. Some of the nozzles in the trickle irrigation system may be blocked, and resulting in nonuniform water distribution in the field. When flooding and furrow irrigation are used and the water continues to have a high or relatively high amount of suspended material, the permeability of the soil may be affected, especially if the irrigated area is characterized by fine-textured soil. Under these conditions the suspended material can also cause crust, which may reduce seedling germination in addition to its adverse effect on infiltration.

Principal Recommendations for Applying Irrigation Water of a Specific Quality

To assess the suitability of a given water source for irrigation purposes, one must proceed in three steps. The first step includes chemical and physiological analyses, the second step deals with pedological and engineering aspects, and the third step considers the economic point of view.

1) The first step concerns the composition of the water to be used in the irrigation system. One must also know the tolerance limits of the plants to be irrigated with respect to total salinity, the effect of each individual ion, and the effect of biocides. If the total salt concentration, or the concentration of any specific ion, is denoted by C_l, and the tolerance level of a given plant is C_p, then the water in question can be used only when $C_l < C_p$. That is to say, no water can be used to irrigate any specific crop if its salt content is higher then the tolerance limit of that crop.

2) The second step in analyzing the suitability of water for irrigation considers all the possible effects on the soil, and the effect of the soil properties on the rate of salt-pollutant accumulation and leaching. It consists of two groups of requirements:

a) *Water-supply Requirements.* The way the water is applied must be adapted to the climatic conditions and to the type of the soil in the irrigated area, and must take into account the quality of the water as well as the salt tolerance of the crops. The use of saline or sewage water leads to special leaching, drainage, and irrigation-method requirements.

b) *Chemical Requirements.* The chemical requirements refer to the influence to the total salt content, ion composition, microelements, or organic compounds of the irrigation water on the soil and the plant.

3) The third step in analyzing the suitability of water for irrigation involves an economic analysis. The fact that water from a given source is found to be unsuitable for a specific crop, does not mean that the water is "bad" and cannot be used economically. The entire project must be reconsidered on the basis of selecting a more tolerant plant. Subsequently, it may be found that under a different set of conditions, the same water is "good."

The exploitation of water for irrigation purposes is complex and farreaching, and in an economic analysis both local and national interests must be taken into consideration.

Table 7. Levels of Herbicides in irrigation waters (from "Water Quality Criteria", 1968)

Herbicide	Site of use	Type of formulation	Treatment rate	Likely concentration in irrigation water reaching crop or field	Crop injury threshold in irrigation water (mg/l)[a]	Remarks
Acrolein	In water from cylinders under nitrogen gas pressure	Soluble liquid	15 mg/l × 4 hours	10 to 0.1 mg/l	Flood or furrow: beans–60, corn–60, cotton–80, soybeans–20, sugar beets –60	Canals up to 200 cfs concentration reduced to minimum in 10 to 20 miles
			0.6 mg/l × 8 hours	0.4 to 0.02 mg/l	Sprinkler: corn–60, soybeans–15, sugar beets–15	Canals, 200 to 500 cfs concentration reduced to minimum in 20 to 30 miles
			0.1 mg/l × 48 hours	0.05 to 0.1 mg/l		Canals 1,000 cfs and larger concentration to minimum in 30 to 50 miles
Aromatic solvents (xylene)	Emulsified in flowing water	Emulsifiable liquid	6–10 gal/cfs in 30–60 minimum (300–750 mg/l).	700 mg/l or less	Alfalfa > 1,600, beans-1,200, carrots–1,600, corn–3,000 cotton–1,600, grain sorghum > 800, oats–2,400, potatoes–1,300, wheat-> 1,200	Concentration reduced rapidly from point of application within 2 to 6 miles and almost completely in 6 to 10 miles
Copper sulfate	In flowing water canals or in reservoirs	Coarse pentahydrate crystals	0.5 to 3.0 mg/l (continuous) 1/3 to 1 lb/cfs (slug)	0,8 to 0.04 mg/l in 10 miles 9.0 to 0.08 mg/l in 10 to 20 miles	Apparently above concentrations used for weed control	Concentration reduced more rapidly with distance from slug applications
Amitrole-T	On bank weeds along irrigation canals and on cattail in drain canals	Foliage spray	6 to 16 lb/A	Usually less than 0.1 mg/l[c]	Beets (Rutabaga). >3.5, corn->3.5	Registered for use only in drain canals and marshes, but actually used for control of bank weeds along western irrigation canals.
Dalapon	do	Foliage	15 to 30 lb/A	Usually less than 0.5 mg/l[b]	Beets-> 7.0, corn->0.35	Same as amitrole-T

Diquat	In water or over surface of canals and reservoirs	Liquid	3–5 mg/lor 1–1.5 lb/A.	Usually less than 0.1 mg/l	Beans-5.0, corn-125.0	Diquat used in Florida for control of submersed weeds and floating weeds. Do not use for 10 days. Not used in western irrigation systems
Diuron	On bottoms and banks of small canals when no water is in canal	Wettable powder suspension sprayed	64 lb/A	Below crop injury threshold	No data	Used mostly in small farm ditches with intermittent water flow.
Monuron	Same as for diuron	Same as for diuron	64 lb/A	do	do	Same as for diuron except first water through canal after treatment not used for irrigation
Endothall Na and K salts.	In ponds and reservoirs mostly in Eastern States	Liquid or granule	1–4 mg/l	Probably little or none after waiting period	Corn-25, field beans-1.0, alfalfa-> 10.0	Must wait 7 days after treatment before using water for irrigation or domestic purposes
Dimethylamines	In water control canals in Florida. Promising use in western canals	Liquid	0.5–2.5 mg/l	Same as for Na and K salts	Corn->25, soybeans->25, sugar beets-25	Wait for 7 to 25 days (depending on concentration) after treatment before using water
2,4–D	Weeds along canal banks	Liquid spray	1 to 4 lb/A usually as amine.	3.0 to 1.0 µg/l, 2 to 10 miles below treatment	Field beans->3.5, <10 Grapes-0.7–1.5	Registered precaution: Do not contaminate water to be used for irrigation
	Floating and emersed weeds in southern canals	do	do	0.1 mg/l or less to none in 3 weeks	Sugar beets-3.5	A minimum waiting period of 3 weeks before using treated water for irrigation
Silvex	Phreatophytes on floodways, along canals, reservoirs, and streams	Liquid spray as ester	2 to 4 lb/A	No data. Probably less than 0.1 mg/l	No data	Silvex registered only for control of aquatic weeds in nonflowing water at 4 lb/100 gallons of water. Do not use in water to be used for agricultural or domestic purposes

Herbicide	Site of use	Type of formulation	Treatment rate	Likely concentration in irrigation water reaching crop or field	Crop injury threshold in irrigation water (mg/l)[a]	Remarks
	Floating and emersed weeds in southern waterways	Liquid spray over surface	2 to 8 lb / A	From 10 to 1,600 μg/l, 1 day after application, 1 to 70 μg/l, 5 weeks after treatment	do	
Dichlobenil	Promising bottom treatments in canals without water	Granules or wettable powder spray	7 to 10 lb / a		Alfalfa-10, corn->10, soybeans-1.0, sugar beets-1.0 to 10	Registered for control of submersed weeds in lakes, ponds, and drainage ditches where water not used for agricultural or domestic purposes
Fenac	Same as dichlobenil	Same as dichlobenil	10 to 20 lb / A	0.66 to 1.8 mg / l below treated area. 0.007 to 0.100 mg / l 2 hours later	Alfalfa-1.0, corn-10, soybeans-0.1, sugar beets-0.1 to 10	Same as dichlobenil
Pichloram	For control of brush and weeds on watershed areas	Liquid spray or granules	1 to 2 lb / A	No data	Corn->10, field beans-0.1, sugar beets-<1.0	Gives excellent control of Canada thistle and other bank weeds, but use near canals hazardous

[a] Date submitted by F. L. Timmons, Crops Protection Branch, Crops Research Division, ARS, USDA (unpublished).

[b] Data are for flood or furrow irrigation for all herbicides except when sprinkler irrigation is indicated for acrolein. Threshold of injury is lowest concentration that caused either temporary or permanent injury. Often this concentration did not cause final reduction in crop yield or quality.

[c] Estimates based upon very limited data and extensive observations.

Literature

ARANY, S.: VI Congres. Int. Soil Sci. Soc. VI **22**, 615–619 (1956).
BOWER, C. A.: Salinity problems in the arid zones. UNESCO, pp. 215–222 (1961).
BOWER, C. A., MAUSLAUD, M.: Symp. Water Logging and Salinity. W. Pakistan, pp. 49–61, 1963.
BRYSINE, G.: Salinity Problems in the Arid Zones. UNESCO, pp. 245–249, 1961.
DIELMAN, P. J.: Reclamation of Salt-Affected Soil in Iraq, p. 174. The Netherlands: H. Veenman and Zone 1963.
DONNEEN, L. D.: Personal communication, 1963.
DURAND, J. H.: Les Sols Irrigables, p. 187. Alger.: Imp. Inebert 1958.
EATON, F. M.: Significance of carbonates in irrigation waters. Soil Sci. **69**, 123–133 (1950).
FIERMAN, M., KRAUS, Y.: Salinity control in irrigated agriculture, p. 46. Israel: Tahal 1965.
GRILLOT, G.: Utilisation des Eaux Salines. UNESCO, pp. 11–39, (1954).
HANDRA, B. K.: Soil Sci. **98**, 264–270 (1964).
HATCHER, J. T., BLAIR, G. Y., BOWER, C. A.: Soil Sci. **88**, 98–100 (1959).
KOVDA, V. A., YARON, B., SHALHEVET, J.: Quality of irrigation water. In: International Sourcebook on Irrigation and Drainage of Arid Lands in Relation to Salinity and Alkalinity. KOVDA, V. A., HAGAN, A. M., VEN DEN BERG, C. (Eds.). Draft edition. FAO/UNESCO, pp. 246–282, (1967).
LINVINGSTONE, I.: Chemical composition of rivers and lakes. USDI Geol. Survey paper 440 G, p. 64 (1961).
LONGENECKER, D. E. LYERLY, P. J.: Soil Sci. **87**, 207–216 (1958).
MAGISTAD, D. C., CHRISTIANSEN, J. E.: USDA Circ. 707 (1944).
OREN, O. H.: Bull. Res. Counc. Israel. MG, pp. 1–33 (1962).
ROHADES, J. D., BERENSTEIN, L.: Chemical, Physical and Biological Characteristics of Irrigation and Soil Water. In: Water and Water Pollution Handbook, CIACCIO, L. L. (Ed. MARCELL DEKKER. Inc N. Y. (1971)
SCHOLLER, M.: Les Eaux Souterraines, 642 pp. Paris: Masson et Cie. 1962.
SCOTFIELD, C. S.: The salinity of irrigation water. Smithsonian Inst. Ann. Rep., pp. 275–287 (1935).
SHAH SINGH, S.: Soil Sci. **98**, 383–388 (1964).
State of Israel, Minsitry of Agriculture, Salinity Survey, Progress Rep. 1963, Tahal, 81 pp. (1964)
THORNE, O. W., PETERSON, H. G.: Irrigated Soils, Their Fertility and Management, (2nd ed.) 483 pp. New York: Blofiston 1954.
U.S. Salinity Laboratory: Diagnosis and improvement of saline and alkali soils. USDA Handbook **60**, 160 pp.
Water Quality Criteria: Federal Water Pollution Control Adminstration, USDI, 234 pp. (1968).
WILCOX, L. V.: Determining the quality of irrigation water. Agr. Inf. Bull. **194**, 7 pp. (1958).
WILCOX, L. V., DURUM, W. H.: Quality of irrigation water. In: Irrigation of Agricultural Lands. HAGAN, R. M., HAISE, H. R., EDMINSTER, T. W. (Eds.). Amer. Soc. Agron., Madison, Wis., pp. 104–125 (1967).
YAALON, D. H., KATZ, A.: IV. Congr. Israel Assoc. Adv. Sci., pp. 189–190, (1961).
YAALON, D. H.: Limnology and oceanography **92**, 218–223 (1964).

III. Water Transport in Soil-Plant-Atmosphere Continuum

A knowledge of the processes of water transport in soil, into plants and from soil and plants to the atmosphere, is fundamental for understanding, the basis of irrigation practice. An attempt is made in this chapter to define the water transport in the soil-plant-atmosphere continuum.

In the first section the forces and laws governing water retention and transport in soil are discussed. Data on soil-water status are presented and introduce a discussion of the dynamic process of water flow into soil. Some agronomic aspects of water flow in soil, such as infiltration, redistribution, and evaporation, are also discussed. Water transfer within the soil toward the plant is analyzed in the second section by defining the system, the physical laws and parameters that govern water transfer, and uptake by both a single root and an entire root system. The treatment of water flow toward plant roots as a dynamic process contributes to the understanding of the problem of soil-water availability to plants.

Following the discussion of water movement from the soil to the root, a treatment of the water movement in the liquid phase across the root and through the plant, and then in the vapor phase from plant to atmosphere, is presented. The mechanism of water movement in liquid phase through the root cortex, the xylem tubes, and the leaf mesophyll is discussed, followed by a description of the water movement in the vapor phase, from the mesophyll through the stomata to the atmosphere. The driving forces conducting the water loss from plant surfaces and the various plant resistances to transpiration are considered.

Finally, the processes dealing with the total atmospheric losses of water from both soil and plant surfaces that affect the water balance of the irrigated crops are presented. After a basic presentation of the physics of evaporation, the environmental and surfaces parameters that affect evapotranspiration are discussed.

1

Water Retention and Flow in Soils

A. HADAS

Soil water is subjected to various forces originated by the gravitational pull of the earth, the presence of solutes in the soil water, or the interaction between water contained in the soil's pores and the walls of these pores. These forces, acting on the soil water or water brought in contact with the soil, cause the water to be adsorbed, retained, transferred, drained, evaporated, or transpired to the atmosphere.

An attempt is made here to define and formulate the forces and laws governing water retention and transfer in soil, and to describe the processes involved in the agricultural hydrological cycle.

Soil-Water Potential

The force with which water is retained by a soil depends on the soil's water content. The drier the soil, the more tightly the water is held, and thus more work is required to extract water from the soil. This work is called total soil-water potential, which can be defined according to the ISSS (1963) as follows: "Soil-water potential is the amount of work that must be done per unit quantity of pure water in order to transport reversibly and isothermally an infinitesimal quantity of water from a pool of pure water at a specific elevation at atmospheric pressure to the soil water at the point under consideration." Since work means, in physical terms, force times length of path covered, one can transform this particular definition into a more generalized definition, as follows: "Soil-water potential is the negative integral of the forces over the path taken when transferring an infinitesimal body of water from a defined standard location to the point in the soil under consideration." This latter definition, when expressed per unit mass or volume, defines the potential of soil water Φ_i in the particular force field i and is given by

$$\Phi_i = -\int F_i \cdot dl \tag{1}$$

where F_i is the force per unit water due to force field i, and dl is the length element.

Two important properties should be noted here. The first is that this potential is a scalar, which indicates that an algebraic summation with respect to different force fields gives the total potential of the soil water:

$$\psi = -\sum_i \int F_i \cdot dl \tag{2}$$

The second property is that in order for the system to be at equilibrium, the sum of all forces must equal zero, or, in other words, the net force acting on an infinitesimal unit of water is zero throughout the system. Thus at equilibrium the total potential is the same at any point in the system.

The various force fields acting on the soil water are the gravitational, pressure, matric, and osmotic forces. Thus, total soil-water potential as defined by Eq. (2) is the algebraic sum of potential components, each of which is caused by a different force field acting on the soil water:

$$\psi = Z_{gravitational} + S_{pressure} + M_{matric} + O_{osmotic} \tag{3}$$

The total soil-water potential can be expressed either as energy per unit mass, unit weight, or unit volume; however, it is common to express it in energy per unit volume, which may be reduced to units of pressure such as bars, atmospheres, or dynes per cm^2. It is now possible to define each potential component separatly.

Gravitational Potential (Z)

This potential component is attributable to the gravitational force field in which a force of 980 dynes acts on a gram mass of any material. In other words, this force is the weight of the particular gram mass. This soil-water potential component is given in energy per unit volume in Eq. (4):

$$Z = \int_{z_o}^{z} \varrho g dz = \varrho g(z - z_o) \tag{4}$$

where z is the elevation of the unit volume of water with respect to an arbitrary datum plane z_o, ϱ is the density of the liquid, and g is the acceleration of gravity. According to the ISSS (1963) the gravitational potential is defined as: "The amount of work that must be done per unit quantity of pure water in order to transport reversibly and isothermally an infinitesimal quantity of water from a pool containing a solution identical in composition to the soil solution at a specified elevation at atmospheric pressure, to a similar pool at the elevation of the point under consideration."

The Z will be negative or positive depending on the location of z with respect to z_o, i.e., if z is above z_o then z is positive.

Pressure Potential (S_p) (also Piezometric or Submergence)

This soil-water potential is due to hydrostatic pressure acting on the soil-water, which is simply the weight of a water column pressing on the unit volume under consideration. This potential, which is usually encountered in saturated soils, is given by

$$S = p_{z_o} + \varrho g(z - z_o) \tag{5}$$

where p_{zo} is the pressure acting on the soil-water at the reference plane z_o. This potential is determined by measuring the depth of a point below the free-water surface at equilibrium by lacing a tube to a point adjacent to the one under consideration.

Matric Potential (M)

This potential component (known also as capillary potential or soil-water suction) originates by the interactions between the solid soil particles and the soil water. It acts in partly saturated soil where air-water-solid interfaces interact with each other. The manifestation of these interactions is the water surface tension σ and the curved air-water interfaces within the soil pores. The soil-water pressure under the air-water interface is lower than the air pressure in the soil pores and is given by the Laplace equation:

$$M = \sigma \cos \alpha / _r = -\varrho g h \tag{6}$$

where α is the solid-water contact angle, r is the pore radius, and h is the equivalent water column height at equilibrium above free-water level in a capillary tube of radius r. The right-hand side of Eq. (6) has a negative sign since the water in the capillary was raised against gravity. If h were designated as z and the surface of the water into which the capillary tube were inserted as z_o, the similarity between Eqs. (6) and (5) would be evident.

The ISSS (1963) definition for this potential component is: "The amount of work that must be done per unit quantity of pure water in order to transport reversibly and isothermally an infinitesimal quantity of water identical in composition to the soil water, from a pool at the elevation and the external gas pressure of the point under consideration, to the soil water." This definition does not refer to possible effects of internal stresses in the soil, expecially in clay, where the overburden weight of the soil layers may influence the matric potential at a given point. The definition given for potential M is actually pertinent to the pressure potential S, the only difference between them is that for S the absolute gauge soil-water pressure is greater than the atmospheric pressure whereas for M it is smaller.

Recently it has become customary to incorporate the pressure potential and the matric potential into a generalized hydrostatic pressure potential P, which is positive for saturated soil $P = S$ and negative for unsaturated soils $P = M$. Equation (6) will hold as long as the soil does not shrink. If it does shrink, the retention of water bound by the clay particles expressed as pressure will have to be incorporated into this equation.

Osmotic Potential (O)

This potential component is attributable to dissolved ions in the soil water. The potential in energy units per unit volume is given by

$$0 = -\pi \tag{7}$$

where π is the osmotic pressure caused by the dissolved solute in the soil water.

The ISSS definition for the osmotic potential is: "The amount of work that must be done per unit quantity of pure water in order to transport reversibly and isothermally an infinitesimal quantity of water from a pool of pure water at a specified elevation at atmospheric pressure, to a pool containing a solution identical in composition with the soil solution, at the point under consideration, but in all other respects identical to the reference pool."

The negative sign indicates that the potential of the water of a given solution is lower than that of pure water at the same conditions, provided a semipermeable membrane separates the two solutions.

Table 1. Potentials describing water state in the soil: reference states, equivalent names, discipline of usage and units. (Modified after ROSE, 1966 and HAGAN *et al.*, 1967)

Potential	Reference	Pool	Names for the units and disciplines of usage		
	Composition	Elevation	Energy / Volume $[ML^{-1}T^{-2}]$	Energy / Mass $(L^2 T^{-2}]$	Energy / Weight $[L]$
Total $\Phi = Z + B$ $M + 0$	Pure water at the given elevation	Arbitrarily chosen	Water potential 1.2 Total suction 2 Total stress 2 Soil-water stress 2 Turgor deficit 1 net osmotic pressure 1 Diffusion pressure Deficit 1	Water potential 1,2 Total water potential 2 Specific free Energy of soil Water 1, 2	Chemical potential of water 1, 2, 3 Partial molar free Energy of water 1, 2
Gravitational Z	Same as the soil solution	As above	Gravitational potential 1, 2, 3 ϱgh	Gravitational potential 1, 2, 3 gh	Elevation head 2, 3 h
Pressure S or M	Same as the soil solution	Same as point in soil where potential is defined	Static pressure 3 Solution pressures 1 Neutral pressure 3 Pore pressure 3 Stress 2 Hydrostatic pressure 1, 2, 3 p	Pressure potential 2, 3 $p/\varrho = gz$	Pressure head 2, 3 $p/\varrho g$
Pressure Piezometric S	Same as the soil solution	Same as above	Piezometric potential 3 $p + \varrho gh$	Piezometric potential 1, 2, 3 Hydraulic potential 1, 2 $p/\varrho + gh$	Piezometric head 2, 3 Hydraulic head 2, 3 $p/\varrho g + h$
Matric M	Same as the soil solution	Same as above	Capillary pressure 2, 3 Suction 2 Matric suction 2 Soil-water suction 2 Soil-water stress 1 Suction pressure 1 Tension 1 $p_a - p$	Capillary potential 2, 3 Suction potential 2 Matric potential 2 $(p_a - p)/\varrho$	Capillary pressure head 2, 3 Suction head 2 Matric suction head 2 2 Tension 2 $(p_a - p/\varrho g$

Potential	Reference	Pool	Names for the units and disciplines of usage		
	Composition	Elevation	Energy / Volume $[ML^{-1}T^{-2}]$	Energy / Mass $(L^2 T^{-2}]$	Energy / Weight $[L]$
Osmotic O	Pure water	Same as above	Osmotic pressures 1, 2, 3 Osmotic suction 2 Osmotic value 1 Osmotic concentration 1	Osmotic potential 1, 2 Solute potential 1	

Note: (1) plant physiologists; (2) soil scientists; and (3) engineers.

Table 2. Conversion factors for the units expressing potential pressure and suction of soil water. (After COREY *et al.*, 1967)

Pressure Energy per volume					Potential Energy per mass		Head Energy per weight	
To Convert from To	dyne / cm^2	bar	atm	erg / g	joule / kg	cal / g	cm H_2O^a	hg
dyne / cm^{-2}	1	10^6	1.013×10^6	1.00	10^4	4.18×10^7	9.8×10^2	1.33×10^4
mbars	10^{-3}	10^3	1.013×10^6	10^{-3}	10^{-5}	4.18×10^{-3}	9.8×10^{-1}	1.33×10^1
bar	10^{-6}	1	1.013	10^{-6}	10^{-2}	4.18	9.8×10^{-4}	1.33×10^{-2}
atm	0.987×10^{-6}	0.987	1.00	0.99×10^{-6}	0.99×10^{-2}	41.4	9.7×10^{-4}	1.32×10^{-2}
erg.g^{-1}	1	10^6	1.01	1	10^4	4.18×10^7	9.8×10^2	1.33×10^4
joule.kg^{-1}	10^4	10^2	1.01	10^{-4}	1	4.18×10^3	9.8×10^{-2}	1.33×10^4
cal.g^{-1}	2.39×10^{-8}	2.39×10^{-2}	2.42×10^{-2}	2.39×10^{-8}	2.39×10^{-4}	1	2.34×10^{-5}	3.18×10^{-4}
cm H_2O^a	1.02×10^{-3}	1.02×10^3	1.03×10^3	1.02×10^{-3}	10.2	42.6×10^4	1	13.6

[a] ϱ water = 1 gr cm^{-3}, g = 980 cm / sec^{-2}. To convert atmospheres to dynes / cm^{-1} multiply by 1.013×10^6.

Hydraulic Potential Φ

This soil-water potential component is usually used by engineers and is the sum of the gravitational and the hydrostatic potentials. Its value is given in Eq. (8):

$$\Phi = Z + P \text{ or } Z + M \tag{8}$$

Actually this potential is more commonly known when expressed in units of energy per weight as the hydraulic head H, which may be defined as: "The elevation with respect to a specified reference level at which water stands in a piezometer or in a tensiometer connected to the point under consideration in the soil." This concept will be used when water transfer is discussed.

Nomenclature and Comparative Terminology

Soil water can be described with respect to its content in the soil and its energy status. In Tables 1 and 2 the synonyms, units and conversion factors are given for the various soil-water potential components.

The methods and devices used in measuring soil-water potential are listed in Table 3, and a more detailed treatment of these methods and devices is given on p. 222.

Table 3. Methods and devices for measuring soil-water potential. (After HOLMES *et al.*, 1967)

Measuring Device	Potential Component	Opertional range	Advantages and disadvantages
Suction table	M	Very wet soils near saturation (0–0.1 bar)	Requires a long time to equilibrate. Inexpensive and simple, operational only in laboratory
Pressure plate or membrane	M	From saturation up to wilting point and beyond (0–100 bar)	Covers field range of soil water. Requires long time to equilibrate. Operational only in laboratory
Tensiometers	M	Wet soil (0–0.85 bar)	Placed in the field. Sensitive to temperature and roots. Fairly accurate. Limited to $M \leqslant 0.85$ bar
Psychrometer	$M + 0$	Saturation to air dryness	Precise laboratory method. Applicable to plants as well. Requires special facilities
Electrical resistance blocks	M	Field capacity to air dryness (0.5–50 bar)	Needs special calibration. Inaccurate, especially at the wet end of soil-water content. Sensitive to salts and temperature

One unit in common usage, proposed by SCHOFIELD (1935), is the pF, where F denotes the "free energy" of the soil water expressed as $\lg_{10}F$. Usually it is used as a measure of $\lg_{10}M$ (where M is in cm water). Some examples may be useful. Let us assume the soil surface to be the references level. At a depth of 100 cm below the surface, the soil solution is subjected to a matric potential of $M = -1.0$ bar and the osmotic potential of the soil solution equals 2.24 atm. The total water potential is given by using Eq. (3) and Table 2 for converting the units (to units of head). The gravitational component

(100 cm) remains the same 100 cm. The matric component value of −1.0 bar should be multiplied by 1.02×10^3 to convert it to cm water (1.02×10^3cm), and the osmotic component value of 2.24 atm converted to cm should be multiplied by 1.03×10^3 (2.307×10^3 cm). The total potential is therefore the sum of all these components and is equal to -3.427×10^3 cm of H_2O.

Soil-Water Retention and Hysteresis

The function relating the soil-water matric potential and the water content is called the soil-water-retention curve or the soil-moisture-characteristic curve. The soil-water-retention curve depends on the soil pore-size distribution. Therefore, any change in the pore-size distribution due to the soil's swelling or shrinking, soil compaction, and change of the soil structure will cause a change of the retention curve. Schematic retention curves are given in Fig. 1.

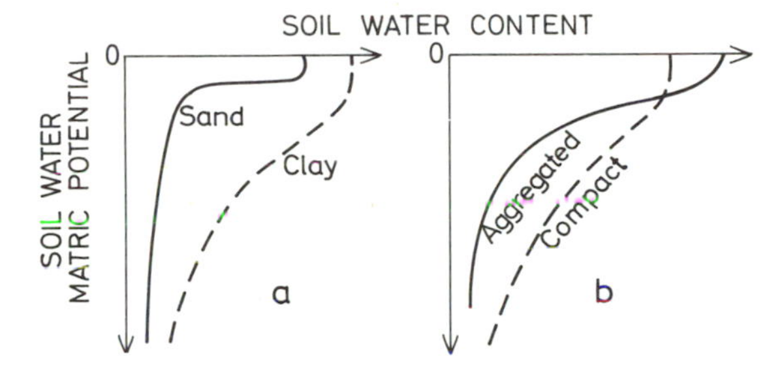

Fig. 1. Schematic retention curves: soil-water matric potential/soil-water content relationship, as affected by (a) soil texture and (b) soil structure

The soil-water-retention curve is not unique and depends on whether it was obtained for a soil-water system that was being drained or wetted. There is a difference between the retention curves for wetting or drying, and for a given soil-water content there are

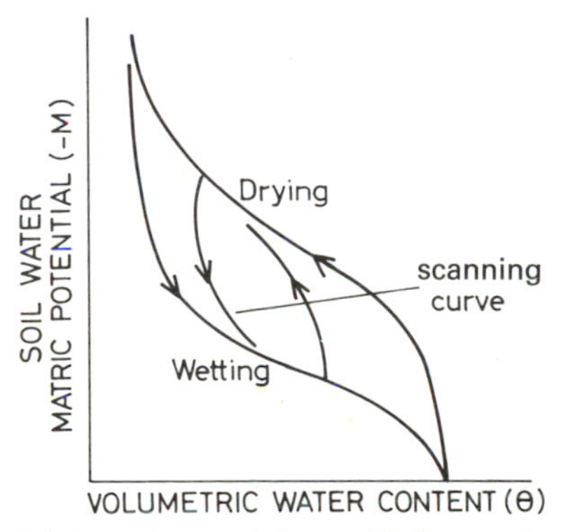

Fig. 2. Schematic hysteresis loop with the scanning curves

many values of matric potential, within a certain interval of potential values, for the particular water content in question. This phenomenon is usually referred to as soil-water hysteresis. By plotting the retention curves for both wetting and drying, one obtains the "hysteresis loop," which represents the extreme changes in the matric potential for any given soil-water content. A schematic hysteresis loop is given in Fig. 2. As one can see, for any given soil-water content the matric potential will be more negative (i.e., lower) for a drying soil than for a wetting soil. This is the case for a nonswelling soil. If the soil does change its volume during water uptake or water loss (swelling or shrinking), the problem is more complicated.

The hysteresis phenomenon can be qualitatively explained by

1) The difference in contact angles for advancing or retreating menisci,

$$\alpha \text{ adv} > \alpha \text{ ret}.$$

2) Air entrapment due to changing mode of wetting or drying.

3) Irreversible changes in soil-particle arragement caused by water-content changes.

4) The "ink bottle effect" where a soil pore with a varying radius may have two points of equal radius, one of which may be filled and the other empty at the same matric potential (see Fig. 3).

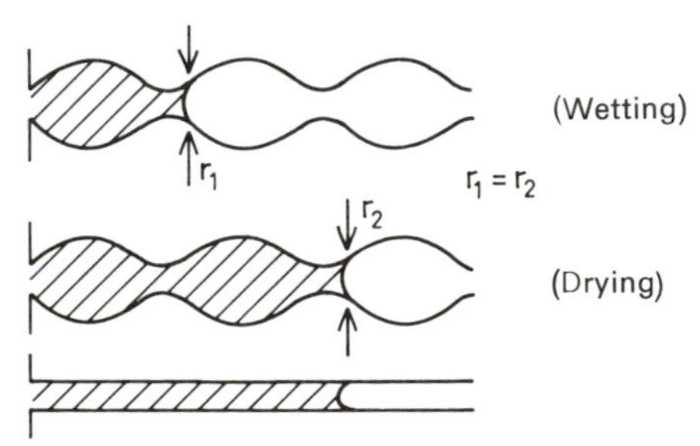

Fig. 3. Schematic presentation of the "ink bottle effect" of hysteresis

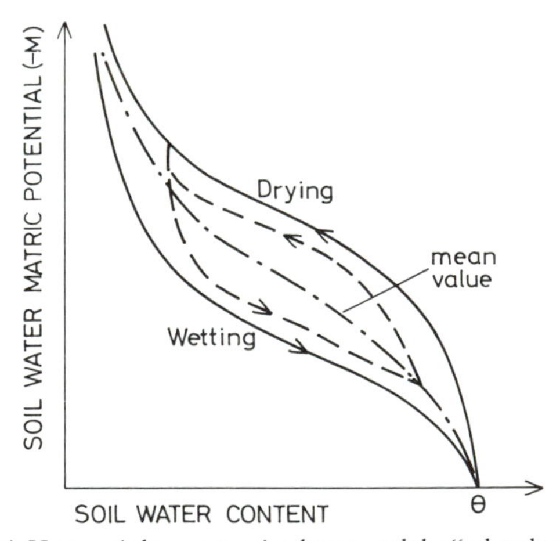

Fig. 4. Hysteresis loop, scanning loops, and the "related valued curve"

The external branches of the hysteresis loop given in Fig. 3 represent the envelope loops for all the intermediate curves, called "scanning curves."

Soil-water hysteresis imposes severe difficulties in analyzing flow systems in the soil as well as the soil-water energy status in the soil under field conditions. It is common to work either on the drying or the wetting branch of the soil-water potential functions.

However, because of the frequent fluctuations in the soil-water potential caused by water transfer, the hysteresis loop may tend to shrink or degenerate into a "relaxed valued" curve (Fig. 4).

Taking the value of the rate of change of water content with the soil-water matric potential has been defined as the "differential water capacity," $C = \partial \theta / \partial M$. This derivative of the retention curve, C is taken as a measure of water yielded by the soil per unit work done, and is a multivalued function of the soil-water content. Therefore, one may expect the hysteretic behavior of the soil-water matric potential to affect the amounts of soil water distributed, stored, evaporated, and absorbed by plant roots.

Water Flow in Soils

Soil water moves in presponse to soil-water potential differences. This movement in the liquid phase is caused by the pressure (positive or negative) gradient as long as there are no semipermeable membrances, such as plant roots, present in the system. If water movement through such membranes is under consideration, the osmotic potential gradient must also be taken into account.

Water will move from points of higher water potential toward points of lower potential. Soil-water potential has been defined as the work expended on or by the soil water during the transfer of an infinitesimal quantity of water from point *A* to a reference pool or point *B* in the soil. Hence, by differentiating the amount of work with respect to the distance from *A* to *B*, one obtains the soil-water potential gradient, which is the driving force that causes water movement.

Soil-water potential differences give rise to water flow, which can be either: (1) a viscous flow of liquid water in water-filled pores, or vapor flow in air-filled pores; or (2) a diffusive flow, namely, diffusion of water molecules as vapor. For water movement the first form is predominant in wet soils, and the second is responsible for water transfer in dry soils.

The analysis of flow may be carried out by using either a *microscopic* or a *macroscopic* approach. However, since the first approcah requires a detailed knowledge of the tortuous flow path through channels and pores, something which cannot be acquired except by micro methods and gross assumptions, it is a common practice to treat water transfer in soils on a macro scale. In this way, flow velocity is averaged with respect to distance, pore-size distribution, and energy gradients, and is treated as a gross flux.

Driving Forces

The forces causing soil-water movement may be divided into classes based on the characteristic form of such movement: *Mechanical forces* are those forces, such as gravitational, pressure, and matricial forces, that cause mass flow. *Molecular forces* include the osmotic, adsorptive, and vapor-pressure gradient forces that cause diffusional transfer. *Other forces* such as electrical or thermal gradients also may be involved.

In this treatment of water movement in soil only the mechanical forces will be dealt with (unless otherwise specified) and thus the soil-water potential will be the sum of the gravitational, pressure, and matric potential components. In the following discussion, as in practice, the dimensions of length will be used to describe soil-water potential; i.e., centimeters of water (the potential units used are energy per unit weight or equivalent water head H).

For energy case considered, the datum plane or reference elevation is arbitrarily chosen in order to obtain the total water head H:

$$H = z + (h \text{ or } m) \tag{9}$$

where h is the pressure head (saturated soil), z is the gravitational head, and m is the negative pressure, or the matric head (for unsaturated soil).

Flow Rate and Energy Relations

The observed relationship based on many experimental data between the saturated flow rate or water flux (v) and the total head gradient (∇H) is linear. This is given by the Darcy equation:

$$v = -K(\nabla H) \tag{10}$$

where K is the proportionality coefficient, known as the hydraulic conductivity, and the minus sign denotes the fact that the flow direction is against the increased potential. The hydraulic conductivity is defined as "the volume flux of water resulting from unit gradient in the hydraulic potential in the particular soil and water situation under consideration." (See ISSS, 1963.)

The dimensions of the hydraulic conductivity depend on the units used for specifying the potential. Table 4. Gives the units of the hydraulic potential gradient and the respective hydraulic conductivity units.

Table 4. Hydraulic potential gradient units and the respective units of the hydraulic conductivity (after ISSS)

Potential	Gradient dimension	Hydraulic dimensions	Conductivity units
Hydraulic potential gradient (mass basis)	LT^{-2}	T	[sec]
Hydraulic pressure gradient (volume basis)	$M^{-1}L^3T$	$M^{-1}L^3T$	[$gr^{-1}/cm^{-3}/sec$]
Hydraulic head gradient (weight basis)	$L\,L^{-1}$	$L\,T^{-1}$	[cm/sec^{-1}]

The hydraulic conductivity may be considered as a combination of two factors: (1) *the intrinsic permeability k*, which is a function of the solid-phase geometry, and (2) the fluid factor, which is known as the dynamic viscositiy $\varrho g/\eta$. This combination is represented by the relation:

$$K = k\left(\frac{\varrho g}{\eta}\right) \tag{11}$$

where ϱ is fluid density and η is viscosity. The dimensions of the intrinsic permeability k are L^2 and the units are (cm^2). As a measure of the geometrical configuration of the soil voids where water flow occurs, the intrinsic permeability, and through it the hydraulic conductivity, depends on the stability of the porous media, and its porosity, pore shape, and tortuous flow paths through which water flows.

Equation (10) applies to a saturated soil where K is presumably constant, whereas for an unsaturated soil K is dependent on soil-water content (θ), as shown in Fig. 5.

The conductivity can also be stated as a function of matric potential M, a relation that is highly hysteretic as compared with the $(K - \theta)$ relationship, which is practically unique Fig. 6).

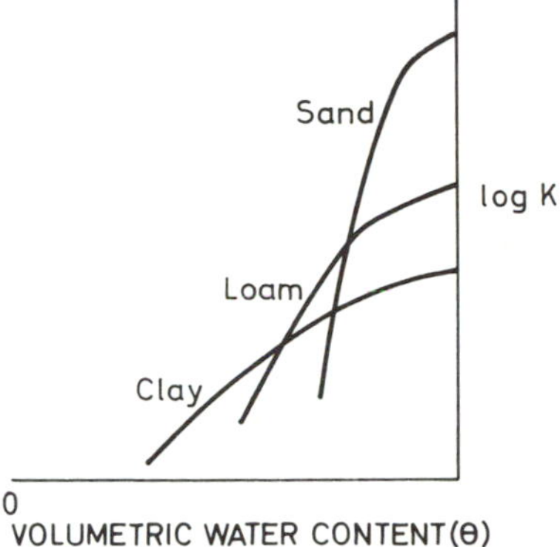

Fig. 5. Hydraulic conductivity K as a function of water content θ, as affected by soil texture

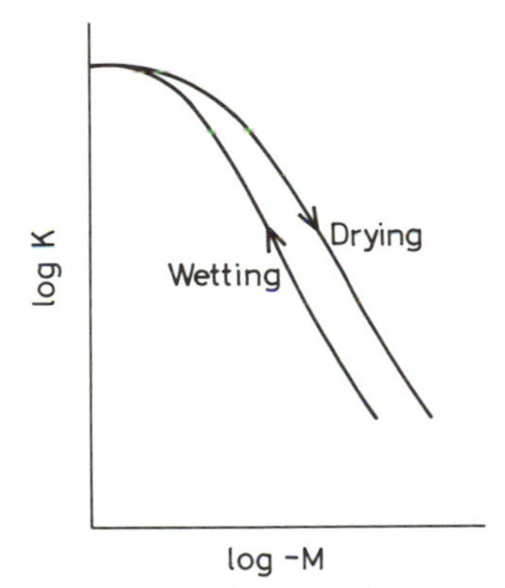

Fig. 6. Hydraulic conductivity as a function of soil-water matric potential (–M)

The drastic reduction of $K(\theta)$ with a decrease in soil-water content is attributed to the fact that the larger pores empty first and the water is forced to flow through even smaller pores and more tortuous flow channels. The smaller the pore radius, the smaller $K(\theta)$ will be, roughly by a factor of the pore radius raised to the fourth power. There are some empirical expressions for the $(K - M)$ or $(K - \theta)$ relations, such as $K = a(b + M)^{-1}$ or $K = a\exp(-nM)$ where a, b, n are constants derived empirically for a given soil.

Equation (10) for an unsaturated soil will then become, using one of the above expressions, for $K(M)$:

$$v = K(M)\,\nabla H \tag{12}$$

If the soil-water-flow system is at steady state – i.e., while water is flowing through the soil there is no change in the soil-water content the potential distribution – Eq. (12) can be solved by using the proper $K(M)$ function. However, if the soil-water content and potential distribution change with respect to time during water flow, the solutions of Eq. (12) become more complicated.

Flow Equations

Regardless of its form [Eq. (10) or Eq. (12)], the Darcy equation assumes a steady flow rate under steady total head and moisture content, meaning a *steady state*. Yet, under most conditions, the total potential (and/or the water content) change with time and position. These changes involve the law of the conservation of matter, expressed for vertical flow by

$$\frac{\partial \theta}{\partial t} = - \frac{\partial}{\partial z} (v) \tag{13}$$

Eq. (13) is called the continuity equation and, by combining it with Eq. (12), one obtains the generalized flow equation:

$$\frac{\partial \theta}{\partial t} = - \frac{\partial}{\partial z} (v) = \frac{\partial}{\partial z} \left[K(M) \frac{\partial}{\partial z} (M + z) \right] \tag{14}$$

Eq. (14) includes three interrelated parameters θ, M, and $K(M)$ and therefore it is difficult to solve. It seems easier to treat the water flow by introducing a new soil-water variable soil-water diffusivity $D(\theta)$, which is defined by Eq. (15) assuming a unique relationship of $M(\theta)$:

$$D(\theta) = K(M) \frac{\partial M}{\partial \theta} \tag{15}$$

where $(\partial M/\partial \theta)$ is the slope of the retention curve, denoting the specific water capacity. Because of hysteresis the soil-water diffusivity should be defined separately for a wetting or drying soil. The introduction of the soil-water diffusivity D (θ) into Eq. (14) yields by applying the chain rule of differentiation of M:

$$\frac{\partial \theta}{\partial t} = \frac{\partial}{\partial z} \left[K(M) \frac{\partial M}{\partial \theta} \cdot \frac{\partial \theta}{\partial z} + \frac{\partial z}{\partial z} \right] = \frac{\partial}{\partial z} \left[\left(D(\theta) \frac{\partial \theta}{\partial z} + K(M) \right) \right] \tag{16}$$

Eq. (16) is somewhat easier to solve, especially for horizontal flow where the the gravitational components are zero. Equation (16) resembles the well-known equation of diffusion called the second Fick equation. Soil-water diffusivity is analogous to the thermal diffusivity in the theory of heat flow, by analogy from that theory it is defined as "the flux of water per unit gradient of moisture content in the absence of other force fields." The dimensions of soil-water diffusivity are L^2T^{-1} and the units are cm^2/sec^{-1}. In order to arrive at solutions for actual conditions and flow systems, these equations should be solved with respect to the initial and boundary conditions appropriate for the case under consideration.

Boundary Conditions

These conditions define the interfaces between the soil layers and between the soil and the atmosphere. Certain assumptions must be made, such as that the pressure head is continuous across the interfaces between layers having different properties. This assumption is necessary since the pressure head must be a continuous variable, and a dependent variable in Eq. (16). It also must be assumed that the normal component of the macroscopic flow velocity vector is continuous across the interfaces. This assumption is a prerequisite stemming from the continuity equation.

In addition to these assumptions, there are conditions describing the relations between the properties of the soil layers involved, the changes of the total head and water content, and whether the flow system is a stationary or transient one. Henceforth, the discussion will consider the actual processes that comprise the hydrologic cycle of bare and arable land under dry and irrigated conditions.

Soil-Water Flow Systems

From an agricultural point of view, water movement involves its penetration into the soil, redistribution, deep drainage, evaporation from the soil surface, and uptake by plants. Sometimes two or three of these processes take place simultaneosly.

Infiltration

The source of water entering the soil may be natural or a man-made system of water application. Water may enter into the soil profile through its upper surface, from below, from a water table, or horizontally. Infiltration is the process describing downward water uptake by the soil through its upper surface. Water may be introduced to the soil surface, by flooding the surface, by sprinkling, or as rain.

The infiltration process has been treated experimentally and theoretically, especially for the case of a homogeneous soil profile and flooded soil surface. It is common knowledge that the infiltration capacity (defined as the maximum intake rate, under given conditions, of excess water on the soil surface) decreases with time and tends to reach a constant rate—the final infiltration rate. Several empirical expressions have been developed to describe the time dependence of the infiltration capacity and cumulative infiltration. PHILIP (1957) has solved Eq. (16) for the conditions:

$$\begin{array}{lll} z > 0 & t = 0 & \theta = \theta_{initial} \\ z = 0 & t > 0 & \theta = \theta_n \end{array} \qquad (17)$$

The solution obtrained for the flow equation is given in Eq. (18), where cumulative water uptake I through the soil surface is given as a function of the time t elapsed since initiation of water infiltration:

$$I = \Sigma_n \, a_n(\theta) \; t^{n/2} \simeq St^{1/2} + At \qquad (18)$$

S is the "sorptivity," which is a factor characterizing the soil-water relationship, and A is a factor that represents the effect of gravity. Both A and S are fuctions of the volumetric moisture content distribution and are calculated from knowledge of the $K(M)$, M, and $D(\theta)$ of the soil under consideration.

The infiltration capacity of the soil can be given by the derivation of Eq. (18) with respect to time t:

$$i = \frac{dI}{dt} = 1/2 \; St^{-1/2} + A \qquad (19)$$

Eq. (19) shows that the infiltration capacity decreases with infiltration time, A gives the final infiltration rate, the value of which is very near the saturated hydraulic conductivity of the topsoil. The infiltration capacity decreases as the initial water content increases, an effect that is more pronounced at the beginning of infiltration. Both the sorptivity S and the A factor depend on the soil-surface conditions. In saline arid-zone soils, these two factors decrease with increased salinization or the affinity of the soil-surface crumbs to slaking. These crumbs, when wet, will seal the soil-surface during infiltration and afterward may develop into a hard surface crust.

The process as described takes place under conditions of flooded soil-surface. Where water is applied to the soil by rain or sprinklers, the uptake rate will equal the application rate as long as the infiltration capacity is higher than the application rate. The moment the rate of water application exceeds the infiltration capacity of soil, the excess water will start ponding at the surface, and in more severe cases flooding and water runoff will start. The water-content distribution and infiltration rates for a flooded soil and for given rain intensity are given in Fig. 7 and 8 (after RUBIN *et al.*, 1964).

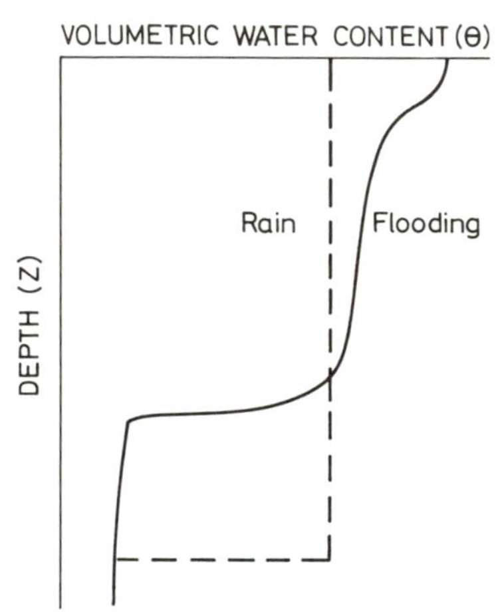

Fig. 7. Water-content distribution as a function of time and mode of water application of an equal amount of irrigation water

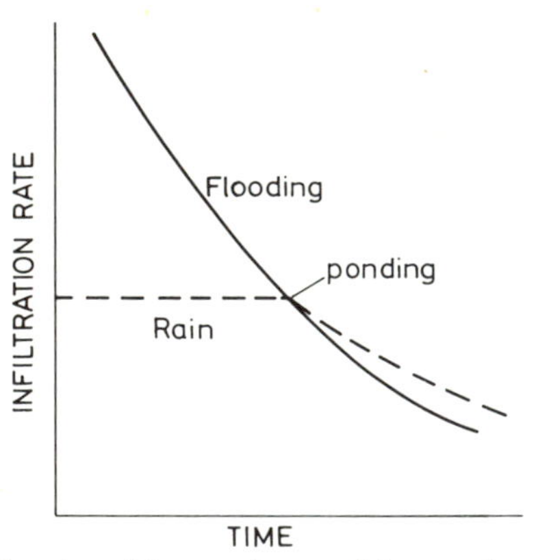

Fig. 8. Infiltration rate as a function of time, surface conditions, and mode of water application

It should be noted that the less intense the rain is, the lower the water content of the wetted soils; thus the infiltration rate of the soil will exceed the rate of water application for a longer time, delaying the appearance of ponded water.

The occurrence of runoff in arid zones, where efficient water use is of paramount importance, means the loss of water for the growing crops, and consequently lower yield and reduced salt leaching. In dry-land farming, a technique is sometimes used whereby runoff from one part of a field is used to irrigate the remainder of the field, and thus an improved water regime is maintained in a part of the total area.

Under field conditions a soil profile is not homogeneous as to either water-content distribution or soil-water properties (i.e., capillary conductivity), and, analytical solutions of Eq. (16) for a nonhomogeneous soil are not yet available. The soil's inhomogeneity may be attributed to gradually changing soil-water properties or to distinct layers that differ from each other in their properties, such as surface crusts.

Some numerical solutions have been developed for layered soils (HANKS and BOWERS, 1962). The solutions for the infiltration rate and soil-water distribution are given in Fig. 9 and 10. As can be seen, the infiltration rates are controlled by the least permeable layer, or

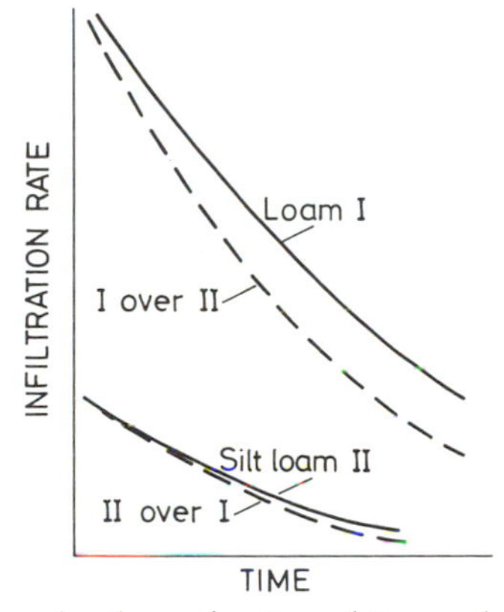

Fig. 9. Infiltration rate of a layered soil as a function of time and order of layering. (Schematically drawn after HANKS and BOWER, 1962)

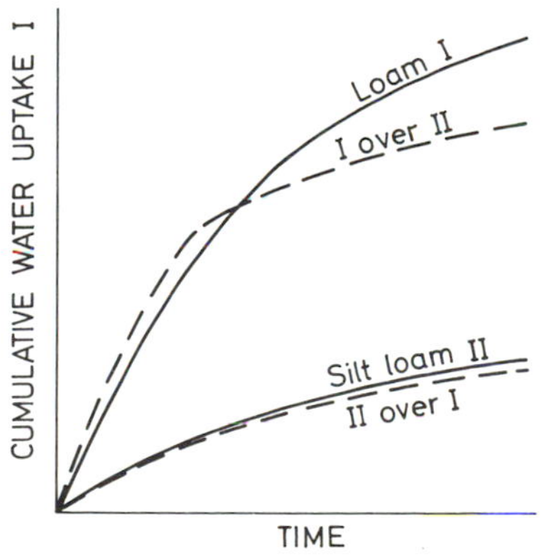

Fig. 10. Cumulative water uptake into layered soil as a function of time and order of layering. (Schematically drawn after HANKS and BOWER, 1962)

by the infiltration capacity of the upper layer in the case of a coarse-textured soil in which the water has not yet reached the second layer. These solutions treat the different layers as homogeneous; however, under natural conditions soil layering is seldom so well defined. Nevertheless, the trends shown in Figs. 9 and 10 yield an interesting insight into the problem of infiltration into a nonhomogeneous soil.

Redistribution and Field Capacity

Redistribution of the infiltrated water starts immediately after it recedes from the soil surface. This process is more complex than infiltration because, in part of the soil profile, the water content decreases due to the downward flow of water, whereas at the bottom of the wetted profile it increases in the absence of a water table or remains saturated if one exists. Hysteresis thus has to be introduced into the analysis of this process.

The rate of the water-content change is considerable at the beginning, but decreases with time. If the soil is so covered as to prevent evaporation from its surface, the water content will tend to reach a "final water content" that is lower than that during the infiltration process. This final water content is commonly known as the "field capacity." It is generally accepted that field capacity is not an equilibrium value and should be considered a transitory value in the soil-water-time-depth distribution, the rate of change of which is very small. Field capacity is usually defined as "the amount of water held in the soil, after excess water has drained away and downward movement has essentially ceased and evaporation has been prevented, this usually takes about two to five days after irrigation has stopped" (Fig. 11). It is quite difficult to determine the point at which the redistribution rate is negligible, and in some soils (e.q., loessial soils) a field capacity cannot be reached. The water content at field capacity depends on soil texture and structure, its water properties [i.e., $K(M)$ and $M(\theta)$], soil layering, depth of wetting, and method of water application.

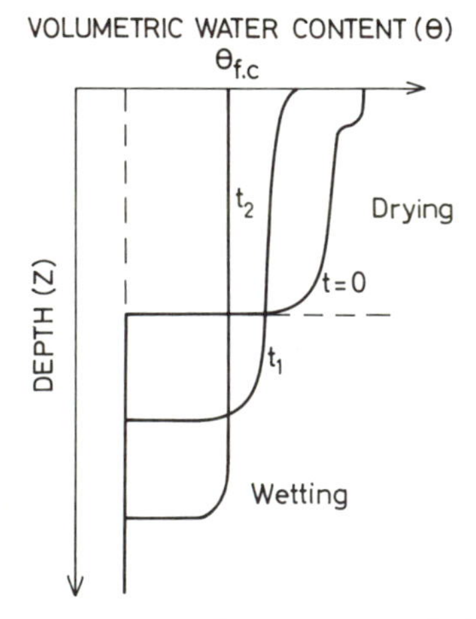

Fig. 11. Redistribution of water content in an homogeneous soil after infiltration (θ_{fc} – moisture content af field capacity)

The redistribution rate decreases with time and moisture content due to a decrease in the hydraulic gradient and in capillary conductivity. The hydraulic-gradient decrease is attributed to an increasing soil-water potential at the wetting end of the soil profile and a decreasing soil-water potential of the draining top of the soil profile (this is a hysteresis effect). The result is a decrease in the water-potential difference, and, at the same time, the distance between the two ends increases, thus decreasing the potential gradient. Based on this factor, combined with the marked reduction in capillary conductivity caused by the decreasing soil-water content, in the decrease in the redistribution rate becomes obvious. If the depth of wetting is small, the field capacity will be observed at a lower moisture content than for soils wetted to a greater depth, because even though the conductivity is very small the hydraulic gradient is still quite large.

Field capacity serves as an estimate of the practical upper limit of available water storage for plant use, since, for transpiring plants, redistribution is often so slow after 2 or 3 days that it is negligible compared with the water uptake by the plants.

The field capacity was found to be correlated with the soil-water content against a pressure differential of -0.1–0.5 bar. The former value is typical for coarse-textured soils and the latter for very fine soils. The capillary conductivity at this moisture content is usually less than 5×10^{-7} cm $\sec^{-1}$ for the corresponding soils. If evaporation is not prevented after infiltration ceases, the soil's-water-content profile will be different than shown in Fig. 11, and for shallow, wetted soils, field capacity may not be reached as defined.

Evaporation and Drying

Water evaporation from the soil is a process involving a change of state. Heat transfer accompanies the water transfer since heat is necessary for vaporization. Mass transfer occurs to move the water vapor to the atmosphere. The physics of the evaporation from a moist surface is discussed in Chap. 3. The present discussion will deal with water transfer within the soil and toward its evaporating surface.

Water loss from the soil can take two forms:

Steady-state Evaporation from a Water Table. In this case the soil-water distribution is practically constant; the soil profile acts as a conveying medium, and the water flux is controlled by the external atmospheric conditions namely, the amount of energy available for the change of state.

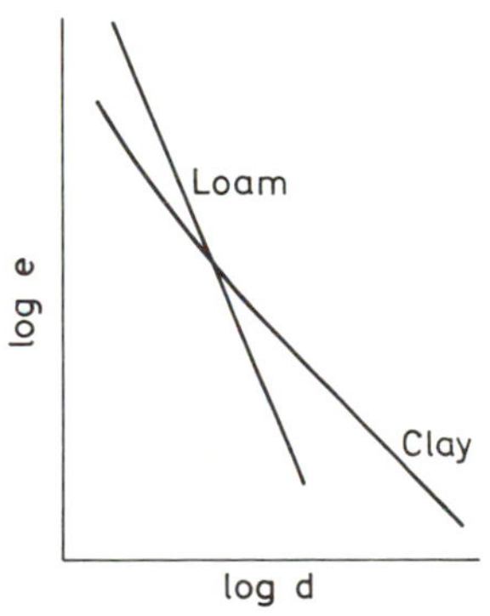

Fig. 12. Steady evaporation flux e as a function of water table depth d

Solutions for steady-state evaporation were developed by GARDNER (1958) from the flow Eq. (12). These solutions present the water potential as a function of the evaporation rate and height above the water table. In another form, these solutions yield the maximal limiting evaporation rate e, which is of the form given in Eq. (20) and in Fig. 12.

$$e = A \cdot d^{-n} \tag{20}$$

where A, and n are constants specific to the soil concerned and d is the depth to the water table. The solutions given here are for homogeneous and isothermal soil conditions; numerical solutions are available for layered soil. The limiting evaporation rate for a layered soil is controlled by the layer with the lowest hydraulic conductivity. The solution for the upward water flow from a water table can be extended to permit an estimate of the upward water movement into the root zone. The rate of water supply to the root zone is of practical value only if the water table is no deeper than 1 meter below the soil surface. However, deeper water-table levels as obtained in drained soils (i.e., at 1.5–2.5 m) cannot maintain an upward rate of water movement sufficient for plant growth, or an appreciable rate of evaporation *e*, wherever saline soil or brackish water are adjacent to the soil surface, steady evaporation may cause a severe saliniation hazard. If the water table lowered by conventional drainage methods, to 2–3 m, the salinization rate is also reduced; however, salinization is not prevented. Saltwater flow from the water table toward root systems will salinize the root zone without salinizing the soil surface.

(2) *Nonsteady-State Evaporation or Drying of the Soil.* In this case there are two well-defined stages. During the first stage the evaporation rate is constant and controlled by atomspheric conditions and changes in the soil-water-content distribution as the soil dries. This stage continues as long as the soil can transport water to the surface at the

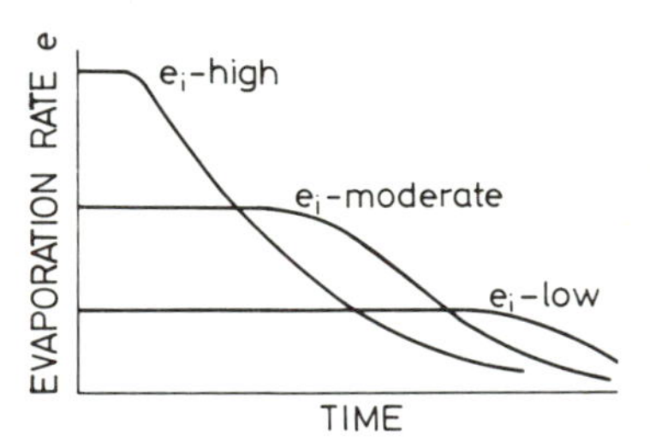

Fig. 13. Time dependence of evaporation rate on the initial rate

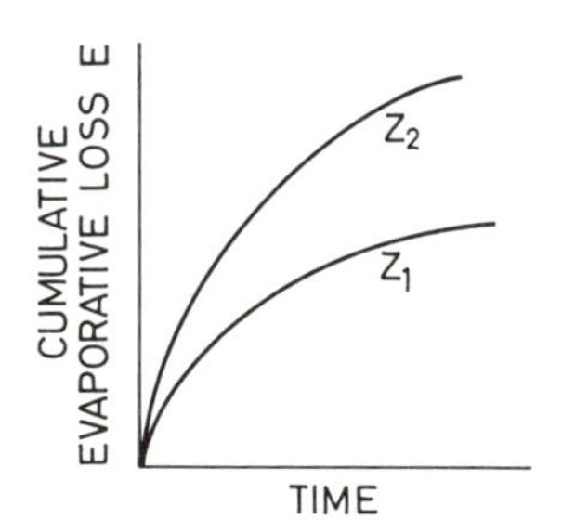

Fig. 14. Cumulative evaporative loss as a function of time and depth of a finite soil system

same rate as the potential evaporation rate. When the soil surface dries out, the rate of water flux decreases – the second stage of drying in which the evaporation rate decreases with time. The two stages are illustrated in Fig. 13.

The falling rate of evaporation can be described by an equation similar to Eq. (18), and thus the rate of evaporation and the cumulative evaporation loss are proportional to the reciprocal to the square root of time. Under field conditions the evaporative losses from soil wetted to different depths will be different, as shown in Fig. 14.

For a layered soil, the cumulative evaporative loss and the evaporation rate depend on the water-conducting properties of the soil layers, on their thickness, and on their sequence in the profile (HANKS and GARDNER, 1966). If a coarse-textured upper layer occurs on top of a finer-textured layer (such as sand over clay, or plowed soil clods over a homogeneous lower profile), the total evaporative loss will be lower from the layered profile than from the homogeneous profile, provided the layering was produced prior to evaporation. The reduction of the total evaporative loss depends on the mean aggregate or particle diameter and on the depth of the top layer. This phenomenon is illustrated schematically in Figs. 15 and 16.

These facts imply that tilling the soil to a proper depth and producing the proper aggregate size may conserve water by reducing the evaporative loss. Practically, this is true only for a period of time, the duration of which depends on the soil layer's water properties, the depth of the top layer, and the wetting depth. The deeper the soil is wetted, the greater is the effect of the top layer in reducing evaporation and preserving the water, since more water can be moved downward due to redistribution. The same effect will be observed if the capillary conductivity of the top layer is drastically reduced. However, over a long period of time the cumulative water loss by evaporation may be practically independent of the agrotechnical measures taken to conserve water.

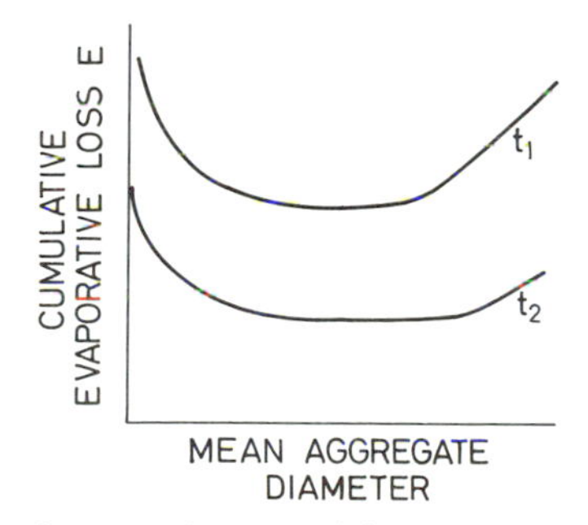

Fig. 15. Cumulative evaporative loss as a function of the mean aggregate diameter for different time periods

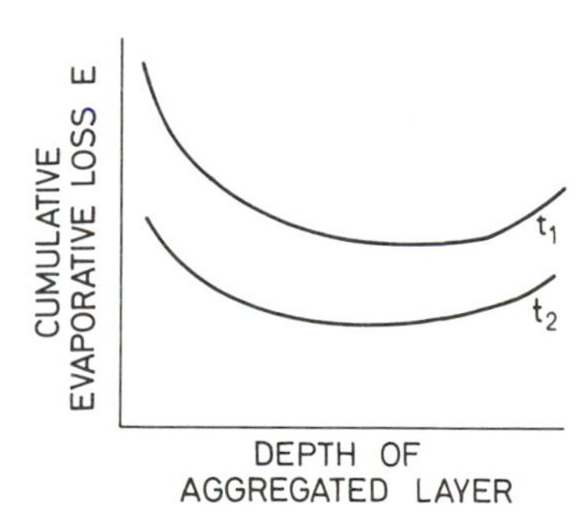

Fig. 16. Cumulative evaporative loss as a function of the aggregated layer depth for two different time periods

Water Flow to Roots

Water flow toward active roots is an important aspect of water transport in soils. A detailed approach to this process is given in Chap. 3.

Concluding Remarks

The preceding paragraphs dealt with the definition and formulation of the forces and laws govering soil-water behaviour under field conditions. For the time being, the physical processes as described for uniforms soils or idealized layered soil explain the behaviour of soil water in the field. To improve our understanding to a state where we can deal with these processes in a quantitative way, the conditions prevailing in the field should be taken into account in our calculations.

Field application of the theoretical treatment and conclusions are further complicated under arid-zone conditions due to greater soil profile variability, the presence of soluble salts, sporadic rainfall, and soil tillage and management practices. However, even though the present knowledge is inadequate, it has contributed considerably to our comprehension of the principles controlling soil-water retention, movement, and management in field practice.

Literature

COREY, A. T., SLATYER, R. O., KEMPER, W. D.: Comparative terminology for water in the soil-plant-atmosphere system. In: Irrigation of Agricultural Lands. Agronomy 11. HAGAN, R. M., HAISE, H. R., EDMINSTER, T. W. (Eds.). Amer. Soc. Agron., Madison, Wis., pp. 427–445 (1967).

DAY, P. R., BOLT, G. H., ANDERSON, D. M.: Nature of soil water. In: Irrigation of Agricultural Lands. Agronomy 11. HAGAN, R. M., HAISE, H. R., EDMINSTER, T. W. (Eds.). Amer. Soc. Agron., Madison, Wis., pp. 193–208 (1967).

GARDNER, W. R.: Some steady solutions of the unsaturated moisture flow equation with application to evaporation from a water table. Soil Sci. **85**, 228–232 (1958).

HANKS, J. R., BOWERS, S. A.: Numerical solution of the moisture flow equation for infiltration into layered soils. Soil Sci. Soc. Amer. Proc. **26**, 530–535 (1962).

HANKS, J. R., GARDNER, W. H.: Influence of different diffusivity-water content relations on evaporation of water from soils. Soil. Sci. Soc. Amer. Proc. **29**, 495–498 (1965).

HOLMES, J. W., TAYLOR, S. A., RICHARDS, S. J.: Measurements of soil water. In: Irrigation of Agricultural Lands. HAGAN, R. M., HAISE, H. R., EDMINSTER, T. W. (Eds.). Amer. Soc. Agron., Madison, Wis. (1967).

International Society of Soil Science, Soil physics terminology. Bull. **23**, 5 (1963).

PHILIP, J. R.: The theory of infiltration: 1. The infiltration equation and its significance. Soil Sci. **83**, 345–357 (1957).

PHILIP, J. R.: The theory of infiltration: 4. Sorptivity and algebraic infiltration equations. Soil Sci. **84**, 257–264 (1957).

RICHARDS, L. A.: Physical condition of water in soil. In: Methods of Soil Analysis. Agronomy 9. BLACK, C. A., *et al.* (Eds.). Amer. Soc. Agron, Madison, Wis., pp. 128–152 (1965).

RICHARDS, S. J.: Soil suction measurements with tensiometers. *Ibid.*, pp. 153–163 (1965).

ROSE, C. W.: Agricultural Physics. London: Pergamon Press 1966.

ROSE, C. W., STERN, W. R.: Determination of withdrawal of water from soil by crop roots as a function of depth and time. Aust. J. Soil Res. **5**, 11–19 (1967).

RUBIN, I., STEINHARDT, R., REINIGER, P.: Soil water relations during rains of low intensites. Soil Sci. Soc. Amer. Proc. **28**, 1–5 (1964).
SCHOFIELD, R. K.: The pF of water in soil. Trans. 3rd. Internatil Cong. Soil Sci. **2**, 37 (1935).

Water Transfer from Soil to Plant

A. HADAS

Introduction

Plant activity responds to the availability of and the water rate of uptake within the root zone. Thus, the water status in a plant's tissues, governed by the water balance, is controlled by the relative amounts of water uptake and loss. Water uptake by plants depends on soil-water and plant-water properties, and on the efficiency of the active root system. The treaty presented here attempts to review water transfer within the soil to the plant by definining the physical laws and parameters governing water flow and uptake by a single root and root systems.

Root Structure

Roots, especially young ones, act as the sole organs for water uptake. It is, thus of interest to review briefly the root's structure and charcteristic features.

A typical young active root has three concentric regions: first, the epidermis, then the cortex, and finally the centrally placed vascular cylinder (see Fig. 1).

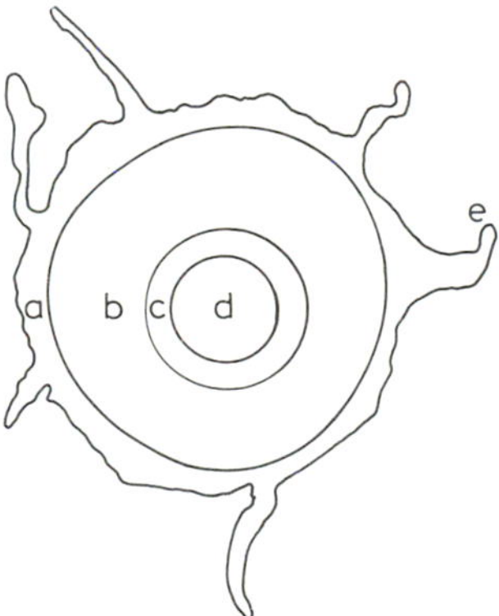

Fig. 1. Schematic cross section of a root showing epidermis (a), cortex (b), endodermis (c), vascular cylinder (d), and root hair (e)

The epidermis usually consists of closely packed, elongated cells. These cells constitute the external layer through which water and nutrients enter the root. Root aging cause these cell walls to become less and less permeable; however, when young, many of these cells produce thin-walled protuberances called root hairs.

Te cortex is made up mostly of parenchimal cells packed in such a way that large voids are left between them, thus providing an extensive airspace throughout the root.

The innermost critical layer is the endomermis. It consists of thin-walled cells with lengthwise so-calles casparian strips that are considered impermeable and supposedly influence the transfer of water or nutrients toward the vascular cylinder.

The vascular cylinder has as its peripherical layer a meristematic cell layer called the pericycle, which produces most of the root branching. Within the vascular cylinder are the phloem tubes, near the pericycle, and the xylem tubes, near the core or the center of the root. The xylem vessles convey the water and the minerals absorbed from the roots towards the stem and plant canopy, and the phloem vessels act as conveyors and distributors of synthesized material within the plant.

Looking along young root from its root tip one can distinguish the root cap, which protects the root's meristematic center. Behind the root cap, is the zone of root-cell elongation where fastest water uptake by the root is supposed to occur. This part gradually merges into a region where the cells differentiate and mature; here the vascular vessels are already fully developed and root hairs are discernable (see Fig. 2). Older parts gradually tend to suberize and thus become less permeable to water.

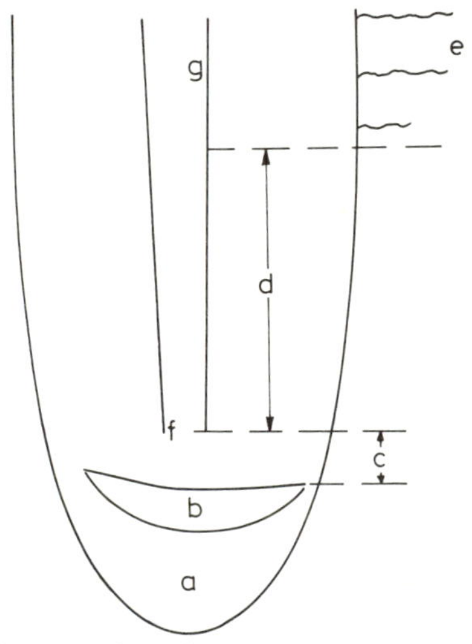

Fig. 2. Schematic longitudinal section through a root showing root ca, (a) meristematic zone, (b) zone of elongation, (c) differentiation and maturation zone, (d) root hairs, (e) initiation of vacular cylinder (f), and mature vascular cylinder (g)

Plant seeds usually emerge with a single root, but this lasts only for a short period and then the root systems start to develop in accordance with their particular genetics, environmental conditions, and the cultural practices.

Many field studies have been made that include mapping of the root system's extent and density distribution with respect to depth for various plant species. Most of the techniques used are laborious and time-consuming and yet the information acquired is incomplete as it does not prove which of the roots excavated, washed, or core sampled actively take up water and which merely act as pipelines. An attempt will be made in the following paragraphs to evaluate rootsystem activity; the reader is referred to EASAU's book (1953) for further information concerning root structures.

Root-System Development

The characteristic shape of the root system of any plant species is programmed within its genetic code, yet this specific root development will be attained only if soil factors, water availability, and climate are adequate.

The soil-solid matrix affects root development by mechanically impeding root penetration and elongation (GILL *et al.*, 1959; BARLEY and GREACEN, 1967). The denser the soil matrix is, the higher this impedance will be. Moreover, because denser, the less oxygen can penetrate into the soil profile, and thus root elongation and branching are further impaired.

Soil-water availability also affects root development since the root has to exert pressure at its tip to penetrate the soil effectively. If water is less available, the pressure at the root tip, and thus root elongation, will be reduced. On the other hand, high soil-water content may also reduce root development because of the oxygen deficiency.

Soil fertility is of paramount importance. A nitrogen deficiency or an excess of it reduces root development whereas in the mid-range maximum development is encouraged.

Data concerning the influence of phosphorus and potassium on root development are meager, but, in general, favorable responses have been reported wherever improved fertility was found.

Soil temperature affects root development through its influence on the chemical processes involved with the plant-root tissues, soil water, and nutrient availability. Root development is enhanced as the temperature increases, until an optimal temperature is reached, after which development decreases.

Cultural practices, such as irrigation and tillage methods, are instituted to improve the soil-water regime and soil matrix compaction, and thus may improve environmental conditions, and consequently root development.

Plant-water Energy Status

The soil-water energy status, a comprehensive definition of plant-water status is needed to account for all the forces operating in the plant cell.

A plant cell can be considered as an ideal osmometer of minute dimensions, made of an elastic cell wall, a very thin semipermeable membrane, and a vacuole. The solutions within the cell behave according to the laws of an ideal gas. Even based on such an oversimplified approach, the general principles of cell-water equilibrium and exchange can be explained.

According to these simple assumptions, the water potential of a plant cell compared with that of free pure water is

$$\psi = P - O \tag{1}$$

where ψ is the total water potential, P is the hydrostatic potential, and O is the osmotic water potential. These potentials are in units of energy per volume (joules/kg) or pressure (bars or dynes cm^{-2}). The total potential ψ is negative under most conditions, except perhaps during gutation, which can take place when a positive pressure exists.

The hydrostatic potential P is usually identified in the plant as the turgor pressure that the cell walls exert on the cell content and O is the total osmotic potential of the cell content. The total water potential of plant cells depends, through its osmotic component, on the ionic strength of the internal and external ionic solutions, the sugars and organic-acid contents, and the proteins and polysacharides contained in the cell. The turgor pressure depends on the elasticity of the cell walls.

Driving Forces for Water Transfer

It is commonly assumed that the total water-potential gradient is the driving force F that causes the water to move (see p 89). Water movement in the soil toward the root occurs mainly in the liquid phase and is thus governed by the total water-potential gradient between the soil water and water at the root surface, given in Eq. (2), where ψ_p is the total plant water potential at the root surface and ψ_s is

$$F = \text{grad}\,[\psi_p - \psi_s] \tag{2}$$

the total soil-water potential. The total root-water potential ψ_p is governed by the root cell contents, the transpiration rate, and elasticity. The total soil-water potential ψ_s depends on soil-water content and solute concentrations. (The reader is referred on p. 90 for a more detailed treatment.)

For a nonsaline soil the osmotic component of the soil-water potential is negligible and ψ_s is approximated by the matricial component. However, for a soil where the osmotic component is of the same order of magnitude as the matricial component or higher, and under high transpirational rates, solutes excluded from entering the root may cause a buildup of osmotic potential at the root surface accompanied by a reduced matricial potential gradient.

Water Flow Toward a Single Root

Water moves in the soil toward a single root along a gradient of decreasing water potential. If water is to enter the root, the root should be in contact with the soil and at a lower water potential. Hence, as the root draws water it should readjust its water potential to keep water flowing in. If the uptake rate equals the rate the soil can supply, the root's-soil-water flow system is said to be at a steady state. Under these conditions the potential gradient across the root surface may be constant, but usually changes with time and rate of water uptake.

GARDNER (1960) analyzed the water flow toward a single root by assuming an infinitely long, thin root, extracting water at a constant rate per unit length. The soil-water conductivity was assumed to be constant. His solution to the flow equation for a radial system, Eq. (3), subjected to the foregiving conditions

$$\frac{\partial\theta}{\partial t} = \frac{1}{r}\,\frac{\partial}{\partial r}\left[rD\,\frac{\partial\theta}{\partial r}\right] \tag{3}$$

is given by Eq. (4).

$$M_r\text{–}M_o = \Delta M = \frac{q}{4\pi k}\left[ln\,\frac{4Dt}{r^2} - 0.5722\right] \tag{4}$$

where θ is volumetric soil-water content, t is time, r ist radius, D, k is the soil-water diffusivity and capillary coductivity, respectively, and M_r, M_o, ΔM are the soil-water matric potential at radius r, at the soil bulk that is unaffected by the root, and the matric potentiol drop between $r = r$ and the root surface. Equation (4) yields the matric potential distribution around a root, assuming no solute concentration at the root surface. A schematic graphical presentation of results computed by Eq. (4) are given in Figs. 3 and 4.

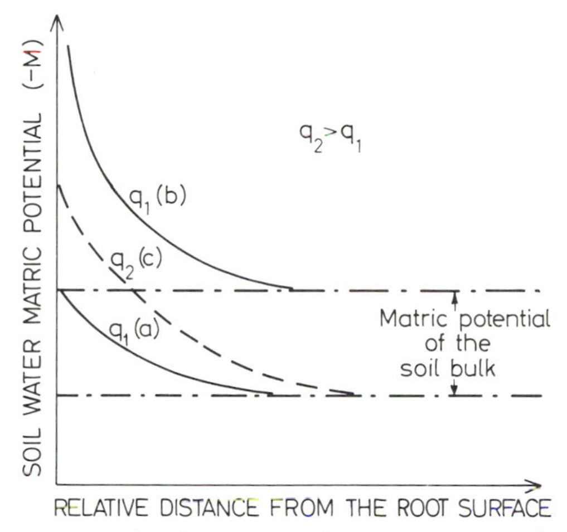

Fig. 3. Soil-water matric potential distribution around an active root as a function of distance from root surface for two water uptake rates q_1, q_2

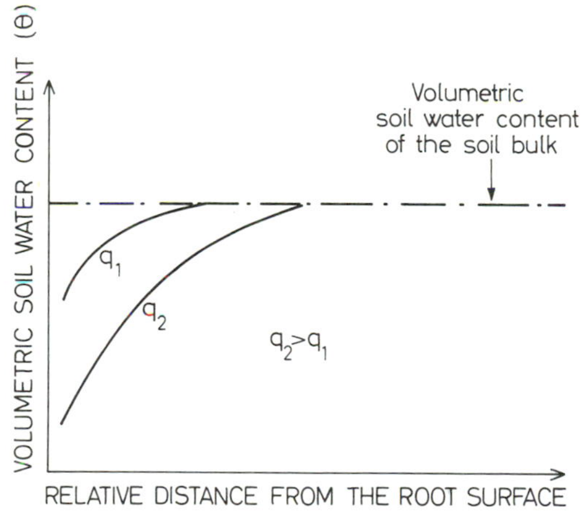

Fig. 4. Soil-water content distribution around an active root as a function of distance from root surface for two water uptake rates q_1, q_2

Equation (4), which is an approximation, yields an interesting insight into the concept of the wilting point. The water-potential drop between that of the unaffected soil bulk and that at the root surface will increase directly proportional to an increase of the uptake rate q or inversely proportional to a decrease in the capillary conductivity k. The effect of D, t, and r being in a logarithmic term is small. When the soil matric potential is high and the k value is comparatively large, ΔM will be small, but as the soil dries and k diminishes very fast, ΔM increases. At high values of M_o e.g. -10 to -15 bars ΔM become very large for a given q (see Fig. 4, curves a, b). If q increases due to environmental drought, ΔM may become very large even at moderate values of M_o (-5 to -8 bars(see Fig. 4, curves a, c). Therefore one is led to conclude that wilting point is a function of environmental conditions, soil properties to water e. g. water retention and conduction, and extent and activity of the plants root system.

If one tries to project these considerations into whole root systems, it can be concluded that q depends upon the transpiration rate and extent of the effective root length. The higher the transpirational rate is, the larger q will become for a given root system.

Gardner's model (1960) assumes a constant rate q; however, under normal conditions, there is no steady state. Several models were tested and yielded practically the same qualitative results (COWEN, 1965, and PASSIURA and COWAN, 1968).

Even though the simple models of the type shown cast some light on water flow to a root, they are not very useful in analyzing the behavior of a root system, since an active root system is not a stationary system imbedded in a uniform soil having a constant water content. Actually, root system's continuous extension is exploring new wet-soil regions and the younger roots are absorbing most of the water taken up. It seems, therefore, that a somewhat different approach should be undertaken, one that considers the whole soil profile and the full extent of the root system.

Water Uptake by an Active Root System

The extent of a root system is great and variable since the rooting depth and total root length change with time and environmental conditions. The number and length of the roots of a given plant cast some doubts whereas the soils hydraulic properties such as suction and water conduction may limit the water uptake. Root lengths of 200–4000 cm/cm^2 of soil surface area are reported for annual cereal plants, and 15–200 cm/cm^2 for nongraminaes (NEWMAN, 1969). Assuming the depth of rooting to reach 100–120 cm, this means that some roots are very near each other and the models of single roots as presented are not very useful in describing the actual situation prevailing in the soil.

An active root system extends new roots most of the time, and since water uptake is fastest near the tips of new roots, it may lead to different amounts of water uptake at various depth within the root zone and thus to different water contents and soil-water potentials at various positions within the root zone. The root system integrates the range of external water potential values and it seems that the ψ_{root} is close to the value found in the wettest regions of the root zone (SLATYER, 1967).

GARDNER (1964) proposed a model predicting water uptake by a nonuniform root system. Assuming water uptake flux per unit volume soil q to be given by

$$q = \frac{\psi_{root} - M_{soil\ bulk}}{I_p + I_s} \tag{5}$$

where I_p and I_s are the impedance to water flow in the plant roots and the impedance to water flow in the soil, respectively and M its water matric potential. By assuming $I_p \ll I_s$ and ψ_{root} to be uniform throughout the root system and $I_s = 1/BkL$ where B is a constant, k is soil-water capillary conductivity, and L root length per unit soil volume, on one can derive for each layer of the root zone i the water uptake q_i for the particular layer. This is given by Eq. (6):

$$q_i = Bh\,(\psi_{root} - M_i - Z_i).\, k_i L_i \tag{6}$$

where h and Z_i are the layer thickness and depth below the soil surface. By integrating q_i with respect to the number of soil layers, one can evaluate q or the transpiration rate T from Eq. (1), provided B, $k_i L_i$ and ψ_{root} are known.

$$q = T = \sum_{i=1}^{n} q_i = Bh \sum_{i=1}^{n} (\psi_{root} - M_i - Z_i)\, k_i L_i \tag{7}$$

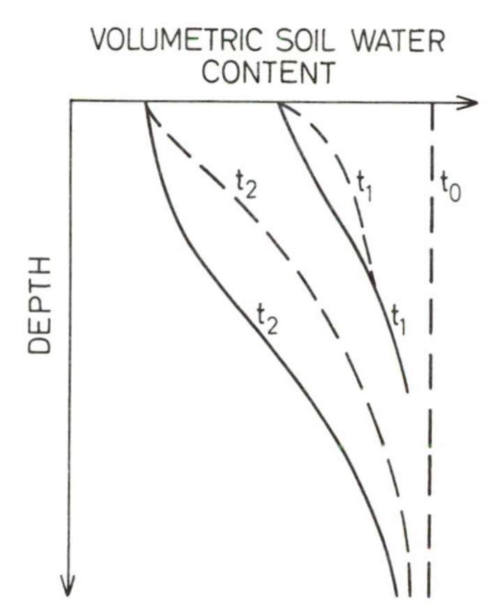

Fig. 5. Volumetric soil-water content distribution with respect to soil depth as a function of time t for a given uptake rate and two values of B. ——B = 1; (– – –B = 100.) (Schematically redrawn after GARDNER, 1964)

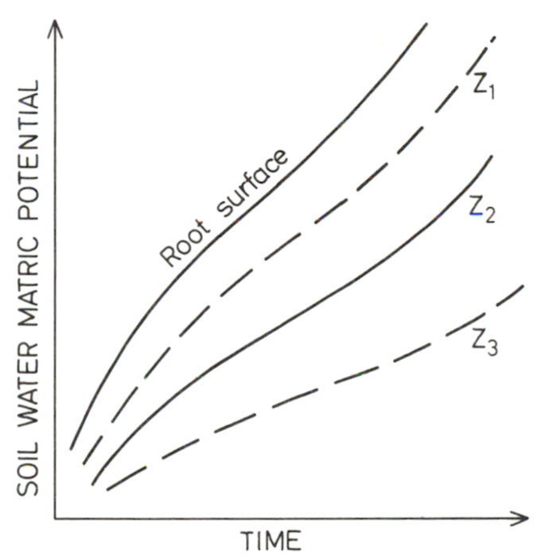

Fig. 6. Soil-water matric potential pattern at three locations as a function of time for a given uptake rate. (Schematically redrawn after GARDNER, 1964)

Some results using this approach are given in Figs. 5 and 6; however if B and Li change with time, this sort of an analysis may lead to erroneous results, depending on the assumed values of B, L_i, k_i.

A different approach was taken by ROSE *et al.* (1967), where, by knowing the changes with time of the water content, matric potential, and the hydraulic conductivity of the root zone, the water uptake by roots can be calculated without knowing B and L_i.

According to ROSE *et al.* (1967), the soil-water conservation equation for the root zone is given for the period $(t_2 - t_1)$, the times of observing consecutive soil-water content distribution, by Eq. (8):

$$\int_{t_1}^{t_2} (I - V_z - E)dt - \int_{o}^{z}\int_{t_i}^{t_2} \left(\frac{\partial\theta}{\partial t}\right) dz\, dt = \int_{0}^{z}\int_{t_1}^{t_2} r_z\, dz\, dt \tag{8}$$

where I is rate of water application, E is evaporation rate at the soil surface, V_z is the vertical water flow rate at depth z (positive downward) and $z = 0$ at the soil surface, and r_z is water withdrawal by roots at depth z. The vertical down flux V_z is given by

$$V_z = K_z \left(\frac{\partial M}{\partial z} + 1\right) \tag{9}$$

where K_z and M are the vertical hydraulic conductivity at z and the soil-water matric potential at z, determined by tensiometers or from-water contents via the retention curve. Integration of Eq. (9) yields

$$\int_{t_1}^{t_2} V_z\, dt = \left[K_z\right] \left(\left[\frac{\partial M}{\partial z}\right] + 1\right) (t_2 - t_1) \tag{10}$$

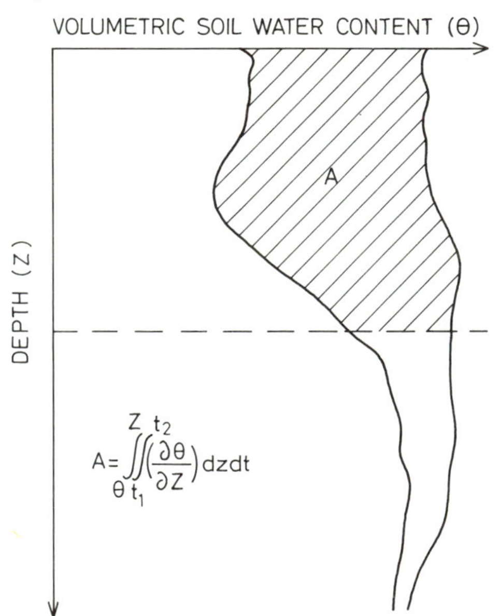

Fig. 7. Schematic presentation of the change in soil-water storage for a soil layer of thickness z and time interval $(t_2 - t_1)$

where angular brackets signify time averages for the period $(t_2 - t_1)$. The evaporative loss $\int E\, dt$ is given by

$$\int_{t_1}^{t_2} E\, dt = -\int_{t_1}^{t_2} V_{z\ o}\, dt \tag{11}$$

or estimated from meteorological data. The term $\int_z \int_t \left(\frac{\partial\theta}{\partial t}\right) dz\, dt$ which actually denotes soil-water storage change for the time interval (t_2-t_1) and soil volume to depth z, is obtained graphically by calculating the area enclosed between successive profiles of water contents at t_2 and t_1 and between the soil surface to a depth of z (see Fig. 7). Substitution of all these calculated values into Eq. (7) yields $\int_o^t \int_{t_1}^{t_z} r_z \, dz \, dt$. By repeating these calculations

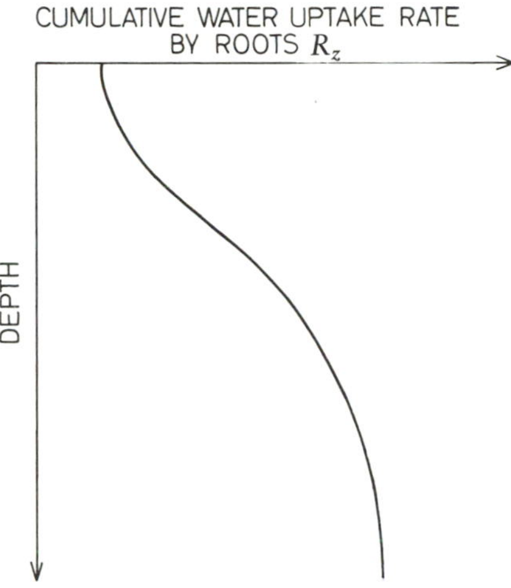

Fig. 8. Cumulative water uptake R_z as a function of soil depth. (Schematically redrawn after ROSE *et al.*, 1967)

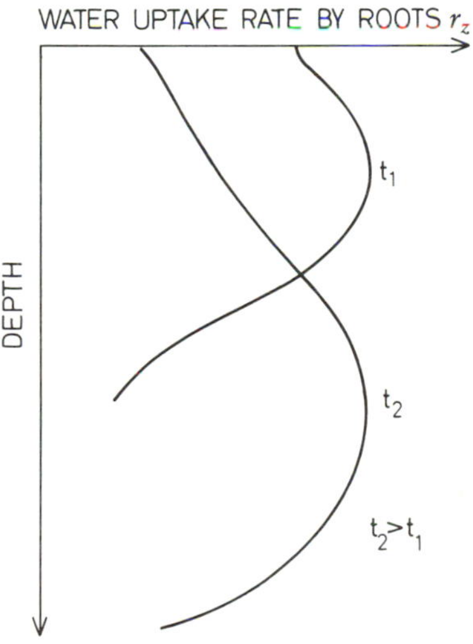

Fig. 9. Water uptake rate by roots r_z as a function of soil depth z. (Redrawn schematically after ROSE *et al.*, 1967)

to any desired depth and time interval, the pattern of water uptake by the active roots can be computed for depth z. The total water uptake for the whole root zone R_z is given by Eq. (12):

$$R_z = \int_o^z r_z \, dz \tag{12}$$

Schematic patterns of r_z and R_z are given in Figs. 8 and 9. It should be noted here that the assymptotic value of R_z should equal the transpiration rate. When using this method, one has to be aware of the possible errors involved; for example, taking short time periods may interoduce less accurate $\left(\frac{\partial\theta}{\partial t}\right)_z$ values whereas large time values may cause unrealistic $(\partial M/\partial z)_z$ values. In addition, errors are introduced in determining the hydraulic conductivity under field conditions. The ability to calculate the amounts of water uptake at various layers within the root zone provides a useful estimate of water consumption for a given crop and climate conditions, and thus may help determine the correct leaching requirements for saline soils, timing of water application and water quotas in precise irrigation schedule, and may help determine the water balance of any given crop.

Soil-Water Availability

The treatment of water flow toward plant roots as a dynamic system may help in understanding the dynamic problem of soil-water availability to plant growth. It is difficult to define water availability because of the number of factors involved and incomplete knowledge of the interactions between these factors. However, some points can be clarified and the crux of the problem explored in view of the previous paragraphs.

It is clear that the range of soil-water contents (between field capacity and wilting point) or soil matric potential (between water contents held at-0.33 and-15 bars of matric potential) can not be used as a definition of available soil water which depend on the extent of and root activity, on one hand, and on the soil-water properties and transpiration demand on the other. For low water uptake rates and a given root system, the plant will not show any water stress signs even if the soil-water matric potential is low, yet for high water uptake rates, signs of water stress will shows up at high matric potential values. Such behavior was shown by DENMEAD and SHAW (1962) and is given in Fig. 10.

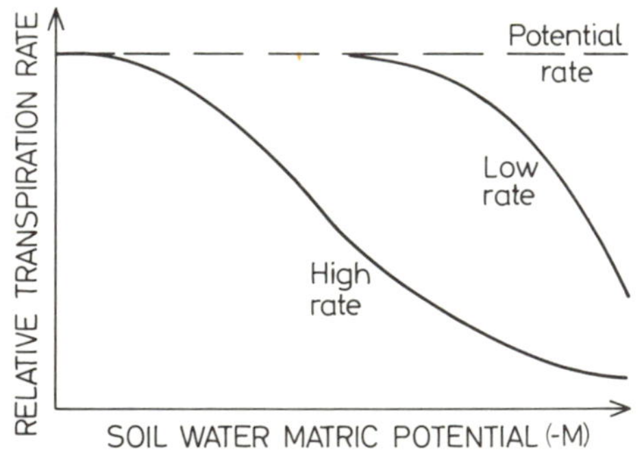

Fig. 10. Relative transpiration rate as a function of soil-water matric potential. (Schematically redrawn after DENMEAD and SHAW, 1962)

These date show that a low transpiration rate, and hence a low root-water uptake rate, can be maintained even at low soil-water matric potential, whereas a high transpiration rate cannot be maintained at high soil-water matric potential values. Thus the problem arises that even though enough water at high matric potential may be held within the root zone, unless the whole volume is fully explored by active roots, not all the water will practically speaking, be available. Such an analysis was presented by COWAN (1965) and is given in Figs. 11 and 12. COWAN (1965) states that the transpirational rate falls at a soil-water matric potential that depends on soil and crop properties (e.g., root density) and that the value is higher under fluctuating conditions than under constant transpiration rates.

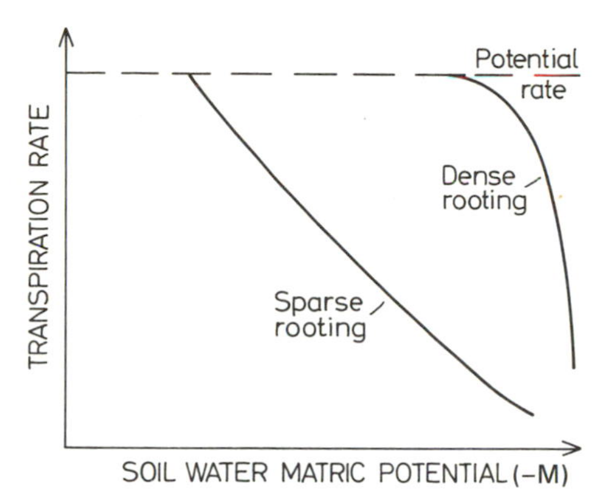

Fig. 11. Transpiration rate as a function of soil-water matric potential for two rooting patterns. (Redrawn schematically after COWAN, 1965)

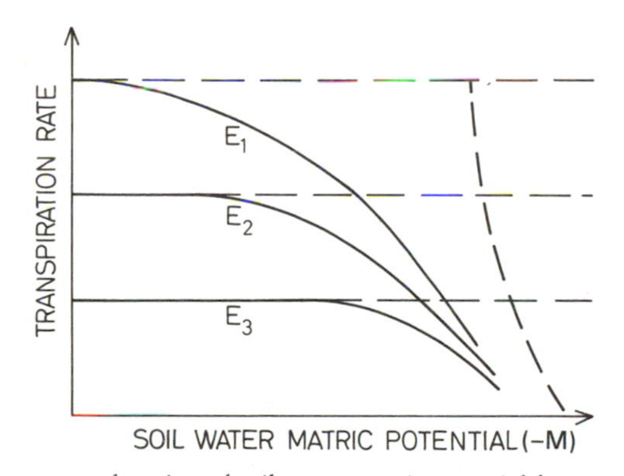

Fig. 12. Transpiration rate as a function of soil-water matric potential for two transpirational modes. Solid line (——) for fluctuating diurnal rate; broken line (– – –) for constant diurnal rate. (Redrawn schematically after COWAN, 1965)

If one integrates the variations in the transpiration rates (Fig. 12) with respect to time, assuming a constant root-zone depth, different root densities, and an initial soil matric potential, one gets a family of curves (see Fig. 13) illustrating the effect of rooting density on the transpiration rate at a given soil water matric potential. For a sparse root density, given transpiration rate cannot be maintained for long and water stress builds up in the plant canopy when the water content is still high. However, if the root density is high, the same rate is maintained for a longer period and the soil water content is lower, but then the rate falls drastically and the stress is more severe. One is led to conclude that for a given soil and crop having a very dense root system, and a moderate or low transpiration rate, the soil waters between "field capacity" and "wilting percentage" are equally available; however, sparse rooting and/or high transpiration rate will result in progressive reduction in soil-water availability.

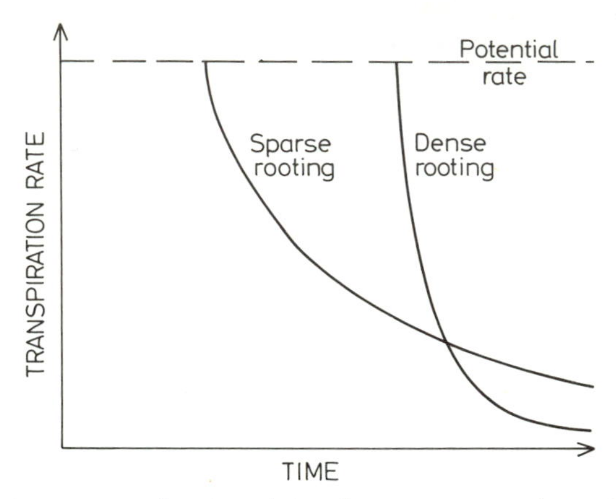

Fig. 13. Transpiration rate as a function of time for two patterns of root densities. (Schematically redrawn after COWAN, 1965)

Literature

BARLEY, K. P., GREACEN, E. L.: Mechanical resistance as a soil factor influencing the growth of roots and underground shoots. Adv. Agron. **19**, 1–43 (1967).

COWAN, I. R.: Transport of water in the soil-plant-atmosphere system. J. Appl. Ecol. **2**, 221–239 (1965).

DENMEAD, O. T., SHAW, R. H.: Availability of soil water to plants as affected by soil moisture content and meteorological conditions. Agron. J. **54**, 385–390 (1962).

ESAU, K.: Plant Anatomy. New York: John Wiley 1953.

GARDNER, W. R.: Dynamic aspects of water availability to plants. Soil Sci. **89**, 63–73 (1960).

GARDNER W. R.: Relation of root distribution to water uptake and availability. Agron. J. **56**, 41–46 (1964).

GILL, W. R., HAISE, H. R., and HAGAN, R. M.: Annotated bibliography on soil compaction. Amer. Soc. Agric. Eng., 32 pp. (1959).

NEWMAN, E. I.: Resistance to water flow in soil and plant. 1. Soil resistance in relation to amounts of root: Theoretical estimates. J. Appl. Ecol. **6**, 1–12 (1969).

PASSIURA, J. B., and COWAN, I. R.: On solving the non-linear diffusion equation for the radial flow of water to roots. Agric. Meteorol. **5**, 129–134 (1968).

ROSE, C. W., and STERN, W. R.: Determination of withdrawal of water from soil by crop roots as a function of depth and time. Aust. J. Soil Res. **5**, 11–19 (1967).

SLATYER, O. R.: Plant-water relationships. New York: Academic Press 1967.

VEIHMEYER, F. J., HENDRICKSON, A. H.: Soil moisture conditions in relation to plant growth. Plant Physiol. **2**, 71–82 (1927).

3

Transport of Water in Plant-Atmosphere System

Z. PLAUT and S. MORESHET

A green plant may be considered as a miniature oasis, absorbing water from the soil that maintains the plant tissue saturated and losing water by evaporation into the surrounding air.

The function of water uptake is conducted in higher plants by the root system. Water movement from the soil to the root surface and the interrelationship between this movement and the characteristics of the root system were discussed previosly. The loss of water from plants, known as transpiration, is a process of water-vapor diffusion from high concentration to lower concentration in the air.

We shall restrict this discussion to water movement in the liquid phase across the root and throughout the plant and then in the vapor phase from plant to atmosphere. Several resistances to water movement were shown to be located along this water pathway. By analogy to the flow of current in an electrical conductor, van den HONERT (1948) wrote an expression for steady-state flow:

$$\text{Flow rate} = \frac{\psi_{soil} - \psi_{root\ surface}}{r_{soil}} = \frac{\psi_{root\ surface} - \psi_{xylem}}{r_{root}} = \frac{\psi_{xylem} - \psi_{leaf}}{r_{xlem} + r_{leaf}} = \frac{\psi_{leaf} - \psi_{air}}{r_{leaf} + r_{air}} \qquad (1)$$

where ψ is the water potential at various sites of the system and r is the corresponding resistance.

Although this model was criticized by PHILIP (1966) and SLATYER (1967), it may be a valuable tool for looking at water relations of the entire plant. It says that the resistance at any segment is related to the potential difference at the two sides of this segment. For instance, the resistance at the leaf air interface is the greatest as the potential difference at this site is the highest.

Water Movement in the Liquid Pase

Water movement throughout the plant demands transfer through two regions of living cells: the root cortex and the leaf mesophyll. The xylem tubes, which can be regarded as relatively empty tubes, connect these two regions. As the mechanism of water movement in these three regions is different, they shall be discussed independently.

Movement Across the Root

A general type of the internal root structure is shown in Fig. 1. The major part of the cross section is the cortex, which includes the external layer – the epidermis – a layer of closely packed cells. Part of these cells develop root hairs that increase surface area of the root. Root hairs develop only in a specified zone 1 to 3 cm long, behind the meristematic region of the root tip and ahead of the region where suberization develops. This is the zone where water absorption is most rapid. The major part of the cortex consistes of large parenchymatous cells and many intercellular spaces. The innermost layer of cells is differentiated and is known as the endodermis. In dicotyledonous roots part of

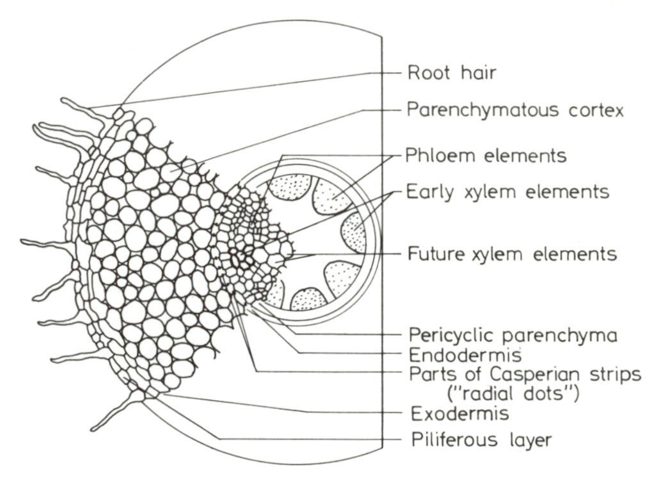

Fig. 1. Structure of a dicotyledoneous root in the regions of maximum water absorption. (From "The Plant in Relation to Water" by R. O. KNIGHT, Heinemann Educational Books, Ltd. London, England, 1965)

the radial walls of these cells are lignified and suberized, forming a strip known as the *casparian strip*. This kind of endodermis usually remains in roots of dicotyledonous plants, even when they develop secondary growth. In this case the cells either divide in the radial direction or are stretched and crushed. In some other plants the endodermis is discarded during secondary growth together with the rest of the cortex. In plants that have no secondary root growth, such as most of the monocotyledons, a suberin lamella develops at the inside of the endodermis cell walls after the casparian strip has already been formed. A thick layer of cellulose is then formed over all cell walls, excluding the outer tangential wall. The walls may then become lignified. This indicates that the endodermis plays an important role in the movement of water and solutes and may be an important barrier, particularly in the monocotyledons. It should therefore be mentioned that the development of endodermis is very often unequal, leaving cells with only casparian strips next to thick-walled cells. The thin-walled cells, which may be regarded as passage cells, are found adjacent to the xylem, whereas those with thick walls are adjacent to the phloem.

The stele inside the endodermis consists of the pericycle and the vascular system. The pericycle is concerned with meristematic activities in the formation of lateral roots. The vascular system consists of xylem, which is primarily responsible for the upward movement of water and ions, and of phloem, which mainly conducts organic solutes. These tissues are located in a radial arrangement.

Water movement from the root surface to the xylem is due to a driving force that is the difference in water potentials ($\Delta\psi$) between the two sites. Since the two systems are separated by a membrane that is permeable to solutes and not only to water, the effect of the moving solutes and water movement has to be taken into account. This was done by DAINTY (1971), who distinguished between permeating and nonpermeating solutes having different effects on $\Delta\psi$. The differnece in water potential can thus be expressed as

$$\Delta\psi = \Delta P - \Delta\pi_{imp} - \sigma\Delta\pi_p - \Delta J \quad (2)$$

where ΔP is the difference in hydrostatic or turgor pressure between the root surface and xylem, $\Delta\pi_{imp}$ and $\Delta\pi_p$ are the differences in impermeating and permeating solutes between the two sites, and σ is the reflection coefficient of the membrane for the particular solutes, expressing the relative osmotic potential exerted by the permeating solutes. σ is always less than 1, and becomes 0 when the membrane is as permeable to the solutes as to water. ΔJ is the difference in matric potential between the two sites.

Assuming, for simplicity, that the hydrostatic pressure at the root surface is zero, that matric potential of the xylem sap is negligible, and that the reflection coefficient is close to 1, then the gradient in water potential will be

$$\Delta\psi = -P_{xylem} - \Delta\pi - J_{soil} \quad (3)$$

As outlined, water is transferred against a resistance located inside the root. It is generally accepted, mainly on the basis of anatomical evidence, that the endodermis constitutes a physical barrier to water transfer (ARISZ *et al.*, 1951, and SLATYER, 1967.). HOUSE and FINDLAY (1966 a, b) have suggested, on the basis of physiological evidencc, that the barrier is located just beyond the xylem vessels. They immersed excised maize roots in aqueous solutions and studied the rate of exudation from the top of those roots as a function of water-potential differences. When their external solution was changed by the addition of a nonpermeable solute, a delay was found in the exudation rate, reaching a

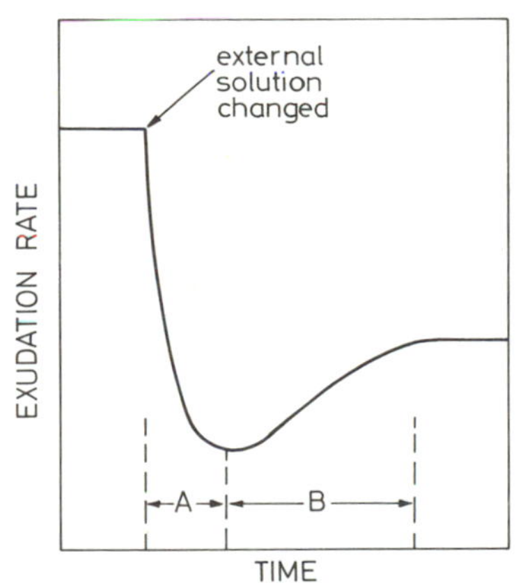

Fig. 2. Effect of osmotic potential changes in external root medium due to a nonpermeating solute on exudation rates. *A* The new solute concentration is diffusing toward the somotic barrier. *B* Readjustment of the osmotic flow as a result of concentration changes within the osmotic barrier (From J. DAINTY, 1971)

minimal value after some time (Fig. 2). This indicates that the barrier to water movements is not located at the epidermis but must be further inside the root. HOUSE and FINDLAY showed a rise in solute concentration inside the root compartment surrounded by the barrier when nonpermeating solute was added to the external solution. This rise was attributed to pumping of ions into the stele. The rate of the rise was determined by the volume of the inside compartment. Since the higher concentration of ions led to an increase in water-potential difference and in rate of exudation, the volume could be estimated. The authors claimed that the volume of this compartment was roughly equal to the volume of the xylem vessels containing protoplasts, suggesting that the barrier for water movement is located at the surface of those elements.

Concerning the nature of this barrier, ARISZ *et al.* (1951) considered it to be an osmotic barrier. If this is so, then a catena of three links is located between the root surface and the xylem: (1) mass flow up to the endodermis, which is in agreement with the "free space" in the cortex; (2) osmosis across the endodermis dominating the overall movement; and (3) probably again mass flow beyond the endodermis. The question arises whether, under active transpiration conditions, pathways in the endodermis may enable mass flow of water through nonsuberized regions, so that the rate of water transport will increase. The possibility of the existence of diffusional and mass-flow pathways in parallel was investigated by WEATHERLEY (1963). If the barrier permits only diffusion of water, then the flux will be at an equal rate when osmotic or pressure gradients are exerted. If, however, mass flow can occur in addition, it will take place as a result of a pressure gradient only. Experiments were carried out with tomato stem stumps attached to the root system. They were placed in a pressure canister with their cut stems protruding. The medium could be rapidly changed, and thus the osmotic potential differnce between medium and root xylem varied. The pressure on the medium surrounding the roots could also be raised, by using compressed air.

Exudation rates were measured before and immediately after changing the osmotic potential of the medium; with hydrostatic pressure remaining zero (atmospheric pressure). The exudation rates can thus be written:

$$J_{v1} = L_o \Delta\psi = L_o(\pi_x - \pi_{m_1}) \tag{4}$$

$$J_{v2} = L_o(\pi_x - \pi_{m2}) \tag{5}$$

where J_{v1} and J_{v2} are the exudation rates before and after the change in osmotic potential respecitvely, L_o is an osmotic permeability coefficient of the barrier, $\Delta\psi$ is the gradient in water potential, and π_x, π_{m1}, π_{m2} are osmotic potentials of the xylem sap and of the medium before and after the change, respecitvely. L_o can thus be estimated from the change in flux ΔJ_v:

$$L_o = \frac{\Delta J_v}{\Delta\pi_m} \tag{6}$$

Similar measurements were conducted at different hydrostatic pressures keeping osmotic potentials constant and estimating L_p, which is the pressure permeability coefficient:

$$L_p = \frac{\Delta J_v}{\Delta \mathrm{p}} \tag{7}$$

It was shown that L_p was always greater than L_o. Weatherley concluded from the ratios between the two coefficients that about 75% of the flux at a pressure gradient of 2 atm was osmotic and only 25% due to mass flow.

The diffusional and mass-flow pathways could also be differentiated by the effect of low temperatures (Fig. 3) and by the use of metabolic inhibitors, such as cyanide. It was shown that 75% of the water movement was sensitive to these two factors, indicating that three quarters of the total flux passes through living cells.

In the whole plant in contrast to excised root systems, low water potentials in the xylem element result from suction caused by transpiration, which enhances the rate of water flow.

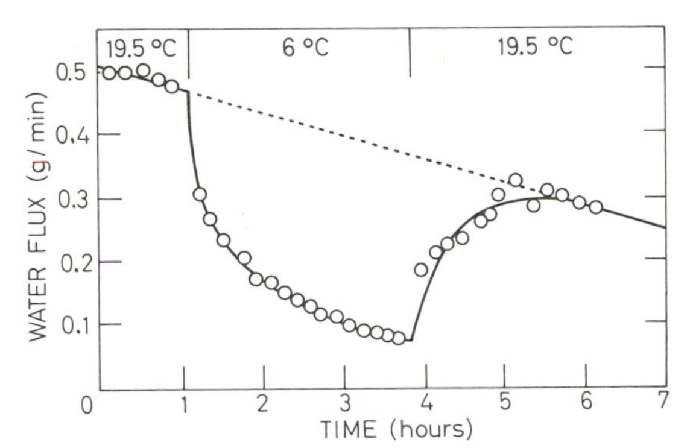

Fig. 3. The effect of temperature on water flux through a tomato root system with the continuous application of a pressure difference of 2 atm across the cortex. (From P. E. WEATHERLEY, 1963)

The Flow of Water Through the Xylem

The path of water from the roots to the leaves is mainly through tracheids in gymnosperms, which serve for both conduction and support, and through xylem vessels in angiosperms, which are specified for conduction. The problem of sap ascent is especially difficult in high trees under intensive transpiration conditions, since hundreds of liters must be transported upward a distance of 100 m or more to the transpiring top. This problem was a subject of debate for many years, and is still not entirely solved. (A more intensive review on the subject is given by KRAMER, 1969).

The driving force for ascent of sap is produced by the evaporation of water from the surface of leaf parenchyma cells, thus reducing their water potential, which causes movement of water from the xylem to the surface cells. This lowers the water potential at the top of the xylem columns, which falls well below zero (suction) and pulls the water up the stem. Continuous water columns in the pathway obviously are needed in order to maintain the water flow. DIXON (1914), who measured the cohesive force of water in sealed capillaries, found that it was in the range of 100–200 atm, which will ensure continuation of the liquid columns in the xylem and their rise, even in high trees.

Measurements of the tension existing in xylem vessels of transpiring trees are very few in number. SCHOLANDER *et al.* (1965) redeveloped the pressure-chamber method for measuring the xylem tension in a twig shortly after being detached. In spite of the criticism of this method, it provides evidence that an increase in tension corresponded to an increase in height of two different trees that were investigated (Fig. 4). The measured tensions were considerably greater during the day than at night; they ranged from −5 atm in plants of moist habitats to −80 atm in desert plants.

One of the main objections to the cohesion theory is that the continuity of the water columns at those high tensions must be rather unstable. Breakage of columns may occur either by dissolved gases in the water, which will form bubbles, or by winds, which will produce cavitations. The formation of such cavitations was observed in individual vessels in petioles under increasing water stress (MILBURN and JOHNSON, 1966). Although the air bubbles may develop mainly in large xylem elements, leaving the smaller ones filled with water, formation of bubbles due to freezing and thawing may

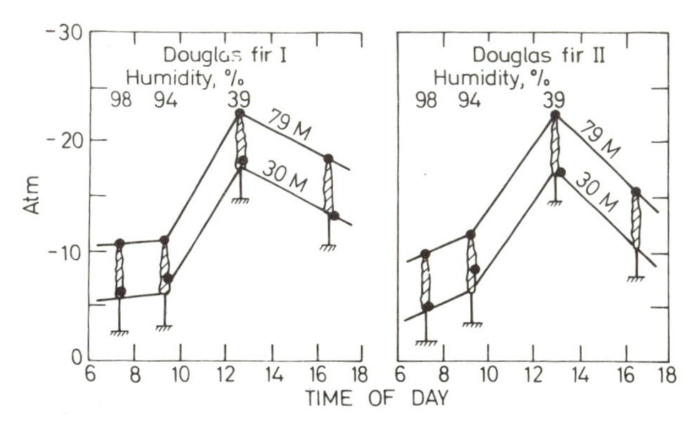

Fig. 4. Diurnal gradients in hydrostatic pressures in upper and lower parts of crowns of Douglas fir trees. The true elevations are indicated by lines. Hydrostatic pressures were measured on excised twigs with a pressure chamber. (From P. E. SCHOLANDER *et al.*, 1965)

block most of the xylem. It seems, however, that the disturbances due to gas bubbles are not as critical as once thought, since the air pockets cannot spread from one vessel to another, still leaving part of the vessels functioning. Secondly, the root pressure developing during the night when water deficits are minimal may cause refilling of the vessels. This pressure may be as high as 2 atm, thus driving water up to about 20 m. Another factor responsible for the operating of the xylem tube is the secondary xylem elements that are formed early in the growing season. These probably replace many of the older vessels that may not be functioning, mainly in tall trees.

As far as other possible resistances in the stem are concerned, low resistances to water flow were usually observed. In the large bundles of tomato plants for instance, only negligible resistance to water flow were found; the water flow through the smaller bundles of the petioles was, however, partly restricted (DIMOND, 1966).

Movement of Water in Leaves

The vascular bundles in the leaves split into many branches, known as veins, which are arranged specifically at various species. In most conifers, a single vein extends through the center of the leaf. In dicotyledons, the veins are usually arranged in a reticulated pattern, and in monocotyledons in a "parallel" pattern. Leaves with a reticulated venation have the largest vein along the medium longitudinal leaf axis – the midvein. It is connected to smaller lateral veins that are connected to smaller ones, from which still smaller veins diverge. In the parallel-veined leaves an alternation is usually found of larger and smaller longitudinal veins. These veins are interconnected by considerably smaller veins. The larger veins are enclosed in ground tissue that rises above the leaf surface and forms ribs, mostly on the lower side of the leaf. The smaller bundles are located in the mesophyll, but are also enclosed in special cells forming the bundle sheath. The bundle sheath extends generally to the end of the vascular tissue, so that the xylem elements are not exposed to the intercellular air spaces within the leaf. Exceptions are hydathodes, in which water is released directly from xylem elements into intercellular spaces. In many leaves the small veins form a very dense network so that most cells of the leaf are only a few cells away from a vein.

Two directions of water movement – vertical and lateral – are found in the leaf outside the vascular system. The direction of the vertical movement is from the end of the vascular bundle to the epidermis, and the lateral movement is within the mesophyll, to the evaporating cell wall adjacent to the stomata. In many leaves the bundle sheaths are connected with the epidermis by extensions of the bundle sheath, which may transfer water to the epidermis cells. Lateral movement of water in the mesophyll may also occur through these extensions, due to water flow back from epidermis to mesophyll. This may enable water conductance to cells that have limited lateral contacts.

The question concerning lateral movement of water is whether the movement is a diffusional movement, i.e., from vacuole to vacuole crossing cell membranes and walls (route d-c-b-a in Fig. 5), or mass flow round the cells by passing the vacuoles (route *d-b'* and *d-a'* in Fig. 5). The driving force for a vacuole-to-vacuole movement will be the gradient in water potentials along the water pathways the potential at the cell adjacent to

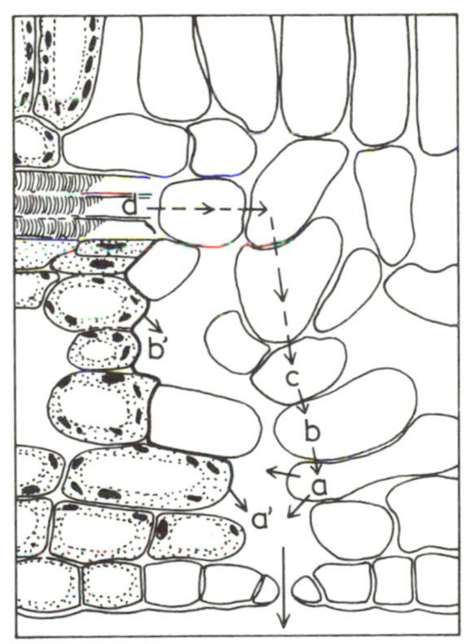

Fig. 5. Diagram of the lower part of a leaf seen in a vertical section. (From "Water and Plant Life" by W. M. N. BARON, Heinemann Educational Books, Ltd., London, England 1967)

the pore *(a)* being the lowest and that at the top of the xylem vessel *(d)* the highest. It was concluded by LEVITT (1966) that as the diffusion rate of the water is much too low to support the diffusion rate of water vapor from the stomates, movement must be a mass flow. The driving force in this case will then be the gradient between the terminal cell (*a*) and the xylem vessel (*d*), the distance between them is short and the resistance to flow propably low.

RUSSELL and WOOLEY (1961) estimated that the resistance to water flow through the walls is only 5% of the water flow through the root cortex. WEATHERLEY (1963) investigated experimentally as to which pathway contributes most to the water movement in the mesophyll. He constructed two hypothetical models of leaf cells, and also tested his hypothesis with leaves of several plant species. A detached leaf was fitted into a potometer as shown in Fig. 6a, and water uptake was measured by following the movement of a meniscus in the horizontal tube. After a few hours of maintaining a steady transpiration rate, the leaf was immersed in water to stop transpiration at once. The initial leaf-water deficit was known from the total amount of water absorbed to retain saturation of the leaf. The change in water deficit during recovery («die-away») was recorded and is shown in Fig. 6b. The initial steep decline in rate of water uptake is attributed to the fast decay in water uptake by the pathway, which was being saturated very fast, whereas the subsequent slow fall represents the decay in uptake by the inner-space

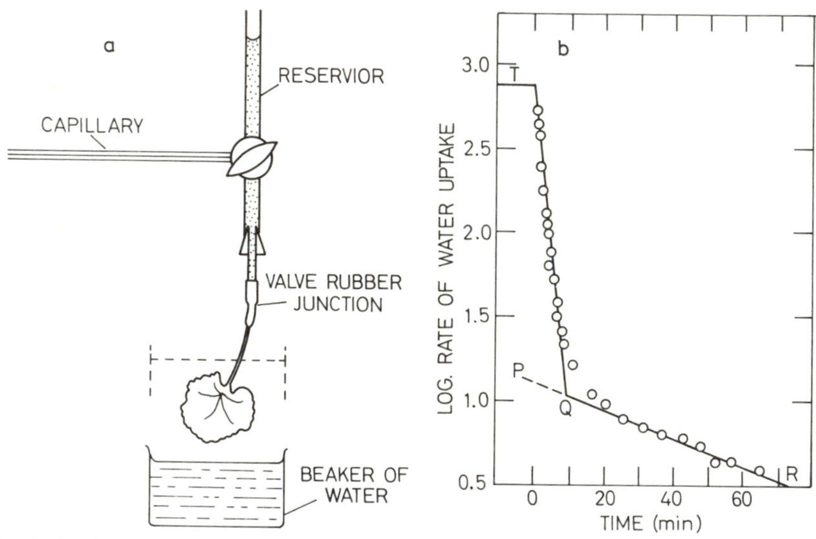

Fig. 6. a A simple potometer with arrangement for stopping of transpiration by raising the beaker to submerge the leaf. b The change in rate of water uptake by a geranium *(Pelargonium zonale)* leaf on stopping of transpiration. (From WEATHERLEY, 1963)

vacuoles, and probably cytoplasm. Extrapolation of the line *P-R* until the time when the experiment was started, giving the line *P-Q*, will show the rate of water uptake by the inner space during the time period when the major water uptake was by cell walls. WEATHERLEY (1963) concluded from these findings that the rate of uptake into the inner space is only 1/50 to 1/60 of the transpiration rate, that is, that water flow through the walls is 50–60 times faster than through the protoplasts. Additional evidence that most of the water movement is around the cells was obtained from the fact that the steep phase of the «die-away» curve in Fig. 6b was hardly affected when leaves were immersed in water at 3 °C whereas the flat phase was stopped completely.

BRIGGS (1967) concluded also that water moves mostly through cell walls rather than across vacuoles. This conclusion was mainly based on a calculation that for similar rates of water movement the potential gradient required for movement from cell to cell was 20 times higher than movement in the walls. The relative lower resistance in the walls determines that this must be the more prominent path for water flow.

Water Movement in the Vapor Phase

Mesophytic plants (i.e. most crop spps.) growing under dry conditions transpire more than 99% of all the water they absorb from the soil. This loss is an unavoidable result of the exposure of large areas of moist surfaces to solar energy, which is necessary for the fixation of CO_2. Efficient absorption of radiant energy during the day implies high potential rates of transpiration, but the actual rates depend on plant and environmental factors.

Table 1. Midsummer transpiration rates of various species of potted seedlings of trees, growing in soil near field capacity, in Durham, N. C. (Extracted from KRAMER, 1969)

Species	Avg. transpiration g cm^{-2} day^{-1}
Pinus taeda	4.65
Clethra alnifolia	9.73
Myrica cerifera	10.80
Liriodendron tulipifera	11.78
Quercus rubra	12.02
Quercus alba	14.21
Ilex glabra	16.10

Table 1 lists midsummer transpiration rates of potted seedlings of trees. The variation in the rates of transpiration per unit leaf areas was about 1:3 despite the similar atmospheric conditions for all the seedlings, which emphasizes the importance of these factors.

Ohm's law, the electrical analogue of Fick's diffusion law, is often used to analyze the complicated pathway from the leaf to the atmosphere (see Fig. 7A):

$$E = \frac{\Delta c}{r} \tag{8}$$

where E is the rate of water loss or transpiration from the leaf to the atmosphere, Δc is the difference in water concentration between the evaporating surface within the leaf and the free atmosphere above the leaf, and r is the leaf resistance. The differences in transpiration, such as those shown in Table 1, can be attributed to differences in the driving forces Δc, or in resistances r. These factors will now be described in detail.

Driving Force

Eq. (1) describes water movement as a function of water-potential differences $\Delta\psi$ along resistances from the root to the atmosphere. Water-potential difference between the atmosphere and the leaves can be of the order of -1000 bars (see Fig. 7B), whereas the difference between leaf and soil is only of the order of -10 to -20 bars (see Table 2). This might lead to the conclusion that the resistance in the vapor phase is two orders of magnitude greater than the other resistances in the liquid phase. Actually, vapor movement from the leaf to the air is a function of differences in vapor concentration Δc or partial vapor pressure Δe. Vapor pressure is related to water potential by the equation

$$\psi = \frac{RK_a}{V} \, ln \, (e_a/e_s) \tag{9}$$

where e_a is actual and e_s is saturated vapor pressure at absolute temperature K_a, R is the gas constant, and V is the specific volume of liquid water. As shown by COWAN and MILTHORPE (1968) and by DAINTY (1971), the expression $\Delta\psi$ may be used as the driving force, but the resulted resistance in the gaseous phase will vary with $\Delta\psi$. The resistance in Eq. (8) is more or less constant, independent of the driving force Δc, and the magnitude of the vapor phase resistance is reduced.

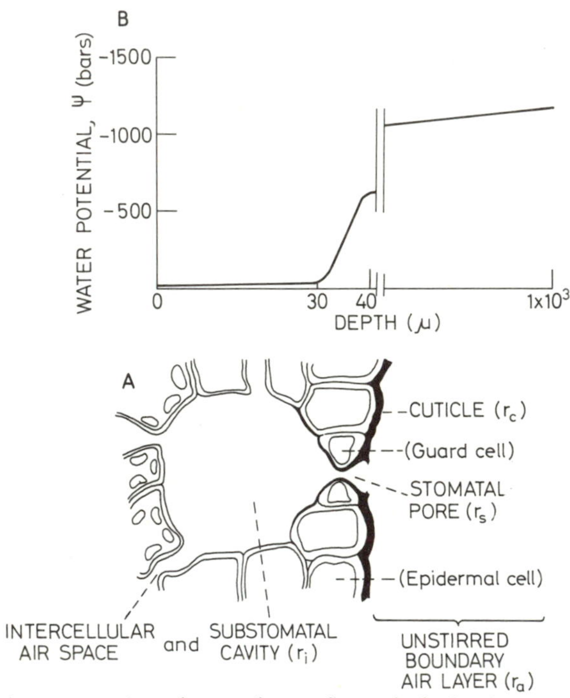

Fig. 7. A Schematic cross section of part of a sunflower leaf, showing a stoma and other leaf resistances. B The drop of water potential (ψ) from the evaporating surfaces inside a wellwatered sunflower leaf to the external atmosphere

Table 2. Approximate magnitudes of water potential in the Soil-plant-atmosphere system

Component	Water potential (bars)
Soil	−0.1 — −20
Leaf	−5.0 — −50
Atmosphere	−100 — −2000

1) Vapor Concentration in the Leaf. The leaf tissue has many air spaces (see Fig. 7 A). In mesophytic plants the ratio of internal free surfaces to external leaf surface can reach 30 : 1 and the volume of these internal air spaces may range from 10 to 60% of the total volume of the leaf (TURRELL, 1965). The tissue that contains these air spaces (the mesophyll layer) is enclosed between two epidermal layers that have small pores (stomates). The cross-sectional area of these pores ranges from 0.15 to 2% of the leaf surface area (MEIDNER and MANSFIELD, 1968). Most of the liquid water moves in the cell walls with no interference from living membranes. As a result, the vapor-pressure difference between these cell-wall surfaces and the stomates is very small. The effect of moderate changes in leaf-water potential has little effect on vapor pressure at the evaporating surfaces, and it may be taken as the saturation vapor pressure of water, which is a simple function of temperature (Smithsonian Meteorological Tables, 1966, pp. 381–383).

2) Vapor concentration in the atmosphere. The vapor concentration in the free atmosphere can be calculated from wet and dry bulb temperatures of the air, using the following equations. For aspirated psychrometers:

$$e_a = e_w - 0.66\ (T_d - T_w) \tag{10}$$

where e_a and e_w are the partial pressures of vapor in free air and of saturated air at wet bulb temperature T_w, respectively (for more details see Smithsonian Tables, 1966); e_w can be found in the tables referred to, and T_d is the dry bulb temperature. Partial vapor pressure in the air is related to water concentration c_a by Eq. (11):

$$e_a = c_a \frac{R}{M} K_a \tag{11}$$

where M is the molecular weight of water, 18, and R/M = 4620 mbr $cm^3\ g^{-1}\ {}^\circ K^{-1}$. K_a is absolute temperature. The units of vapor pressure are mbr.

Table 3 illustrates some examples of vapor concentration differences between the leaf and the ambient air, assuming the evaporating surfaces to be saturated. Under desert conditions and low air temperatures, the difference may be larger than the difference under

Table 3. Effect of leaf temperature, air temperature, and relative humidity of the air on water-vapor concentration ($g\ cm^{-3}\ 10^{-6}$)

Air temperature (°C)	Leaf vapor concentration at saturation	Ambient vapor concentration	Vapor concentration difference
	Humid conditions (leaf = air temperature, RH = 80%)		
10	9.40	7.52	1.88
20	17.30	13.83	3.47
30	30.38	24.26	6.12
40	51.19	40.85	10.34
	Desert conditions (leaf = air temperature + 5°C, RH = 10%)		
10	12.83	0.94	11.89
20	23.05	1.73	21.32
30	39.63	3.03	36.60
40	65.50	5.11	60.39

humid conditions, with high air temperatures. In dry and hot desert conditions (40°C and 10% relative humality), the vapor concentration difference can be so high that mesophytic plants will dessicate. Only very high resistances will be able to compensate for such high vapor concentration differences. Low leaf-water potentials such as -100 bars with water activity of 0.934 will reduce this vapor concentration difference by not more than 7% (from 60.39 to 56.07 gr $cm^{-3}\ 10^{-6}$).

Resistances

The most difficult terms of Eq. (8) to define and evaluate are the various resistance components. Fig. 7A demonstrates a schematic cross section of part of a sunflower leaf whose structure may be taken as representative of most mesophytic plants. Water evaporates from the cell walls into the adjacent intercellular air spaces and moves through the substomatal cavity and the stomatal pore out to the ambient air. Most crop plants have stomates on both surfaces of the leaf so that vapor can transpire from the two surfaces. Water can also transpire directly from the epidermal cells through the cuticle. Fig. 8 describes an electrical analogue for the various resistances involved in the transpiration process. The two parallel meshes represent the two leaf surfaces. The resistances will be described in the following section.

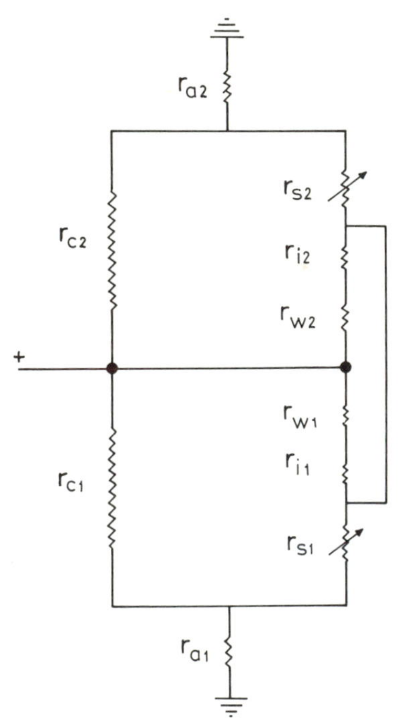

Fig. 8. An electrical analogue of leaf resistances to diffusion of water vapor. r_c = cuticular, r_a = boundary layer, r_s = stomatal, r_i = intercellular, r_w = cell wall resistances. 1 and 2 refer to upper and lower epidermis

Evaporating Surfaces (r_w) and Intercellular Air Spaces (r_i). The density of the cells in the mesophyll between the two epidermal layers is low. The mesophyll contains many intercellular air spaces, mainly in the very loose part called the spongy layer. The other part of the leaf mesophyll, which is called the palisade layer, is more compact, but still contains free air spaces. These air spaces are interconnected to the substomatal cavities. Water evaporates from the cell walls to the air spaces. The cell walls are almost saturated and can hardly offer significant resistance to water movement. As the ratios of internal free surfaces to external leaf surface and of internal air volume to total leaf volume are so large, it can be assumed that the intercellular resistance to vapor movement r_i is small.

This has been shown to be the case for mesophytic plants at least, although the substomatal cavity is covered by a thin layer of waxy deposits (COWAN and MILTHORPE, 1968).

Stomates (r_s). Stomatal pores are the connection between the internal leaf atmosphere and the external free atmosphere. Each stomatal pore is surrounded by two guard cells (see Fig. 7A). As a result of unequal thickening and elasticity of their walls, they can contract and expand nonuniformly, which causes opening and closing of the stomatal pore. This dominant type of movement is called active movement. Passive movement results from turgor changes in the neighboring epidermal cells that might contract or expand, causing similar movement of adjacent guard cells.

Many plant surfaces have stomates, but they are active and most numerous in leaves. The ranges of distribution, densities, and dimensions of stomates are given in Table 4.

Table 4. Distribution and dimensions of stomates in leaves[a]

Leaf type	Location of stomates	Plant types leaves	Range of stomatal density		length of pore[b] (μ)	
			upper surface	lower surface	upper surface	lower surface
Amphistomata	Both surfaces	Herbaceous	7–182	15–270	9–49	13–52
Hypostomata	Lower side	Most trees and shrubs	–	60–370	–	10–38

[a]Extracted from MEIDNER and MANSFIELD (1968), Table 1.1.

[b]Average pore width when fully open and pore depth are aprox. 10 μ.

The stomatal size changes markedly in response to changes in a number of physiological factors and environmental conditions. Recent reviews of the subject have been given by MEIDNER and MANSFIELD (1968) and by HEATH and MANSFIELD (1971). Here we shall list only the major factors: (1) Stomatal aperture is inversely proportional to light intensity. The opening reaction is slower (about 15 to 30 minutes) than the closure (less

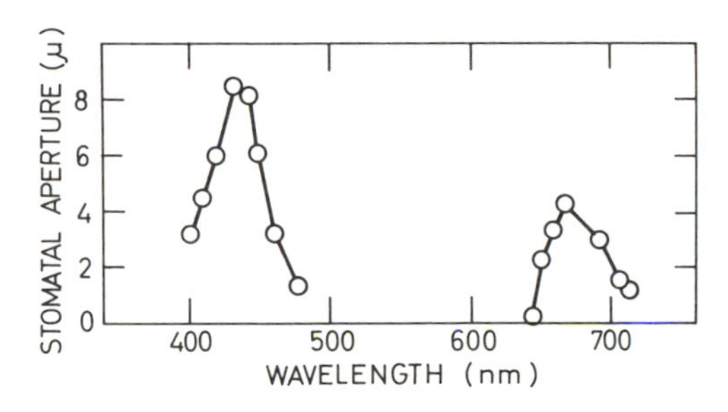

Fig. 9. Effect of wavelength of illumination on stomatal opening in *Senecio odoris.* (From KUIPER, 1964)

than a minute to few minutes). There is a clear spectral sensitivity; only blue and, to a lesser extent, red light cause opening (Fig. 9) (exceptions to this light response are plants that are especially adapted to arid conditions, such as members of the Cactaceae and

Crassulaceae, whose stomates open at night and close during the day). (2) Increasing CO_2 concentration causes stomatal closure, and decreasing concentration causes opening. There is an interaction between light intensity and CO_2 concentration: the higher the light intensity, the higher the CO_2 concentration required to produce a given degree of stomatal closure. (3) Increases in temperature sometimes cause stomatal opening up to an optimum degree above which additional increase in temperature may cause stomatal closure. (4) Water supply greatly affects stomatal movement. Water stress causes stomatal closure. Fig. 10 shows diurnal changes in stomatal opening in plants exposed to a wide range of soil-water content. The lower the water content in the soil, the lower the degree of stomatal aperture as shown by the infiltration index (infiltration of liquids with low

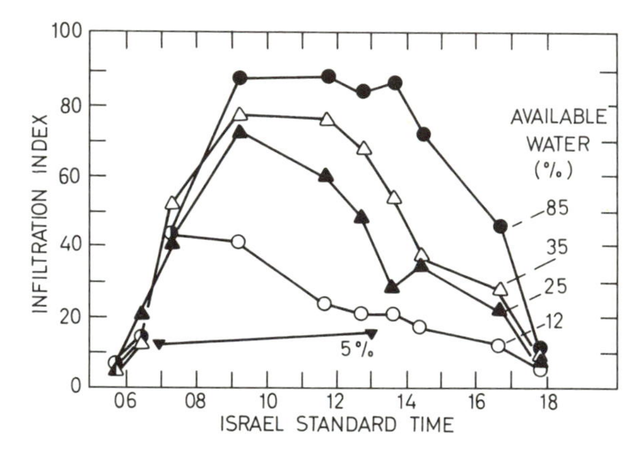

Fig. 10. Diurnal changes in infiltration measurements under different percentage of available soil water. (From OFIR *et al.*, 1968)

surface tension into stomates is often used as a measure of their degree of opening). There is an aftereffect of wilting on stomatal opening and full recovery from stress may take up to five days. (5) Stomates exhibit endogeneous rhythmic movement, even under constant environmental conditions.

The resistance of stomates to water-vapor diffusion may be computed from their dimensions, making a number of simplifying assumptions. Thus, the resistance of a narrow cylindrical tube is given by l/Da where l is the length of the tube, a is its cross-sectional area, and D is the diffusion coefficient of water vapor. Two end corrections must be added to this value (BROWN and ESCOMBE, 1900), each of them $\pi d/8$, where d is the tube's diameter. In the case of n parallel tubes the resistance will be

$$r_s = \frac{l + \pi d/4}{nDa} \tag{12}$$

This equation is oversimplified since none of the stomates of the plant kingdom has a cylindrical shape, and the stoma in Fig. 7A can be taken as an example. Thus, most promising ways of evaluating resistances to water movement in the gaseous phase are by direct measurements, as described by SHIMSHI (see p. 249).

Since stomates may open and close, their resistance to water-vapor diffusion may range from almost infinity to very low values. These lowest values may vary from species to species. Table 5 shows some of these variations.

This table gives only an order of magnitude of stomatal resistance values, which can change with environmental and growth conditions.

Cuticle (r_c). There is no direct method for measuring cuticular resistance. The usual technique has been to calculate it from nighttime fluxes of transpiration, assuming

Table 5. Minimum values of stomatal resistance (sec cm^{-1}).

Species (Mesophytes)	r_s	Species (Mesophytes)	r_l	Species (Xerophytes)	r_l
Beta vulgaris	0.5	Brassica sp.	1.6	Kochia indica	8
Helianthus annuus	0.7	Gossypium hirsutum	1.8	Zygophyllum dumosum	8
Medicago sativa	0.8	Alocasia sp.	2.4	Reaumuria hirtella	14
Solanum tuberosum	0.9	Hyoscyamus niger	4.8	Pinus halepensis	17
Pinus resinosa	2.3	Lycopersion esculentum	4.8	Atriplex sp.	24
Triticum aestivum	3.0	Phaseolus vulgaris	4.8	Haloxylon articulatum	26

r_s values were calculated from viscous-flow porometer and from pore dimension.
r_l values were obtained from flux measurements. They are called r_l (leaf resistance) since all leaf resistances were included in the measurement. Table adapted from COWAN and MILTHORPE (1968).

complete closure of stomates in the dark (an assumption difficult to substantiate) or from transpiration measurements from leaf surfaces containing no stomates (HOLMGREN *et al.*, 1965). A method of calculating cuticular resistance of amphistomatous leaves (see Table 4) was described recently (MORESHET, 1970). It showed r_c to be variable and dependent on environmental factors, in particular, the relative humidity of the ambient air (Fig. 11). It was suggested that changing relative humidity of the air changes the physical attraction

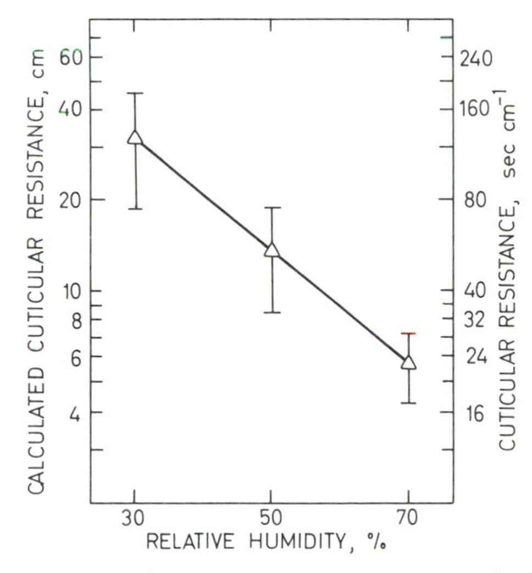

Fig. 11. Relationship between cuticular resistance to transpiration and relative humidity (vertical lines represent double standard errors). (From MORESHET, 1970)

forces of the cuticle. However, the impotance of cuticular resistance in controlling transpiration is significant mainly under arid conditions where water supply is limited and vapor concentration difference between the leaf and the atmosphere is high.

Atmospheric Boundary Layer (r_a). The external boundary layer resistance of a single leaf is provided by the layer of stationary air at the leaf surface. The resistance may be defined as the time required for a unit volume of air to exchange water vapor with a unit area of evaporating surface. The thickness of this layer depends on the rate of air movement in the free air and leaf dimensions, primarily its characteristic length, or length in the direction of air movement. Leaf angle has a much smaller effect on thickness. Many empirical equations have been developed for the calculation of this resistance, using wind speed and leaf dimensions. MONTEITH (1965) suggested Eq. (13) as the most representative for leaves under natural conditions

$$r_a = 1.3 \sqrt{b/u} \tag{13}$$

where b is the leaf width and u is wind speed. With fluttering leaves and moving plants, with air streams that do not flow parallel to the leaf suface, and with the difficulties of measuring air movement near the leaves, equations of this type are of limited practical application. The best empirical method of determination is by measuring water loss from artificial, wet surfaces such as blotting paper, which have no internal resistance. STANHILL and MORESHET (1965) used this method for measuring boundary layer resistance under field conditions. Figure 12 illustrates a profile of boundary layer resistances in a field of cotton in Israel in the middle of the day. MONTEITH (1965)

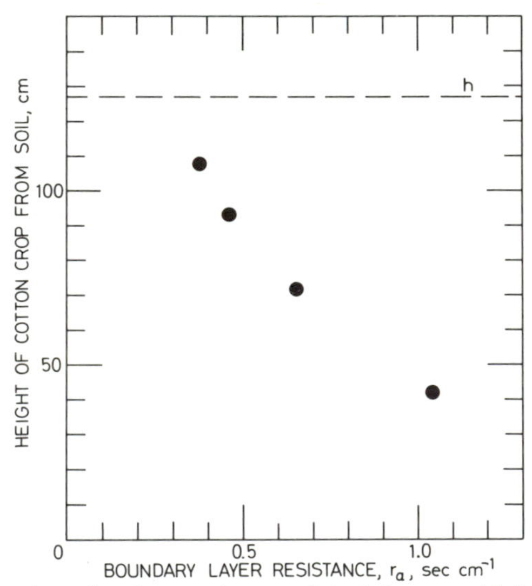

Fig. 12. Relation between boundary layer resistance of single leaves and height of measurement in a field of cotton *(Gossypium hisutum* var Acala 15–17) measured at Bet-Dagan, Israel. (Each point is the mean of three replicates with mean standard error = ± 0.057; h = crop heigth. Redrawn from STANHILL and MORESHET, 1965)

calculated resistances not exceeding 0.2 sec cm^{-1} for 8-cm-wide leaves and a wind speed of 3 m sec^{-1} a low value compared with the range 0.5–5.0 sec cm^{-1} minimum stomatal resistance reported in Table 5 for mesophytic plants. However, this low value may double or triple at low wind speeds, or inside the canopy as shown in Fig. 12.

The Leaf (r_t). Eq. (14) (MORESHET *et al.*, 1968) combines the various leaf resistances as presented in the electrical analogue (Fig. 8) neglecting r_w and r_i, which are very small:

$$r_t = \frac{[(r_{s1}^{-1} + r_c^{-1})^{-1} + r_a]\,[(r_{s2}^{-1} + r_c^{-1})^{-1} + r_a]}{(r_{s1}^{-1} + r_c^{-1})^{-1} + (r_{s2}^{-1} + r_c^{-1})^{-1} + 2r_a} \tag{14}$$

where r_t is the resistance of a unit leaf area, assuming that r_c and r_a on the two leaf surfaces are the same (multiply by 2 to get average r_t per unit leaf surface).

The Canopy (R_a and R_t). An interaction between leaves in the canopy and diffusion by atmospheric turbulence governs the process of exchange between the atmosphere and the surface of the crop. Eq. (15) has been developed (MONTEITH, 1965) to calculate crop boundary layer resistance R_a over uniform surfaces:

$$R_a = \frac{[\ln(z-d)/z_o]^2}{k^2 u} \tag{15}$$

where u is average wind speed measured at height z above the soil, d is the height above the roughness length z_o where wind speed is theoretically zero, and k is a constant of proportionality. A more detailed explanation of these parameters can be found on p. 143).

A similar treatment can be used to calculate internal canopy resistance, which is directly related to net radiation energy gained by the canopy surface and inversely to rate of evapotranspiration. For a detailed analysis see MONTEITH (1965) and COWAN and MILTHORPE (1968).

The relative importance of internal and external resistances in controlling water loss from plant surfaces has been debated for more than two decades. BANGE (1953) showed that in still air changing stomatal resistance of a single leaf has little effect on rate of transpiration as compared with moving air, where stomatal aperture controls rate of

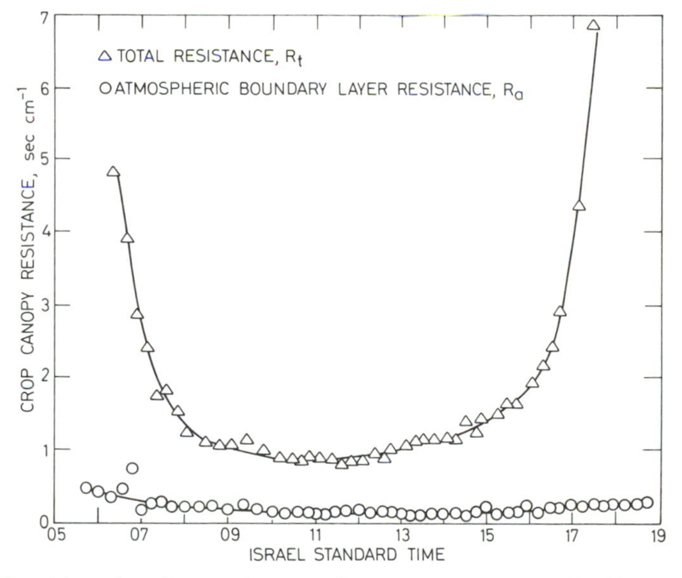

Fig. 13. Total and boundary layer resistances of cotton crop canopy at Bet Dagan, Israel. (From STANHILL and MORESHET, 1967)

water vapor loss. Under natural conditions in Israel, the canopy boundary layer resistance of lawn, cotton, and corn did not exceed 10 percent of total crop resistance to water vapor loss (STANHILL and MORESHET, 1967). An example can be seen in Fig. 13, which shows a diurnal course of crop boundary layer and total crop resistances of a cotton field in Israel. It may be concluded that, at least under similar conditions, crop resistances are more important than boundary layer resistances in controlling water loss. But it should be kept in mind that water loss is governed not only by resistances, but also by the difference in water-vapor concentration between the leaves and the ambient air.

Literature

ARISZ, W. H., HELDER, R. J., VAN NIE, R.: Analysis of the exudation process in tomato plants. J. Expl. Bot. **2**, 257–297 (1951).

BANGE, G. G. J.: On the quantitative explanation of stomatal transpiration. Acta Bot. Neerl. **2**, 255–297 (1953).

BERNSTEIN, L., NEIMAN, R. H.: Apparent free space of plant roots. Plant Physiol. **35**. 589–598 (1960).

BRIGGS, G. E.: Movement of Water in Plants. Philadelphia, Pa.: F. A. Davis 1967.

BROWN, H. T., ESCOMBE, F.: Static diffusion of gases and liquids in relation to the assimilation of carbon and tranlocation in plants. Phil. Trans. Roy. Soc. B. **193**. 223–291 (1900).

COWAN, I. R., MILTHORPE, F. L.: Plant factors influencing the water status of plant tissues. In: Water Deficts and Plant Growth. KOZLOWSKI, T. T. (Ed.) Vol. **I**, pp. 137–193. New York: Academic Press 1968.

DAINTY, J.: Water relations of plant cells. In: Advances in Botanical Research. PRESTON, R. D. (Ed.). Vol. **1**, pp. 279–326. New-York: Academic Press 1963.

DAINTY, J.: The water relations of plants. In: Physiology of plant Growth and Development. WILKINS, M. B. (Ed.) pp. 419–452. New York: McGraw-Hill 1971.

DIMOND, A. E.: Pressure and flow relations in vascular bundles of tomato plants. Plant Physiol. **41**, 119–131 (1966).

DIXON, H. H.: Transpiration and the Ascent of Sap in Plants. New York: McMillan 1914.

HEATH, O. V. S., MANSFIELD, T. A.: The movement of stomata. In: The Physiology of Plant Growth and Development. WILKINS, M. B. (Ed.) pp. 302–332. New York: Mc Graw-Hill 1971.

HOLMGREN, P., JARVIS, P. G., JARVIS, M. S.: Resistance to carbon dioxide and water vapor transfer in leaves of differnet plant species. Physiol. Plant. **18**, 557–573 (1965).

VAN DEN HONERT, T. H.: Water transport as a catenary process. Faraday Soc. Discuss. **3**, 146–153 (1948).

HOUSE, C. R., FINDLAY, N.: Water transport in isolated maize roots. J. Expl. Bot. **17**, 344–354 (1966a).

HOUSE, C. R., FINDLAY, N.: Analysis of transient changes in fluid exudation from isolated maize roots. J. Expl. Bot. **17**, 627–640 (1966b).

KRAMER, P. J.: Plant and Soil Water Relationships: A Modern Synthesis. New-York: McGraw-Hill 1969.

KUIPER, P. J. C.: Dependence upon wavelength of stomatal movement in epidermal tissue of Senecio adoris. Plant Physiol. **39**, 952–955 (1964).

LEVITT, J.: The physical nature of transpirational pull. Plant Physiol. **31**, 248–251 (1956).

MEIDNER, H., MANSFIELD, T. A.: Physiology of Stomata, 179 pp. London: McGraw-Hill 1968.

MILBURN, J. A., JOHNSON, R. P. C.: The conduction of sap. II. Detection of vibrations produced by sap cavitation in *Ricinus* xylem. Planta **69**, 43–52 (1966).

MONTEITH, J. L.: Evaporation and environment. In: The State and Movement of Water in Living Organisms. FOGG, G. E. (Ed.) Symposia of the Soc. Expl. Biol. **19**, 205–234 (1965).

MORESHET, S.: Effect of environmental factors on cuticular transpiration resistance. Plant Physiol. **46**, 815–818 (1970).

MORESHET, S., KOLLER, D., STANHILL, G.: The partitioning of resistances to gaseous diffusion in the leaf epidermis and the boundary layer. Ann. Bot. **32** 695–701 (1968).

OFIR, M., SHMUELI, E., MORESHET, S.: Stomatal infiltration measurement as an indicator of the water requirement and timing of irrigation for cotton. Expl. Agric. **4**, 325–333 (1968).

PHILIP, J. R.: Plant water relations: some physical aspects. Ann. Rev. Plant Physiol. **17**, 245–268 (1966).

RUSSELL, M. B., WOOLEY, J. T.: Transport processes in the soil plant system. In: Growth in Living Systems. ZARROW, M. X. (Ed.). pp. 695–722. New York: Basic Book Inc. 1961.

SCHOLANDER, P. E. HAMMEL, H. T., BRADSTREET, Edda D., HEMMINGSEN, E. A.: Sap pressure in vascular plants. Science **148**, 339–346 (1965).

SLATYER, R. O.: Plant Water Relationships. New-York: Academic Press 1967.
Smithsonian Meteorological Tables: Prepared by R. J. LIST: Smithsonian Misc. Coll. Vol. 114. Smithsonian Institution, Washington, 527 pp. (1966).
STANHILL, G., MORESHET, S.: Atmospheric resistance to transporation and photosynthesis. Ann. Rep. USDA Project A10-SWC-29, pp. 28–45 (1965).
STANHILL, G., MORESHET, S.: Atmospheric resistance to transpiration and photosynthesis. Ann. Rep. USDA Project A10-SWC-29, pp. 39–86 (1967).
TURRELL, F. M.: Internal suface intercellular space relationships and dynamics of humidity maintenance in leaves. In: Humidity and Moisture II: Applications. WEXLER, A. (Ed.) pp. 39–53. New York: Reinhold 1965.
WEATHERLEY, P. E.: The pathway of water movement across the root cortex and leaf mesophyll of transpiring plants. In: The Water Relations of Plants. RUTTER, A. J., WHITEHEAD, F. H. (Eds.). London: Blackwell 1963.

4

Water Transfer from the Soil and the Vegetation to the Atmosphere

M. FUCHS

Defining the Problem

The water balance of irrigated crops teaches us that a major portion of the water in the root zone is lost as vapor to the atmosphere. Water losses occurring directly from the soil surface are usually called *evaporation,* whereas *transpiration* refers to the water losses from plants. Physically, the processes are identical, because both involve a change of phase of water from liquid form to vapor form and transport of water into the atmosphere. The concept of *evapotranspiration* refers to the total atmospheric losses of water from soil and plant surfaces. However, as used in this chapter the word evaporation will mean all water losses in vapor form, regardless of the nature of the surface.

The purpose of discussing the physics of evaporation is to define the environmental factors that determine the evaporation. Two physical models describe evaporation in nature. First, evaporation is a process whereby water is transformed from the liquid phase to the vapor phase. The change of phase of one gram of liquid water at 20 °C into vapor requires 585 calories. This energy is known as the *latent heat of vaporization,* λ; consequently, the study of the *energy balance* of the surface can be used to calculate evaporation. Second, evaporation is a process of transporting water vapor from the evaporating surface into the atmosphere. This description of evaporation is based on the *mass balance* and the *momentum balance* of an air layer above the surface. Energy, mass, and momentum balances are classical techniques used by engineers to solve transport problems. Meteorologists apply the same techniques to describe evaporation.

Dimensions and Units

Because we are treating evaporation as a physical problem we must define the dimensions of the terms in the balance equations. The most convenient dimensions are those of flux densities, i.e., the flow of a physical quantity across a unit area per unit of time. To homogenize the dimensions and units in the mass and energy balance equations, we choose to use energy as our basic physical quantity, so that all the terms of our balance equations will be energy flux densities expressed in cal cm^{-2} min^{-1}. Dividing by the latent heat of vaporization converts the energy flux density into water flux density.

Example: To convert an energy input of 0.8 cal $cm^{-2}min^{-1}$ at 20°C into mm of water evaporated per hour, we proceed as follows:

$$0.80 \text{ cal cm}^{-2}\text{min}^{-1} \times \frac{1 \text{ g water}}{585 \text{ cal}} \times \frac{1 \text{cm}^3 \text{ water}}{1 \text{ g water}} \times \frac{60 \text{ min}}{1 \text{ hr}}$$

$$= 0.082 \text{ cm water hr}^{-1} = 0.82 \text{ mm water hr}^{-1}$$

A useful rule of thumb to remember is that 1 cal cm^{-2} min^{-1} represents approximately an evaporation of 1 mm of water per hour.

The Energy Balance

As the primary source of energy at the evaporating surface is solar radiation, we shall start with a brief survey of the main features of the radiation balance.

Direct solar radiation R_s provides on the average approximately 60% of the total energy reaching the surface during daytime. A typical, average value for a summer day in Israel is 0.6 cal cm^{-2} min^{-1}. Diffuse solar radiation R_d represents 10% of the total incoming radiation. The remaining 30% is emitted by the heated atmosphere, R_L. Direct and diffuse solar radiation are in the wavelength range from 0.3 to 3 microns. Atmospheric radiation is characterized by wavelengths larger than 6 microns.

Only a portion of the total incoming radiation is absorbed by the surface. A fraction of the solar radiation is reflected by the surface and defines the surface albedo r. The fraction of atmospheric long-wave radiation absorbed by the surface defines the *thermal absorptivity*, α. Because of its own absolute temperature T, the surface emits long-wave radiation proportionally to T^4, where T is in °K. The proportionality factor is made up of the product of a physical constant, the Stefan-Boltzmann constant, $\sigma = 0.817 \times 10^{-10}$ cal cm^{-2} min^{-1} $°K^{-4}$, and the *thermal* emissivity ε_T, which is a property of the surface. It can be proved that for most natural surfaces and under most conditions α equals ε_T so that the net radiation flux density R_n absorbed by the surface is given by

$$R_n = (1-r)(R_s + R_d) + \varepsilon_T (R_L - \sigma T^4) \quad (1)$$

In Eq. (1) the term $(R_s + R_d)$ accounts for most of the diurnal and seasonal variations of the radiation balance, whereas $(R_L - \sigma T^4)$ is a relatively conservative quantity. For a clear summer day $(R_s + R_d)$ gradually increases from zero at sunrise to reach a maximum of + 1.5 cal cm^{-2} min^{-1} at midday. The term $(R_L - \sigma T^4)$, which is almost always negative, typically ranges from –0.2 to –0.5 cal cm^{-2} min^{-1}. This indicates that the net radiation is strongly correlated with the solar radiation during daytime, and also that it is negative during the night.

Another consequence of Eq. (1) is that net radiation strongly depends upon the surface albedo and the surface thermal absorptivity. Characteristic values of these two parameters listed are in Table 1. From them we can predict that net radiation will be larger over a pine forest than over grassland.

Although few measurements of thermal absorptivity have been made, they all indicate that the value is high and that there is little variation between different vegetative surfaces.

As a surface has no heat capacity, the net radiation absorbed by the surface must be totally dissipated. There are three major dissipating mechanisms: increasing the temperature of the soil and of the plants underneath the surface or *ground heat flux density G,* heating of the air above the ground, or *sensible heat flux density H;* and phase transformations of water, or *latent heat flux density* E. Evaporation and condensation are the processes described by the latent heat flux density that are the most relevant to

Table 1. Radiation properties of agricultural and natural surfaces

Surface type	Albedo r	Thermal absorptivity[4] α
Pine forest[1]	0.12	–
Oak forest, open[1]	0.18	–
Desert vegetation[1]	0.37	–
Orange grove[1]	0.17	–
Cotton[2]	0.16–0.19	–
Corn[3]	0.16–0.17	0.97
Wheat[3]	0.14–0.27	–
Sugar beets[3]	0.14–0.24	–
Potatoes[3]	0.17–0.27	–
Alfalfa[3]	0.16–0.22	0.98
Grass[3]	0.24–0.26	0.98
Meadow[3]	0.25	0.97
Water[1]	0.11	0.96

1. STANHILL *et al.,* 1966.
2. STANHILL and FUCHS, 1968.
3. GATES and HANKS, 1967
4. FUCHS, unpublished.

irrigation problems in arid lands. For completeness the radiation used by plants in photosynthesis should also be included in the energy balance, but since it represents only 1 or 2% of the net radiation, it can be neglected.

The energy balance equation of the surface is now written:

$$R_n = G + H + E \tag{2}$$

Evaporation easily derives as

$$E = R_n - G - H \tag{3}$$

When water is readily available at the surface, E is the largest term on the right-hand side of Eq. (2). For this condition, we can say that to a first approximation evaporation is proportional to the net radiation. The importance of this conclusion in field practice is that evaporation of extended areas is nearly impossible to measure whereas R_n, is routinely measured with commercially available *net radiometers* (also called radiation balance meters or pyrradiometers).

According to Eq. (3), factors that modify the net radiation will also affect the evaporation. Consequently, the surface albedo that has been shown to affect the radiation balance provides useful information of the expected evaporation of various forms of land use. In this respect a pine forest is likely to evaporate more water than grassland.

For more detailed information on evaporation, we need to investigate the ground heat flux density and the sensible heat flux density. The flow of heat into the ground depends upon the heat capacity and the thermal conductivity of the substrate. As the heat capacity of the plants is two orders of magnitude smaller than of the soil, its contribution is neglected. The prediction of the soil-heat flux density from a physical model is difficult because the soil is a heterogeneous, multiphase porous medium. Nevertheless, the difficulty is overcome experimentally by simple calorimetric measurements and the use of heat flux transducers. An excellent treatment of the problem is given in VAN WIJK (1963). We should mention here that the theory as well as experimental results show that if the energy balance is made over a period of one or two weeks the net ground heat flux is extremely small and can be neglected altogether.

The laws of heat transport in gases cannot be applied to predict the sensible heat flux density because the atmosphere is an open system. The air flow over the surface continuosly brings new air in contact with the surface and mixes this air with the air stream at higher levels. Experimental techniques cannot resolve the difficulty because surface measurements of the sensible heat flux density are presently impossible. Consequently, the energy balance cannot provide further information on evaporation.

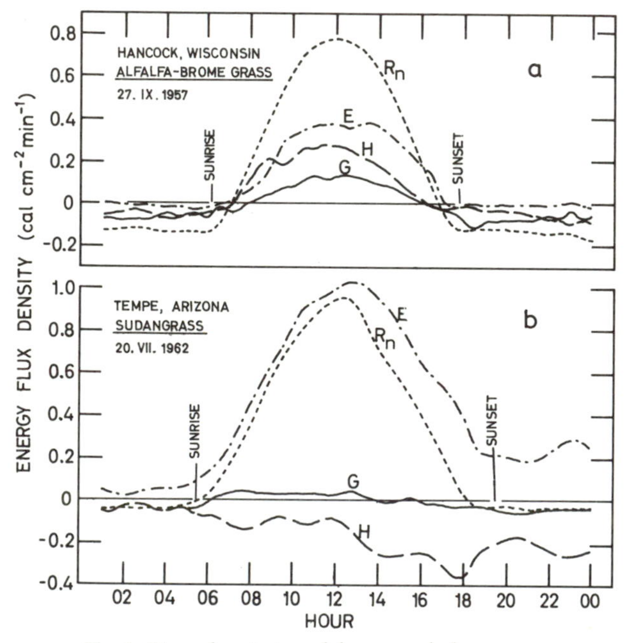

Fig. 1. Diurnal variation of the energy balance terms. (After SELLERS, 1965)

Two classical examples of the diurnal trend of the components of the energy balance are shown in Fig. 1 (SELLERS, 1965). Part a is typical of large irrigated areas, or subhumid conditions. Evaporation here is about half of the net radiation; however, near sunrise and sunset evaporation is larger than net radiation. During these periods, the evaporation process extracts energy from the air and the soil. Negative nighttime values of E indicate condensation and dew formation. The energy balance plotted in part (b) was measured over a small irrigated field in a dry surrounding. Here, evaporation is larger than the net radiation. The additional latent heat is provided by hot dry air flowing over the wet field. This feature of the energy balance describes the oasis effect, or local advection of sensible heat.

The Mass and Heat Transport Approach

Evaporation can take place only if a water vapor pressure gradient exists in the air layer above the surface. We can therefore consider evaporation as analogous to a diffusion process and write:

$$E = (\lambda \varrho \varepsilon / P)\, K_E de/dz \tag{4}$$

where λ is latent heat of vaporization: 585 cal (g water)$^{-1}$ at 20°C; ϱ is air density, 1.22×10^{-3} g cm^{-3} at 20°C; ε is the relative molecular weight of water with respect to air, 0.622 dimensionless; p is barometric pressure, millibars; K_E is eddy diffusivity of water vapor, cm^2 sec^{-1}; and de/dz is vertical water vapor pressure gradient, mb cm^{-1}

We can also express the sensible heat flux density as proportional to the air temperature gradient by analogy to thermal diffusion:

$$H = \varrho c_p K_H dT/dz \tag{5}$$

where c_p is specific heat of air, 0.24 cal (g air)$^{-1}$ °C^{-1}; K_H is eddy diffusivity of sensible heat, cm^2 sec^{-1}; and dT/dz is vertical temperature gradient, °C cm^{-1}.

Although theoretically interesting, Eqs. (4) and (5) have no practical applications because K_E and K_H are not constants. They vary strongly with wind speed, and may range from 10^{-1} to 10^4 cm^2 sec^{-1}. It has been shown experimentally that the magnitude of the eddy diffusivites is related to the turbulence level of the air layer in contact with the surface.

Momentum Transport and Turbulent Mixing

The wind velocity at a given point in the atmosphere is a vector that continuously fluctuates in magnitude and direction. The motion of a flag on top of a mast demonstrates this fact very well. We can, however, define a time-averaged value of the wind speed and estimate a time-averaged wind direction. If we resolve the instantaneous wind velocity vector V into its components and express each component as an average plus a deviation, we obtain:

$$V = \begin{cases} u = \overline{u} + u' \\ v = \overline{v} + v' \\ w = \overline{w} + w' \end{cases} \tag{6}$$

The barred terms indicate the time-averaged value and the primed terms the instantaneous deviation. If u is the horizontal component parallel to the surface in the average direction of the wind, v is the horizontal component perpendicular to this direction, and w is the vertical component, the following simplifications result:

$$\overline{v} = 0 \text{ and } \overline{w} = 0 \tag{7}$$

Another experimental observation is that $\overline{u}$ increases with height, and is zero at the interface between the air and the land surface. This means that the land surface absorbs momentum and that momentum is transferred from one air layer to the next. By analogy with the definition of molecular kinematic viscosity, the mean momentum flux density

$\overline{\tau}$, i.e., the horizontal momentum transferred from one air layer to the next per unit horizontal area, is given by:

$$\overline{\tau} = K_M \, d\overline{u}/dz \tag{8}$$

Where K_M is the eddy diffusivity for momentum or eddy viscosity, $cm^2 \, sec^{-1}$; and $d\overline{u}/dz$ is vertical wind shear, sec^{-1}.

The fluctuating nature of the wind vector expressed by Eq. (6) indicates that the horizontal momentum per unit volume ϱu is vertically transferred at a rate proportional to w, so that the average momentum flux is given by:

$$\overline{\tau} = \overline{\varrho u w} = \overline{\varrho u' w'} \tag{9}$$

The second equality in Eq. (9) is obtained by developing the product $(\overline{u} + u')(\overline{w} + w')$, recalling that the choice of the reference system imposes that $\overline{w} = 0$, and that $\overline{u}' = 0$ and $\overline{w}' = 0$ by definition.

A third experimental fact is that the temperature and the water vapor pressure at a given point in the air fluctuate continuously. As previosly, we can here write the instantaneous value in terms of a time average plus a deviation:

$$T = \overline{T} + T' \tag{10}$$

$$e = \overline{e} + e' \tag{11}$$

If a unit volume of air at temperature T and vapor pressure e moves vertically, it carries sensible heat and water vapor at a rate proportional to w. If we average the instantaneous transfer over a certain time lapse, we should obtain the sensible heat flux density and the evaporation:

$$H = \overline{\varrho c_p T w} = \varrho c_p \overline{T' w'} \tag{12}$$

$$E = \overline{(\lambda \varrho \varepsilon / P) \, e w} = (\lambda \varrho \varepsilon / P) \, \overline{e' w'} \tag{13}$$

Note that $\overline{T'w'}$ and $\overline{e'w'}$ are the cross products of the correlation coefficient between w and T, and e respectively. If Eqs. (12) and (13) are true, the correlation coefficient between w and T and between w and e should be high when sensible heat flux density and evaporation are large. This fact is experimentally verified and leads to an accurate measurement of evaporation known as the *eddy correlation method.*

For Eq. (13) to be representative of the evaporation, all the fluctuations of w and e should be included in the averaging process. Measurements have shown that at a height of 100 cm above the crop canopy, significant fluctuations may have a frequency of about 20 Hz. This imposes the need for instruments with fast response and for formidable sampling and data processing problems. Advances in electronics and computer technology have permitted development of the evapotron (DYER and MAHER, 1965). The instrument is presently limited to research applications, but future improvements should make its use in field work possible.

The turbulent nature of the transfer properties in the atmosphere as described in Eqs. (9), (12), and (13) shows that the modes of transport of momentum, heat, and water vapor are similar. We can therefore assume that the eddy diffusivities in Eqs. (8), (5), and (4), are equal:

$$K_M = K_H = K_E \tag{14}$$

This assumption leads us to the aerodynamic model for evaporation.

The Aerodynamic Approach

A lengthy mathematical derivation using mixing length theory combines Eqs. (8) and (9) to show that the vertical distribution of the average wind velocity is logarithmic (SUTTON, 1953, pp. 75–83 and pp. 232–240):

$$\bar{u} = (u_*/k)\ln[(z + z_o - d)/z_o] \qquad (15)$$

where $\bar{u}$ is average wind velocity at height z, cm sec^{-1}; u is friction velocity, $(\tau/\varrho)^{1/2}$ cm sec^{-1}; k is Karman constant, 0.42 dimensionless; z is height above the soil surface, cm; z_o is roughness length, cm; and d is zero-displacement height, cm.

The logarithmic profile is only valid under an adiabatic condition; i.e., when the sensible heat flux density is equal to zero. This condition does not occur frequently. However, in the case of crop surfaces well supplied with water the dominant term of the energy balance is E; consequently, H is small and its effect on the wind profile in the air layer 1 or 2 meters above the ground can be neglected. In the case of the strong surface heating or cooling that pervails over unirrigated arid-land surfaces the curvature of the wind profile is modified by the diabatic effect. Under these conditions, the analysis of the wind profile is more difficult. For details consult TANNER (1968).

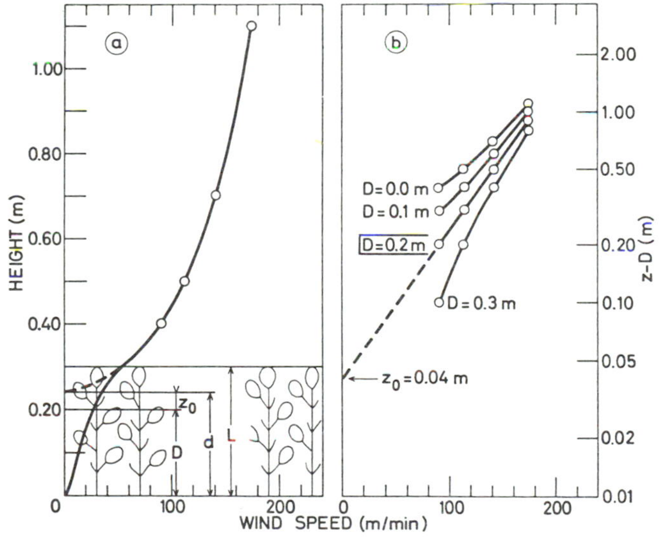

Fig. 2. Wind profile representation on linear (a) and semilogarithmic (b) scales. z = height above the soil surface, z_o = roughness length, d = zero-displacement height, D = height adjustment term, L = vegetation height above the soil surface

The paramaters z_o and d are determined by the nature of the surface. The roughness length z_o is a measure of the irregularity of the surface. Because momentum is absorbed not only by the soil surface but also by the vegetation, the height at which the wind speed is zero should be somewhere between the soil surface and the top of the vegetation. The height of theoretical zero wind speed is given by d. The role of d is thus simply to define the origin of the wind profile.

A physical representation of z_o and d is given in Fig. 2a. The height d, for which the wind profile extrapolates to zero wind speed, in principle can be measured from any arbitrary reference level. Whenever possible, it is recommended to choose for this the average soil surface, because it enables fixing d and z_o in relation to the crop height. The base line of Fig. 2a represents the average soil surface from which we also measure L, the crop height. The length z_o is often interpreted as the thickness of the layer of air near the plant surface, in which there is laminar flow. It can also be regarded as the contribution of the plant surface irregularities to the scaling of the mixing length.

Fig. 2b shows a practical method of determining d and z_o from wind profile measurements. One first defines $D = d - z_o$ and chooses a series of arbitrary values for D. The wind profile is plotted on semilog paper using $z - D$ as the ordinate. If the chosen D is too small, the profile will have a positive curvature; if D is too large, the curvature is negative. The correct value of D is obtained when the plot is linear. The value of z_o is found at the interception of the straight line at zero wind velocity.

These two parameters can also be computed from the following empirical relations (KUNG and LETTAU, 1961; quoted by TANNER, 1968, and STANHILL, 1969).

$$z_o = 0.058 \; L^{1.19} \tag{16}$$

$$d = 0.66 \; L^{0.98} \tag{17}$$

where L is the average height of the vegetation.

Using z_o and d from Eqs. (16) and (17), we can obtain τ by applying Eq. (15) to wind speed measurements at a single height.

$$\tau = \varrho k^2 \bar{u}^2 / [\ln(z + z_o - d)/z_o]^2 \tag{18}$$

Comparing Eq. (18) and Eq. (8) indicates that

$$\int_o^z dz/K_M = [\ln(z + z_o - d)/z_o]^2 / k^2 \bar{u} \tag{19}$$

Consequently, the equality of eddy diffusivities in Eq. (14) transforms Eq. (4) into:

$$E = (\lambda \varrho \varepsilon / P) k^2 \bar{u} (\bar{e}_o - \bar{e}_z) / [1\mathrm{n}(z + z_o - d)/z_o]^2 \tag{20}$$

where $\bar{e}_o$ is average water vapor pressure at height d; and $\bar{e}_z$ is average water vapor pressure at height z.

The practical usefulness of Eq. (19) is limited by the fact that e_o is very difficult to measure. However, Eq. (19) tells us that for a given vapor pressure gradient, evaporation increases if the wind speed increases. We can also deduce that a rough surface will evaporate at a higher rate than a smooth surface. Because of the relation between roughness length and crop height we can conclude that: (1) an irrigated orchard is likely to evaporate more water than an irrigated cotton field, and (2) changes in wind speed will affect the evaporation from an orchard more effectively than that from a cotton field.

Equation (20) can be applied to measurements of $\bar{e}_1$, $\bar{u}_1$ and $\bar{e}_2$, $\bar{u}_2$ at heights z_1 and z_2 respectively. This enables us to eliminate $\bar{e}_o$ from Eq. (20) and to obtain the Thornthwaite-Holzmann equation for evaporation:

$$E = (\lambda \varrho \varepsilon / P) k^2 \; (\bar{u}_2 - \bar{u}_1) \; (\bar{e}_1 - \bar{e}_2) / \; [\ln(z_2 + z_o - d)/(z_1 + z_o - d)]^2 \tag{21}$$

This formula has mainly been used to determine evaporation from open water surfaces. STANHILL (1969) has devised an apparatus that directly measures the product $(\bar{u}_2 - \bar{u}_1)$ $(\bar{e}_1 - \bar{e}_2)$, which considerably simplifies data collection and computation.

Bowen Ratio Method

The Bowen ratio method enables accurate determinations of the hourly evaporation. The Bowen ratio β is defined as:

$$\beta = H/E \tag{22}$$

By applying Eqs. (4), (5), and (14), we can rewrite Eq. (22) as:

$$\beta = (c_p P/\lambda\varepsilon)\ dT/de = \gamma dT/de \tag{23}$$

Here $\gamma = c_p P/\lambda\varepsilon$ is the psychrometric constant. At 1000-mb atmospheric pressure, γ equals 0.66 mb/°C.
Combining Eq. (23) with the energy balance. Eq. (2), we obtain:

$$E = (R_n - G)/(1 + \beta) \tag{24}$$

Eq. (24) confirms the statement on p. 145 that evaporation is directly related to the net radiation.

Comparing the curves E and H in Fig. 1 indicates that there is a strong diurnal correlation between $(R_n - G)$ and β. Therefore it is not possible to compute directly the average daily evaporation from the daily average $(R_n - G)$ and daily average β. Measurements of R_n, G, the temperature, and water-vapor pressure gradients over time intervals of 30 to 60 min yield the best results.

Conclusions

From the review of the environmental and surface parameters that affect evapotranspiration, it appears that the most significant meteorological factors are incoming radiation, wind speed, and water vapor pressure gradient in the air. However, it may be seen that the effects of these factors depend upon the nature of the surface. The albedo and the absolute temperature of the surface modify the radiation balance. The roughness length affects the rate of transfer of water vapor from or to the surface.

If we compare a 250 cm high corn field ($r = 0.16$, $z_o = 17$ cm) with a 50 cm high meadow ($r = 0.25$, $z_o = 6$ cm), growing under similar environmental conditions, we can predict from strictly physical considerations that the evaporative loss from the corn field will be larger than from the meadow. If we measure the meteorological parameters involved, we can quantitatively assess the evaporation from both surfaces.

Literature

DYER, A. J., MAHER, F. J.: Automatic eddy-flux measurement with the Evapotron. J. Appl. Meteorol. 4, 622 – 625 (1965).

GATES, D. M., HANKS, R. J.: Plant factors affecting evapotranspiration. In: Irrigation of Agricultural Lands. HAGAN, R. M., HAISE, H. R., EDMINSTER, T. W. (Eds). Amer. Soc. Agron., Madison, Wis., pp. 506–521 (1967).

SELLERS, W. D.: Physical Climatology. 271 pp. Chicago Univ. Chicago Press: 1965.

STANHILL, G.: A simple instrument for field measurement of turbulent diffusion flux. J. Appl. Meteorol. 8, 509–513 (1969).

Stanhill, G., Fuchs M.: The climate of the cotton crop: physical characteristics and microclimate relationships. Agric Meteorol. **5**, 183–202 (1968).

Stanhill, G., Hofstede, G. J., Kalma, J. D.: Radiation balance of natural and agricultural vegetation. Quart. J. Roy. Meteorol. Soc. **92**, 128–140 (1966).

Sutton, O. G.: *Micrometeorology*, 333 pp. New York – Toronto – London: McGraw-Hill 1953.

Tanner, C. B.: Measurement of evapotranspiration. In: Irrigation of Agricultural Lands. Hagan, R. M., Haise, H. R., Edminster, T. W. (Eds.). Amer. Soc. Agron., Madison, Wis., pp. 534–574 (1967).

Tanner, C. B.: Evaporation of water from plants and soil. In: Water Deficits and Plant Growth. Kozlowski, T. T. (Ed.). New York-London: Academic Press 1968.

Van Wijk, W. R.: Physics of Plant Environment, 383 pp. New York: Wiley 1963.

IV. Chemistry of Irrigated Soils – Theory and Application

The chemistry of irrigated soil must be viewed as a dynamic relationship between the properties of soil colloids and the composition of irrigation water.

In the first part of this chapter, our discussion concerns the exchange properties of arid-zone soils that are characterized primarily by the predominance of montmorillonite in the clay phase. The origin of electrical charges in the clay particles, as related to the location of adsorbed ions, is described, and the reaction causing bivalent ions to be replaced by monovalent ions during the irrigation process is explained.

Water-soluble salts occurring in the root zone during irrigation constitute part of the plant environment. The manner in which such salts accumulate in the soil during irrigation and are leached down under the root zone is described in the second part of this chapter. In this latter part, the reader will encounter tools capable of forecasting the dynamics of solute movement during the irrigation-postirrigation cycle, and an understanding of the physicochemical processes of diffusion and convective (or viscous) flow governing solute transport in porous media.

Nutrient supply to crops is different under dry and irrigated agriculture, and therefore special attention must be given to the behavior of important nutrients such as nitrate, phosphorus, and potassium under conditions of irrigation. Here, the availability of nutrients to the crop is related to the exchange properties of thc soils and to the solute movement in the soil profile.

An important factor to be considered when one deals with the planning and management of arid-zone agriculture is the hydraulic conductivity, which depends on the physicochemical properties of the soil as well as the liquid. Solutes in the irrigation water may change the composition of the soil solution and, consequently, of the soil-adsorbed phase. This, in turn, affects the swelling and dispersion properties of the clay material, produces changes in the pore geometry, and changes the soils hydraulic conductivity. The latter problems are discussed in the last part of this chapter.

In this chapter, the general theoretical considerations are presented, together with specific cases showing examples of forecasting salt accumulation and nutrient supplies under irrigation.

1

Ion Exchange Properties of Irrigated Soils

I. SHAINBERG

Introduction

The solid paricles of the soils are classified, according to their size, as sand (> 0.02 mm), silt (0.02 mm > silt > 0.002 mm), and clay (< 0.002 mm). The clay particles of the soils usually carry a net negative charge. This charge is neutralized by adsorbed cations than can be exchanged for a stoichiometrically equivalent amount of other ions occurring in the solution. The reaction whereby a cation in solution replaces an adsorbed cation is called cation exchange.

Calcium and magnesium are the principal cations found in a soil solution and on the exchange complex of the irrigated soils from arid regions. When a soil is irrigated with low-quality water, an excess of soluble salts accumulates in it and sodium frequently becomes the dominant cation in the soil solution. Under such conditions, part of the calcium and magnesium originally adsorbed is replaced by sodium.

The physical and mechanical properties of the soils (such as dispersion of the particles, infiltration, permeability, soil structure, stability of the aggregates,) are very sensitive to the type of exchangeable ions. Divalent ions (mainly calcium) are responsible for many of the physical properties characterizing a "good" soil. The deleterious effect of adsorbed sodium on agricultural soil is well known (RICHARDS, 1954). Moreover, it is generally recognized that even an amount of 15 % sodium in the exchange complex is sufficient to interfere with the growth of most crop plants. Our purpose in this chapter is to discuss the origin of the charge present in clay particles, show where the adsorbed ions are located, explain why monovalent ions cause dispersion of the clay particles, and discuss the reaction whereby adsorbed Ca is being replaced by Na occurring in irrigation water.

Nature of Clay Minerals and Origin of Charge

Two structural units are involved in the atomic lattices of most clay minerals. One unit consists of two-dimensional arrays of siliconoxygen tetrahedra. In each tetrahedron, a silicon atom is centered at equal distances from four oxygens located at the four corners of the tetrahedron. Three of the four oxygen atoms are shared by three neighboring tetrahedra to form a tetrahedral sheet or a silical sheet (Fig. 1 a)

The second unit consists of sheets of aluminum, or iron- or magnesium-oxygen-hydroxyl octahedra. In each octahedron, the Al or Mg atoms are coordinated with six oxygen atoms or OH groups located around the central atom at the six corners of the octahedron. The sharing of oxygen atoms by neighboring octahedra results in a sheet (Fig. 1 b), which is called an octahedral sheet or an alumina or magnesia sheet. When magnesium is present, all the octahedral positions are filled to balance the structure, which has the formula $Mg_3(OH)_6$. When aluminum is present, only two-thirds of the possible positions are filled to balance the structure, which has the gibbsite formula $Al_2(OH)_6$

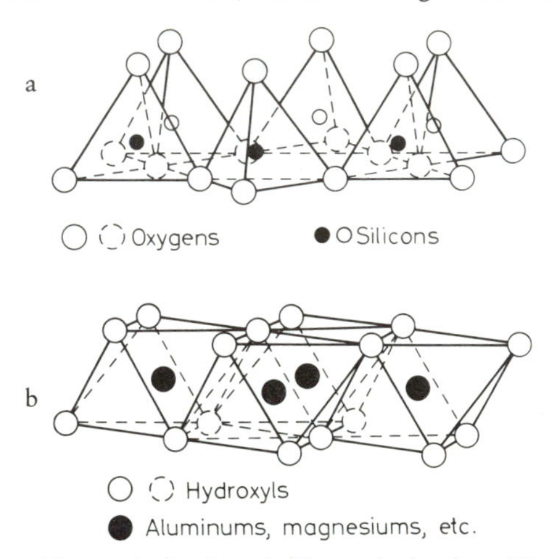

Fig. 1. a The tetrahedra sheet. b The octahedra sheet. (GRIM, 1968)

The analogous symmetry, and the almost identical dimensions in the tetrahedral and octahedral sheets, allow the sharing of oxygen atoms between various sheets. The main clay minerals in the soil are classified according to the way these two sheets are stacked to form a unit layer.

GRIM (1968) classified the clay minerals as follows:

1) *Two-Layer Types.* The unit layer is composed of one sheet of silica tetrahedra and one sheet of alumina octahedra. The two sheets are tied together by shared oxygen atoms. As shown for kaolinite in Fig. 2, the surface of the layer on the alumina side is composed

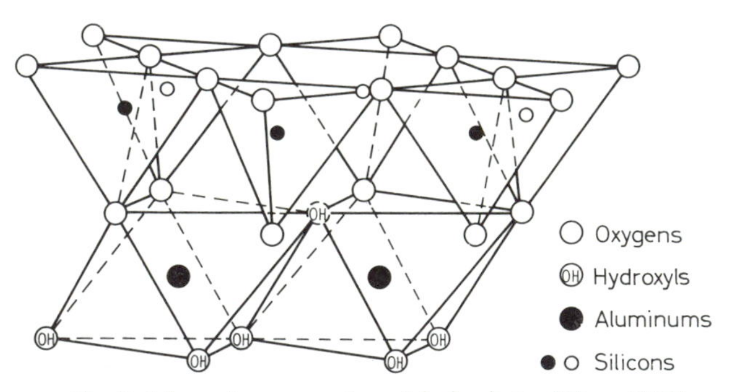

Fig. 2. Schematic presentation of the kaolinite. (GRIM, 1968)

of hydroxyls and on the silica it is composed of oxygens. The crystals consist of superimposed unit layers with the hydroxyl and oxygen surfaces adjacent to each other. The layers show little tendency to separate from each other, probably because of the hydrogen bonds between the adjacent oxygen and hydroxyl layers. Thus, kaolinite clays have relatively low specific surface area.

2) *Three Layer Types.* The unit layers are composed of two sheets of silica tetrahedra

and one central octahedral layer. The three sheets are held together by shared oxygen atoms. Individual layers are stacked to form the crystalline particles. Depending on the magnitude of attraction forces between the layers in a crystal, the three-layer minerals are subdivided as:

a) Expanding lattice—the smectite group: montmorillonite, nontronite, hectorite, etc. When montmorillonite clays are in contact with water, molecules of the latter penetrate between the unit layers, and the lattice expands. The extent of this expansion depends on the exchangeable ion.

b) Nonexpanding lattice – illite. These clays are distinguished from the smectites primarily by the absence of interlayer swelling with water.

The charge on the clay particles may be created in two ways:

1) Imperfections within the interior of a crystal lattice. Substitution of trivalent aluminum for quadrivalent silicon in the tetrahedral layer, and magnesium or ferrous ions for aluminum in the octahedral layer, results in a negative charge in the lattice. The charge due to this isomorphous substitution is uniformly distributed within the plate-shaped clay particles.

2) Preferential adsorption of certain ions at specific sites, on the particle surface. At the edges of the plates, the tetrahedral silica sheets are disrupted, and primary bonds are broken. These surfaces resemble those of silica and alumina soil particles. That part of the edge surface where the octahedral sheet is broken may be compared with the surface of an alumina particle, such particles being charged either positively or negatively, depending on the pH of the solution. Similarly, wherever the tetrahedral sheet is broken, the charge is also determined by the pH, as in the silica particles. Hence, under appropriate conditions, the edge surface area may carry a positive charge.

The particle charge is compensated by the accumulation of an equivalent amount of ions of opposite sign in the liquid immediately surrounding the particle, keeping the whole assembly electroneutral. The particle charge, and the equivalent amount of counter-ions, both form the electric double layer. The total amount of compensating cations may be determined analytically. This amount, expressed in milliequivalents per 100 gm of dry soil, is known as the cation-exchange capacity (CEC) of a soil. Consideration of the origin of clay charges shows that the measured CEC depends on the experimental method (e.g., the pH and concentration of solutions used, the types of replacing cations, the method of leaching out free electrolytes, the method of preparing soil samples for measurement). H. D. CHAPMAN (1965) summarized the experimental difficulties in determining the cation-exchange capacity of a soil sample.

Configuration of the Diffuse Double Layer

The dominant clay in arid-zone soils is montmorillonite. Most of the negative charge on this clay is due to isomorphous substitution in the crystal lattice and is uniformly distributed within the plate-shaped clay particles. The negative charge of the clay particles is compensated by the exchangeable cations forming a double layer.

Knowledge of the double-layer structure is important for understanding the colloidal behavior of soil systems. Fig. 3 shows the ionic concentration distribution near a flat surface.

The counter-ions of the double layers are subject to two opposing tendencies.

Electrostatic forces attract them to the charged clay particle surfaces, whereas diffusion tends to repel them as far from these surfaces (where their concentration is high) as possible, into the bulk of the solution (where their concentration is low). A mathematical description for these two tendencies was given by GOUY (1910), and a detailed discussion and derivation were given by OVERBEEK (1952). A complete discussion of the mathematics is beyond the scope of this chapter. We shall only mention that the derivation is based on the following two equations, which take into account the tendencies.

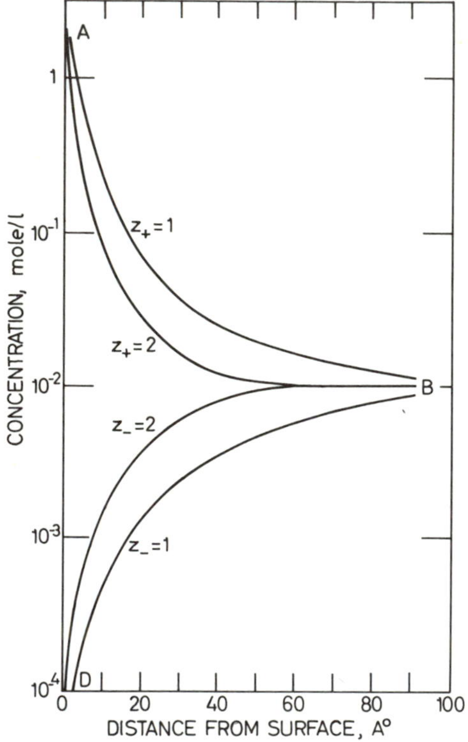

Fig. 3. Concentration distribution of mono- and divalent ions in the diffuse layer formed on montmorillonite in equilibrium with 0.01 *M* solutions

Attracted by the surface charge, counter-ions accumulate near the particle surface. The coulombic interaction between the charges present in the system is described by Poisson's equation

$$\frac{d^2\psi}{dx^2} = \frac{-4\pi\varrho}{\varepsilon} \tag{1}$$

where ψ is the electric potential in electrostatic units (esu), x is the distance from the particle surface, ε is the dielectric constant of the medium, and ϱ is the net volume charge density, as given by

$$\varrho = z_+ en_+ - z_- en_- \tag{2}$$

Here, n_+ and n_- are the local concentration of the positive and negative ions (number per cm^3), e is the electronic charge, and z_+ and z_- one the valences of the ions.

In addition to being electrically attracted to the surface, the counter-ions also have a tendency to diffuse away from the surface toward the bulk of the solution where their concentration is low. The distribution of ions in the solution is given by the Boltzmann equation:

$$n^{+} = n_o^{+} \exp(-ze\psi/kT)$$
$$n^{-} = n_o^{-} \exp(ze\psi/kT) \tag{3}$$

where n_o is the concentration of the ions in the equilibrium solution (where $\psi = 0$), k is the Boltzmann constant (which is equal to the molar gas constant for one molecule), T is the absolute temperature, and z (unsubscripted) is the absolute value of the valencies of the ions. Since electric potential near the clay surface is negative, Eq. (3) indicates that the concentration of cations increases as the negative clay surface is approached, where as the concentration of anions decreases (as in in Fig. 3).

Introducing a dimensionless parameter for electric potential y, affecting an ion at the dimensionless distance ξ, the mathematical description of potential and charge distribution is easily generalized (Overbeek) as

$$y = \frac{ze\psi}{kT} \tag{4}$$

and

$$\xi = Kx = \sqrt{\frac{8\pi z^2 e^2 n_o}{\varepsilon kT}}\, x \tag{5}$$

For water suspensions at 25 °C, the value of K is

$$K = \left(\frac{8\pi z^2 e^2 n_o}{\varepsilon kT}^{1/2}\right) = 0.326\, z \sqrt{m_o}^{-1} \tag{6}$$

where m_o is concentration in the bulk solution in mol/I.
Substituting Eqs. (2)–(5) in Eq. (1) gives the differential equation

$$\frac{d^2y}{d\xi^2} = \sinh y \tag{7}$$

Integrating once, subject to the boundary conditions that for $\xi = \infty$,

$$\frac{dy}{d\xi} = 0, \text{ and } y = 0, \text{ one obtains}$$
$$\frac{dy}{d\xi} = -2 \sinh y/2 \tag{8}$$

Additional integration, subject to the boundary conditions that at the clay surface $\xi = 0, \psi = \psi_o$ or $y = y_o = Z$, yields

$$e^{y/2} = \frac{(e^{Z/2}+1) + (e^{Z/2}-1)\, e^{-\xi}}{(e^{Z/2}+1) - (e^{Z/2}-1)\, c^{-\xi}} \tag{9}$$

Eq. (9) enables one to calculate the electric potential in the solution phase y as a function of the dimensionless distance ξ, provided that the dimensionless potential at the surface Z is known.

The elctric potential at the clay surface, as a function of the clay charge density and the concentration of the equilibrium solution, can easily be calculated from Poisson's Eq. (1). The final equation is

$$Z/2 = \sinh^{-1}[(\pi/2\varepsilon k T n_o)^{1/2} q_o] \quad (10)$$

where q_o is the total charge of the double layer, esu/cm^2, and all the other terms are as before.

If one plots $q_o/\sqrt{m_o}$ as a function of Z (Fig. 4), the value of Z can be easily found.

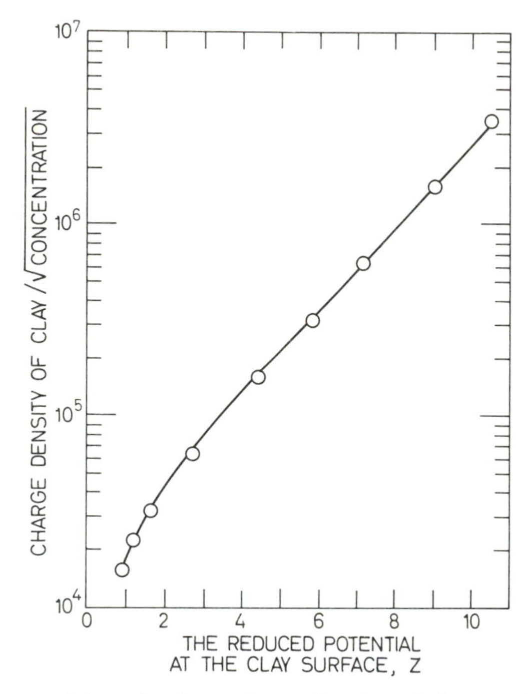

Fig. 4. The reduced potential at the clay surface a function of the clay charge density and the concentration of the eqlibrium solution

Numerical Calculations

Ionic distributions in Na and Ca montmorillonite at equilibrium with 0.01 M solutions of NaCl and $CaSO_4$, respectively, are presented in Fig. 3. The numerical details for this example are as follows:

1) The surface charge density is q_o. The cation-exchange capacity of montmorillonite is 0.9 meq/g (VAN OLPHEN, 1963). The specific surface area is 760 m^2/g (VAN OLPHEN, 1963). Thus, the charge density is 0.9 meq/760 $\times$ 10^4 cm^2 = 1.18 $\times$ 10^{-7} meq/cm^2. Since there are 6.02 $\times$ 10^{20} electronic charges in meq, and the charge on an electron is 4.8 $\times$ 10^{-10} esu, then

$$q_o = 1.18 \times 10^{-7} \times 6.02 \times 10^{20} \times 4.8 \times 10^{-10}$$

$$q_o = 3.42 \times 10^4 \text{ esu/cm}^2$$

2) The concentration in bulk solution is 0.01 mole/1, so that from Eq. (6) one has $K =$

0.326 zm_o, thus $K = 0.0326$ Å^{-1} for NaCl and 0.0652 Å^{-1} for $CaSO_4$. $1/K$, which is a measure of the 'thickness' of the diffuse layer, is equal to 30.6 and 15.3 Å, respectively. The dimensionless distance from the clay surface ξ is given by Eq. (5) as $0.0326x$ and $0.0652x$ for Na and Ca systems, respectively, where x, the distance from the clay surface, is in angstroms.

The quotient of charge density over the square root of the solution concentration may now be calculated as

$$\frac{q_o}{\sqrt{m_o}} = \frac{3.42 \times 10^4}{\sqrt{10^{-2}}} = 3.42 \times 10^5$$

3) The dimensionless potential at the clay surface is found from Fig. 4, or Eq. (10), and is equal to $Z = 6.0$. The electric potential in volts may be calculated from Eq. [4].

$$\psi = \frac{kT}{ze} \cdot y + \frac{1.38 \times 10^{-16} \times 298}{z \times 4.8 \times 10^{-10}} y = 8.55 \times 10^{-5} y \text{ esu of potential}$$

$$= 25.6 \frac{y}{z} \text{ mv}$$

Thus, the electric potential at the clay surface is 153 and 76 mv for Na and Ca systems, respectively.

The dimensionless potential distribution is calculated from Eq. (9). Concentration of the ionic species is calculated from the calculated potentials and the Boltzmann equation (3), and is summarized in Fig. 3.

Fig. 3 shows that the concentration of the counter-ions is high near a particle surface and decreases with increasing distance from this surface. The diffuse layer in divalent cation systems is compressed because these ions are more strongly attracted electrostatically to the charged surface. The bulk solution salt concentration also causes the diffuse layer to compress, due to the diminished tendency of counter-ions of diffuse away from the surface, toward the bulk of the solution. This phenomenon is not expressed in Fig. 3 but can be easily calculated from double-layer theory.

Fig. 3 and the relevant equations also indicate that a diffuse layer does not consist only of excess counter-ions but that there also is a deficiency of ions of similar sign near the negative clay surface, because these ions are electrostatically repelled by the particle. The deficiency of ions of similar sign in the neighborhood of the surface is called negative adsorption or anion exclusion.

Ion-Exchange Equilibria

The distribution of adsorbed ions near the clay surface is very important in determining the swelling and dispersion properties of clay particles, thus very strongly affecting the hydraulic conductivity of the soil. This propblem is discussed on p. 189

Composition of the adsorbed ions is determined by the composition of the soil solution.

Ion-exchange reactions and equations describe the distribution of cations between the adsorbed and solution phases. In this section, a short discussion of ion-exchange reactions and equations is presented.

The ionic equilibrium, and the preference of clay particles for one of the two counter-ions, may be described by the law of mass action, which gives the selectivity coefficient. Thus, for the heterovalent exchange reaction

$$2\overline{\text{Na}} + \text{Ca} = 2\text{Na} + \overline{\text{Ca}} \tag{11}$$

where bars indicate ions in the adsorbed phase, the selectivity coefficient is given by

$$K_{Na}^{Ca} = \frac{(\overline{\text{Ca}})\,(\text{Na})^2}{(\text{Ca})\,(\overline{\text{Na}})^2} \tag{12}$$

If the concentrations of ionic species in both the adsorbed and solution phases are given in moles, then Eq. (12) defines the molar selectivity coefficient. More often, adsorbed ion concentrations is expressed in units of ionic equivalent fraction $\overline{X}$ and concentration in the solution phase in moles per liter, so that Eq. (12) becomes

$$K = \text{const.} = \frac{\overline{X}_{Ca}(m_{Na})^2}{(\overline{X}_{Na})^2\, m_{Ca}} \tag{13}$$

Values of the apparent constant can be estimated for various soils from data published by LEVY and HILLEL (1968). They found that for a wide range of equivalent fractions of exchangeable sodium (between 0.1 and 0.7) in a typical Israeli soil, the value of this constant is 4.0. Similar observations on Na-Ca exchange reactions in montmorillonite were made by LAUDELAUT *et al.* (1968) and LEWIS and THOMAS (1963). As will be shown later, this constant is identical with the "Gapon" constant recommended for use by the U.S. Salinity Staff (1954). If, instead of molar concentration in the solution phase, mole fraction units were used in the solution, namely,

$$N_{Na} = \frac{m_{Na}}{m_o} \quad \text{and} \quad N_{Ca} = \frac{m_{Ca}}{m_o} \tag{14}$$

where m_o is the total molar concentration in solution, then substituting these values in Eq. (13) would give

$$K = \frac{(\overline{X}_{Ca})\,(N_{Na})^2 \cdot m_o}{(\overline{X}_{Na})^2\,(N_{Ca})} = 4.0 \tag{15}$$

For a soil in equilibrium with a solution of 0.01 M and $N_{Na} = N_{Ca} = 0.5$, the ionic composition of the exchanger is

$$\frac{\overline{X}_{Ca}}{(\overline{X}_{Na})^2} = \frac{4.0\,(N_{Ca})}{m_o\,(N_{Na})^2} = 800$$

Since $\overline{X}_{Na} + \overline{X}_{Ca} = 1.0$, solving for X_{Na} one obtains

$$\overline{X}_{Na} = 0.035, \overline{X}_{Ca} = 0.965$$

and the affinity of the clay for the divalent ion is clearly demonstrated.

The effect of solution concentration may be demonstrated by substituting the values $m_o = 1.0$, 0.1, 0.01, and 0.001 M in eq (15). The results are summarized in Table 1, which shows that clays prefer counter-ions of higher valence and that this preference increases as the solution becomes more diluted. This effect is readily explained by the diffuse layer potential. Electric potential attracts the counter-ions with a force that is proportional to the ionic charge. Hence the divalent ion is more strongly attracted and is preferred by the ion exchanger. The absolute value of the electric potentials increases with dilution of the solution, thus increasing the affinity of clay for the divalent ion with increasing dilution.

Table 1. Effect of solution concentration on the equivalent fraction of adsorbed cations from solutions of a constant mole fraction ($N_{Na} = 0.5$ $N_{Ca} = 0.5$)

Solution Concentration (mole/1)	Equivalent fraction of adsorbed Na	Ca
1.0	0.353	0.647
0.1	0.106	0.894
0.01	0.035	0.965
0.001	0.011	0.989

Gapon Equation

Eq. (13) is simmilar to Gapon's equation, which was applied extensively to soils by the staff of the U.S. Salinity Laboratory (RICHARDS, 1954). Rearranging Eq. (13), and taking the square root – which means dividing the reaction of Eq. (11) by 2-gives

$$\frac{\bar{X}_{Na}\sqrt{m_{Ca}}}{\sqrt{\bar{X}_{Ca}}\,m_{Na}} = \sqrt{\frac{1}{K}} = \sqrt{\frac{1}{4}} = 0.5 \tag{16}$$

or

$$\frac{X_{Na}}{\sqrt{\bar{X}_{Ca}}} = 0.5\,\frac{m_{Na}}{\sqrt{m_{Ca}}} \tag{17}$$

The Salinity Staff defined a quantity called the sodium adsorption ratio (SAR),

$$\text{SAR} = \frac{m'_{Na}}{\sqrt{m'_{Ca}}} = 31.6\,\frac{m_{Na}}{\sqrt{m_{Ca}}} \tag{18}$$

where m' is the solution concentration, mmole/1. Combining Eqs. (18) and (17) gives

$$\frac{X_{Na}}{\sqrt{X_{Ca}}} = \frac{0.5}{31.6}\,\text{SAR} = 0.0158\,\text{SAR} \tag{19}$$

At a low equivalent fraction of Na in the exchange phase X_{Ca} is close to unity, and $\sqrt{X_{Ca}}$ can be replaced by $\bar{X}_{Ca}$. Consequently, Eq. (19) takes the familiar form

$$\text{ESR} = \frac{X_{Na}}{X_{Ca}} = 0.0158\,\text{SAR} \tag{20}$$

An analysis of a large number of soil samples by the Salinity Staff led to the empirical equation

$$\text{ESR} = -0.0126 + 0.01475\,(\text{SAR}) \tag{21}$$

More recently, BOWER (1959) published a similar empirical equation in which

$$\text{ESR} = 0.0057 + 0.0173\,(\text{SAR}) \tag{22}$$

It is quite obvious that the last three equations are, in effect, modifications of the mass-action approach, with $K = 4.0$.

Concluding Remarks

In most soils, divalent calcium and magnesium ions are dominant in the adsorbed phase. Calcium ion is probably responsible for many of the physical and mechanical properties characterizing a "good" soil. Very little is known about the behavior of Ca clay, despite its importance. Ca montmorillonite is known to exist in packets or tactoids, consisting of several clay platelets, each having a film of water 4.5 Å thick on each internal surface (BLACKMORE and MILLER, 1961). However, there is still no explanation for the limited swelling of Ca montmorillonite (even in distilled water) and the forces stabilizing the Ca tactoids. Even the properties of the water and the divalent ions enclosed between these closely adjacent platelets are relatively unknown. These factors need to be further investigated.

The behavior of Na clay is best described by the diffuse-double-layer theory. However, some of the assumptions of the Gouy theory are too crude and are currently being modified. Since no complete theory exists for either of the homoinic clay systems, even less is known about a mixed system of mono- and divalent ions, and therefore many related questions remain unanswered: Are the adsorbed ions mixed at random in a mixed system, or does "demixing" occur by which the monovalent ions concentrate on certain surfaces and the divalent ions on others? What causes demixing? What is the effect of demixing on the measured properties of clays and soils? The deleterious effect of adsorbed sodium, even at low percentages in the adsorbed phase (10 or 15 percent), may be mainly the result of this demixing phenomenon.

Literature

BLACKMORE, A. V., MILLER, R. D.: Particle size and osmotic swelling in Ca-montmorillonite. Soil Sci. Soc. Amer. Proc. **25**. 169–173 (1961).

BOWER, C. A.: Cation exchange equilibrium in soils affected by Na salts. Soil Sci., **88**. 32–35 (1959).

CHAPMAN, H. D.: Cation exchange capacity. In: Methods of Soil Analysis (Pt. 2). BLACK, C. A., (Ed.), 1965.

GRIM, R. E.: Clay Mineralogy (2nd ed.). New York: McGraw-Hill 1968.

LAUDELOUT, H., van BLADEL, R., BOLT, G. H., PAGE, A. L.: Thermodynamics of heterovalent cation exchange reactions in a montmorillonite clay. Trans. Faraday Soc. **64**. 1477–1488(1968).

LEVY, R., HILLEL, D.: Thermodynamics equilibrium constants of Na/Ca exchange in some Israeli soils. Soil Sci. **106**, 393–398 (1968).

LEWIS, R. J., THOMAS, H. C.: Adsorption studies on clay minerals VIII. A consistency test of exchange sorption in the systems Na/Cs/Ba montmorillonite. J. Phys. Chem. **67**. 1781 (1963).

OVERBEEK, J. Th. G.: In: Colloid Science. KRYT, H. R. (Ed.). Vol. **I**. Chaps. 4, 5, and 6. Amsterdam: Elsevier 1952.

RICHARDS, L. A. (Ed.): Diagnosis and improvement of saline and alkaline soils. USDA Handbook 60, 1954.

VAN OLPHEN, H.: An Introduction to Clay Colloid Chemistry. New York: Interscience 1963.

2

Solute Movement in Soils

E. BRESLER

Water-soluble salts entering the soil profile through the process of irrigation constitute an important part of the plant environment. These salts may accumulate in the root zone or may be leached out of this zone, depending on the convective diffusive transport processes and solute interactions in the soil. Understanding these processes is important in establishing management practices directed toward preventing the hazardous effect of salt on the plant and the soil. Theories based on macroscopic considerations provide a satisfactory means for describing such transport phenomena. In these theories, the soil is considered to be a continuous porous medium, and the governing equations are derived for a representative elementary volume that is large enough so that its properties can be expressed in terms of statistical averages.

Theory

Solute Movement by Diffusion

The terms diffusion and thermal motion refer to the movement of individual particles of molecular size as a result of their thermal energy. This energy causes the particles to move at random within the phase that contains them (for solute movement in soil, this phase is the soil-water solution). As a result of such random motion, more particles tend to move from points of high concentration to points of low concentration than from points of low concentration to point of high concentration.

The average macroscopic flow rate of the particles is proportional to the concentration gradient (dc/dx) and to the cross-sectional area. This is known as Fick's first law,

$$\frac{dQ}{dt} = -DA\,\frac{dc}{dx} \tag{1}$$

where Q is the amount of diffused material in units of mass or volume, t is time, A is the cross-sectional area available for diffusion, c is the concentration of the material in mass or volume of material per unit volume of solution, x is the space coordinate parallel to the flow, and D is a proportionality constant known as the diffusion coefficient. Values of D for various ions in a pure-water system at isothermal conditions have been published in the literature.

Since the soil constitutes a complex charged porous medium, the effective diffusion coefficient (D_p) for any given ion is less than the equivalent coefficient (D_o) in a pure-water system. The relationship between D_p and D_o may be expressed as

$$D_p = \theta\,(L/L_e)^2\,\alpha\gamma\,D_o \tag{2}$$

where θ is the volumetric water content of the soil (cm^3 water/cm^3 bulk soil). When a solute moves by diffusion in the water phase of the soil, the cross-sectional area available for flow is only a fraction, θ, of the macroscopic cross-sectional area, A, of Eq. (1). Therefore, θ represents the reduction in the area available for flow in a soil, as compared with the area available in a body of water. L is the macroscopic average path of diffusion, whereas L_e is the actual tortuous path along which a particle moves. Since the tortuous path is longer, dx must by multiplied by L_e/L. Furthermore, since the actual path is likely to form an angle with the macroscopic direction, the cross-sectional area (θA) is decreased by L/L_e. The resulting tortuousity factor, arising from the porous nature of the soil, is therefore $(L/L_e)^2$; α accounts for reduction in water viscosity due to the presence of charged particles in the soil matrix, and γ takes into account the retarding effect of anion exclusion on flow in the vicinity of negatively charged soil particles.

The factors θ, $(L/L_e)^2$, α, and γ are less than unity, and therefore the effective diffusion coefficient (D_p) is less than the equivalent diffusion coefficient (D_o) in a free-water solution. However, in a given unsaturated soil, reducing the soil-water content results in a smaller cross-sectional area available for diffusion, a longer flow path, and more significant values of water viscosity and negative adsorption. Thus, one can express D_p in terms of D_o and θ. It has been shown that, in a clay-water system, the solute diffusion coefficient is a positive exponential function of water content (θ) and, for practical purposes, is independent on salt concentration. This simply means that

$$D_p\,(\theta) = D_o a e^{b\theta} \tag{3}$$

where a und b are empirical constants. OLSEN and KEMPER (1968) stated that data collected on soils fit Eq. (3) reasonably well with $b = 10$ and a ranging from 0.005 to 0.001, depending on the surface area of the soil studied (sandy loam to clay). This simplified $D_p\,(\theta)$ relationship was found to apply to water contents corresponding to a range of suctions between 0.30 and 15 bars, which is enough for most cases where diffusion is important. Thus, taking into account the effect of water content in the soil on ion diffusion, Fick's modified first law can be written as

$$J_d = \frac{dQ}{Adt} = -D_o\theta\,(L/L_e)^2\,\gamma\alpha\,\frac{dc}{dx} = -D_p\,(\theta)\,\frac{dc}{dx} \tag{4}$$

where c is the solute concentration in the soil solution, and $D_p\,(\theta)$ may be calculated from Eq. (3).

Solute Movement by Convection

In studying transport phenomena, it is convenient to picture the water as an imperfect crystal, which is easily deformable by forces such as those generated by pressure gradients. The resulting movement of the "crystal" is known as viscous flow. Dissolved ions that are carried by the moving water are said to undergo convective, viscous, or mass flow. When the solution flows through the soil, the convective transport of the solute depends on the flow velocity. Owing to the soil's porous nature, the actual pore water velocity is distributed about an average value, in a manner that depends on the distribution of pore sizes and shapes. Flow in large pores is faster than in small pores, and is much faster at the center of a pore than near the periphery. In this way, the soil matrix with its porous structure causes the solute to disperse.

The macroscopic convective transport of a solute is usually described by an equation that takes into account two modes (or components) of transport: (a) the average flow velocity component, and (b) the mechanical dispersion component (the latter resulting from local variations in the flow velocities). This mechanical dispersion effect is similar to diffusion in the sense that there is a net movement of the solute from zones of high concentration to zones of low concentration. Therefore, it is commomly agreed that an equation similar to (1) provides a good first-order description of the despersion component of convective flow, provided that the diffusion coefficient is replaced by the coefficient of mechanical dispersion (D_h).

Experiment as well as theory indicate that the magnitude of the mechanical dispersion coefficient (D_h), in a given porous material, depends on the average flow velocity. In a simple nonaggregated porous medium, D_h may be considered proportional to the first power of the average flow velocity,

$$D_h(\overline{V}) = D_m\overline{V} \tag{5}$$

where $\overline{V}$ is the average interstitial flow velocity (cm/sec^{-1}), and D_m (cm) is an experimental constant depending on the characteristics of the porous medium.

Assuming that average and local velocities are additive, the total amount of solute transported across a unit area, in the direction of flow (x) by convection, is obtained from

$$J_h = -D_h(\overline{V})\frac{dc}{dx} + \overline{V}\theta c \tag{6}$$

where J_h represents the total amount of solute moving by convection across a macroscopic area, in unit time (g, or mole, cm^{-2} sec^{-1}). The first term represents solute flow due to dispersion and the second term is due to the average flow velocity.

Combined Effects of Diffusion and Convection on Solute Movement

The joint effects of diffusion and convection are described by combining Eqs. (4) and (6),

$$J_s = -[D_h(\overline{V}) + D_p(\theta)]\frac{dc}{dx} + \overline{V}\theta c = -D(\overline{V},\theta)\frac{dc}{dx} + qc \tag{7}$$

where D is the so-called hydrodynamic dispersion coefficient (cm^2 sec^{-1}), and q is the volumetric water solution flux (cm^3 · cm^{-2} · sec^{-1}). The right hand side terms in Eq. (7) are only an approximation because the macroscopic quantities D, V, θ, and c are spatial averages. Nevertheless, Eq. (7) is very useful in predicting solute transport in soils.

Changes in soil-water content as a result of infiltration, redistribution, evaporation, and transpiration bring about the simultaneous movement of water and salt. Steady flow conditions, such as those described in Eq. (7), are rare in nature. A mathematical expression for the more general transient conditions is obtained by combining Eq. (7) with the equation of continuity, or mass conservation. The latter equation states that the rate of change of solute within a given soil element ($\partial q'/\partial t$) must be equal to the difference between the amounts of solute that enter and leave the element. By equating the difference between outflow and inflow to the amount of dissolved salt that has accumulated in the soil element, and considering the case of one-dimensional vertical flow, one obtains the expression

$$\frac{\partial q'}{\partial t} = \frac{\partial(c\theta)}{\partial t} = \frac{\partial}{\partial z}\left\{[D_h(\bar{V}) + D_p(\theta)]\frac{dc}{dz}\right\} - \frac{\partial(\bar{V}\theta c)}{\partial z} \tag{8}$$

Eq. (8) is the mathematical statement of transient vertical diffusive-convective solute transport under the conditions described. It applies only when the solute does not interact chemically with the soil and when there is no gross loss or gain of salt inside the flow system. Mathematical solutions of Eq. (8) may be developed for problems involving salinity and fertility under fallow soil conditions, and c may represent either the total salt concentration or a specific anion, provided, of course, that neither salt precipitation nor adsorption and exclusion take place.

In steady-state horizontal flow of water, $\bar{V}$ und θ are constant, and Eq. (8) becomes

$$\frac{\partial c}{\partial t} = D'\frac{\partial^2 c}{\partial z^2} - \bar{V}\frac{\partial c}{\partial z} \tag{9}$$

where $D' = (D_h + D_p)/\theta$ is a dispersion coefficient obtained by fitting the breakthrough data to an analytical solution of Eq. (9). A comprehensive review of steady-state miscible displacement was published by BIGGAR and NIELSEN (1967).

In many problems of salinity and soil fertility, the investigator is interested in the concentration of a specific ion in the soil solution. Often, the ion may interact with the soil material: cations may be adsorbed by the clay fraction of the soil and, in some cases, anion exclusion may occur. When ions in the exchange phase and in the soil solution reach equilibrium very rapidly, the amount adsorbed or excluded may be assumed to depend only on the equilibrium concentration of the solution. Applying the mass conservation principle to the solute in a small soil element gives

$$\frac{\partial[(c\theta) + Q'(c)]}{\partial t} = \frac{\partial}{\partial z}\left[D(\theta, \bar{V})\frac{\partial c}{\partial z}\right] - \frac{\partial(qc)}{\partial z} \tag{10}$$

where $Q'(c)$ is the amount adsorbed or excluded per unit volume of soil, $c\theta$ is the amount dissolved, $q = \bar{V}\theta$ is the volumetric water flux, and D is the coefficient of hydrodynamic dispersion.

Applying the chain rule, $dQ'(c)/dt = (dQ'/dc)(dc/dt)$ to Eq. (10), rearranging, and remembering that $\partial\theta/\partial t = -\partial q/dz$, it gives

$$[\theta(z, t) + b(c)]\frac{\partial c}{\partial t} = \frac{\partial}{\partial z}\left[D(\theta, \bar{V})\frac{\partial c}{\partial z}\right] - q(z, t)\frac{\partial c}{\partial z} \tag{11}$$

where $b(c) = d'Q(c)/dc$ is the lope of the adsorption or exclusion isotherm. Note that Eqs. (11) and (8) are identical except for the term $b(c)$. The left-hand side and the last term on the right-hand side, of Eq. (8) were expanded and the conservation of mass principle was used in converting Eq. (8) into (11). Also note that the function $b(c)$ is given in terms of volumetric water content and is positive for adsorption and negative for exclusion.

Applications

Estimation of Salt Distribution in the Soil

Estimating the distribution of salt in depth and time is important in many problems involving irrigated agriculture. Among these are problems concerning soil salinity and alkalinity, placement and leaching of plant nutrients, soil pollution, etc.

Fallow Field Conditions

Under fallow field conditions, or when the effect of vegetation is negligible, one can estimate the salt concentration from numerical solution of Eq. (8) or (11). Such solutions are restricted to total salt concentration whenever there is no salt precipitation or salt uptake by plants. They may also be useful in estimating the distribution of a specific ion whose adsorption or exclusion functions are either known [Eq. (11)] or negligible [Eq. (8)].

To obtain estimates of salt distribution, $c(z, t)$, at various time and depth values, water content and flux as functions of time and depth [$\theta\ (z, t)$ and $q(z, t)$, respectively] must be known. These are obtained by solving the Darcy'-type water-flow equation (see p. 89), subject to appropriate boundary conditions depending on the flow process, whether one speaks of infiltration, redistribution, drainage, or evaporation. Once $\theta\ (z, t)$ is known, the diffusion coefficient $D_p\ (z, t)$ may be calculated from Eq. (3), provided that a and b are given (OLSEN and KEMPER, 1968). Knowing $q(z, t)$, one may calculate $\overline{V}(z, t) = q(z, t)/\theta\ (z, t)$ to evaluate the hydrodynamic dispersion coefficient (D_h) from Eq. (5). The coefficient D_m in Eq. (5) may be approximated by fitting an experimental breakthrough curve to the solution of Eq. (9) for constant $\overline{V}$, θ, and D', or from other theoretical considerations.

Using such approximations, Eq. (8) may be solved numerically with the aid of a computer. The following initial and boundary conditions may be considered to be appropriate for many practical situations:

1) During the infiltration process, the solute flux (J_s in Eq. (7) at the soil surface is equal to the product of the infiltration rate (surface water flux) and the salt concentration of the water applied.

2) During the redistribution process, both solute and water fluxes at the soil surface are zero.

3) During the evaporation process, solute flux at the soil surface is zero, and the surface water flux is negative.

4) At an arbitrary depth Z well below the root zone, where a unit hydraulic gradient is considered, the salt concentration gradient is practically zero.

5) The initial salt concentration profile is prescribed independently of the numerical solution.

These initial and boundary conditions may be expressed mathematically as follows:

$$c(z, 0) = c_n(z) \tag{12a}$$

$$-\{ D_p\,[\theta(0, t)] + D_h[\overline{V}(0, t)]\}\ \frac{\partial c}{\partial z}\bigg|_{z=o} + q(0, t)c(0, t) = q(0, t)C_o\,(t) \tag{12b}$$

for any $t > 0$ during infiltration

$$\left.\frac{\partial c}{\partial z}\right|_{z=0} = 0 \tag{12c}$$

for any $t > 0$ during redistribution

$$-\{D_p[\theta(0,t)] + D_h[\bar{V}(0,t)]\} \left.\frac{\partial c}{\partial z}\right|_{z=0} + q(0,t)c(0,t) = 0 \tag{12d}$$

for any $t > 0$ during evaporation

$$\left.\frac{\partial c}{\partial z}\right|_{z=Z} = 0 \text{ for any } t > 0 \tag{12e}$$

where $C_o(t)$ is the salt concentration of the water applied $c_n(z)$ the prescribed initial salt concentration, $q(0,t)$ the volumetric water flux at the soil surface, and all the other symbols are the same as before. It should be noted that the volumetric water flux at the soil surface $[q(0,t)]$ is negative during the evaporation stage.

To estimate the salt distribution $c(z,t)$ by solving Eq. (8), subject to Eq. (12), the functions D_p (θ) and $D_h(V)$ must be known. However, one may assume for many practical purposes that, under transient conditions, the overall hydrodynamic dispersion term contributes very little compared with the macroscopic-average viscous-flow term (i.e., $D(\bar{V}, \theta)\ dc/dz \ll qc$). Applying this assumption to Eq. (8) and integrating the latter from the soil surface ($z = 0$) down to any depth (z), one has

$$\frac{\partial}{\partial t} \int_0^z [c(z,t)\,\theta\,(z,t)]dz = q(0,t)\,C_o(t) - q(z,t)\,c\,(z,t) \tag{13}$$

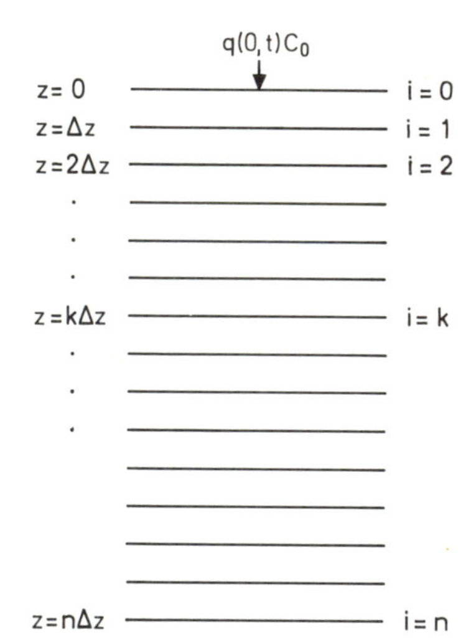

Fig. 1. Schematic division of soil profile into segments

This equation may be solved for $c(z, t)$ if $\theta\ (z, t)$ and $q(z, t)$ are known.
Eq. (13) can be expressed in finite difference from as

$$c^j_k = \{1/(\Delta z\,\theta^j_k + q^{j-1/2}_k\,\Delta t)\}\Big\{2\Delta t(qc)^{j-1/2}_o - 2\Delta z\,[1/2(c\theta)^j_o + \sum_{i=1}^{k-1} (c\theta)^j_i - 1/2(c\theta)^{j-1}_o - 1/2(c\theta)^{j-1}_k - \sum_{i=1}^{k-1} (c\theta)^{j-1}_i] - q^{j-1/2}_k\,c^{j-1}_k\,\Delta t\Big\} \tag{14}$$

where j is a time index defined by $t = j\Delta t$ (j–1/2 refers to the period Δt between $t = j\Delta t$ and $t = (j-1)\Delta t$), and i is an index of soil depth as illustrated schematically in Fig. 1.

If q_i^j, θ_i^j, the initial concentration c_i^{j-1}, and the concentration of the applied water c_o^j are known, Eq. (14) can be solved explicitly for c_k^j corresponding to arbitrary values of k and j. To do this, one starts with the first Δt when $j = 1$ ($t = \Delta t$) and with $k = 1$ ($z = \Delta z$) and solves for c_i^1. The procedure is repeated for $k = 2,3, \ldots, m$, then c_k^j is calculated for $j = 2$ and $k = 1,2, \ldots, n$, and so on for $j = 3,4, \ldots, J, k = 1,2 \ldots, n$, where J and n correspond to the desired time and depth.

Crop-growing Conditions

The assumption that salt uptake by plants may be neglected can be applied also in the presence of plants. However, water uptake by plants is a dominant factor that must be taken into account when salt distribution under crop-growing conditions is considered. Under such conditions, a macroscopic water extraction term should be added to the water flow equation to account for the loss of water due to transpiration. Since a satisfactory theoretical formulation of water extraction by plants is not yet available, one must often rely on empirical models. Introducing an empirical or theoretical extraction term into the water flow equation and solving it numerically, one can estimate the salt distribution in space and in time.

The degree of accuracy obtainable with the described methods in estimating soil salinity variables such as $c(z,t)$, $\theta(z,t)$, and $q(z,t)$, depends primarily on (a) the proper choice of a mathematical model to provide adequate description of the physical system, (b) the proper specification of boundary conditions, (c) the accuracy and stability of the numerical procedure, and (d) the accuracy of the soil and plant parameters used in the computation. Out of these factors, the accuracy of the soil and plant parameters is most critical.

Owing to difficulties in obtaining all the necessary soil water plant salt parameters that are required for these numerical methods, they are sometimes difficult to apply to real problems. In these cases, it is useful to consider a simplified numerical procedure, based on the mass-balance Eq. (13), where the time increment Δt is set equal to the time interval between two successive water applications (rain or irrigation). The method (BRESLER, 1967) is restricted to downward flow, that takes place in a limited range of water contents ($\overline{\theta}$), between saturation and an assumed "field capacity" of the soil. In addition, the amount of water passing any depth $z \neq 0$ during Δt, ($q^j(z)\Delta t^j$), is taken to be the difference between the amount of water applied ($Q_j = q^j \Delta t^j$) and the amount consumed by the crop and lost through evaporation from the soil surface down to the depth z during Δt^{j-1}. With these approximations, Eq. (13) becomes:

$$\int_0^z [c^j(z) - c^{j-1}(z)]\overline{\theta}(z)dz = Q^j C_o^j - \left[Q^j - \int_0^z E^{j-1}(z)dz\right] c^{j-1/2}(z) \tag{15}$$

where Q is the amount of water applied to the soil surface by irrigation or rain (cm) j is the index of water applied by rain or irrigation; and E is the amount of water per unit volume of soil ($cm^3\ H_2O/cm^3$ soil) consumed by the crop and lost through evaporation from the soil during the time interval Δt^{j-1}. If $\overline{\theta}(z)$ and $E(z)^{j-1}$ can be evaluated, and if the initial salt concentration $c_n(z)$, as well as the amount of water applied and its salt concentration are measured, Eq. (15) can be solved for the salt concentration. Values of $E^{j-1}(z)$ may be obtained from the large number of water-requirement experiments that have been conducted in many countries (particularly in Israel). If we denote $E_i^{j-1} =$

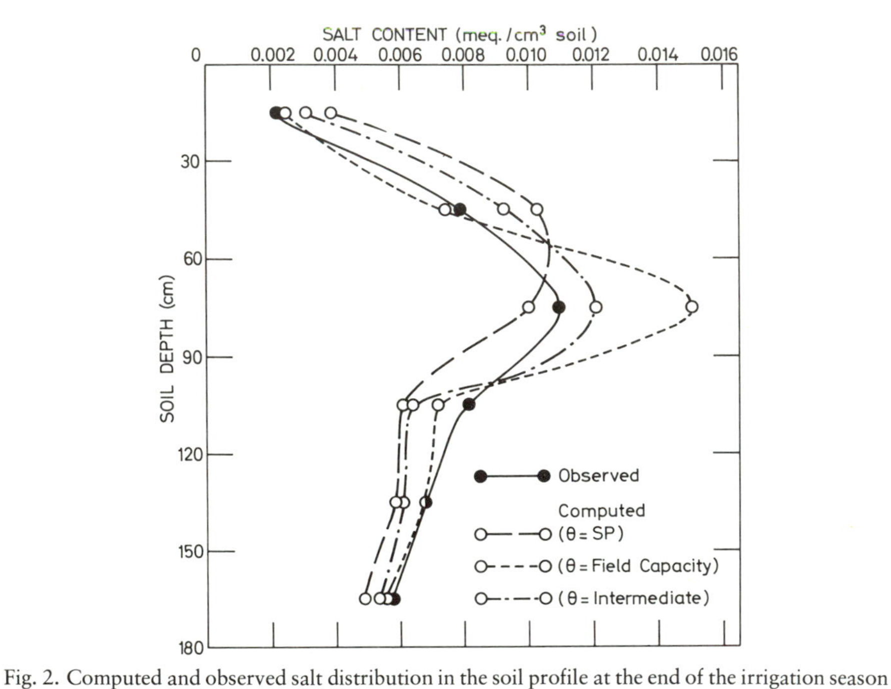

Fig. 2. Computed and observed salt distribution in the soil profile at the end of the irrigation season

$= E(z)^{j-1}$ as the water deficit in the ith increment before the jth rain or irrigation, then $q_k^{j1/2}\,\Delta t$ in Eq. (14) may be substituted for $q_o^j\Delta t^i - \left[1/2(E_o^{j-l} + E_k^{j-l}) + \sum_{i=1}^{k-1} E_i^{j-1}\right]$, so that now the solute distribution at any soil depth and after each water application may be calculated.

Table 1. Data from citrus irrigation experiment (conducted by H. BIELORAI at GILAT)

Soil Depth (cm)	Depth Index (i)	Initial Salt Concentration (c_i) (meq/1.) $\theta^a = 0.45$	$\theta^b = 0.36$	$\theta^c = 0.27$	Water Deficit (E_o^{j-1}) $J = 1$	2	3	4	5	6
0–30	1	3.7	4.0	5.4	35	42	55	43	50	48
30–60	2	7.2	8.7	12.8	30	37	51	39	47	42
60–90	3	11.1	14.2	20.7	6	8	12	16	16	15
90–120	4	14.1	17.6	27.0	1	2	4	5	4	5
120–150	5	13.0	16.4	25.5	0	1	2	3	4	2
150–180	6	10.2	14.4	20.6	0	1	1	2	1	0

Water applied (Q^j) (mm)						Salt concentration of irrigation water (C_o^j) (meq/1)					
j = 1	2	3	4	5	6	j = 1	2	3	4	5	6
69	87	118	98	113	105	7.5	7.5	7.5	7.6	7.6	6.7

[a] SP (saturation percentage). [b] Between SP and field capacity. [c] Field capacity.

Table 1 contains the data necessary for such a calculation. They were taken from a citrus irrigation experiment conducted at the Gilat Experimental Station in Israel by H. BIELORAI, where they were used to compute salt distribution in the soil profile after each irrigation. Results of the computation at the end of the irrigation season, after six irrigations, are presented in Fig. 2, which shows the effect of $\bar{\theta}(z)$ on salt distribution and on the way that the numerical results compare with measured data. D. YARON (1971)[1] compared several irrigation experiment data with calculations performed by the above method and found good agreement in many practical cases.

Note that the salt-content scale (Fig. 2) is in terms of milliequivalents of salt per unit volume of soil and not in terms of concentration in the soil solution. This is so to enable comparison between computations based on different assumptions as concern $\bar{\theta}$. If one is interested in getting salt distribution data in terms of meq./l of soil solution of saturated soil paste, the data of Fig. 2 are simply multiplated by $1000/\theta_s$, where θ_s is the volumetric water content of the saturated paste (SP).

Evaluating the Optimal Combination of Quantity and Quality of Irrigation Water

Knowing the solute concentration as a function of time (or number of irrigation cycles) and depth, and measuring the crop response pattern to some soil salinity index, it is possible to determine the most efficient combination of amount of irrigation water (Q) and its solute concentration (C_o). The crop response salinity index may be taken as the specific concentration of an ion, or as the electrical conductivity of the soil solution (or of extract obtained from the saturated soil paste).

Some investigators maintain that yield response to salinity is significant only above a certain critical threshold concentration. Below this threshold, the salinity effect is negligible. The associated critical salinity index is unique for any crop under given growing conditions.

Use of linear programming and computer simulation to estimate the efficient Q-C_o combination enables one to consider the effect of varying the critical threshold values in time and space (YARON and BRESLER, 1970). In this way, changes in crop response during different stages of growth can be taken into account. The problem may be formulated in many different ways. To illustrate, let (a) the amount of water applied be prescribed as $Q(t)$, and (b) the soil salinity index at any time and depth be restricted to remain below a specified threshold function, i.e., $c(z,t) \leq c_r(z,t)$. The problem is to find the maximum level of C_o that will not exceed the threshold values $C_r(z, t)$, subject to the finite difference equations in (15). This leads to a linear programming problem where the objective function is to be maximized, subject to a set of linear equality (finite difference equations) and inequality [restrictions on values of $c(z,t)$] constraints. By systematically varying the values of $Q(t)$, a set of Q_t-C_o combinations is obtained, maintaining the soil salinity index below the prescribed critical values. Another way to solve the same problem is by using a computer simulation model. Here the "best" combination of the man-controlled variables Q_t-C_o, subject to similar constraints, is found by an automated trial-and-error (YARON and BRESLER, 1970) procedure.

---

1 Personal communication

Any optimization process of the type considered here must be judged, finally, from the point of view of economic efficiency. It is therefore appropriate to conclude this section with an illustration of an economic analysis of the problem.

The economic problem is to find the least costly combination of water quality (C_o) and quantity (Q) ensuring that a given critical threshold salinity index (SS_i) is not exceeded. To further illustrate this problem, let the critical salinity index be represented by a relevant part of the isosoil salinity line, estimated from Eq. (15), say SS_1 in Fig. 3. Such a line may be obtained by any of the estimation methods described previously. For the analysis to be realistic, one must assume that both Q and C_o may be controlled, and that the cost per unit amount of water increases when the salt concentration decreases. To derive the optimal $Q - C_o$ combination for the system represented in Fig. 3, the total cost of water must be minimized. Consider the Lagrangian expression

$$L = QP_q + (C_s - C_o)P_cQ - \lambda[g(C_o, Q) - SS_1] \tag{16}$$

where L is the total cost per unit water, P_q is the cost per unit amount of water of standard salinity, C_s is the standard water salinity in terms of a salinity index (SS), p_c is the price paid when in one water unit C_o deviates from the prescribed standards value by one SS unit, and g (C_o,Q) is the Q-C_o function in Fig. 3. To minimize the Lagrangian in Eq. (16), each partial derivative of L with respect to Q, C_o, and λ should be equal to zero. Thus, one has

$$\begin{aligned} \frac{dC_o}{dQ} &= [P_q + (C_s - C_o) \cdot P_c]/P_cQ \\ SS_1 &= g(Q, C_o) \end{aligned} \tag{17}$$

Eq. (17) gives the necessary conditions for the desired least-cost quantity and quality (Q–C_o) combination. This least-cost combination is illustrated graphically in Fig. 3. The curve marked SS_1 represents threshold concentration at a certain $SS = 1$ level, and the

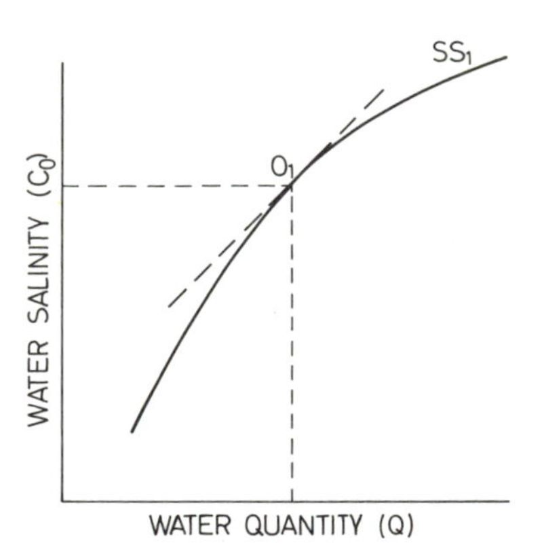

Fig. 3. Treshold isosoil salinity curve and determination of the optimal quantity (Q)–quality (C_o) combination (BRESLER, 1972)

dashed straight line represents the cost of quanity and quality ratio. The point O, at which the straight line in tangent to the isothreshold salinity line, represents the optimal Q-C_o combination. Obviously, an optimal combination can be derived in a similar manner for any given level of threshold conditions.

Literature

BIGGAR, J. W., NIELSON, D. R.: Miscible displacement and leaching phenomenon. In: Irrigation of Agricultural Lands. HAGAN, R. M., HAISE, H. R., EDMINSTER, T. W. (Eds.). Amer. Soc. Agron., Madison, Wis., pp. 254–274 (1967).

BRESLER, E.: A model for tracing salt distribution in the soil profile and estimating the efficient combination of water quality and quantity. Soil Sci. 104, 227–233 (1967).

BRESLER, E., HANKS, R. J.: Numerical method for estimating simultaneous flow of water and salt in unsaturated soils. Soil Sci. Soc. Amer. Proc. **33**, 827–832 (1969).

BRESLER, E.: Control of soil salinity. In: Opimizing the Soil Physical Environment Toward Greater Crop Yields. HILLEL, D. (Ed.), pp. 101–132. New York–London: Academic Press 1972.

BRESLER, E., YARON, D.: Soil water regime in economic evaluation of salinity in irrigation. Water Resources Res. 8, 791–800 (1972).

HANKS, R. J., KLUTE, A., BRESLER, E.: A numerical method for estimating infiltration, redistribution, drainage and evaporation of water from soil. Water Resourcess Res. **5**, 1064–1069 (1969).

OLSEN, S. R., KEMPER, W. D.: Movement of nutrients to plant roots. Adv. Agron. **20**, 91–151 (1968).

PERKINS, T. K., JOHNSON, O. C.: A review of diffusion and dispersion in porous media. Soc. Petroleum Eng. J. **3**, 70–84 (1963).

YARON, D., BRESLER, E.: A model for the economic evaluation of water quality in irrigation. Aust. J. Agric. Economics **14**, 53–62 (1970).

3

Nutrient Supply to Irrigated Crops

U. KAFKAFI

Plants receive their nutrients in the form of various sources in the soil, as well as from chemicals supplied to the crop directly in the form of fertilizers and manure. The choice of a specific fertilizer, and the amount to be used in a given case of irrigated agriculture, depend on the kind of crop, the state of plant development, the time of application, the kind of soil and nutrient residues in the soils, the price of unit nutrient in the field, and the irrigation technology.

Although investments in fertilizers represent less than 10% of the overall cost of irrigated crop production, misuse of fertilizers may lead to a decreased yield and a total failure of the irrigation effort. For example, in a crop like cotton, where the fertilizer investment represents only 6–8% of the total production cost, excess use of nitrogen fertilizers will promote vegetative growth at the expense of lint.

In general, farmers should be able to control their yields under field conditions by knowledgeably regulating the amount and timing of irrigation, as well as the fertilization.

Nitrogen Fertilizers

Most organic manures and chemical fertilizers contain nitrogen in the form of ammonium. Nitrogen fertilizers containing only nitrate are usually more expensive. In the soil, nitrogen is continuously consumed and excreted by soil organisms and plants

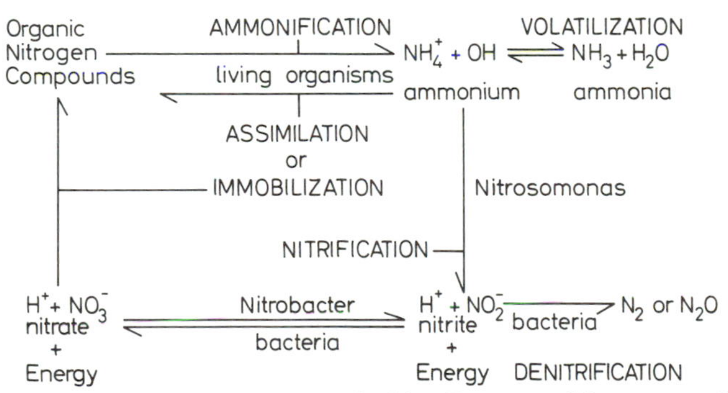

Fig. 1. Pathways and products of nitrogen in soil. (After FREDRIC and BROADBENT, 1966. Biological interactions pp: 198. Agricultural anhydrous ammonia, technology and use. Edited by M. H. MCVICKAR, W. P. MARTIN, I. E. MILES and H. H. TUCKER. Published by A. A. I., Memphis, Tenn. American Society of Agronomy, Madison, Wis. Soil Science Society of Amer. Madison, Wis.)

(Fig. 1). Most of the ammonium is held in the form of exchangeable ions on the clay and organic material. In dry soils with a pH above 7.5, some may be lost as ammonia gas by volatilization.

Nitrate is the product of nitrification reactions, and all of it is available for plant uptake. In the plant, it transforms into organic nitrogen compounds. It is readily transported in the soil. Irrigation may cause either leaching of nitrates below the root zone, or denitrification when a shortage of oxygen occurs due to flooding or sprinkling.

Nitrogen Mineralization

The term "nitrogen mineralization" is used to describe the transformation of soil nitrogen by microorganisms from organic to mineral form. Irrigation may influence the rates of mineralization. As an example, Table 1 shows the influence of soil moisture on the rate of mineralization during the first 10 days of incubation.

Table 1. Mineralization rate as a function of soil moisture during ten days' incubation. (After ROBINSON, 1957)

Moisture level, %	10.5	17.5	21.0[a]	25.0	42.0[b]
Mineralization rate:					
(1) μg N/g/day	0.2	0.5	0.9	1.0	1.4
(2) Kg N/ha/day in 0–20-cm layer	0.5	1.0	1.8	2.5	2.8

[a] Permanent wilting percentage.
[b] Field capacity.

In a field experiment a rate of nitrogen uptake by corn of 1.4 kg N/ha/day was found (KAFKAFI and BAR-YOSEF, 1971). The rate of mineralization of the soil mentioned in Table 1 is such that it alone could account for the entire demand of a corn field producing 80 ton/ha of forage corn. However, worldwide observations of yield response to the addition of nitrogen fertilizers suggest that the demands of high production cannot be satisfied only by nitrogen mineralization.

Nitrification and Denitrification

Nitrification means the oxidation of ammonium to nitrate by microorganisms. The reverse process, whereby nitrate is reduced to molecular nitrogen or ammonium, is known as denitrification. One of the main factors influencing the direction and magnitude of these oxidation reduction processes is the soil-moisture content (BLACK, 1968).

Although there is practically no nitrification in air-dry soils, it may take place at wilting point. The largest rate of nitrification occurs when the moisture content corresponds to a suction of 0.5 bar. With further increase or decrease in the rate of matric suction, nitrification decreases (MILLER and JOHNSON, 1964). The rate is slower at low oxygen concentrations, such as those occurring under certain irrigation conditions. GREENWOOD (1962) found that the presence of 3 micromoles of elemental oxygen in the water slows down the nitrification process to 50% of that found in the presence of water in equilibrium with air oxygen.

Most of the published data on rates and amounts of nitrifications and denitrification were obtained in the laboratory, in the absence of a growing plant. The effect of root excretions on denitrification was demonstrated by WOLDENDORP (1963); Fig. 2. The roots serve as a source of assimilable material such as sugar and amino acids. When there is an oxygen shortage and high activity of microorganisms in the rhizosphere, nitrate is used instead of oxygen as a hydrogen acceptor. Irrigation of a standing crop, during hot periods, can produce conditions favorable for denitrification. In a field experiment with irrigated corn on a montmorillonitic clay soil (KAFKAFI and BAR-YOSEF, 1971), 330 kg NO_3–N/ha were lost within 35 days; most of this was lost during the first two weeks following 68 mm of sprinkler irrigation (Table 2).

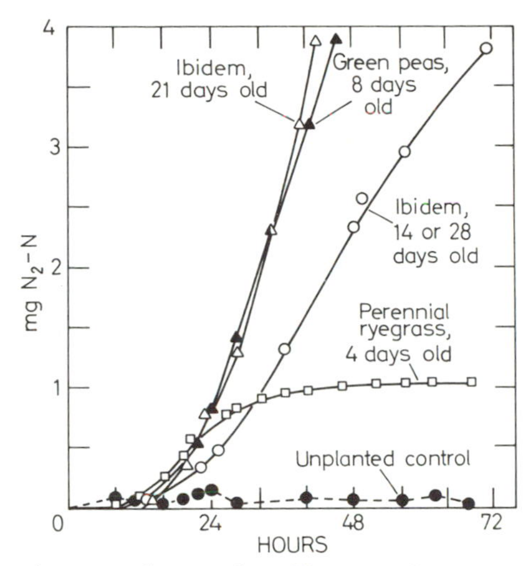

Fig. 2. Effect of growing plants on denitrification. (After WOLDENDORP, 1963)

Table 2. Nitrogen amounts in the soil and uptake by plant in irrigated cornfield. (After BAR-YOSEF and KAFKAFI, 1971)

Amount of Fertilizer (kg N/ha)	Amount of N at seeding (kg N/ha in 0–40 layer)		Amount of N 35 days later (kg N/ha in 0–40 layer)		Plant uptake (kg N/ha in 35 days)
	NO_3^-–N	NH_4–N and organic N	NO_3–N	NH_4–N and organic N	
0	50	3400	35	3210	26
400	480	3900	150	3620	36

On the 24th day, 68 mm of water were applied.

Nitrate Leaching

Irrigation influences the movement of nitrates in the soil. Water movement can be predicted and nitrate movement calculated in the same way as was done for chlorides (see p. 166). Agreement with calculation of nitrate downward displacement is expected during cold winter periods or on fallow soils when low activity of soil microorganisms is anticipated. The general pattern of nitrate leaching by percolating water is identical to that of chloride. However, because of microbial attack, part of the nitrate may be lost during leaching. Denitrification has been demonstrated to occur during leaching of soil columns (COREY *et al.*, 1967). The addition of sugar to the percolating water may increase nitrate losses by denitrification, the sugar being the energy source for the microorganisms conducting denitrification. Under field conditions, organic matter and root exudates such as sugars, carbohydrates, and amino acids are the energy sources. At soil temperatures below 13° C, the rate of microbial activity is low, and irrigation may favor the leaching of nitrates. When the soil temperatures are around 30° C, microbial activity is at its peak, and therefore more denitrification and less nitrate leaching may be expected. Summer temperature favors rapid nitrification. Any top dressing of ammonium fertilizer will convert to nitrate within a week, and would be leachable in the next irrigation.

A well-planned irrigation must assure that the water reaches the depths of the root zone. Under such conditions, the hazards of nitrate losses by leaching are minimized.

Since it is difficult to predict the rate of denitrification, the depth of nitrate movement with irrigation water can only be calculated approximately. Such a calculation is of practical importance, since it indicates whether one has to add top dressing of nitrogen fertilizer. LEVIN (1964) calculated the downward leaching of nitrate that had been spread on the soil surface for a soil initially at field capacity, using the following equation:

$$d = \frac{Q}{Pv}$$

where d is depth where maximum nitrate concentration is expected in soil, in mm; Q is the quantity of irrigation, mm; and Pv is volume fraction of water at field capacity.

Potassium

Potassium in the soil is commonly divided into three phases (BLACK, 1968): (1) in feldspars, micas, and illite clay minerals (fixed K); (2) at exchange sites on surfaces of clay minerals (exchangeable K); and (3) in the soil solution (dissolved K). Phases 2 and 3 can radily be taken up by plants.

When one considers a time scale of weeks in an irrigated field, there is a reduction in potassium concentration of the soil solution and of the exchangeable fraction as a rescult of crop removal. The long-term effect of moisture becomes significant when the crop is removed. Potassium diffuses from its position inside the clay minerals and regenerates the exchangeable and soluble pool (BLACK, 1968).

The relationship between potassium and bivalent cations in the soil solution under equilibrium is defined as (WOODRUFF, 1955):

$$\Delta F = RT \times 2.303 \log \frac{(K)}{(\mathrm{Ca} + \mathrm{Mg})^{1/2}}$$

where R is the gas constant in cal/mol = 1.99; T is absolute temperature; and K and Ca + Mg are the activities (or, approximately, the concentrations) of the cations in the soil solution of a saturation extract, mol.

Similar experimental data, but without constants, were used (U.S. Salinity Lab. Staff, 1954) to determine the potassium adsorption ratio (PAR) according to the equation:

$$\mathrm{PAR} = \frac{\mathrm{K}}{\sqrt{\frac{\mathrm{Ca} + \mathrm{Mg}}{2}}}$$

where the concentrations of K, Ca, and Mg are expressed in milliequivalents per liter.

Figure 3 shows a nomogram of ΔF and PAR values as functions of cation concentration in the soil solution. This nomogram is useful in the absence of gypsum because Ca then enters the soil solution independently of the exchange complex. By plotting a straight line between the concentrations of K and Ca + Mg, one obtains the corresponding ΔF or PAR values. In a particular experiment with irrigated cotton (HALEVY, 1966),

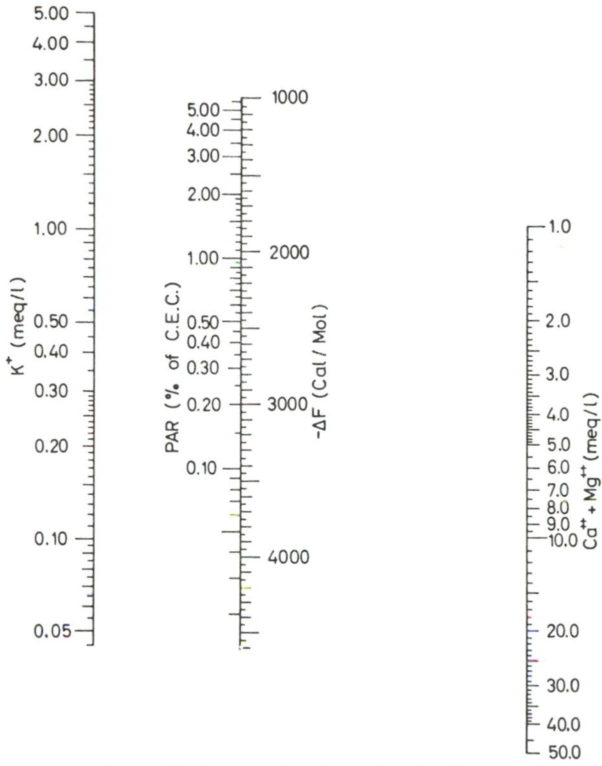

Fig. 3. A nomogram to estimate PAR and ΔF values of soils. (Prepared by Div. of Field Service, Extension Service, Ministry of Agriculture, Israel)

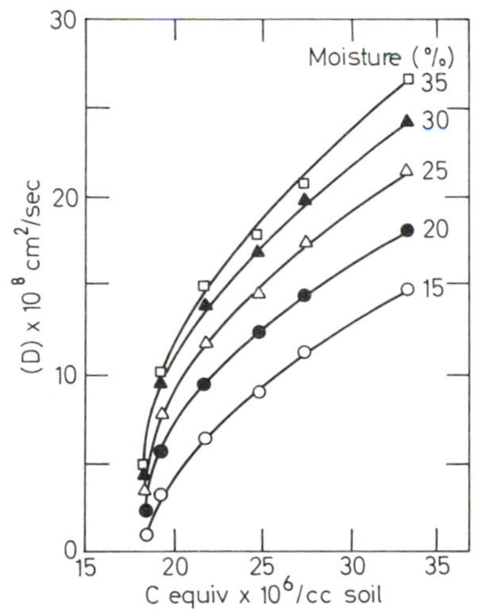

Fig. 4. Effect of soil-moisture content on the diffusion coefficient of potassium in soil. (After VAIDYANATHAN and NYE, 1968)

a substantial reduction in yield was observed near ΔF of about – 4000 cal/mol deg. This value corresponds to a PAR of about 0.04 in the exchange complex. A decrease in this value is associated with a shortage of potassium. Values obtained from the nomogram in Fig. 3 are useful in estimating the need for potassium fertilization of irrigated soils.

The equilibrium concentration of potassium is merely an index for the availability of potassium. BARBER *et al.* (1962) estimated that only about 9% of the potassium inside the plant can be attributed to uptake by mass flow of the water entering the root as a result of transpiration. The remaining 91% must reach the root, and enter the plant, by some other mechanism (probably diffusion). The effect of soil-moisture content on the diffusion coefficient of potassium in a soil, for various values of potassium concentration, is shown in Fig. 4. However, the question of how this affects the amount of potassium taken up by the plant still remains unanswered.

Leaching of Potassium

The amount of potassium leached out of the soil under irrigation depends on the cation-exchange capacity of the soil, as well as on the cation concentration of the irrigation water. A quantitative prediction of the depth to which potassium moves is important for nontillage agriculture, such as in irrigated orchards. Under normal irrigation and fertilization conditions, the downward migration of potassium is restricted to a few centimeters per year. Mechanical mixing of the soil by tillage ensures that potassium is present in the active root zone. Unlike nitrate and chloride, potassium lags behind the wetting front. No appreciable losses of potassium are likely to occur due to leaching in irrigated clay soils with high cation exchange capacity.

Phosphorus

Most of the phosphorus in the soil occurs in the solid matrix, in the form of slightly soluble calcium phosphate, or as adsorbed material on the clay particles. In general, all of these forms are in equilibrium with the soil solution. However, unlike potassium, concentration of phosphate in the solution does not change significantly when the moisture content is lowered. This is so as long as there are open sites available for phosphate adsorption, as has been demonstrated on a drying kaolinite-phosphate system (BAR-YOSEF *et al.*, 1969).

When phosphate is applied in heavy doses (KAFKAFI *et al.*, 1968), the concentration of phosphorus in the soil solution increases for a short period, and then falls off due to precipitation and slow reactions with the soil particles. The latter reactions involve aluminium from the broken edges of clay particles, and are probably enhanced by phosphate replacing silica on the clay particles (KAFKAFI *et al.*, 1970). The leaching of clay with water causes heavy losses of silica, thereby opening new sites for phosphate adsorption and bringing about an increased affinity of phosphate to clay. Since phosphate is highly adsorptive, there is little loss due to leaching. The main loss of phosphorus is due to erosion of the soil surface (BLACK, 1968).

The equilibrium concentration of phosphate in the soil solution is low. It has been estimated (BARBER *et al.*, 1962) that only 3% of the total amount of phosphorus in a corn plant could be attributed to mass flow. The rest must reach the plant root by diffusion. However, a unit surface of corn root does not participate in phosphate adsorption for more than a week (BAR-YOSEF, 1971). Any shortage of water, caused by insufficient irrigation, may reduce root propagation and decrease the amount of phosphate moving toward the root.

Due to its limited mobility, most of the phosphorus is concentrated in the plow layer. One expects that since irrigation prevents drying of the upper soil layer, it should ensure proper conditions for the maximum utilization of soil and fertilizer phosphorus.

Excess amounts of phosphatic fertilizers are less harmful to crops than excess amounts of nitrogen. However, to reduce the cost of crop production, a fertilizatiòn advisory service can operate successfully by testing the soil for phosphate. An empirical method developed by OLSEN *et al.* (1954) in the U.S. has been found applicable to calcareous soils in other parts of the world, such as Australia (COLWELL, 1963) and Israel (SADAN *et al.*, 1969). The numerical values obtained by this method, and their relationship to crop response, varies according to the soil type and the kind of crop, and therefore must be calibrated in each case (KAFKAFI *et al.*, 1968). OLSEN (1954) gives a scale according to which 0–5 ppm = low P; 5–10 ppm = medium P; and above 10 ppm = high P.

Effect of Fertilization on the Use of Water by Plants

The relationships between crop yield, water supply, and soil fertility were reviewed by C. A. BLACK (1965) and F. G. VIETS (1962).

The biological response of a plant to any given growth factor is nearly linear, provided that this factor is the only one that is in short supply and is therefore yield-determining (MCCANTS and BLACK, 1957). The results presented on p. 366 demonstrate such a response to water under field conditions. When no nitrogen was applied, the corn plant transpired the added water without producing a yield over 2500 kg/ha. When 200 kg N/ha were applied, the yield was 8500 kg/ha kernel, and the response to water was linear between 400 and 700 mm of consumptive use.

Table 3. Effect of fertilization composition on transpiration ratio[a] of wheat (var. N-652) in Bet Dagan. Total rainfall 32.7 cm. (Calculated from results of the Permanent Plot Team, 1968)

	Fertilizer Combinations (kg/ha)								
	N_0P_0	N_0P_{48}	N_0P_{96}	$N_{120}P_0$	$N_{120}P_{48}$	$N_{120}P_{96}$	$N_{240}P_0$	$N_{240}P_{48}$	$N_{240}P_{96}$
Grain, kg/ha	850	740	770	1490	1870	1930	1470	1700	1890
Straw, kg/ha	880	740	800	2470	3040	3450	2760	3630	4130
Total, kg/ha	1730	1480	1570	3960	4860	5380	4430	5330	6020
Transpiration ratio	1560	1820	1710	682	555	500	610	510	450[b]

[a] Evaporation was calculated from meteorological data to be 5.7 cm. Total amount of available water for transpiration was 27.0.

[b] Close to values reported by BRIGGS and SHANTZ (1914).

The total yield of dry matter is important in evaluating crop production. For small grains and vegetable crops, only the marketable product is important in the economic evaluation of water. The values of the ratios mentioned above must be regarded with reservations, since the production of dry matter is not only a function of water amount. Changes in nutrient levels in the soil may significantly affect the values presented in Table 3.

This table demonstrates that under a given amount of irrigation water, the transpiration ratio under field conditions may change by a factor of three as a result of fertilization. In cases where there is ample amount of water, as in hydroponics, potassium and nitrogen fertilizers have a detrimental effect on the yields of tomatoes, but less so on their transpiration ratio. The results of such experiments are summarized in Table 4. In these experiments, the total amount of nutrients (except calcium) for each level of nutrition is the same. Three potassium sources are compared.

Table 4. Effect of nutrition ions on dry-matter production, transpiration ratio, and water requirement for fresh fruit of a single tomato plant grown in circulating nutrient solution. (U. KAFKAFI–first published results)

Nitrogen fertilizer Potassium fertilizer	NH_4NO_3 K_2SO_4		KNO_3 KNO_3		NH_4NO_3 KCl	
Fertility level	1	2	1	2	1	2
Conductivity, mmhos/cmat 25° C	1.8	3.3	1.6	3.1	2.5	4.6
Dry-matter yield, g/plant	169	127	252	346	180	62
Transpiration ratio	259	281	224	213	270	266
Water transpired/fresh fruits g/g	27	29	25	20	35	24

When only nitrate is present in the solution, increasing the conductivity to 3.1 mmhos by raising the concentration of KNO_3 salt caused an increase in the production of dry matter and a decrease in the transpiration ration. At a given salinity level, when ammonium concentration is equal to that of nitrate, the yield of dry matter using potassium sulfate is equal to that using potassium chloride at the low salinity level, and is twice that using potassium chloride at the high salinity level.

The "transpiration ratios" in this experiment are all below 300, as opposed to field conditions where, for wheat, they exceed 450. These relatively low values indicate a very high efficiency of water use by the plants (RUSSELL, 1950). Clearly, the proper choice of a nitrogen fertilizer may increase the efficiency of water use.

Practical Problems

In dealing with a particular irrigation and fertilization problem, one must first estimate all the pertinent soil and plant parameters. A practical example will illustrate the ways one may take to reach a decision.

The problem: What fertilization and irrigation practice should be followed for silage corn on a clay soil?

Data needed: Soil information; plant data such as rate of nutrient consumption by the plant and estimation of transpiration ratio and rate of evaporation.

a) *Soil data*	*Laboratory data*
1. Depth of sampling	40 cm
2. Bicarbonate soluble P	16 ppm
3. Potassium adsorption ratio (PAR)	0.06
4. Mineral nitrogen(NH_4–N + NO_3–N)	100 kg/ha
5. Organic N	0.06%
6. Field capacity by volume	40%
7. Wilting point: 1/2 field capacity (according to U.S. Salinity Lab. Staff)	20%

8. Bulk density		1.2
9. Mineralization rate		2%/year
10. Average soil temperature		22° C

b) *Plant data*		*Values known*		
1. Age of plants (days)		30	60	90
	N	4.0	2.0	1.2
2. Nutrient content (%)	P	0.45	0.22	0.19
	K	6.0	3.5	1.6
3. Rate of dry-matter production of good yields (kg/ha / day)		78	215	215
4. Consumption of nitrogen		Monthly nutrient consumption[1] kg/ha/month		
		93.6	129	77.6
5. Transpiration ratio		400	400	400
6. *Evaporation rat from the soil* (mm/day)		2	2	2

Assumptions: Transpiration ratio and evaporation rate remain constant; (2) No addition of micronutrients is necessary; (3) The upper 40 cm of the soil is allowed to dry to within 75% of the available water before irrigation starts.

1 The monthly nutrient consumption per hectare is obtained by the following calculation:

$$\frac{\text{Nutrient content (\%)}}{100} \times \text{rate of dry-matter production (kg/ha/day)} \times 30 \text{ (days per month)}.$$

Calculations and considerations

Soil data:

As a first approximation based on experience (see values given above), one may assume that, in this case, there will not be any response to potassium or phosphate fertilization, due to the large amounts already present in the soil, as indicated by the results of the soil-extracting method. Soil nitrogen data show that 100 kg N/ha are in a readily available (mineral) form. The crop is capable of utilizing at least 300 kg N/ha^{-1}. Can the organic soil supply the rest of the plant demands?

Amount of organic N in upper 40-cm soil layer = weight of the soil layer × total nitrogen concentration / 100 = bulk density × depth area × % N/100.

4,8000,000 × 0.006 = 28,800 kg N/ha

Amount of nitrogen that soil can supply during plant growth by mineralization = amount of organic N × annual mineralization rates × time fraction of the year occupied by crop =

$$28{,}800 \times 0.02 \times \frac{90}{365} = 142 \text{ kg N/ha}$$

Total potential soil supply of N = 142 + 100 = 242 kg N/ha.

Conclusion. More nitrogen fertilization is needed. What remains to be answered is: How much, when, and what kind of fertilizer should be used? Assuming a constant mineralization rate, the soil is capable of supplying 47 kg N/ha/month from the organic source. The mineral soil nitrogen is completely available. Knowing the possible growth rate and the desired nitrogen concentration in the plant, it is seen that nitrogen consumption needed during the first 60 days of growth is 270 kg N/ha. At the same time, the soil can supply only about 200 kg N/ha. In the last month, 30 kg/ha is needed to supply nitrogen demands. The above considerations assumed that the plant can utilize all of the nitrogen that is present in ammonium and nitrate forms.

Experience indicates that corn plants suffer from nitrogen deficiency whenever the concentration of NO_3–N is below 10 ppm N, on a dry-soil basis. This amount is equal to 48 N/ha in the upper 40 cm of the soil.

Keeping the concentration of NO_3–N to above this value, it can be assumed that the mineralization rate of soil organic N is enough only to maintain a basic N level and the remaining 2/3 of the plant demand in this case should be supplied by N fertilizers.

Conclusions: Apply 200 kg N/ha.

When to apply?

Soil and plant data considerations. The initial amount of mineral nitrogen present in the soil (100 kg N/ha) is sufficient for the first month of crop demand.

Conclusions: Apply N fertilizer just before the end of the first month.

The second month of growth requires the highest amount of N (130 kg N/ha^{-1}). The conclusion would be to apply all 200 kg N before the end of the first month.

What kind of fertilizer should be applied? The considerations here involve combinations of nitrogen in the form of ammonium or nitrate, nitrification rate, amount and timing of irrigation water, and cost of fertilizer.

Case 1: Ammonium-type fertilizer

Method of application: Three possible ways exist when the plants are only one month old. (1) surface application; (2) subsurface application with soil cultivator; and (3) application with irrigation water.

In all of these methods, the ammonium stays in the upper few cm of the soil that are subject to drying. Assuming a rate of nitrification of 10 kg N/ha/day, the total amount applied will be converted to nitrate, under optimum conditions, within 20 days. If the soil becomes dry, more time will be required. Therefore, the moisture in the upper layer must be frequently controlled. One must remember that ammonium fertilizer are subject to volatilization on dry basic soils. Therefore, it is essential to irrigate immediately after fertilizer application.

Case 2: Nitrate-type fertilizer

Can be applied by all the above-mentioned methods. No volatilization expected. Depth of penetration depends on amount of irrigation. Timing of irrigation: In every case, top dressing must be immediately followed by irrigation. Since top dressing is calculated to take place about 28 days after seeding, the amount of water needed at this time can be calculated as follows:

Irrigation considerations. Assumption: Irrigation given immediately after seeding and the soil is wet down to a depth of 60 cm.

$$\text{Evaporation losses} = \text{evaporation rate} \times \text{time} = 2 \times 28 = 56\ \text{mm}$$

$$\text{Crop demands} = \text{dry-matter rate of production} \times \text{time} \times \text{transpiration ratio} =$$

$$78 \times 28 \times 400 = 8.7 \times 10^5\ \text{kg/ha} = 87\ \text{mm}$$

$$\text{Total water loss expected} = 56 + 87 = 143\ \text{mm or } 5.3\ \text{mm/day}$$

The total available water in the upper 40 cm, according to the above assumptions, is equal to 15% by volume (field capacity – lower permissible moisture), i.e.,

$$40 \times 0.15 = 6\ \text{cm} = 60\ \text{mm}$$

At a loss rate of 5.3 mm/ha/day, the amount of 60 mm available water will be depleted in 12 days (60/5.3). This means that irrigation is required every 12 days during the first month. If each irrigation supplies 60 mm, the *maximum possible* downward displacement of nitrate due to each irrigation in such a clay soil is 12 cm (according to Levin's equation).

Should one choose ammonium or nitrate fertilizer? When frequent irrigation occurs, and the rate of nitrification is high, the benefit from a nitrate fertilizer is no greater than that from ammonium.

Economic considerations. These depend on current prices and may change in time and place. One must estimate the extra yield obtained from one kind of a fertilizer as compared with another. The fertilizer expected to give the highest return per unit of fertilizer is the one to use. When the form of the nitrogen in the fertilizer is irrelevant, the obvious choice is the cheapest fertilizer available.

When no nitrification can take place (e.g., hydroponics or crops after soil fumigation), or when quality of the product is the main feature (e.g., no chlorine in tobacco), and the plant is known to prefer nitrate over ammonium (e.g., tomato), one has no other choice despite the higher price of nitrate fertilizer.

Further irrigation considerations. The consideration for irrigation demands are discussed by Bielorai (Chap. 8.1). From a plant-nutrition point of view, the upper 40 cm of the soil should be kept moist during the main uptake period; e.g., until heading in silage corn. During the rest of the growing period, water can come from deeper layers without causing a decline in yields, provided that the root system has penetrated to deep layers. Drying the upper 50-cm layer near harvest time is especially important if heavy machinery is to be utilized.

Taking into consideration all the given soil and plant data, one may formulate the following irrigation and fertilization schemes:

Amount of water allowed to be depleted at the end of the growing season = total available water in the upper 50 cm $= (0.40-0.20) \times 50 = 10$ cm $= 100$ mm

$$\text{Time of consumption of 100 mm} = \frac{\text{amount of water}}{\text{transpiration ration} \times \text{rate of dry matter production}} = \frac{100 \times 10^4}{400 \times 215} = 12 \text{ days}$$

Therefore, the last irrigation can be given 12 days before harvest. The considerations for the first month were shown earlier. During the second month, the possible growth rate is 215 kg day matter/ha/day. With transpiration ratio of 400, the water demand is

$$215 \times 30 \times 400 = 89{,}000 \text{ kg water/ha} = 8.9 \text{ mm/day}$$

Allowing the soil to dry only down to 60 mm, and assuming that transpiration alone is responsible for water loss, irrigation every 7 days is required,(60/8.0). The irrigation fertilization scheme is

Days from sowing	Irrigation (mm)	Water front Depth (cm)	Fertilization
–30	–		Soil test for N, P, K.
–20	–		Add P and K if soil test low; prepare seed bed
–10	120	60	
0	20	60	
12	60	60	
24	60	60	Apply 200 kg N/ha
36	60	60	As ammonium salt
43	60	60	
50	60	60	
57	60	60	
60	60	60	
70	60	60	
77	60	60	
90 harvest	60	60	
Total	680		Total potential dry-matter production (98 + 215 + 215) = 16 ton/ha

Mantell (see p. 388) quotes results where 18 ton/ha of corn were obtained with only 450 mm (including preplanting) of what is called "highly fertile field." This means a higher growth rate and lower transpiration ratio than those assumed in our example. A value of 400 for the transpiration ratio indicates insufficient efficiency; deficiency of nutrients is suspected. An average value of 200 under field conditions for the total growth period is a very good value, indicating that no nutrient or water shortage is suspected. If the last water-use efficiency is assumed, the total amount of water used would be cut almost in half.

Trickle irrigation, or a fixed sprinkler system, allows frequent irrigation, which enables drying of only the upper 20 cm of the soil. Theoretically, such frequent irrigation results in optimum water conditions for plants, approaching the conditions of hydroponics. In practice, the upper soil should be allowed to dry, to enable weed control in the first month. Too frequent irrigation can also cause unaerobic conditions. The practice to be followed in each case needs special consideration.

Literature

BARBER, S. A., WALKER, J. M., VASEY, E. H.: Principles of ion movement through the soil to the plant root. Internl. Soc. Soil Sci. Trans. Joint Meeting Comm. IV: **80,** (New Zealand 1962) 121–124 (1962).

BAR-YOSEF, B.: Fluxes of Ca and P into intact corn roots and their dependence on solution concentration and root age. Pl. Soil **35,** 589–600 (1971).

BAR-YOSEF, B., KAFKAFI, U., LAHAV, N.: Relationships among absorbed phosphate, silica and hydroxyl during drying and rewetting of kaolinite suspension. Proc. Soil Sci. Soc. Amer. **33,** 672–677 (1969).

BLACK, C. A.: Crop yields in relation to water supply and soil fertility. In: Plant Environment and Efficient Water Use. PIERRE, W. H., KIRKHAM, D., PESEK, J., SHAW, R. (Eds.). Amer. Soc. Agron., Madison, Wis., pp. 177–206 (1965).

BLACK, C. A.: Soil Plant Relationships (2nd ed.). New York: Wiley 1968.

BRIGGS, L. J., SHANTZ, H. L.: Relative water requirement of plants. J. Agr. Res. **3,** 1–65 (1914).

COLWELL, J. D.: The estimation of the phosphorus fertilizer requirements of wheat in southern New South Wales by soil analysis. Aust. J. Expl. Agric. Anim. Husb. **3,** 110–113 (1963).

COREY, C. Y., NIELSEN, D. R., KIRKHAM, D.: Miscible displacement of nitrate through soil columns. Soil Sci. Soc. Amer. Proc. **31,** 497–501 (1967).

GREENWOOD, D. J.: Nitrification and nitrate dissimilation in soil. II. Effect of oxygen concentration. Pl. Soil **17,** 378–391 (1962).

HALEVY, J.: The relationship between potassium fertilizer and leaf browning in corn. Potash Rev. 23–33 (1966).

KAFKAFI, U., BAR-YOSEF, B.: Long-term effects of cumulative application of fertilizers on irrigated soil and crops. Project A-III/6, Final Report. Volcani Inst. Agric. Res., Bet Dagan, Israel (1971).

KAFKAFI, U., GISKIN, M., HAGIN, J.: Phosphate and silica adsorption and desorption from soils. Israel J. Chem. **8,** 373–381 (1970).

KAFKAFI, U., HADAS, A., HAGIN, J.: The effect of soil surface area and time of contact on the extraction of phosphate from superphosphate by soil test and test plant. Agrochimica **12,** 231–238 (1968).

LEVIN, I.: Movement of added nitrates through soil columns and undisterbed soil profiles. Trans. Intern. Cong. Soil Science. IV **4,** 1011–1021 (1964) (Bucharest).

MCCANTS, C. B., BLACK, C. A.: A biological slope ratio method for evaluating nutrient availability in soils. Soil Sci. Soc. Amer. Proc. **21,** 296–301 (1957).

MILLER, R. D., JOHNSON, D. D.: The effect of soil moisture tension on carbon dioxide evolution, nitrification and nitrogen mineralization. Soil Sci. Soc. Amer. Proc. **28,** 644–647 (1964).

OLSEN, S. R., COLE, C. V., WATANABE, F. S., DEAN, L. A.: Estimation of available phosphorus in soils by extraction with sodium bicarbonate. USDA Circ. 939 (1954).

ROBINSON, J. B. D.: The critical relationship between soil moisture content in the region of wilting point and the mineralization of natural soil nitrogen. J. Agric. Sci. **49,** 100–105 (1957).

RUSSEL, E. J.: Soil conditions and plant growth. London: Longmans, Green 1950.

SADAN, D., MACHOL, D., KAFKAFI, U.: Soil tests for fertilizer recommendations under extensive conditions. In: Transition from Extensive to Intensive Agriculture with Fertilizers. Proc. VII Colloquium of IPI in Israel (1969).

U.S. Salinity Lab. Staff: Diagnosis and improvement of saline and alkali soils. USDA Handbook 60 (1969).

VAIDYANATHAN, L. V., DREW, M. C., NYE, P. H.: The measurement and mechanism of ion diffusion in soils. IV. The concentration dependence of diffusion coefficient of potassium in soils at a range of moisture contents and a method for the estimation of the differential diffusion coefficient at any concentration. J. Soil Sci. **19,** 94–107 (1968).

VIETS, F. G. Jr.: Fertilizers and the efficient use of water. Adv. Agron. **14,** 223–261 (1962).

WALDENDORP, J. W.: The influence of living plants on denitrification, **63** (13), 1–100 (1963). Mededelingen van de Landbouwhogeschool, Wageningen.

WOODRUFF, C. M.: The energies of replacement of calcium by potassium in soils. Proc. Soil Sci. Soc. Amer. **19,** 167–169 (1955).

4

Electrolytes and Soil Hydraulic Conductivity

B. YARON and I. SHAINBERG

Hydraulic conductivity is one of the main factors to be considered in arid-zone irrigation planning and management, as it depends on the physicochemical composition of both the soil and the permeating liquid.

Hydraulic conductivity is usually expressed as

$$K = \frac{k \varrho g}{\eta} \tag{1}$$

where ϱ is the fluid density, g is gravity, η is the fluid viscosity, and k is intrinsic permeability, which depends on the size and shape of the soil pores. A relationship between permeability and the properties of the porous medium is given by the well-known Kozeny-Carman equation:

$$k = \frac{\theta^3}{mt^2 S_o^2 (1-\theta)^2} \tag{2}$$

where θ is porosity, m is the pore shape factor (~ 2.5), t is tortuosity ($\simeq(2)^{1/2}$), and S_o is the specific area per unit volume of particles.

Large discrepancies between measured flow and flow predicted by the Darcy and Kozeny Carman equations were observed. One possible reason for these discrepancies is the fact that both of these equations neglect the chemical composition of the adsorbed phase and of the soil solution. The composition and concentration of the salts in irrigation water may produce significant changes in the chemical composition of the soil solution. This, in turn, affects the composition of the adsorbed phase, and swelling and dispersion of the clay material produces changes in the geometry of the pores, thereby affecting the hydraulic conductivity.

The present chapter deals with the dynamic relationship between those physical properties of the soil and its colloidal material that have a bearing on hydraulic conductivity, and the electrolytes present in the permeating water. Changes in the swelling and dispersion properties of clay material, as affected by the presence of electrolytes, are discussed in light of the pore size and volume dynamics. The theory is first described for pure clays, and is later applied to practical cases concerning saturated soils.

Swelling in Clays

It has been mentioned previously that the diffuse layer is more compressed in divalent adsorbed systems than in Na-adsorbed systems, because divalent ions are attracted to the charged surface with a larger electrostatic force. The salt concentration of the bulk solution also tends to compress the diffuse layer, due to the smaller diffusion of counter-ions from the surface toward the solution. The first phenomenon was previously described in Fig. 3 of contribution 1, chap. IV. The "thickness" of the diffuse double layer is directly related to the swelling pressure. The present section deals with the effect that electrolyte concentration and the type of adsorbed ion have on the swelling pressure of clays.

Homoionic Systems

Studies on montmorillonite pastes, saturated with monovalent ions, showed that swelling can be predicted by the diffuse-double-layer theory (WARKENTIN *et al.*, 1957). The presence of diffuse ion layers around clay particles leads to interparticle repulsion, as a result of osmotic activity of the ions. The repulsive force decreases with increasing spacing between the clay particles, with electrolyte concentration, and with the valence of the exchangeable ions. However, repulsion is not the only factor determining the properties of montmorillonite saturated with divalent ions. For example, calcium-saturated montmorillonite consists of packets, particles, or tactoids, each of which include several (4 to 9) clay platelets arranged parallel to each other, at distances of 9 Å (AYLMORE and QUIRK, 1959; BLACKMORE and MILLER, 1961). If a diffuse ion layer is present, it acts only on the outside surface of each tactoid (BLACKMORE and MILLER, 1961), thereby reducing the effective surface area of the clay system. As a result, particle interaction causes relatively small volume changes. The diffuse double layers in sodium and calcium montmorillonite clays are shown schematically in Fig. 1.

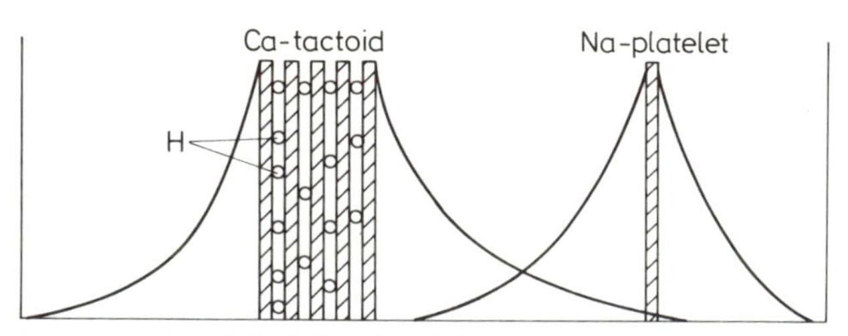

Fig. 1. Schematic presentation of the diffuse double layers on Na and Ca-montmorillonite systems

The difference in osmotic pressure between the midplane of the interacting diffuse layers of two clay platelets and the bulk solution serves as an approximate measure for the swelling pressure of the clay. This pressure difference is given by the van't Hoff equation:

$$\pi = RT \sum_i (C_{id} - C_{io}) = 2RTC_o (\cosh u - 1) \quad (3)$$

where C_{id} is the concentration of the ionic species i at the midplane; and C_{io} is its concentration in bulk solution.

The half-spacing between clay plates in Na systems, d, is estimated (assuming uniform distribution of particles in parallel orientation) from the relationship

$$d = \frac{v}{w.\,s}$$

where v is volume of water, w is dry weight of the clay, and s is the specific surface area of the clay.

The swelling pressure of Na-montmorillonite, as a function of moisture content, is shown in Fig. 2. Theoretical as well experimental data are included.

As mentioned earlier, calcium-saturated montmorillonite consists of tactoids rather than single platelets. According to the osmotic model, these tactoid formations reduce

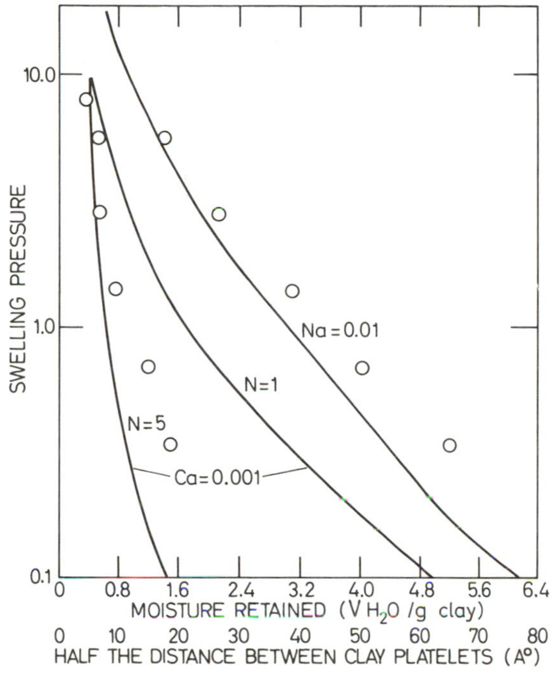

Fig. 2. The swelling pressure of Na and Ca montmorillonite as a function of the moisture retained (theoretical curves and experimental points). (SHAINBERG *et al.*, 1971)

the active interacting surface of the clay, thereby reducing macroscopic swelling. This is indicated in Fig. 2, where the experimental points for Ca-montmorillonite lie much below the classical diffuse layer curve ($N = 1$). Assuming that only the external surfaces of the tactoids are osmotically active, another curve is shown, corresponding to 5 platelets in a tactoid ($N = 5$). The experimental points appear to fit this particular curve.

Bi-Ionic Systems

Whereas in Na-clays the platelets are separated, in Ca-clays only a fraction of the particle surfaces are active in swelling. The swelling characteristics of a clay water system, saturated with a mixture of mono and divalent ions, are illustrated in Fig. 3. Here, the volume of retained water is given as a function of percent exchangeable sodium. It is seen that a slight adsorption of Na by clay, saturated mainly with Ca, has little effect on the amount of solution retained by the clay at any given pressure. SHAINBERG *et al.* (1968, 1971) concluded from these results that replacing 10% of the adsorbed Ca by Na does not affect the basic structure of Ca-clay tactoids. This means that the size of the tactoids remains constant, and the external surface area, which is osmotically active, remains fixed.

Hydraulic Conductivity of Clay

It has been shown that the swelling of Ca-clay is not affected by the addition of a small amount of exchangeable Na. This means that swelling is not the main cause for the decrease of hydraulic conductivity of clays in the low ESP range. Dispersion of the

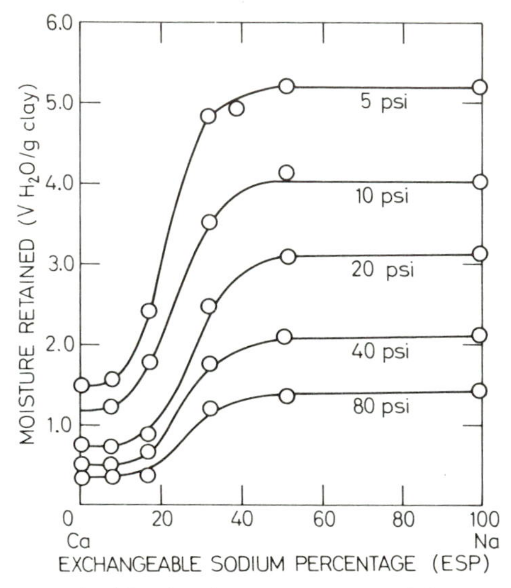

Fig. 3. The volume of water retained by the clay at various consolidation pressures as a function of the exchangeable sodium percentage (SHAINBERG *et al.*, 1971)

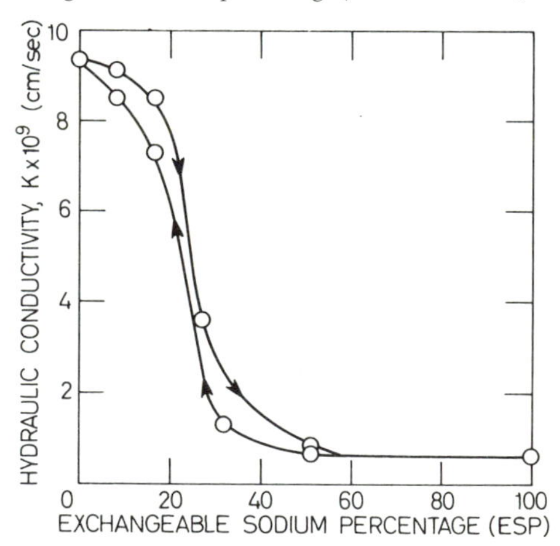

Fig. 4. The hydraulic conductivity of the clay membranes as a function of ESP at total salt concentration of 0.01 *M*. (SHAINBERG and CAISERMAN, 1970)

clay particles may be another cause for the reduction in the HC of clays, since it results in rearrangement and blocking of the pores through which flow takes place. SHAINBERG and KAISERMAN (1971) measured the movement of water through clay systems when the hydraulic conductivity was affected only by swelling (Fig. 4), and when the water movement was affected by swelling as well as by dispersion and movement of the particles (Fig. 5). The HC curve in Fig. 5 is very similar to the swelling curves in Fig. 3. In both cases, the addition of a small amount of exchangeable Na to a Ca system had only a moderate effect on the swelling properties (Fig. 3) and HC (Fig. 4) of the system. The picture is very different when the clay particles are free to move with the solution flowing through the clay system, since here the addition of a small amount of exchangeable Na to Ca-saturated clay has a considerable effect on the movement of water in the clay paste (Fig. 5). When the ESP reaches a value of about 25, the HC of the clay systems is identical to that of pure Na montmorillonite. A similar behavior was observed by MARTIN *et al.* (1964).

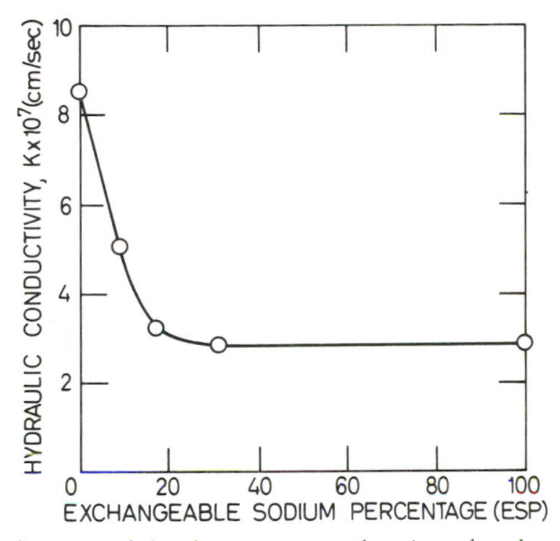

Fig. 5. The hydraulic conductivity of the clay pastes as a function of exchangeable sodium percentage. (SHAINBERG and CAISERMAN, 1971)

The electrophoretic mobility of montmorillonite particles is also very sensitive to the ESP (BAR-ON *et al.*, 1970). Introducing a low percentage of Na into the exchange complex of Ca-montmorillonite increases the mobility of the clay particles considerably. The fact that a low ESP has no effect on the size of the particles and the swelling pressure, but causes an increase in the electrophoretic mobility and a very rapid decrease in the water movement, is explained by the arrangement of Ca-montmorillonite in packets of platelets. When only about 10% of the adsorbed Ca is replaced by Na, most of the adsorbed Na concentrates on the external surfaces of the tactoids, and there is no swelling of the platelets in the packet. Conversely, the zeta potential and dispersion of the tactoids increase very rapidly because the external surface is saturated mainly with Na. When a solution flows through the clay paste, the dispersed clay packets can easily move and block the big pores through which most of the flow takes place.

Hydraulic Conductivity of Soil

Once the influence of electrolytes on the clay material has been established, their effect on the hydraulic conductivity of the soil is more easily understood.

The Hydraulic Conductivity of Soils in Homoionic Systems

The hydraulic conductivity of a soil (K) at equilibrium with irrigation water of a given quality is determined by the concentration of electrolytes and the exchangeable cations in the soil complex. This fact was established by QUIRK and SCHOFIELD (1955), whose work included saturating soil samples with Na, K, Mg, and Ca, and percolating through them a solution characterized by a single-ion system similar to that used in the initial treatment. It is seen from Fig. 6 that, after 3 hours of flow, the hydraulic conductivity decreases by 88% for the Na-saturated soil, and by 40% for the K-saturated soil.

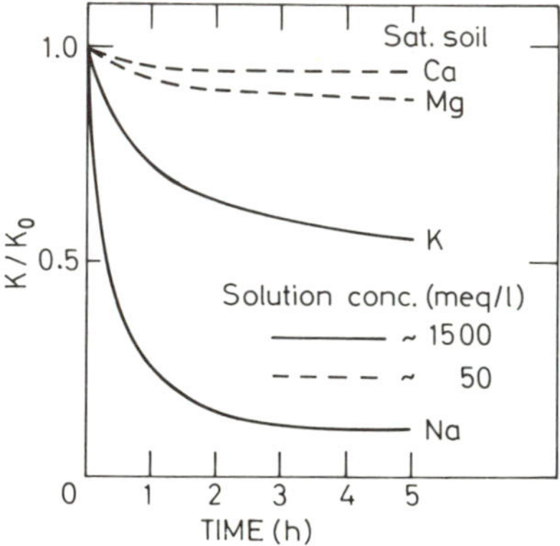

Fig. 6. Hydraulic conductivity as affected by soil cation saturation. (From the data of QUIRK and SCHOFIELD, 1955)

However, when the exchange capacity is saturated with the divalent cations Ca and Mg, there is practically no change in hydraulic conductivity. The beneficial effect of the divalent ions becomes further evident when one considers that the concentration of the permeating solution in the divalent cation systems in 30 times lower than in the monovalent cation system.

However, since homoionic systems seldom exist in nature, the influence of a single ion on hydraulic conductivity is mainly of theoretical interest.

The Hydraulic Conductivity of Soils in Mixed Systems

In a mixed system under equilibrium, the percentage of exchangeable sodium (ESP) is the most important soil characteristic influencing hydraulic conductivity. This fact was pointed out by QUIRK (1957), who showed that permeability decreases with increasing percentage of exchangeable sodium (Fig. 7).

The U.S. Salinity Lab. (1954) has tentatively established an ESP of 15 as the upper limit above which the adverse effect of Na becomes evident.

In mixed systems, increasing the concentration of electrolytes in the solution results in larger flow rates through the soil as compared with that corresponding to the initial cation composition. The influence of electrolyte concentration on the hydraulic conductivity of soils whose complex is characterized by different percentages of exchangeable Na is also shown in Fig. 7.

It is seen that the effect of electrolytes on hydraulic conductivity is more pronounced in soils with a high ESP than in those with a low ESP.

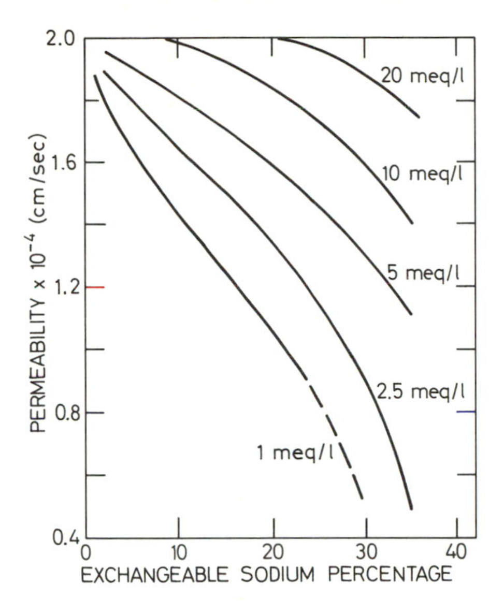

Fig. 7. Effect of the degree of sodium saturation on the permeability of Sawyers soils. (QUIRK, 1957)

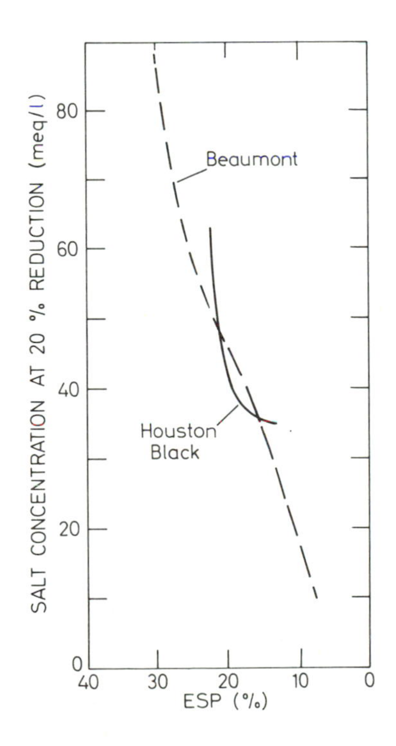

Fig. 8. Salt concentration needed at a given ESP to maintain the hydraulic conductivity at 20% reduction. (NAGHSHINEH-POUR *et al.*, 1970)

A quantitative relationship between solution concentration and ESP, when the latter is expressed as threshold concentration, is given by NAGHSHINEH-POUR *et al.* (1970). The threshold concept was first proposed by QUIRK and SCHOFIELD (1955). It is defined as the salt concentration needed in order to maintain the hydraulic conductivity in a Na-saturated system 20% below that existing in an equivalent Ca-saturated system. Fig. 8 shows threshold concentrations for various values of ESP. It is seen that, in general, there is a relationship between sodium adsorption ratio, water solution concentration, and soil hydraulic conductivity.

Under dynamic conditions, when a soil that is initially nonsodic is irrigated with sodic water, the hydraulic conductivity decreases continuously until the exchangeable sodium in the entire soil profile reaches equilibrium with the permeating solution. YARON and THOMAS (1968) showed that this decrease in hydraulic conductivity is characterized by a curve whose shape depends on the nature of the soil, the cation composition, and the total salt concentration of the permeating solution. The hydraulic conductivity depends on the mean ESP of the soil profile and is independent of the volume of the effluent, a constant only when exchangeable sodium content is achieved along the soil column.

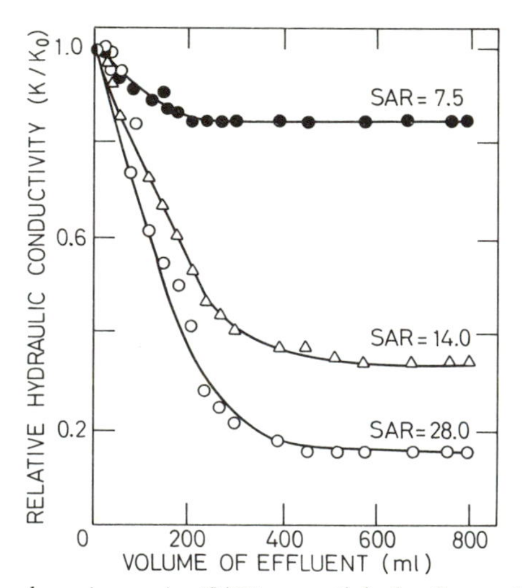

Fig. 9. Effect of sodium adsorption ratio (SAR) on soil hydraulic conductivity. (Soil: Burleson; total electrolyte concentration of irrigation water = 11 meq/1; THOMAS and YARON, 1968)

Composition of the irrigation water has a considerable effect on the exchange process and the hydraulic conductivity of the soil. In dealing with the effect of irrigation water on hydraulic conductivity, one must consider the sodium adsorption ration (SAR). For a given concentration of irrigation water, an increase in SAR results in a decrease in hydraulic conductivity. Fig. 9 shows the reduction in hydraulic conductivity when three solutions with one value of total concentration (11 meq/1) and three SAR values (7.5, 14.0, and 28.0) were passed through columns of a Burleson loamy clay vertisol. At equilibrium, an SAR value of 7.5 resulted in a 15% decrease in hydraulic conductivity, whereas an SAR of 28 reduced the initial hydraulic conductivity by 84% (THOMAS and YARON, 1968).

Influence of Soil Mineralogy

The relationship between hydraulic conductivity and electrolyte type and concentration is affected by the mineralogical composition of the porous medium. McNeal and Coleman (1966) indicated that in soils where the clay minerals are dominated to the extent of 2 to 1 by layer silicates, and where there are moderate amounts of montmorillonite, ESP values in excess of 15 can often be maintained without causing a serious reduction in hydraulic conductivity (provided that the concentration of the permeating

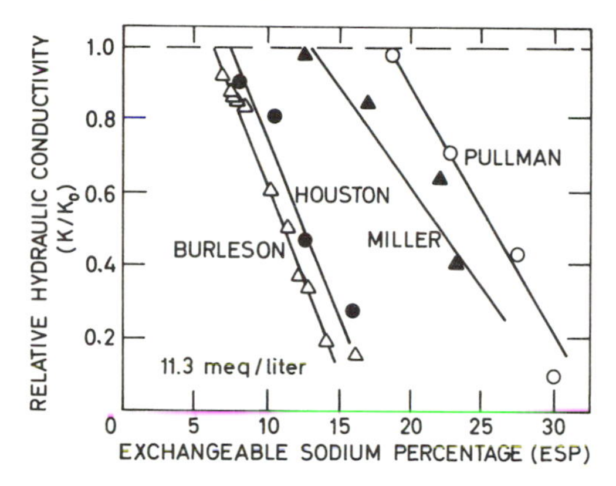

Fig. 10. The relation between the exchangeable sodium percentage and the relative hydraulic conductivity for soils with different mineralogy: Burleson and Houston montmorillonite; Miller mixed; Pullman illite, kaolinite. (Yaron and Thomas, 1968)

solution is 3 meq/1). A 20% decrease inhydraulic conductivity was produced by an ESP of 10 in montmorillonitic soils, by an ESP of 16 in a soil with mixed mineralogy, and an ESP of 23 in a kaolinitic soil (Fig. 10, after Yaron and Thomas, 1968). This behavior is similar to that previously discussed in connection with clays that are saturated with different cations. A relationship between hydraulic conductivity, ESP, and clay mineralogy of vermiculite soils has also been observed. Rhoades and Ingvalson (1969) mentioned that the ESP value needed in order to reduce appreciably the permeability of such soils is higher than in montmorillonitic soils.

Prediction of Decrease in Hydraulic Conductivity

One of the main problems with irrigated agriculture in arid zones is to regulate the water-soil interaction so as to avoid reduction of the hydraulic conductivity, and to increase the conductivity of sodic-affected soils. The common presence of sodium in irrigation water causes problems whose solution requires predicting the expected changes in hydraulic conductivity. During the past few years, several methods of computing changes in hydraulic conductivity during irrigation have been developed. One such method (Yaron and Thomas, 1968) is based on forecasting exchange processes during the flow of saline water (dynamic conditions), and another (McNeal *et al.*, 1968) relies on changes in the bulk volume of the soil due to swelling (equilibrium conditions).

Recently, MCNEAL *et al.* (1968) proposed a procedure for predicting the hydraulic conductivity of a soil, relative to that in a mixed solution, using calculated swelling values of montmorillonite as the basis. This prediction is expressed by the function

$$1-y = cx^n/(1 + cx^n) \tag{4}$$

where y is hydraulic conductivity relative to that in a mixed solution; x is the swelling factor (the calculated interlayer swelling of soil montmorillonite); and c and n are constants for a given soil within a specified range of ESP values. Good agreement between measured and calculated hydraulic conductivity was obtained by MCNEAL using this model.

YARON and THOMAS (1968) used a semi empirical relationship to predict the decrease in relative hydraulic conductivity (K/K_o) due to the presence of sodium in the water. Their relationship is expressed as

$$K/K_o = 1-b\,(K'_G)\,(\text{SAR}) \cdot \frac{(\text{meq salt added})^{1/3}}{\text{meq CEC soil}} - k \tag{5}$$

where the value of b is obtained from the slope of K/K_o versus the mean ESP and the value of k from the intercept of the curve at $K/K_o = 1.0$. K_G' is the modified Gapon constant,

$$K_G' = \frac{(\overline{\text{Na}})\,(\text{Ca})^{1/2}}{\text{CEC (Na)}} \tag{6}$$

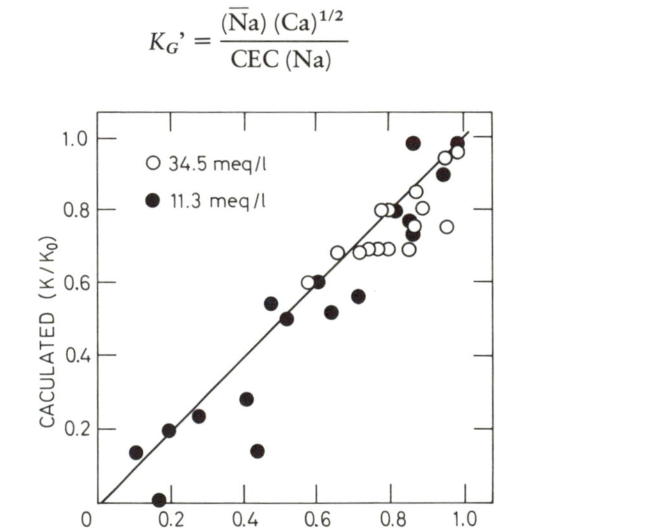

Fig. 11. Plot of estimated versus experimental K/Ko values for four soils and two salt solutions. (Diagonal line indicates perfect agreement; YARON and THOMAS, 1968)

where the bar indicates concentration of the cation in the absorbed phase (meq/100 g), and concentration of ions in the solution is expressed in mmole/l.

The agreement between measured and calculated relative hydraulic conductivity (K/K_o) for four soils and two salt solution using Eq. (5) is seen in Fig. 11.

Eqs. (4) and (5) were applied only to a few soils, and so little can be said about their applicability to other soil conditions.

Literature

AYLMORE, L. A. G., QUIRK, J. P.: Swelling of clay water systems. Nature **183.** 1752–1753 (1959).

BAR-ON, P., SHAINBERG, I., MICHAELI, I.: The electrophoretic mobility of Na/Ca monmorillonite particles. J. Colloid and Interf. Sci. **33.** 471–472 (1970).

BLACKMORE, A. V., MILLER, R. D.: Tactoid size and osmotic swelling in Ca-montmorillonite. Proc. Soil Sci. Soc. Amer. **25.** 169–173 (1961).

CARMAN, P. C.: Fluid flow through granular beds. Trans. Inst. Chem. Engs. London **15.** 150–166 (1937).

MARTIN, J. B., RICHARDS, S. J., PRATT, P. F.: Relationship of exchangeable Na percentage at different soil pH levels to hydraulic conductivity. Proc. Soil Sci. Soc. Amer. **28.** 620–622 (1964).

MCNEAL, B. L., COLEMAN, N. T.: Effect of solution composition on soil hydraulic conductivity. Proc. Soil Sci. Soc. Amer. **30.** 308–312 (1966).

MCNEAL, B. L.: Prediction of the effect of mixed salt solution on soil hydraulic conductivity. Proc. Soil Sci. Soc. Amer. **32.** 190–193 (1968).

NAGHSHINEH-POUR, B., KUNZE, G. W., CARSON, C. D.: The effect of electrolyte composition on hydraulic conductivity of certain Texas soils. Soil Sci. **110.** 124–127 (1970).

QUIRK, J. P., SCHOFIELD, R. K.: The effect of electrolyte concentration on soil permeability. J. Soil Sci. **6.** 163–178 (1955).

QUIRK, J. P.: Effect of electrolyte concentration on soil permeability and water entry in irrigated soils. Internat. Conf. on Irrig. and Drainage, 3rd Congress R6 Question **8.** pp. 115–123 (1957).

RHOADES, J. D., INGVALSON, R. D.: Macroscopic swelling and hydraulic conductivity properties of four vermiculite soils Proc. Soil Sci. Soc. Amer. **33.** 364–369 (1969).

SHAINBERG, I., OTOH, H.: Size and shape of montmorillonite particles saturated with Na/Ca ions. Israel J. Chem. **6.** 251–259 (1968).

SHAINBERG, I., BRESLER, E., KLAUSNER, Y.: Studies on Na/Ca montmorillonite systems. I. The swelling pressure. Soil Sci. **111.** 214–219 (1971).

SHAINBERG, I. CAISERMAN, A.: Studies on Na/Ca montmorillonite systems. II. The hydraulic conductivity. Soil Sci. **111.** 276–281 (1971).

THOMAS, G. W., YARON, B.: Adsorption of sodium from irrigation water by four Texas soils. Soil Sci. **106.** 213–219 (1968).

U.S. Salinity Lab. Staff: Diagnosis and improvement of saline and alkaline soils. USDA Handbook No. 60, 1954.

WARKENTIN, B. P., BOLT, G. H., MILLER, R. D.: Swelling pressure of montmorillonite. Proc. Soil Sci. Soc. Amer. **21.** 495–499 (1957)

YARON, B., THOMAS, G. W.: Soil hydraulic conductivity as affected by sodic water. Water Resources Res. **4.** 545–552 (1968).

V. Measurements for Irrigation Design and Control

Irrigation design and control impose the need to study the soil and to measure water status in the continuous soil-plant-atmosphere system. It was therefore decided to discuss the methodology concerned with the selection of soils to be irrigated, the determination of water in soils and plants, and the measurement of water losses from soil and plant to atmosphere together, to provide a satisfactory background for those dealing with irrigation problems.

The selection of a soil for irrigation requires knowledge of its properties: the soil's genetic formation and relation to irrigation water. With regard to the irrigation requirement as a function of soil properties, it is shown that the soil profile must be considered in determining the need for water and the irrigation system to be employed. A soil survey for irrigation purposes should be undertaken in conformity with the specific requirements of irrigation planning and management and should be based on a knowledge of specific parameters. Some examples of soil surveys on different scales are presented. In the management of irrigation systems, the control of soil evolution is very important in order to avoid soil degradation, and therefore it is recommended that the soil of an irrigated area be surveyed periodically.

Detailed discussions on techniques for measuring water status in soils and plants and evapotranspiration may be found in the literature. However, emphasizing the application to irrigation problems, the present chapter deals only with selected methods appropriate to the requirements of irrigation science and technology.

The dynamics of water status in irrigated soils may be followed by periodically measuring the soil-water content or soil-water potential. Special measurements for calculating the water requirement of irrigated land, such as field capacity and wilting point, are described, together with the measurement of hydraulic conductivity and infiltration data necessary in irrigation design and management.

Water balance control for designing a properly irrigated system may also be carried out by relating the actual evapotranspiration to standard climatological data. In this chapter the critical presentation of some formulas—i.e., solar radiation, pan evaporation, and temperature—will help the irrigationist in providing data to forecast irrigation needs.

The knowledge of water status in plant tissue (defined by the ratio of the actual water content to what it can hold when fully turgid, or by the water potential) may give an indication of the water requirement and the timing of irrigation for many irrigated crops. In this light we present methods for measuring tissue-water content, water potential, osmotic potential of plant sap, stomatal aperture, transpiration, and resistance to vapor transfer that may be used for irrigation purposes.

1

Soil Survey for Irrigation

B. YARON and A. P. A. VINK

Soils are three-dimensional bodies. They have profiles, consisting of specific kinds and combinations of horizons, as well as specific surfaces and relief or landscape features. They are formed by a combination of natural processes under the interrelated influence of climate, vegetation, relief (including hydrology), parent material, and time. In the drier areas of the world–the arid, semiarid, and subhumid zones–soil productivity is usually limited by lack of water. Therefore, if sustained agricultural production with satisfactory yields is desired, irrigation must supplement the natural water supply. The quantity of water to be applied, as well as the frequency and the techniques used in its application, depend on the nature of the soils, as well as on the prevailing climate and the system of agriculture. Furthermore, in most dry areas, water is a scarce commodity. Also, its quality, in terms of contents of salt and silt, is very variable. The kind and amount of water applied not only influence the yields of the crop, but also modify the properties of the soils, and this may essentially affect productivity in later years. Some problems may be partially overcome by a drainage system to leach excess salts. A soil's suitability for irrigation therefore cannot be assessed without due attention to its suitability for drainage.

Selection of soils for irrigation depends on the predicted interactions between water, soil, and crop. It also depends on the land-utilization system envisaged and on the position of the soils in the landscape. Soil suitability for irrigation is one aspect of land appraisal and should always be regarded in a multidisciplinary context, involving economic and social as well as scientific and technical considerations. It is clear, therefore, that no simple, general recipe can be given. Irrigation engineers should consult with soil scientists, agronomists, and others on the establishment of soil suitability at a given moment within a given area. Preferably, this should be done in an integrated multidisciplinary survey covering at least climate, landscape, and soils, as well as the study of the appropriate land-utilization systems.

Irrigation Requirement as a Function of Soil Properties

Given a certain landscape and one or more well-defined land-utilization systems within the terms described, the composition and sequence of the soil horizons provide basic data for determining the suitability of the soil for irrigation. These data should be regarded with due attention to the effective depth of the soil in relation to the root zone of the crop. This depth is partly determined by the nature of the soil profile itself, but also by the properties of rooting of the crops and by the techniques of land use, including irrigation and drainage, used. The characteristics of the horizons within the root zone, in particular their structure and porosity (distribution of the various pore-size classes), determine the intensity and effectiveness of the root system.

With regard to irrigation, the contents and movement of soil water are of primary importance; these are determined by the physical properties of the various soil horizons, their interactions, and distance to the groundwater. These, in turn, are determined by the kind of landscape in which a soil is situated and by its position in the landscape: in depressions, on slopes, or on more or less elevated hills. As these factors influence soil formation, it is clear that many interrelations exist between the various aspects indicated above (see p. 20). It is the task of the pedologist to describe these natural phenomena in soil maps and reports and to interpret them to the irrigation specialists in terms of soil suitability for irrigation. This so-called "soil survey interpretation" is gradually developing into a separate subdiscipline of soil science.

As an example, Fig. 1 shows the variation of the volumetric water content (θ) at "field capacity" to a depth of 100 cm for a toposequence of four soils from the coastal plain of Israel, all of which are used for citrus growing. The field capacities of these four soils differ greatly. They require different amounts of water to replenish their water deficits, thus necessitating different irrigation intervals.

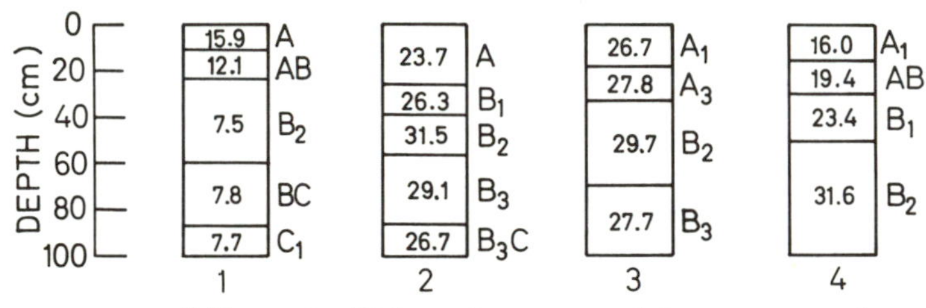

Fig. 1. Water content at field capacity (Volumetric percentages) for a sequence of topographically related soils in the coastal plain of Israel, see Fig. 9 p. 25: 1 – sandy red luvisol; 2 – sandy clay loam reddish-brown luvisol; 3 – sandy clay brown vertic luvisol; 4 – clayey vertic luvisol with pseudogley (the sequence is given from the top of a hill to the bottom of the adjoining depression). (After DAN, MARCU and HAUSENBERG, 1967)

The quantitiy of water required to reach field capacity is different for coarser-textured soils and for heavy vertisols. Other factors also contribute to this difference, in particular, permeability-which is often, although not always, greater in coarser-textured soils. The hydraulic conductivity of a solonetz soil with a natric B-horizon (textural B-horizon with a high sodium adsorption ratio) may be 95% lower than that of a sandy regosol (without B-horizon) and 60% lower than that of a brown luvisol (with argillic-textural B-horizon). The quality of the irrigation water also influences the permeability of different soil profiles differently. Irrigation water with a sodium adsorption ratio of 10, for example, will cause a decrease in the hydraulic conductivity of a soil with an argillic B-horizon but will have little effect on a sierozem (soil with calcic horizon and little or no clay illuviation) and little or no effect on some sandy regosols.

The natural drainage of the C-horizon and of the deeper substrata has much more influence on the suitability for drainage and irrigation of a soil than is often understood. In many cases, due to the shape of the natural drainage curves, hydraulic conductivity to a certain depth below the surface must be known, at least approximately. Furthermore, the interactions between soil drainage and subsoil drainage differ considerably in different landscapes. The direct influence of the deeper horizons within the soil profile itself is shown in the case of an impermeable C-horizon. In a flood-irrigation system this may cause excessive accumulation of water within the root zone, with the resulting lack of air for the roots and grave danger of salinization. The hydraulic conductivity of the calcareous layers underlying a rendzina soil may have a serious effect on its suitability for irrigation. Finally, there is a difference between soils with regard to their suitability for the various kinds of irrigation systems: flood, furrow, sprinkler, and trickle.

Criteria for Selection of Irrigable Land

Land is a broader concept than soil. It embraces not only the soil itself but also its position within a certain landscape, including its climate, vegetation, rocks, and surface and groundwater hydrology. Land, in most cases, has been influenced by past human activity and today is largely influenced by such activity within the scope of an irrigation project. To be complete, any analysis of the irrigability of a particular tract of land must take into consideration the cost of structures to be built for irrigation purposes and related activities as well as how the land will be utilized after the project is completed. The nonrecurrent (capital investments) and recurrent inputs for a project can then be taken into account and the future payment capacity of the project, with or without amortization of the capital invested, can be used to calculate the land's payment capacity. To a large extent, this capacity depends on the productivity of the various soils under an appropriate irrigation system, and on the landscape, which largely determines the nonrecurrent inputs to be made.

In recent years, the land classifications developed by the U.S. Bureau of Reclamation have often been used for this purpose. In this system, on the basis of soil surveys and other investigations, six land classes are recognized: Class 1 has the highest level of irrigation suitability, and hence the highest payment capacity. Class 2 has intermediate suitability and payment capacity, Class 3 has the lowest suitability and payment capacity. Class 4 designates land with strong limitations that is irrigable only under special technical and economic conditions. Class 5 is a temporary designation due to incomplete knowledge of the land. Class 6 includes all soils that are unsuitable for irrigation (see also MALETIC and HUTCHINGS, 1967).

For practical purposes it can be said that the following main properties of land and soil must always be considered in the selection of irrigable land (modified after STORIE, 1964):

1. Topography and microrelief.
2. Depth and quality of groundwater.
3. Quantity, quality, and seasonal periodicity of available irrigation water.
4. Permeability and water-retention capacity of the soil profile and soil drainage conditions.
5. Salinity, alkalinity, and other toxic conditions.
6. Depth of soil to hardpan, bedrock, or cobble.
7. Texture, mineralogical, and chemicophysical characteristics of the soil profile.

The interrelationships between these factors must be carefully considered within the scope of each project.

Soil Survey for Irrigation

Soil surveys provide indispensable data for the planning, execution, and management of irrigation projects. In these surveys not only the essential data on the soils, but important knowledge as to the general conditions of landscapes and groundwater, are collected (VINK, 1963). In many arid and subhumid countries this includes data on soil salinity and on the salinity of shallow groundwater (to a depth of less than 2 m below ground level).

The soil-map scale and, in accordance with this, the survey intensity and the details required of the maps, depend on the various stages of irrigation, planning, and management. These include (1) a reconnaissance soil survey during the "preinvestment" stage, (2) a semidetailed soil survey for the "feasiblity" studies, and (3) a detailed soil survey for the "design and execution" of the project.

Table 1. General indications on the detail required of soil surveys for irrigation (modified after FAO)

Level	Scale	Type of Survey	Type of Mapping Unit
Preinvestment	1:250,000 to 1:100,000	Reconnaissance	Physiographic units including associations of great soil groups and of soil families; individual soil families; phases of associations and families
Feasibility	1:50,000 to 1:10,000	Semidetailed	Associations of soil series; physiographic units (including identified soil series); soil series; soil complexes; soil types; phases of associations
Design and execution	1:20,000 to 1:2,000	Detailed	Soil series; soil types; phases of soil series and of soil types

Table 1 gives general guidelines on the detail required for soil surveys in the various stages. The cost of the surveys, if executed by competent personnel, never exceeds 1 or 2% of the total costs of the project for which they are made, particularly if systematic interpretation of aerial photographs precedes or accompanies the essential field observations. The use of air-photo interpretation is particularly useful for the first two stages, when relatively large areas have to be surveyed within limited periods of time. In the detailed surveys required in the third stage, many field observations and laboratory analyses have to be made. However, more time is available for this purpose; it must be done in advance of, but not more quickly than, the execution of the various parts of the project ("irrigation blocks").

Reconnaissance Soil Survey (Preinvestment Level)

This kind of soil survey is used for preliminary studies in order to obtain general information about a large project area or for a regional survey. As is mainly used to decide which parts of the area will be included and to obtain a broad outline of the major and the minor investments to be made within the project, it is also called the "project identification stage." The publishing scales of the soil maps produced in this stage usually are in the range of 1:250,000 and 1:100,000, but in general include more detailed studies of representative sample areas for between 1 and 10% of the total surface area.

The soil units represented on these maps in general are physiographic units; i.e., they combine the essential data on soils and landscapes in such a manner that the relevant information necessary for planning as given is sufficiently detailed but also easily understandable. In terms of taxonomic pedology (soil classification) these units are,

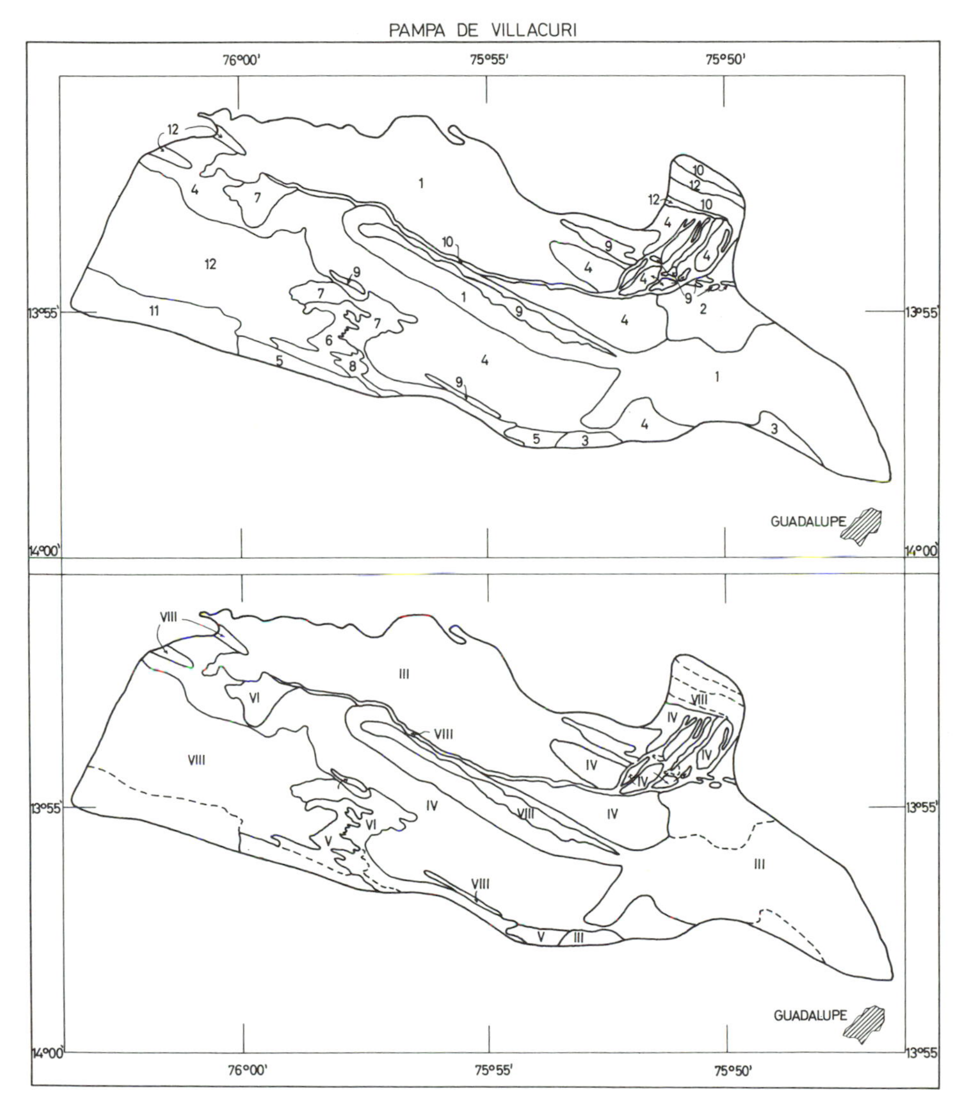

Fig. 2. The soil map (A) and suitability for irrigation map (B) at the reconnaissance level of the Pampa Villacuri area. (After MARCU, Tahal Consulting Eng., 1969). *Soil units:* 1–eolian sandy regosol; 2–complex of sandy and saline sandy regosol; 3–saline and sandy regosol; 4–saline slightly calcareous sandy regosol; 5–sandy regosol with saline hardpan; 6–sandy regosol with remains of solonchak; 7–eolian solonchak; 8–glay solonchak; 9–sandy regosol with active dune formation; 10–alluvial and eolian soils; 11–geological formations with saline accumulations; 12–sandy alluvial regosol with active dune formation. *Irrigation suitability units:* (a) Soils good for irrigation (capability classes III, IV, soil units included – 1, 2, 3). (b) Soils able to be irrigated with special precaution measure (capability classes IV, VI; soil units included – 5, 6, 7, 8. (c) Soils not suitable for irrigation (capability classes VIII, soil units included – 9, 10, 11, 12)

in general, associations of great soil groups with additional information on landscapes and parent materials. Wherever necessary for the project, phases of relief, salinity, stoniness, and soil depth are given. With regard to the soil profile, the essential information, albeit in broad categories, of the nature and sequence of soil horizons is given. It includes orientative data on their texture, as well as on the lime and/or gypsum contents, on soil salinity, and on some physicochemical characteristics.

Each mapping unit has to be characterized by a morphological description of a representative soil profile and by data on its main physical and chemical characteristics (texture, organic matter, pH, lime and/or gypsum content, soluble salts, exchangeable salts).

This survey, together with additional hydrologic information (in particular, on groundwater below 2 m in depth and on available quantities and quality of irrigation water), provides the main criteria essential for the selection of irrigation areas. In all cases where extensive displacement of soil materials, i.e., land leveling or the digging of main canals, is foreseen, more detailed and quantitative information on the topography is required. Experience indicates that this is often obtainable by intensive topographical surveys, in particular, those including contour lines at intervals of not less than 1 m, of small sample areas in representative parts of the overall area. The selection of these sample areas may be based on the data of the soil survey. Soils considered to be suitable for irrigation only after execution of soil improvement are indicated separately on the soil-suitability map.

Semidetailed Soil Survey (Feasibility Level)

This kind of survey is used for geographically well-defined areas. The size of the areas may vary between approximately 30,000 and 5,000 ha. It is understandable that for the larger areas a smaller scale is used than for the smaller areas. For the former, a soil survey with a publishing scale of 1:50,000 is generally found appropriate if approximately 5 to 15% of the area is mapped in more detail for representative sample areas (BURINGH, 1960). For the smaller areas, soil map scales of 1:25,000 to 1:10,000 are often used. Even the latter, if executed in the detail appropriate for this scale, is considered a detailed soil survey. For soil surveys on scales of 1:50,000 to 1:20,000, systematic air-photo interpretation, if appropriately employed within the field survey procedure, may considerably enhance the survey's efficiency.

In a semidetailed soil survey, physiographic soil-map units are used to the extent that they give sufficiently detailed and accurate information for the purpose in hand. However, in cases, where the soil information needed cannot be obtained in this way, separate series and phases of these series are mapped. In some parts of an area the soil pattern may be too intricate to map the soil units separately, even on a 1:10,000 scale and so soil complexes must be mapped. Their delineation, however, has to be carried out in such a manner that the essential information for the irrigation studies is gathered in sufficient detail; this may often be done by a judicious use of phases of these complexes (slope, stoniness, drainage, texture, salinity).

A soil survey at the feasibility level gives information on the sequence of the soil horizons with depth; on the principal physical and physicochemical characteristics of these horizons; on chemical characteristics of particular importance, such as salinity of soils and of groundwater; on the occurrence of calcic and/or gypsic horizons and/or hardpans; and on the essential features of the soil landscape. Soil texture is included, but at this level its use as an approximation of permeability and drainability should be approached with great care; in the past, many mistakes have been made in this respect. For this reason, at least in some representative sample areas, permeability has to be determined in the field by approximative methods, including the auger-hole method for the saturated zone and the determination of infiltration rates for the unsaturated zone.

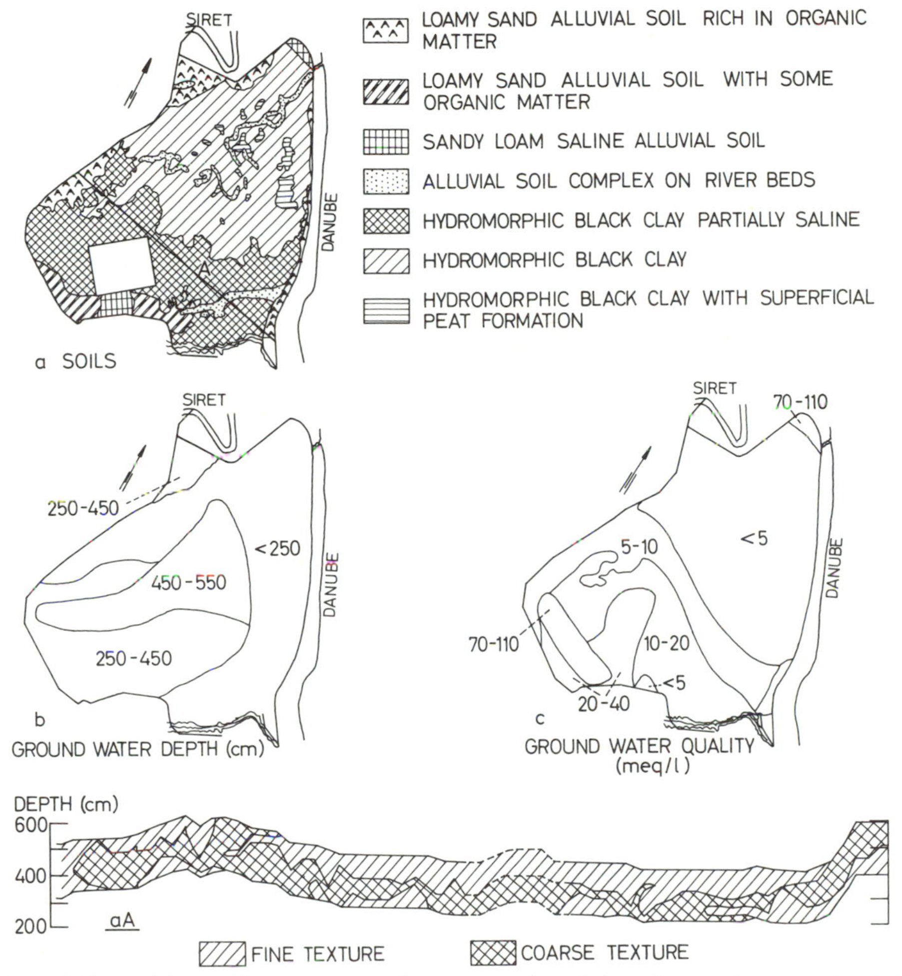

Fig. 3. Soil map (a), complementary groundwater maps (b and c), and textural cross section (aA) of Braila-Danube-Siret area at a feasibility level. (After YARON, 1957)

A complementary groundwater map with information on groundwater quality and flow, also in the depper subsoil, must be prepared in order to group the soils with regard to their suitability for irrigation. In all areas where a considerable displacement of soil materials for land leveling and/or the construction of canals is foreseen, a sufficient amount of quantitative information on the relief must be collected by topographic surveying; in some cases, special aerial photography may be used to expedite this work.

At this level, various soil-survey interpretation maps ("land classification maps") must be derived from the soil map and, whenever necessary, from the additional information given in topographic and hydrologic maps. The interpretation of these maps leads to suitability maps as well as to soil-quality maps, which indicate the ecological land conditions with regard to their agrohydrological and fertility aspects. Also, maps indicating possibilities for soil improvement often are made on the basis of the soil map. In some cases, separate maps showing the suitability of the various soils for the various irrigation methods have to be drawn. And all these maps have to be made with regard to one or more land-utilization systems that may be envisaged within the project. Fig. 3 shows, for example, a soil survey at the feasibility level.

Detailed Soil Survey (Design and Execution Level)

Detailed soil surveys, with a publishing scale of 1:20,000 to 1:2000 are used for the detailed planning of the implementation of a project and for its design and execution. Such surveys, particularly those based on scales of 1:10,000 and larger (1:2000), need not be carried out for an entire area, but only for specific sections where soil improvement is needed or where special irrigation management is recommended. On these maps nearly pure taxonomic units of the lowest classification level, such as soil

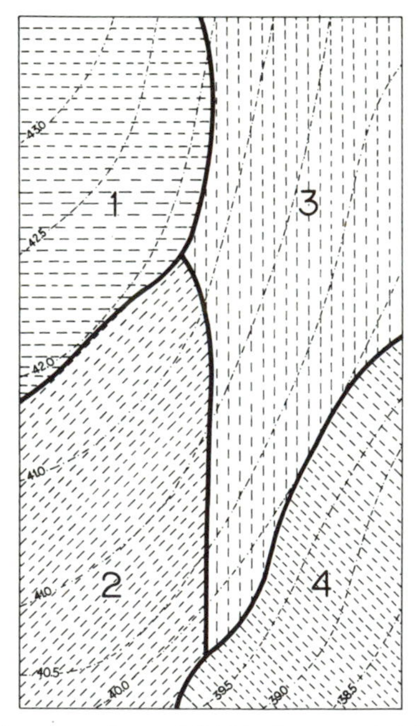

Fig. 4. Detailed soil map of citrus experimental field in the coastal plain of Israel. 1–Mediterranean red; 2–Mediterranean brown; 3–Mediterranean red brown; 4–Mediterranean dark brown. (After YARON, MOR and GOEL, 1968)

series and types, and phases of these, are represented. Detailed information on physical, physicochemical, and chemical characteristics of all mapped soils is given; in some cases, particularly where the construction of irrigation structures is foreseen, mechanical data also have to be gathered. Moreover, detailed agrohydrological information must be amassed–in particular, on the hydrophysical constants of the soils (field capacity, water-retaining capacity, hydraulic conductivity, tilth, and stability under surface wetting) and on the groundwater (flow, levels, and quality).

The interpretation of the soil survey on the basis of these data includes maps of the ecological qualities of the soils (agrophysical and fertility) and of those aspects of soil hydrology and salinity that have a direct bearing on the design of the project. Comprehensive soilsuitability maps at this level may have special importance at an early stage if the cultivation of special crops or the use of special irrigation techniques is envisaged; in other cases, such maps may be needed only after execution of the project, which often includes land leveling and soil improvement.

Example: As an example, consider the very detailed soil survey undertaken for an experimental field at Bet-Dagan, Israel (YARON, 1968). The total area surveyed was a citrus grove. From this detailed survey, four soil types were delineated (Fig. 4). The juxtaposition of many soil types in such a small area is typical of the coastal plain of Israel, where relatively saline water is used for irrigation purposes. In order to predict the soil response to irrigation with regard to salt accumulation and leaching, a test for the uniformity of the experimental plot was deemed necessary. The saturation percentage as defined by the U.S. Salinity Laboratory was selected as a representative index of uniformity. Samples were taken of each 30-cm layer to a depth of 150 cm from the 50 sampling loci in the experimental field, the soil profile was determined, and the results were analyzed statistically. The results show that the coefficient of variation for all soil-mapping units together is about 10% for the upper soil layers (0–90 cm) and increases with depth to 10–13%. In the light of this evidence (the coefficient of variation was considered quite satisfactory for the 0–90 cm layer required for the citrus development), it was felt that the experimental plot could be considered as a uniform unit from the point of view of irrigation response.

Control of Soil Development under Irrigation

Irrigation as a Soil-Forming Factor

Modern technology as used in irrigation modifies the effects of all factors of soil formation. Its intensity may even lead to a considerable change in the rate of soil formation, often resulting in substantial and quantiative changes in the characteristics of a soil. Ultimately, these changes may bring about the development of a soil profile that differs even in its main taxonomic classification. Prior to the introduction of irrigation, there exists in every soil a natural water balance between rainfall, on the one hand, and stream flow, groundwater, evaporation, and transpiration, on the other. Irrigated agriculture disturbs the relatively steady state of soils by acting as a new and continuous soil-forming factor. This man-induced process is called metapedogenesis. Whereas in the natural processes of soil formation the pedogenic factors in general act slowly and over prolonged periods of time, the metapedogenic factors tend to have a strong, and often rapid, effect on the soil properties.

Table 2 shows some examples of soil changes due to metapedogenetic processes, many of which are caused by irrigation. Terracing, land leveling, or dam construction may change a rendzina (rendoll) to a "brown rendzina"(inceptisol) or an alluvial soil (entisol) to a solonchak (aridisol). Drainage may change a gley soil (aqualf) into a "parabraunerde" (udalf) if a sufficiently long period is taken into account. Periodic flooding may gradually transform an alluvial soil (entisol) into a gley soil (aquent, and even aquept). The influence of regular irrigation of paddy fields in the humid tropics causes the formation of plinthite, thus leading to the formation of latosols with duripan (plintaquox); in tropical monsoon climates latosols (oxisols) may even be transformed into grumusols (vertisols) by the continuous addition of water rich in silica.

Table 2. Selected examples of soil formation resulting from metapedogenetic processes in irrigated agriculture[a] (After YAALON and YARON, 1966, and others)

Manipulation	Principal processes Observed in the soil	Soil formation Initial soil	Resulting soil
Terracing or land leveling	Reduction of erosion; humus content increase; rejuvenation of pedogenetic processes; catenary slope differentiation altered	Lithosolic terra rossa (lithic rhodustalf) Rendzina (rendoll)	Terra rossa (rhodustalf) Brown rendzina (inceptisol)
Dam construction on flood plains	Stopping of sedimentation and leaching; water table rise; salt accumulation	Alluvial soil (entisol)	Solonchak (aridisol)
Drainage, lowering of water table	Improved oxidation; structure formation; permeability change	Gleysoil (aqualf)	Parabraunerde (udalf)
Flooding of paddy fields	Hydromorphic water regime; reduced oxidation; gleying; formation of duripan; formation of 2:1 lattice clays	Alluvium (entisol) Latosol (oxisol) latosal (oxisol)	Gley (aquept) Latosols with plinthite (plinthaqox) grumusol (vertisol)
Irrigation with sodic water	Adsorption of sodium; structural degradation; decrease in permeability	Brunizem or chestnut (udoll)	Solonetz (natrustalf) (natrargid)

[a] For reader convenience, 7th Approximation equivalents are included.

Man's manipulations of the soil in irrigated agriculture affect not only one or a few properties but result in simultaneous variations and in interrelated chains of changes of many processes and characteristics of the soil. The use of permanent irrigation makes it necessary to study beforehand the processes induced by man-made changes on comparable soils and under comparable circumstances of climate and irrigation system. The prediction of the evolution of the initial soil under the new conditions of soil formation si an essential aspect of the suitability study for irrigation on a permanent and continuous basis.

Periodic Soil Control

In order to study the kind and rates of soil changes that occur in irrigated agriculture, periodic field and laboratory investigations have to be carried out. These have to start as soon as the project has been partly implemented, particularly for those soil characteristics that are subject to rapid changes: salinization-desalinization, formation of gypsum-accummulation horizons, and hydraulic conductivity. The scheduling of the frequency and intensity of these periodic observations is not rigid, although to check on salinity it may be necessary to carry out yearly samplings of some representative fields for the first five to ten years. Also, specific controlsample areas often can be selected in advance on the basis of the original soil map, and the data from these "control surveys" then are presented on very detailed maps, i.e., a scale of 1 : 2000.

Examples: A. Kibbutz Y'sodot (Israel). Thirty hectares were leveled to facilitate mechanical movement of irrigation laterals in a sprinkler irrigation system. The natural soil was a grumusol with a sodic B-horizon at a depth of about 80 cm. As a result of leveling, the B-horizon appears on the surface in 20 percent of the area. The first step was a very detailed soil survey, which was necessary to locate the soils with a high ESP in the layer. Chemical reclamation of these alkaline surfaces was required (gypsum application and supplementary leaching). A new, very detailed soil survey was then executed to assess the effects of the reclamation.

B. Agricultural Experiment Farm, Gilat (Israel). Five hectares of citrus are irrigated (about 7000 m^3/ha) with water containing 250 mg/l chloride. The soil is a sierozem on loessial material (calciorthid) with a fairly uniform profile and sandy loam texture. The chloride content of a saturated soil paste extract (0–90 cm deep) before irrigation was about 6 meq/l. After 3 years, the chloride content had increased to 15 meq/1 (experiment by H. BIELORAI). Periodic surveys of the electrical conductivity and the chloride content of the soils will give further indications of the leaching requirements and of the effects of irrigation and leaching.

Similar experiments in other arid countries also show the possibilities of leaching of salts with fresh water and brackish water, provided that these are applied with suitable methods of irrigation and drainage to suitable soils (see, for example, UNESCO, 1970). Annual checks on the slinity of the different soil horizons (top soil, 0-25 cm; subsoil, 25–50 cm; and often also deeper layers) are indispensable for determining the effects of the methods used.

Literature

Bureau of Reclamation: Land Classification handboock. U.S. Dept. Interior, Bur. Reclam. Publ. V. Pt. **2**, 53 pp., 1953.

BURINGH, P.: Soils and Soil Conditions of Iraq. 322 pp., 1960.

MALETIC, J. T., HUTCHINGS, T. B.: Selection and classification of irrigable land. In: Irrigation of Agricultural Lands. HAGAN, R. M., HAISE, H. R. and EDMINSTER, T. W. (Eds.). Amer. Soc. Agron., Madison, wis., pp. 125–156 (1967).

Soil Survey Staff: Soil Survey Manual. USDA Handbook **18**, 503 pp., 1951.

STORIE, R. E.: Soil and land classification for irrigation development. Proc. 8th Internt'l. Congr. Soil Sci. Bucharest, vol. **V**, pp. 873–882 (1964).

TYURIN, I. V., GERASIMOV, I. P., IVANOVA, E. N., NOSIN, V. A.: Soil survey, a guide to field investigations and mapping of soils, 356 pp., Israel Program Sci. Translations, Jerusalem, 1965.

VINK, A. P. A.: Planning of soil surveys in land development. Intern. Inst. Land Reclam. Improvement, Wageningen, Publ. **10**, 55 pp. (1963).

UNESCO: Tunisia, Research and Training on Irrigation with Saline Water. 256 pp. and appendixes, 1970.

YAALON, D. H., YARON, B.: Framework for man-made soil changes – an outline of metapedogenesis. Soil Sci. **102**, 272–277 (1966).

2

Measurement of the Water Status in Soils

S. GAIRON and A. HADAS

There is no special preferred method for determining the water status in arid-zone soils, but because of the wide range of soil water contents encountered in these regions, some of the techniques will be more satisfactory than others. Actually, the dry end of the available water range is frequently encountered in the rooting profile of arid-zone soils, and the following will describe primarily methods that are more appropriate for, or adaptable to, dry soils.

Defining the Water Content of Soils

Water in arid soils is encountered as a liquid, and as a vapor. The water distribution within the porous soil mass depends upon the composition and spatial arrangement of the soil particles and on the pore-size distribution. Soil water in the liquid form will cover the surfaces within the soil and fill the pores between the particles. The space free of liquid water is occupied by gases and water vapor. The tightly bound water is sometimes defined as hygroscopic water; the loosely bound water is termed capillary water.

Soil-water content varies in time and with location. These variations are caused by external conditions and depend upon the properties of the soil and their relationship to water. In order to study these properties and relationships, and to establish the basic laws governing the soil-water-content variations, one must define the concept and devise procedures to measure it.

Usually, water content is defined as the dimensionless ratio of the mass of water to the mass of dry soil in a sample.

$$\theta_{dwt} = \frac{m_{wt}}{m_s} \tag{1}$$

where θ_{dwt} is the water content, the subscripts *dwt* indicating that the dry weight of soil is used as a reference basis, m_{wt} is the mass of water; and m_s is the mass of the dry soil. The soil-water content can also be defined as a dimensionless ratio of the volume of water to the bulk volume of the sample.

$$\theta_{vol} = \frac{V_w}{V_b} \tag{2}$$

where the subscript "vol" indicates "volume basis," and V_w and V_b are the volumes of water and sample, respectively. Since it is difficult to measure V_b, it can be calculated, provided the bulk density D_b is known.

When the bulk density of a soil is known, the water content by volume is calculated by the following formula:

$$\theta_{vol} = \frac{D_b}{D_w} \theta_{dwt} \tag{3}$$

where D_w is the density of water, which is unity (1 g/cm^3) in the c.g.s. system, but has to be included in order to satisfy dimensional considerations. Very often water percentages are used by multiplying the foregoing ratios by 100, but in any case the basis (volume or dry mass) should be stated:

$$\text{Water percentage by weight} = P_{wt} = \theta_{dwt} \cdot 100 \tag{4}$$

$$\text{Water percentage by volume} = P_{vol} = \theta_{vol} \cdot 100 \tag{5}$$

When the basis is not stated, it is customary to assume that the dry-weight basis is used.

It is convenient for many purposes to express water content in the field as the volume of water per unit area of land. In this case, the amount of water is averaged over a given depth of soil and the moisture content θ_h has the dimension of length, which represents an equivalent surface depth of water. It is related to the average moisture percentage by weight P_{wt} by the equation:

$$\theta_h = \frac{P_{wt} D_b Z}{10} \tag{6}$$

For the case where θ_h is in mm and Z in cm, 1 mm of surface depth is equivalent to 1 l/m^2, 1 m^3/1000 m^2, or 10 m^3/ha.

This equation is used to calculate the volume of water found in a given layer of soil per unit area of land at the time of sampling, P_{wt} being determined experimentally. The

Table 1. Water-content terminology, definitions, and units

Symbol	θ_{dw}	θ_{vb}	θ_h
Definition	mass H_2O/mass oven-dried soil	vol H_2O/vol soil	height H_2O/area
Formula	$\frac{\text{wet mass-dry mass}}{\text{dry mass}}$	$\theta_{dw} \cdot \frac{D_b{}^a}{D_w{}^b}$	$\frac{P_{wt} D_b Z^c}{10}$
dimensions	dimensionless	dimensionless	length
Units	(g/g)	(cm^3/cm^3)	(cm or mm)

[a] D_b = soil bulk density (g soil/cm^3) $= \frac{m_s}{V_b}$

[b] D_w = water density (g water/cm^3)

[c] Z = total depth

same formula is also used to compute the soil-water deficit before an irrigation, this being the difference between the „field capacity" value and the water content as determined by sampling prior to irrigation. These concepts are summarized in Table 1.

Methods of Determining Water Content

There are many ways of measuring the soil-water content and the literature on the subject is very extensive. Actually, only a few methods are practical and widely used, and all have their limitations. Because of the high variability of soil properties, as well as of water content encountered in the field, it is generally very difficult to find a representative number for a given field (BLAKE, 1965; GARDNER, 1965). Let it be remembered here, that in general, samples are taken from each foot of soil profile (or each genetic layer) at two to four sites a few meters apart, to represent a fairly uniform field that has an area of about 1 ha. For irrigation experiments, the number of samples taken, depends on the statistical experimental layout and size of plots.

Gravimetric Method

The gravimetric method is the basic and standard determination of soil-moisture content to which all other methods are referred. Soil samples are dried until a constant weight is reached. By convention, the samples are dried in an oven at a temperature kept practically constant at 105 °C, and the drying process is continued until the sample does not lose more than 0.1% of its weight during one additional hour in the oven. Thus, drying generally lasts between eight hours for coarse-textured soils to 16 hours for fine-textured soils. The difference in weight before and after drying divided by the weight of dry soil gives the water content by weight, θ_{dwt} or P_{wt}. Because of the very large variability of water content in the vertical as well as the horizontal direction, the sample procedures, and especially the number of samples needed to assess the water content in the field, are critical. The tools used for sampling vary widely and are more or less adapted to the soils in which they are used: The King tube or Veihmeyer tube is suitable for medium-textured and coarse-textured soils, but the spiral auger is easier to use on fine-textured compacted soils. Descriptions of these instruments, many of which are power-driven, can be found in the literature (see GARDNER, 1965).

The main advantage of the gravimetric method is simplicity. Simple and cheap equipment is used: an auger and sample containers for the field work; a balance and an oven in the laboratory (sometimes the soil sample is mixed with absolute alcohol and then burned making the oven unnecessary). The accuracy that can be obtained by the gravimetric method, together with the possiblity it presents of assessing the measurement and sampling errors involved, make it the standard with which all other methods are compared. It can be carried out on all soils without any special skill being required.

The disadvantage of the method lies mainly in the fact that it is destructive and that the measurement of water content at a given site is not made on a single sample. Moreover, relatively hard physical labor is involved (digging up samples in the field), and in most cases a day is needed between sampling and obtaining the results. Nevertheless, the direct gravimetric method is the most suitable and is extensively used in routine agricultural work, especially where relatively cheap semiskilled labor is available for the physical tasks.

Electrical Resistance Methods

The electrical resistance of a stable porous medium is a function of its water content. Measuring directly the electrical resistance of the soil *in situ,* between two electrodes, and calibrating against the water content as determined by gravimetry has been tried with little success, because of the many factors that affect this measurement in soils. However, electrodes that are carefully spaced and embedded in a porous meterial, usually plaster of Paris, nylon, or fiberglass, or various combinations of these, have proved successful in USE. The block so constructed is buned in the soil with the wire leads connected to the electrodes extending to a terminal on the soil surface, where the resistance of the block is measured with an Wheatstone bridge (generally 1000 HZ) to avoid electrolysis or the polarization that would occur with direct current. Moisture equilibrium is usually obtained very quickly when thorough contact between the block and the surrounding soil is assured. Careful calibration of the electrical resistance of the block against the water content of the soil, measured gravimetrically when equilibrium is reached, enables one to get on-the-spot readings of the soil-water content *in situ,* without destroying the site, and with possibility of continuous recording. This type of equipment has several other advantages because of its cheapness, high speed of readout, and the fact that many blocks can be used for replicate measurements. A large number of replicates, used together with the technique of *in situ* measurement, reduces the error due to the large random variations of water content in the field. The ruggedness of the equipment and simplicity of the readings also make this method suitable for use by semiskilled labor. The blocks are compatible with automatic, remote reading and recording devices.

The difficulties of this method stem from the following characteristics: (1) Each block should be calibrated individually, as there is no way to build identical blocks. (2) The calibration may be unstable, as gypsum blocks gradually dissolve in the soil. Partially dissolved blocks will give erroneous readings. The durability of gypsum blocks has been improved by covering them with a thin plastic membrane that is permeable to water, and that affects the response time of the block only a little. (3) Electrical blocks are sensitive to soluble salts in the soil and are generally not recommended under saline conditions, except for the gypsum blocks that are relatively insensitive to salts. (4) The electrical conductivity of the blocks depends upon their temperature. Thermistors are included in some designs to facilitate temperature correction. (5) The blocks are rather insensitive to moisture changes in the wet range (a large change in water content results in a small change in electrical resistance). (6) The main difficulty is probably the hysteresis phenomenon. Blocks reach equilibrium with the soil moisture through matric potential equilibration, and the moisture content of the block (and thus its electrical conductivity) will depend on how equilibrium was reached by drying or wetting.

All these properties limit the use of resistance blocks to areas where accuracy is not a primary requisite.

The Neutron Method

A source of fast neutrons such as Ra-Be, or Am-Be, is lowered into an access tube (usually an aluminium tube of adequate diameter) that has been installed previously in the soil (Fig. 1).

By collisions with the surrounding atoms in the soil, these neutrons are slowed down to the level of thermal neutrons, losing practically all their energy in the process. Hydrogen atoms are most efficient in slowing the fast neutrons, and most of the hydrogen is in the water contained in a mineral soil. A cloud of slow neutrons is thus formed in the vicinity of the fast neutron source. The number of slow neutrons in this cloud rapidly becomes constant (within less than 0.01 sec.), and this steady state depends on the rate at which collisions occur; i.e., mainly on the number of hydrogen atoms or water molecules that exist in the soil medium. A BF_3 proportional counter enriched with B^{10} mounted near the source will detect these neutrons, and, with appropriate electronic equipment (which

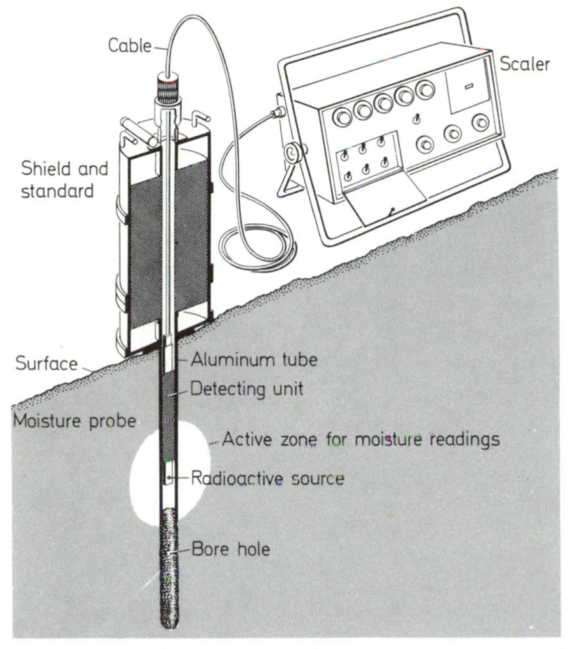

Fig. 1. Neutron moisture probe and portable scales for moisture-content determination in the field

includes preamplifier, discriminator, scaler, and timing unit), they can be counted. With adequate calibration (count rate versus volumetric water content, within the sphere of influence of the probe), this equipment provides a sophisticated, precise, and reliable method for measuring soil water. The linear universal calibration curve furnished by the manufacturer cannot be relied upon, in most cases, and for accurate work a calibration curve should be made for each soil.

This equipment features several important advantages: it is a nondestructive sampling method, allowing one to follow moisture variations *in situ,* at any desured depth; it is a fast method, giving an immediate answer, with each measurement being made in a matter of minutes. Thus, many sites can be checked daily without excessive physical work (because of its weight, the equipment is generally carried on a cart).

Finally, thanks to the linear relationship between count rate and volumetric water content within the sphere of influence of the probe, the method gives the same precision of measurement throughout the entire moisture range, as long as the measuring conditions are similar to the calibration conditions.

However, a number of disadvantages seriously restrict the routine use of the neutron-scattering devices presently available. (1) Careful precautions have to be taken to reduce the health hazards that accompany nuclear equipment. (2) The equipment is very expensive and requires skilled operators and costly maintenance. (3) The sphere of influence of the probe is rather large (ordinarily between 10 and 30 cm in diameter) and varies with the water content, becoming larger as the water content diminishes. As a result, the equipment is unable to measure abrupt changes in the water-content profile, such as wetting fronts or near the soil surface, where it lacks resolution. (4) Lack of uniformity in the soil, like the presence of stones or cracks, will reduce the reliability of the method.

As a result the method seems rather unsuitable for ordinary agricultural work and is used mainly in research projects where it is necessary to follow the changes in soil-water content on a large scale.

The Gamma Attenuation Method

Unlike the neutron-scattering method, where one probe with source and detector is used, the gamma-ray attenuation method is based on the double-probe design: one probe contains the gamma-ray source (usually Cs-137 with a gamma-ray energy of 0.661 MeV); the other contains a scintillation crystal with a photomultiplier and preamplifier, which detects the monoenergetic beam emitted by the radioactive source. The intensity (count rate) of the beam arriving at the detector depend upon the overall density and thickness of the material it has to traverse. In the case of a soil layer, where the distance between source and detector is constant, and there is no change in the bulk density of the soil with time and space, the ratio of the gamma-ray flux (count rate) through the soil when wet to the flux through the soil when dry, is an exponential function of the water content, θ_{vol}. In other words, θ_{vol} is given by the equation

$$\theta_{vol} = \frac{\log_{10} N_m/N_d}{0.4343\ \mu_w x} \tag{7}$$

where θ_{vol} is the soil moisture content by volume, N_m is the count rate through moist soil, N_d is the count rate through dry soil, μ_w is the mass attenuation of water, and x is thickness of the soil.

This equation assumes that attenuation due to the air in the soil column is negligible. The gamma method, which is used successfully in the laboratory with scanning and recording devices, is rather cumbersome in the field, and presents the same difficulties as the neutron-scattering method. In fact, the necessity of knowing the bulk density very accurately for calibration purposes makes the method even more problematic in the field than the neutron-scattering method. Although some double-probe field equipment is already available commercially, it is still unreliable and difficult to calibrate. Only laboratory scanners have given satisfactory results.

The Tensiometer Method

The tensiometer (which is described on p. 225) measures the soil-water matric potential. But with appropriate calibration it can be used as an indicator of soil-water content although the readings are then subject to hysteresis. The reader is referred to p. 94 for a complete discussion of the advantages and limitations of the tensiometer for measuring soil-water matric potential and water content.

Bulk Density Determinations

The bulk density, also called volume weight or apparent density, is the mass of dry soil per unit bulk volume of the sample: $D_b = m_s/V_b$. But volume measurements of porous material are time-consuming and it is generally difficult to get accurate results. As we have seen in Eq. (3), θ_{vol} can be computed from θ_{dw} (the water content by weight) when the bulk density of the sample D_b is known. Actually, the bulk density is a frequently used soil characteristic, for instance, when one needs to calculate the soil porosity knowing particle density, or when one needs to estimate the weight of a volume of soil too large to be weighed conveniently.

On the other hand, bulk density is generally a highly variable quantity in space and time, and it is therefore necessary to determine its value at enough sites, depths, and times in a field to obtain a reliable value of this soil parameter. For many practical problems (as in routine irrigation work), an approximate value of the bulk density can be satisfactory, in which case, the number of samples of sufficient uniformity to be taken from a field will not be excessive, and the variation with time can often be neglected.

In principle, all the methods of determining the bulk density, except the gamma-ray attenuation method, are based on the drying and weighing of a known volume of soil.

The Core Method

In the core method a cylindrical metal sampler is driven into the soil to a predetermined depth and removed carefully so as to preserve a known volume of the soil in a condition as similar as possible to condition it had *in situ*. The sample is then dried to a constant weight at 105 °C. The bulk density in g/cm^3 is the dry mass of the soil sample divided by its volume. This method generally fails when the soil profile is very dry or very hard and compacted. Various types of core samplers have been described in the literature (SHAW and ARBLE, 1959) and some of them are commericially available.

The Clod Method

In this method a sufficiently coherent clod of soil is suspended on a thin wire, coated with a water-repellent substance (paraffin wax, for instance), and then weighed, first in air, and then while immersed in a liquid of known density (Archimedes' principle). Taking into account the weight of the wire, the volume of the paraffin wax, and the moisture content, the mass and volume of the soil clod can be calculated, and thus its bulk density. The soil-moisture content is determined either from one-half of the original clod, or from the entire clod after weighing and removal of the paraffin layer. This method is the most accurate and is considered as the standard method with which the others are compared. It has a few desadvantages: it is slow and time-consuming, and according to various authors it may not give an accurate picture of what happens in the field where cracks and stones are found.

The Nuclear Method

This method is not different in principle from the one that has been described briefly as the double-probe gamma-ray attenuation method for soil-water-content determinations. But in this case, one has to measure the water content independently and the calibration curve will give the bulk density of the soil. The same disadvantages as previosly described exist here, and the method is used mainly in research work.

Methods of Determining Soil-Water Potential

Functional relationships exist between the soil-water content and physical parameters such as conductivity, vapor pressure, and free energy of the water. The most important feature of these functional relationships is the marked decrease in the free energy of water as the water content of the soil decreases.

Table 2. Methods and devices for measuring soil-water potential

Measuring device	Potential component	Operational range	Advantages and disadvantages
Suction table	*M*	Very wet soils near saturation (0–0.1 bar)	Requires a long time to equilibrate; inexpensive and simple; operational only in laboratory
Pressure plate or membrane	*M*	From saturation to wilting point (0–20 bars); can work for 100 bars	Covers field range of soil water; requires long time to equilibrate; operational only in laboratory
Tensiometers	*M*	Wet soil only (0–0.85 bar)	Placed in the field; sensitive to temperature and roots; fairly accurate
Psychrometer	*M* + *O*	Saturation to air dryness	Precise laboratory method; applicable to plants as well as soil; requires special facilities
Electrical resistance blocks	*M*	Field capacity to air dryness (0.5–50 bar)	Needs special calibration; inaccurate, especially in wet soil; sensitive to salts and temperature

Various devices enable us to determine the soil-water potential (Table 2). The determination is based in the assumption that equilibrium is attained between the soil and the measuring device we use. Thus, by applying pressure or suction on the soil and measuring the water content at equilibrium, we may evaluate the soil-water potential water-content relationship.

Total Soil-Water Potential

The total soil-water potential incorporates various components such as the osmotic, the matric, and the adsorptive components. The common methods for determining the total potential are those in which the relative humidity of an atmosphere in equilibrium with the soil is measured. Another method is to measure the freezing-point depression of the moist soil.

The Vapor-Pressure Methods. The equilibrium vapor pressure of a moist soil is very close to the saturation vapor pressure. Accurate methods were developed only recently that employ miniature psychrometers made of wet and dry thermistors or thermocouples. The units are inserted into small sample chambers and the whole system is allowed to reach equilibrium. The principle involved is that the vapor-pressure depression from saturation is proportional to the sum of the osmotic potential and the matric potential of the soil solution.

Because of the small differences in vapor pressure in the field moisture range, the methods involving psychrometric measurements require intricate temperature control and expensive electronic equipment to measure the psychrometer output accurately. The accuracy attained under laboratory conditions is about 0.5 bar; under field conditions it is 1–2 bars.

The Freezing-Point Depression. This method is commonly used whenever there is a need to determine the osmotic pressure of solutions or the molecular weight of the solutes. The data so obtained when applied to a moist soil, are difficult to interpret, since the free energy of soil water is affected by solutes, and soil capillaries filled with water. Corrections have been proposed, but the uncertainty of the data still exists. This method is often used to measure plant-water status.

Matric Potential

It is commonly assumed that the soil-water potential encountered by plant roots is determined by forces that act upon the soil solution and that stem from the soil water-soil matrix interaction. This is true when the osmotic potential due to solute content is negligible.

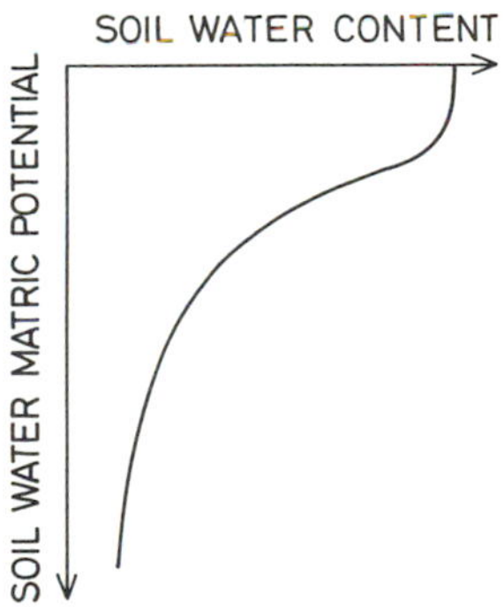

Fig. 2. Schematic retention curve

If one neglects the osmotic potential, the matric potential can be measured by various methods, such as pressure plates and membranes, tensiometers, electrical resistance blocks, etc.

The Pressure-Plate or Pressure-Membrane Method. This method is a standard method used all over the world and the data obtained from it, namely, the soil-water content water potential function, are called the "retention curve" or the "soil-water characteristic." An example is given in Fig. 2.

Soil-water potential may be expressed as work per unit volume, or simply as differential pressure, or the equivalent pressure by which soil water is "held" by the soil. This equivalent pressure is negative when compared with atmospheric pressure. To measure the matric potential, one has to apply differential pressure on a disturbed or undisturbed soil sample and let the system reach equilibrium. The differential pressure is applied across a saturated medium such as ceramic plate, sintered glass, or cellulose membrane, which do not allow air to replace the water contained in them while the pressure is applied.

The apparatus consists of a pressure chamber on the bottom of which is a porous unit – a ceramic plate, a sintered glass, or a cellulose membrane. Soil samples are placed on the porous unit and wetted to saturation and the cavities below the unit filled with water. The differential pressure is produced either by increasing the pressure in the pressure chamber through inlet A or by applying vacuum through outlet B (Fig. 3). The matric potential range covered by this apparatus is 0.8 when the pressure difference is controlled by vacuum, and 20 bars or more when compressed air is used.

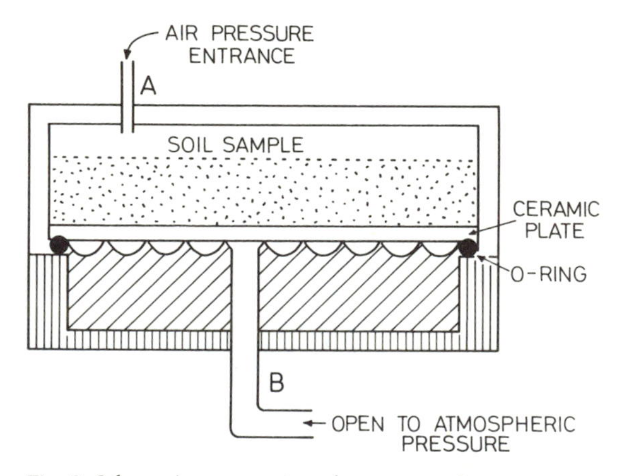

Fig. 3. Schematic presentation of a pressure-plate apparatus

The time needed for equilibrium to be attained increases with the fineness of the porous unit and of the height of the soil sample.

The data obtained by this method are reproducible and may be applied to field work provided the soil samples that are placed in the apparatus are undisturbed cores.

The range of soil-water potential covered by these methods is from saturation to about 50 bars; for the 0–1 bar range ceramic plates are used, and above this pressure difference, cellulose membranes are used. Recently, new ceramic plates were developed that withstand pressure differences up to 15 bars.

The Electrical Resistance Method. On p. 218 this method was described for soil-water-content determinations and its merits and disadvantages were discussed previously. The same units can be calibrated in pressure units on pressure plates and/or pressure membranes and thus, instead of relating the electrical resistance readings to water content, one can relate them directly to the soil-water potential. In fact, it is preferable to use resistance blocks to determine water potential rather than content.

The Tensiometer Method. A porous cup (Fig. 4) is filled with pure deaerated water and closed with a vacuum gauge (or manometer). When the cup is introduced into a free-water reservoir (or a saturated soil), the gauge will indicate a value that is arbitrarily taken to be zero. When the cup is then transferred and brought into contact with a drier soil, water will leave the cup, and a vacuum will develop in the measuring gauge. This vacuum depends upon the soil-water matric potential (which is functionally related to soil-water content), and is higher, the drier the soil. This instrument is called a tensiometer, and it directly measures the matric potential of soil water. Various types of tensiometers are available commercially, varying mainly in their vacuum gauge design and cup material, but they all have the same positive and negative features. The tensiometer

Fig. 4. A tensiometer

is valuable because it gives *in situ,* continuous readings, and also because it directly gives capillary potential values when needed. However, although its use is rather widespread in irrigated areas, the tensiometer has numerous limitations. The most important disadvantage is its narrow measuring range between soil-water saturation and the moisture content corresponding to a capillary potential of about 0.85 bar. Because of their high cost, only a small number of instruments can generally be used, limiting the number of replicates that can be made. Skilled maintenance is needed to keep tensiometers operative: air often leaks into the vacuum system or diffuses through the wall of the porous cup and must be expelled periodically. They are sensitive to temperature variations, and especially to temperature gradients between their various parts.

Finally, most of the devices currently available, influence the water content they measure, because of the rather large volume of water that has to move back and forth from the cup to the soil in order to adjust to the ambient water potential. Thus, the use of tensiometers is restricted today to high-value crops, where high water contents are maintained during the growing season, and the return from the crop can justify the high cost of the instruments and the skilled labor required.

It should mentioned here, that the tensiometer is primarily designed and calibrated to monitor the soil-water matric potential. In order to overcome some of its limitations, such as response time and limited range of measurement, a new type was recently introduced in which the ceramic cup is replaced by a cup covered by a semipermeable membrane and filled with a solution of a known osmotic potential, and its manometer is replaced by an electric transducer. These types are still experimental, but may be useful for measurement through a wider potential range.

Literature

BLAKE, G. R.: Bulk density. In: Methods of Soil Analysis. Agronomy 9. BLACK, C. A. *et al.* (Eds.). Amer. Soc. Agron., Madison, Wis., 1965.

GARDNER, W. H.: Water content. In: Methods of Soil Analysis. Agronomy 9. BLACK, C. A. *et al.* (Eds.), Amer. Soc. Agron., Madison, Wis., pp. 83–127, 1965.

HOLMES, J. W., TAYLOR, S. A., RICHARDS, S. J.: Measurement of soil water. In: Irrigation of Agricultural Lands. Agronomy 11.

HAGAN, R. M. *et al.* (Eds.), Amer. Soc. Agron., Madison, Wis., 1968.

RICHARDS, L. A.: Physical conditions of water in soil. In: Methods of Soil Analysis. Agronomy 9. BLACK, C. A. *et al.* (Eds.)., Amer. Soc. Agron., Madison, Wis., pp. 128–152, 1965.

SHAW, M. D., ARBLE, W. C.: Bibliography on methods for determining soil moisture. Eng. Res. Bull. B. 78, Pennsylvania State University, Pa., 1959.

U.S.D.A.: Diagnosis and improvement of saline and alkali soils. USDA Handbook No. **60.** 1954.

World Meteorological Organization: Practical soil moisture problems in agriculture. Technical Note No. 97, WMO No. 235 TP 128 (G. STANHILL *et al.*, Eds.), 1968.

3

Important Soil Characteristics Relevant to Irrigation

S. GAIRON

Introduction

Water in the soil profile is very seldom under equilibrium conditions. Many forces act upon the soil solution in a continously changing manner, thus causing redistribution of water. As water always moves in the direction of decreasing potential, the rate of flow is determined by the potential gradient and the ability of the soil to transmit water. Mathematically, the water flux is expressed by the Darcy equation:

$$v = -K \nabla \Phi$$

where v is the water flux, K is the soil hydraulic conductivity, and $\nabla \Phi$ is the gradient of the total potential field in which the water is moving. The minus sign expresses the fact that water is flowing in the direction of decreasing potential.

All the important soil characteristics relevant to irrigation are actually related to this equation. It deals with the transmission of water and governs the determination of the soil conductivity to water and the infiltrability of the soil profile. On the other hand, when the flux is nearly zero, because of vanishing K or $\nabla \Phi$ (see p. 104, the conditions of quasi-equilibrium govern the availability of water to plant growth within the range of field capacity to wilting point.

Defining these characteristics and describing their methods of measurement are the objectives of the present chapter.

Soil Conductivity to Water

Principles

We shall limit ourselves here to isothermal flow of liquid water in response to the gravitational and pressure gradients. For practical purposes the gravitational and pressure gradients may be regarded as the driving forces that are responsible for the movement of the water to plants, so that Darcy's equation may be written:

$$v = -K \nabla H \tag{1}$$

where v is the volume flux of water, i.e., the volume of water passing through unit cross-sectional area of soil per unit time; and ∇H is the hydraulic gradient in the direction of flow. The conductivity of the soil to water K, which appears as a proportionality factor in Darcy's equation, is not always constant. First, it depends upon the moisture content i.e., $K = K(\theta)$. When the soil is saturated, K is called the hydraulic conductivity of the soil, and under unsaturated conditions it is referred to as the unsaturated conductivity of the soil. On the other hand, the conductivity is affected by the various physical, chemical, and biological processes that occur in the soil, and so it happens that the conductivity changes as water passes through the soil. The water flow can produce a change in the exchangeable-cation status, especially when its ionic composition and concentration are different from those existing in the displaced soil solution, and, as a consequence, very large changes in the conductivity may occur. The conductivity tends to decrease when the salt concentration is low and/or when the proportion of dispersing monovalent to divalent cations is relatively high in the displacing solution. The physical displacement of small particles or microbial activity within the sample can also be responsible for important changes in the soil conductivity, as can be the entrapment of air bubbles and their subsequent slow dissolution. These have to be assessed and controlled during the measurement.

Methods and Apparatus

Although a very simple setup can be devised to measure the hydraulic conductivity of saturated soil samples (KLUTE, 1965), we shall describe here an apparatus that can be used for unsaturated samples as well. The slight differences in procedures will be stressed in each case.

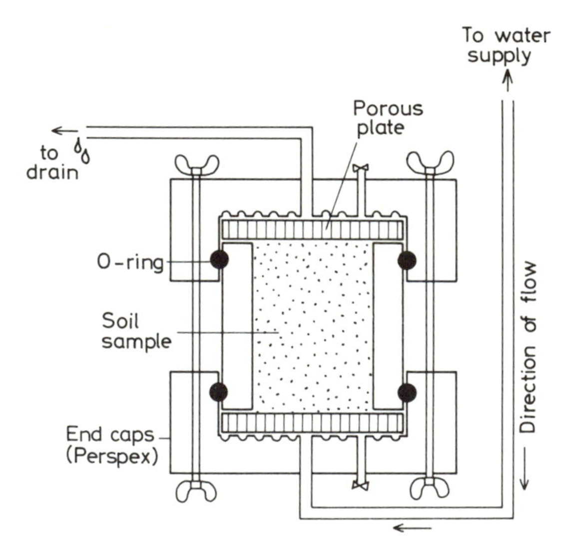

Fig. 1. Permeameter for measuring soil-water conductivity (for saturated samples the O-rings have to be in place; they are excluded for unsaturating the samples)

A schematic representation of a permeameter, which can be used for measuring the water conductivity of soil samples, is presented in Fig. 1. The porous plates are sealed to the end caps. For measurement of the hydraulic conductivity of saturated samples, the O-rings have to be in place, for unsaturated samples, they are excluded. This apparatus can accommodate disturbed soil samples, by filling, in the laboratory, the sample ring with air-dried sieved soil. It can also be used with adequate rings from an undisturbed core sampler thus giving an acceptable approximation of the hydraulic conductivity of the soil in the field.

Constant-Head Method. In this case we use devices that keep the water pressure constant at a certain level (above the permeameter cell for saturated soils, and below it for unsaturated samples) both at the inlet and at the outlet. The device most often used is the Mariotte bottle (with or without a siphon), but any similar overflow or float device can be utilized (Fig. 2).

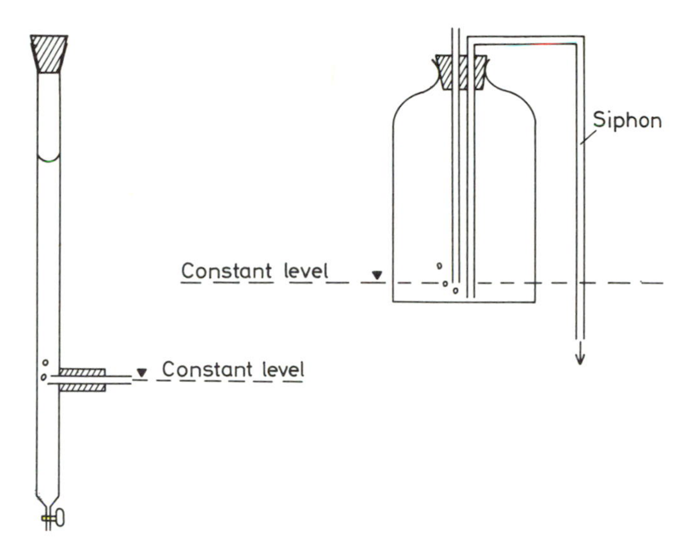

Fig. 2. Constant-level devices (Mariotte bottle type)

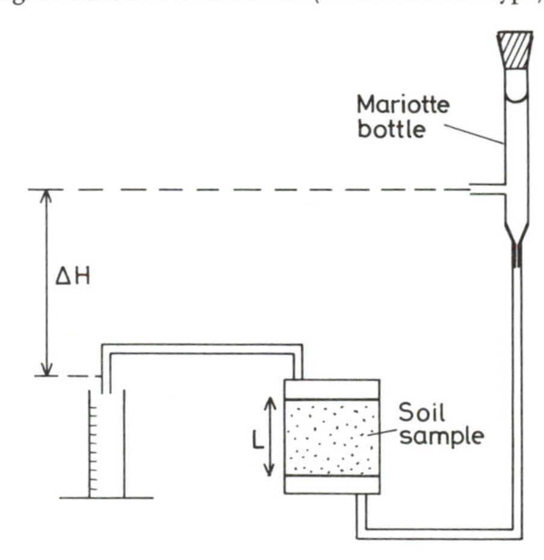

Fig. 3. Permeameter set up for measuring soil-water conductivity (saturated case)

In general, the outlet will be a single overflow tube or drop tube at a fixed position on the outlet side (Fig. 3). In the case of very slow flow, the measuring device may utilize the movement of an air bubble or an air-water meniscus in a capillary tube. These devices may introduce relatively large errors when very small gradients are used (OLSEN, 1965), and when the capillary tube is of small diameter.

In order to obtain a saturated sample, the permeameter has to be thoroughly evacuated (by pumping for several hours) or flushed with CO_2 and then the deaerated water admitted into the sample from the bottom. It is wise to leave the sample soaking overnight before beginning the flow measurement.

In the case of unsaturated samples the procedure is generally to bring the sample to saturation and then desaturate it by lowering both the inlet and outlet reservoirs below the permeameter cell to the mean matric potential desired (Fig. 4). When equilibrium is reached (determined by lack of water flow in either direction), the inlet and outlet reservoirs are brought to the planned elevations (both below the cell), symmetrical as to the position of ψ, thus creating the gradient ∇H responsible for the water movement.[1] Care must be taken to keep the porous plates saturated with water. Therefore, the values of ψ and ∇H will be limited by the bubbling pressure of the porous plates. In both cases, saturated and unsaturated, the amount of water that passes during measured periods of time is recorded, and the constancy of the flow checked several times.

To calculate the conductivity, the following equation will be used:

$$K = \frac{Q}{At} \cdot \frac{L}{\nabla H} = V \frac{L}{\nabla H} \qquad (2)$$

where K is the water conductivity at the given capillary potential or at saturation, Q is the volume of water that has passed through the sample during time t; A is the cross-sectional area of the sample; L is the length, V is the mean velocity of flow, and H is the water head.

When Q is expressed in cm^3, t in sec, A in cm^2, L in cm and H in cm of water, K is obtained in cm/sec (or $cm^3 . cm^{-2} . sec^{-1}$).

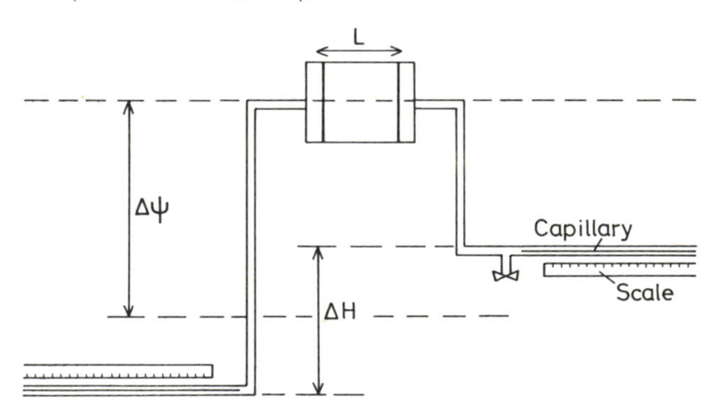

Fig. 4. Permeameter set up for measuring soil-water conductivity (unsaturated case)

In the case of the unsaturated sample (Fig. 4), the moisture content can be reduced by increasing ψ. If a series of measurements is made at various ψ's, the amounts of water flowing into the sample or out of it during the change of ψ must be noted. At the end a gravimetric determination of the moisture content of the sample is made, which allows the water content at each ψ to be calculated, and a curve of K to be drawn, as a function of θ.

---

1 One has to remember that in this case the measured conductivity will correspond to a moisture content reached by desaturation. Because of the hysteresis phenomenon encountered in soils, this moisture content will generally be higher than the one attained by the soil at the same potential by capillary uptake.

Falling-Head Method. A falling-head permeameter is sometimes used with water-saturated samples (Fig. 5). An advantage of this method is that the precision of head measurement depends only on the length of the water-supply column. (In the case of constant-head devices such as Mariotte's bottle, there is always a slight change in head during the formation of each air bubble.) Also, with this method it is very easy to design a system that allows reuse of the rather small volume of flowing water and reversal of the flow direction. This reduces the changes in hydraulic conductivity that may occur because of solute concentration changes in the soil solution, and enables one to check the uniformity and isotropy of the sample.

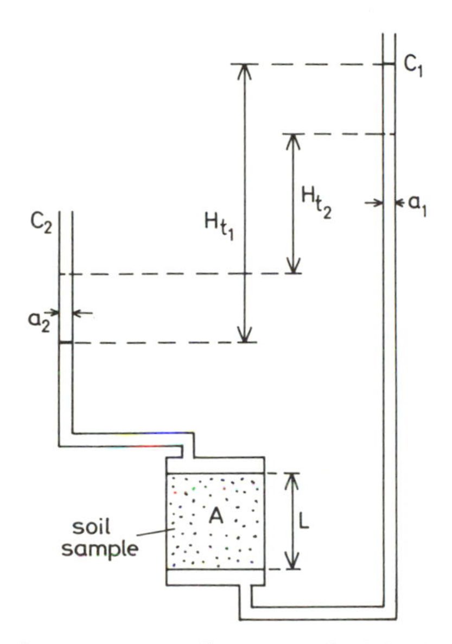

Fig. 5. Permeameter set up for measuring soil-water conductivity (saturated case – falling-head method)

The soil sample is saturated as described above and left to soak for 24 hours with the water levels in columns C_1 and C_2 at the same elevation. Then the columns are displaced so as to separate their water levels symmetrically around their initially identical level. The time $(t_2\text{-}t_1)$ for the head differential to change from H_1 to H_2 is recorded, and the hydraulic conductivity K calculated by the following formula:

$$K = \frac{L}{A(t_2\text{-}t_1)} \frac{a_1 a_2}{a_1 + a_1} \ln \left(\frac{H_{t1}}{H_{t2}}\right) \qquad (3)^2$$

2 This formula is arrived at in the following manner: according to Darcy's equation:

$$(1/A)\, dQ/dt = -(K/L)H,$$

where dQ is the amount of water passing through the sample during time dt, and H is the water-head difference across the sample at time t. If dh_1 and dh_2 are the changes in the water level in each column corresponding to dQ, we can write: $dQ = a_1\, dh_1 = a_2 dh_2$, where a_1 and a_2 are cross-sectional area of columns C_1 and C_2, respectively. But the total change in the water-head

Comments

It is of interest to note here, that the hydraulic conductivity has the units of velocity, and that it represents the combined properties of the flowing fluid and the porous medium. It is sometimes written as the ratio between the intrinsic permeability (k), which ideally describes the conducting medium alone, and the kinematic viscosity $g\varrho/\eta$, which characterizes the fluid,

$$K = \frac{\varrho g}{\eta} \cdot k$$

where η is the viscosity of the water, ϱ is its density, and g is the gravity. (In the CGS system k is in cm^2)

Here we can see how temperature can influence hydraulic conductivity, by changing the viscosity of the fluid (water). In many cases, the changes in η with temperature are taken from tables, but these values would appear to be incorrect as the viscosity of the soil solution is probably different from that of pure water.

In order to give the reader an idea of the range of values that can be encountered for hydraulic conductivity and permeability, the classes as defined by O'NEAL (1952) are presented in Table 1.

Table 1. Conductivity classes for saturated subsoils (after O'NEAL, 1952)

	Hydraulic conductivity (k cm/hr)	Permeability (k cm^2)
Very slow	0.125	3.10^{-10}
Slow	0.125 – 0.5	3.10^{-10} – 15.10^{-10}
Moderately slow	0.5 – 2.0	15.10^{-10} – 60.10^{-10}
Moderate	2.0 – 6.0	60.10^{-10} – 170.10^{-10}
Moderately rapid	6.0 – 12.0	170.10^{-10} – 350.10^{-10}
Rapid	12.0 – 25.0	350.10^{-10} – 700.10^{-10}
Very rapid	25.0	700.10^{-10}

With the described permeameter the hydraulic conductivity value may be influenced by the porous plates. It is of paramount importance to choose plates that have a conductivity to water, several orders of magnitude larger than the one expected for the soil. This may not always be possible, especially for unsaturated samples, in which case corrections should be made to eliminate the error.

---

difference corresponding to dQ is $dH = dh_1 + dh_2 = [(a_1+a_2)/a_2]\ dh_1$ and thus $dQ = a_1a_2/(a_1+a_2)\ dH$. By combining these equations one gets $[(l/A)a_1 \cdot a_2/(a_1 + a_2)]\ dH/dt = -(K/L)H$.

Separating the variables and integrating from time t_1 to time t_2 we obtain:

$$\int_{t_2}^{T_1} K dt = -(L/A).\, a_1a_2/(a_1 + a_2) \int_{H_1}^{H_2} dH/H$$

which gives $K = [(L/A)(t_2 - t_1).\, a_1a_2/(a_1 + a_2)]\ [1n\, H_1 - 1n\, H_2]$, where A is the cross-sectional area of the soil sample, L is its length, H_1 and H_2 are the water-head difference across the sample at times t_1 and t_2, respectively, and K is the hydraulic conductivity. When the length units are in cm, the time in sec, K is obtained in cm/sec^{+1}. When $a_1 = a_2 = a$, the formula becomes $K = (La/2A)\ (t_2 - t_1)$ $(1nH_1 - 1nH_2)$, and if the outlet is kept at constant level, $K = [La/A(t_2 - t_1)].\ [1n\, H_1 - H_2.]$

McNeal (1964) reported that, in some cases, especially when swelling soils are used and the water varies sharply in salt concentrations, flow errors may occur. He suggested a special permeameter, which, in these cases, helped to assess and eliminate these errors.

The hydraulic conductivity of soils can be measured in the field by methods that have been devised for depths above groundwater level or below the phreatic level. Below the phreatic level the auger-hole and piezometer methodes are the most frequently employed. The water naturally present in the saturated zone is used, and an adequate solution of the flow equation for the appropriate boundary conditions obtained in each method, gives a fairly accurate value of the hydraulic conductivity, provided one adheres to the conditions imposed for solving the equations.

Measurements above the water table are made using the double-tube method, the shallow well pump-in method, or the permeameter method, which consists mainly of measuring flow rate of water from a lined or unlined auger hole. They are used when the layer to betested is never, or seldom, náturally saturated. These methods are involved and time-consuming, large amounts of water are needed to saturate the soil, and it is difficult to assess accurately whether the boundary conditions required for solving the flow equation in each case are met.

In all field methods for measuring hydraulic conductivity, the reproducibility of the results is poor, and variations of two to three orders of magnitude may be encountered for measurements made a few feet apart. In order to overcome the high variability, a large number of measurements should be made, and this is the reason why these methods are seldom used in irrigation practice.

Infiltrability

Principles

Infiltration describes the disappearance of water at the surface of the soil into the profile. When there is no time lag between the arrival of a water drop and its penetration through the soil surface, the infiltration rate is equal to the water application rate. If at any moment the rate of application (by sprinkling of flooding) is so high that not all the water disappears at once, the "infiltrability" of the soil is said to have been exceeded. By definition the infiltrability is the maximum rate at which water penetrates into the soil at a given moment, the soil surface being in contact with water at atmospheric pressure. This term has been suggested by D. Hillel (1971) to replace the often-criticized "infiltration capacity" (see L. A. Richards, 1952). It is also sometimes defined as the difference between the rate of water application and the rate of runoff from an irrigated plot. Thus, infiltrability is a most important soil property for planning irrigation systems and schedules. Upon it will depend the rate at which water can be applied, the time necessary to apply an appropirate amount of water, and consequently the size and quantity of irrigation equipment to be used. The infiltrability of a given soil is not a constant. It decreases with time from the beginning of the irrigation; it depends upon the initial water-content distribution within the profile, and changes in the same way as the hydraulic conductivity with changes in the physicochemical properties of the soil such as mineralogical and chemical composition and structure (status and stability of the particle aggregation).

The Time Dependence. Many equations have been proposed. It seems that a reasonable approximation is given by the Philip (1957a) equation:

$$i = 1/2\, St^{-1/2} + A \tag{4}$$

where i is the infiltration rate, S is called the sorptivity and appears to be a fairly good soil characteristic, and A is a function of the soil properties, which occurs as a parameter in the theoretical development of the equation, and approximates the constant value of i for large values of t. Fig. 6 shows the general appearance of the infiltrability-time curve. The infiltrability approaches a rather constant value for large t (PHILIP, 1967b).

Effect of Initial Moisture Content. Fig. 6 also shows that for an initially wet soil the infiltrability curve will remain below the "dry soil" curve. Qualitatively, this can be explained as follows: The hydraulic gradient responsible for the infiltration rate is the one existing between the surface layer (practically zero potential as this layer will generally be nearly saturated) and the layer ahead of the wetting front. The higher the initial moisture content, the closer to zero is the water potential in the soil ahead of the wetting front, and the potential gradient is thus smaller. Although the wetting front progresses at a higher velocity into wet than into dry soil, the infiltrability will be lower.

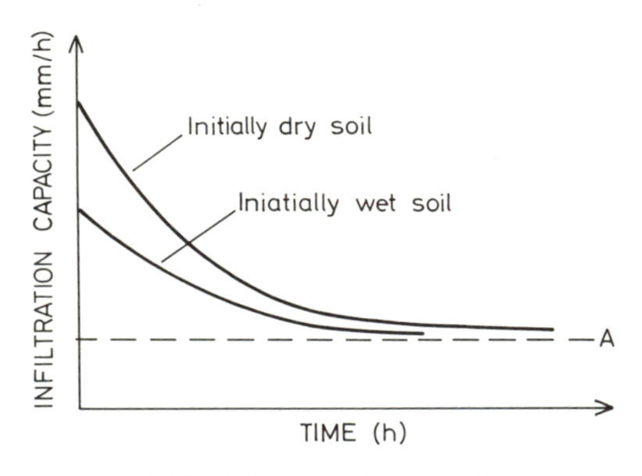

Fig. 6. Infiltrability curves for dry and wet soils

The Effect of Physicochemical Soil Properties. The influence of the soil texture is fairly obvious as the water paths will generally be narrower, and thus resistance to liquid flow larger, when the percentage of small particles is higher. However, because of the tendency of soil particles to stick together, building aggregates held by electrostatic or other forces, very-fine-textured soils may have a rather high hydraulic conductivity or infiltrability. The soil structure thus created may not be very stable, but be subject to breakdown by the leaching of divalent ions (Ca, Mg). These ions will be replaced by monovalent ones (expecially Na), whose efficiency in flocculating the soil colloids is very much lower. In this case the infiltrability may decrease very rapidly because of the dispersion of the soil aggregates. On the other hand, the type of clay mineral present in the soil may also greatly influence the behavior of the infiltrability. Swelling minerals (as for example, the montmorillonitic type) can very strongly reduce the hydraulic conductivity and infiltrabilitiy of soils by restricting the interparticle pathways. However, even if nonswelling clay minerals are present, or if the particles are not well aggregated, they may move around and block pathways, thus having a similar effect of reducing infiltrability.

Methods of Measurement

Flooding Method. The following method, being the simplest and cheapest, is the one most commonly used. A metal cylinder, with a diameter larger than 30 cm and a length of about 40 cm, is driven in the soil, leaving only 5–8 cm above ground[3], if possible. When infiltration to a depth greater than the one to which the ring has been driven is to be measured, this cylinder is placed concentrically inside a larger ring – which may consist of rough planks or of a soil levee – that allows flooding of the soil outsinde the cylinder to the same depth as it is inside. A Mariotte bottle is used to keep the water level constant in the metal cylinder (see Fig. 7). As the water level in the measuring burette is the same as that in the calibrated metal container, it is possible to compute the volume of water that leaves the container by recording the change of water level in the burette with time. This gives the cumulative amount of water that infiltrates into the soil through the metal cylinder where the water level is kept constant. As the area

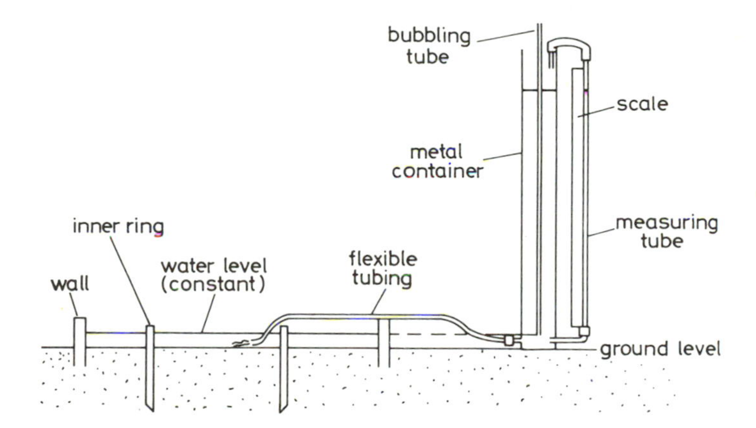

Fig. 7. Double ring infiltrometer setup for measurement in the field

of this cylinder is known, the infiltrability (which is the time derivative of the cumulative curve described above) can be calculated. It is generally given as the volume of water infiltrated per unit area per unit time (e.g., m^3/ha/hr) or in length per unit time (e.g., mm/hr).

It is important to use a cylinder having at least a 30 cm diameter and to drive it into the soil for at least 20 cm, if a significant measurement is to be made.

The radius of the buffer area should be at least 20 to 30 cm larger than that of the inner ring, and should be kept covered with water throughout the measurement. For best results, the water depth should be the same as in the inner ring, as leakage may occur from one area to the other. The buffer area is necessary to insure vertical movement of water in the measured area. When only one ring is used, and if its diameter is relatively

3 It is recommended that the edge of the cylinder be sharpened to facilitate its penetration into the soil.

small, water will move sideways below the depth to which the ring is driven, giving unrealistically high values for the infiltration capacity.

The surface of the ground should be smoothed and leveled. For soils where slaking is likely to occur, it is advisable to protect the ground surface by covering it with coarse cloth or a thin layer of gravel before applying the water. Measurement should be continued until the infiltration rate is practically constant. Adequate water should be used (the same that will be used for irrigating). It is most important to assess the soil and water properties, especially regarding salt concentration.

Artificial Rainfall Method. Sometimes, especially when sprinkler irrigation is to be used, this method is recommended instead of the flooding method. Many rainfall simulators have been described, and the reader is referred to the pertinent literature or reviews (BERTRAND and PARR, 1961). The principle of the method is to irrigate a plot of known area at a rate sufficient to produce runoff. By measuring the rates of water application and runoff, the characteristic curve of infiltrability of the soil is obtained. The rainfall simulator must have characteristics comparable to the sprinkler system to be used: drop-size distribution, uniformity over the test plot, drop velocity, and intensity (or water application rate). It is important to apply the irrigation to an adequate buffer zone surrounding the measured plot in order to insure unidimensional flow. Precautions as to choosing the site or sites, and assessing the soil and water characteristics, are the same as stated for the flooding method.

Comments

Infiltrability is a highly varying property, in time and in space. The reproducibility of any single determination is rather low, and the less uniform the soil, the more determinations there will have to be made. It is also important to take into consideration the purpose of the determination before deciding which method to use. For example, for coarsetextured soils having a rather high infiltrability, it is recommended that the artificial rain method he used and that relatively short runs be made (it is advisable not to wet deeper than about 60 cm of soil). For fine-textured compacted soils, the flooding method is more advisable. However, the infiltrability curve is rather difficult to evaluate. For many practical purposes, the infiltrability after one hour of irrigation or the final infiltrability can be used as a soil characteristic. Table 2 provides a rough classification of soils with regard to their infiltrability.

Table 2. Soil classification in respect to infiltrability after one hour of irrigation

Class	Infiltrability after one hour of Irrigation (mm/hr)	Remarks
Very low	5	Flood irrigation recommended
Low	5–15	Sprinkling possible (low intensity)
Medium	15–25	Flooding possible only on short strips
High	25–50	Flooding impossible
Very high	50	All irrigation rather difficult

Field Capacity

Principles

Field capacity has been defined and described on p. 104 and thus we shall limit ourselves here to a few remarks relevant to its measurement. Although water movement does not strictly cease when field capacity is reached, any further decrease in moisture content is so slow that for limited periods of time (as are encountered in irrigation schedules) it may be neglected, and for most practical purposes the field capacity is considered as a limit. A soil will not remain, under conditions of free drainage, at a moisture content above field capacity, and water will not move out of a soil profile at field capacity when only gravitational forces are applied. In other words, practically speaking, field capacity is considered a reasonably stable equilibrium value. It must also be remembered that very seldom does a uniform profile occur in the field, and we shall not obtain the ideal profile of constant moisture content within the wetted depth when drainage is practically nil. Thus, the volumetric moisture content at field capacity is determined in each predetermined layer (it is advisable to choose the natural genetic layers of the soil profile).

Method of Measurement

The field capacity is determined at a number of sites in the field, according to its uniformity (usually one site per ha). At each site a small plot (2 m × 2 m) is chosen, and irrigated with an amount of water sufficient to wet it thoroughly below the depth to which the field capacity is to be determined. Then the plot, bare of vegetation, is well covered with a plastic sheet or wet sacks and earth to prevent evaporation. The water content within this irrigated plot is determined every day at depth intervals corresponding to the resolution that has been decided upon, until the water content no longer changes appreciably. This is the "field capacity." For nonuniform soils each layer (especially if genetic layers of the profile are chosen) may have a different field capacity. We wish to remind the reader here that for irrigation purposes the water content by volume is required; i.e., it is imperative to know the value of the bulk density for each layer if the water-content determinations are made on a weight basis.

Comments

It has been found experimentally that in many cases, mainly for medium-textured soils, the value of field capacity is very well approximated by the 1/3-bar water content (VEIHMEYER and HENDRICKSON, 1931). The 1/3-bar water content is commonly determined as an alternative laboratory estimate of the field capacity.[4]

The field capacity, in spite of its being a water content whose value has been found to be rather difficult to reproduce for a given soil in a given field, has been recognized as one of the soil-water characteristics and is used in classifying soils, especially when irrigation or drainage practices are involved. Its value is generally higher the finer the soil texture and the more compacted the profile is. In irrigation practice the field capacity

---

4 For coarse-textured soils one usually takes as an estimate the 1/10-bar water content and for very fine textured soils the 1/2-bar water content.

is used to calculate the amount of water one has to apply at a single irrigation, when the average water content by volume $\bar{\theta}$ of the soil profile is known. This amount will be $W_D = 10\,(\theta_{FC} - \theta)\,D$, which is the number of mm of water necessary to bring the soil from $\bar{\theta}$ to $\bar{\theta}_{FC}$ to a depth of D cm, θ_{FC} being the average water content at field capacity within the depth D. W_D is also called the „water deficit“ in the soil to a depth of D cm.[5]

Permanent Wilting Percentage

Definitions

Let us assume that we follow the changes in moisture content in a profile that has been irrigated and where plants are growing. In this case, the water content corresponding to field capacity will be reached in a relatively short period, but because of the presence of plants, it will continue to diminish at an appreciable rate due to evapotranspiration. After a few days, the growing plants will show signs of stress and wilting will occur, first in the older parts (leaves) and then in the plant as a whole. At first the plants will recover their turgor if placed in a water-saturated atmosphere – but as the water content is further reduced, the older leaves wilt permanently, without being able to recover in a saturated atmosphere (the younger parts will recover). The water content of the soil at this stage is the incipient wilting point. If the experiment is continued until the whole plant wilts permanently, the water content determined then is the permanent wilting percentage. Experience shows that plant growth ceases at the incipient wilting point, and the water content in the soil cannot change much more by plant extraction alone. However, there may be an appreciable, though small, change of water content between the two points, especially for fine-textured soils. The definitions thus arrived at are (PETERS, 1965): "The incipient wilting point is the water content at which the lowest pair of true leaves of a particular kind of plant at a particular stage of growth wilt and fail to recover in a saturated atmosphere. (The elongation of plants ceases at about the same water content.) On the other hand, the permanent wilting point is the water content of a soil when plants growing in that soil are first reduced to a wilted condition from which they cannot recover in a saturated atmosphere."

Method of Determination

About 500–600 g of air-dried soil, sieved through a 2-mm sieve, is placed in a weighed watertight container with a hole in its lid. A water-content determination is made on a separate soil sample to get the oven-dried weight of the soil in the container. Nutrient solution is added to the soil, which is thus moistened to a water content corresponding roughly to the 1/3-bar percentage. A few seeds of dwarf sunflower (*Helianthus annus,* large seeded variety) are planted and allowed to germinate. When the seeds have germinated, all but the best one are cut of at the soil-surface level and discarded. The remaining seedling is then led through the hole in the lid and the plant is grown in the green-

5 In practice, the amount of water applied will be higher than this value in order to allow for inevitable losses and uneven distribution of the water, as well as for the lack of uniformity in the soil.

house until the third pair of leaves is completely developed. The water supply is kept adequate for normal plant growth by weighing the cans, and watering them regularly to the same water content (allowance can be made for the weight of the growing plant by estimate). Finally, the container is watered one last time to the original water content, and the space between the stem and the sides of the hole in the lid is filled with cotton in order to reduce direct evaporation from the soil to as low a level as possible (care must be taken to avoid anaerobic conditions for the roots). The plants are left until the lowest pair of true leaves wilts. The container is transferred to a dark humid chamber, and if the leaves recover their turgidity after 10 to 15 hr, the plants are brought back to the greenhouse, and the procedure is repeated until the leaves fail to recover turgor. When the lowest pair of leaves fails to recover (the others should recover), the plant is cut off at the soil-surface level and the soil-water content determined. This corresponds to the incipient wilting point. The procedure is continued until all three pairs of true leaves wilt without recovering.

Comments

The wilting point is generally a well-reproducible characteristic, although it is not an intrinsic soil property. The reason for this is the fact that the water content changes very slowly as the water potential approaches the wilting range. Experience has shown that for very many soils the wilting point is very closely approximated by the 15-bar moisture percentage. This determination is generally made routinely in the laboratory (pressure-plate method), instead of by the cumbersome, time-consuming plant-response method.

The wilting point is the water content at which plants no longer can extract water from the soil for the purposes of growth, and thus it is the lower limit of available soil water to plants. The range of water content between the permanent wilting point and the field capacity is sometimes referred to as the "readily available water capacity of the soil." (RAWC). It should be noted here that this RAWC will vary from one soil type to another, mainly, according, to texture:

China silty clay = 46.2 − 34.3 = 11.9%
Pachappa fine sandy loam = 20.5 7.9 = 12.6%
Hanford sand = 11.9 − 4.1 = 7.8%

In other words, for medium and fine-textured soils the readily available water capacity measured in percent by volume will remain roughly constant in the range of 11–13%, whereas, for coarse-textured soils it will go down to as low as 4–8%, implying frequent, small water applications in the irrigation schedule.

Literature

Bertrand, A. R., Parr, J. F.: Design and operation of the Purdue sprinkling infiltrometer. Purdue Univ. Res. Bull. No. 723 (1961).
Hillel, D.: Soil and Water: Physical Principles and Processes. New York: Academic Press 1971.
Klute, A.: Laboratory measurement of hydraulic conductivity of saturated soils Agronomyg. Amer. Soc. Agron., Madison, Wis., 1965.

MCNEAL, B. L., REEVE, R. C.: Elimination of boundary flow errors in laboratory hydraulic conductivity measurements. Proc. Soil Sci. Soc. Amer. **28.** 713 (1964).

OLSEN, H. W.: Deviation from Darcy's law in saturated clays. Proc. Soil Sci. Soc. Amer. **29.** 135 (1965).

O'NEAL, A. M.: A key for evaluating soil permeability by means of certain field clues. Proc. Soil Sci. Soc. Amer. **16.** 163 (1952).

PETERS, D. B.: Water Availability. Agronomy 9. Amer. Soc. Agron., Madison, Wis., 1965.

PHILIP, J. R.: The theory of infiltration: 4. Sorptivity and algebraic infiltration equations. Soil Sci. **84.** 257 (1957a).

PHILIP, J. R.: The theory of infiltration: 5. The influence of initial moisture constant. Soil Sci. **84.** 329 (1957b).

RICHARDS, L. A.: Report of the Subcommittee on Permeability and Infiltration. Committee on Terminology. Proc. Soil Sci. Soc. Amer. **16.** 85–88. 1952).

VEIHMEYER, F. J., HENDRICKSON, A. H.: The moisture equivalent as a measure of the field capacity of soils. Soil Sci. **32.** 181 (131).

4

The Estimation of Evapotranspiration

M. FUCHS

To determine irrigation requirements and to design irrigation systems, the irrigationist must know the evaporation losses from crops. The physics of evaporation treated in Chapter 3 indicate that these losses could be assessed from purely meteorological considerations. However, the exact meteorological methods require specialized instrumentation that is generally not available to the field irrigationist. Furthermore, these methods are inadequate when it is necessary to forecast the irrigation needs of new areas. A large number of empirical and semiempirical formulas relating evapotranspiration to standard climatological data have been developed to obviate these shortcomings. In this chapter we shall describe the most commonly used formulas. More detailed reviews and complete references are available in HAGAN *et al. (1967)*.

Potential Evapotranspiration

The determination of irrigation requirement, from meteorological data does not take into account plant factors, such as drought resistance. Therefore, these methods estimate *potential* rather than *actual evapotranspiration*. The potential evapotranspiration of a crop under a given set of meteorological conditions is defined as that occurring from a crop surface where the water vapor pressure is at the saturation point. We have seen in Chapter 3 that the radiative and aerodynamic properties of crop surfaces modify the effect of the meteorological factors on evaporation. Consequently, the potential evapotranspiration will vary from crop to crop.

Potential evapotranspiration can be determined experimentally by measuring evaporation from crop surfaces maintained under wet conditions. It can also be computed by combining the energy balance equation and the aerodynamic formula to yield

$$E_p = (R_n\text{-}G)\,\triangle / \lambda\,(\triangle + \gamma) + h\,(e_s\text{-}e)\gamma / (\triangle + \gamma) \qquad (1)$$

$$h \;= k_2\,(\varrho\varepsilon / P)\,u / [\,\ln(z + z_o - d) / z_o]^2 \qquad (2)$$

$E_p =$ Potential evapotranspiration, g water/cm^{-2}min^{-1}
$R_n =$ net radiation, cal cm^{-2} min^{-1}
$G =$ ground-heat flux density, cal cm^{-2} min^{-1}
$\triangle =$ slope of the saturation water vapor pressure curve at air temperature, mb C^{-1}
$\lambda =$ latent heat of vaporizaion of water, 585 cal (g water)$^{-1}$
$\gamma =$ psychrometric constant, mb °C^{-1}
$e_s =$ saturation water vapor pressure of the air at height z, mb
$e =$ water vapor pressure of the air at height z, mb

$k =$ Karman constant, 0.42 dimensionless
$\varrho =$ density of the air, 1.22×10^{-3} g cm^{-3}
$\varepsilon =$ molecular weight ratio, 0.622, dimensionless
$P =$ barometric pressure, mb
$u =$ wind speed at height z, cm min^{-1}
$z =$ height above the soil, cm
$z_o =$ roughness length, cm
$d =$ zero plane displacement, cm
Values of e_s, Δ, and γ are listed in Table 1.

Table 1. Thermodynamic parameters used in the computation of the potential evapotranspiration from Eq. (1). Atmospheric pressure: 1000 mb

T (°C)	e_s (mb)	Δ (mb C^{-1})	γ (mb C^{-1})	$\Delta/(\Delta+\gamma)$	$\gamma/(\Delta+\gamma)$
0	6.108	0.443	0.646	0.406	0,593
5	8.719	0.606	0.650	0.482	0.517
10	12.27	0.819	0.654	0.555	0.444
15	17.04	1.093	0.659	0.623	0.376
20	23.37	1.440	0.663	0.684	0.315
25	31.67	1.875	0.668	0.737	0.262
30	42.43	2.417	0.672	0,782	0.217
35	56.24	3.082	0.676	0.819	0.180
40	73.78	3.898	0.681	0.851	0.148
45	95.86	4.878	0.685	0.876	0.123

Eq. (1) is an improved form of the equation proposed by PENMAN (1948). It includes all the parameters that play a role in evaporative processes. However, its complexity and the accuracy required for the measurements limit its use to research problems and to the determination of diurnal trends of potential evapotranspiration (VAN BAVEL, 1966). It is also of great importance for the investigation of the relationship between potential and actual evaporation.

A simplified version of Eq. (1) is Penman's original equation, which neglects the contribution of the surface roughness and the ground head flux density:

$$E_p = R_n \Delta / (\Delta + \gamma) + E_a \gamma / (\Delta + \gamma) \text{ mm day}^{-1} \tag{3}$$

where R_n is expressed in equivalent mm day^{-1} of evaporation (an example of the transformation cal cm^{-2} min^{-1} into equivalent mm hr^{-1} of evaporation is given in Chapter 3), and E_a is an aerodynamic function, which is given as

$$E_a = 0.35\ (0.5 + 0.0062u)\ (e_s\text{-}e) \text{ mm day}^{-1} \tag{4}$$

where u is the wind speed at 200cm height in km day^{-1}, and $(e_s\text{-}e)$ is in mb. This equation correctly predicts evaporation from open water surfaces and from well-irrigated short turf, over weekly periods.

Solar Radiation Formulas

The strong dependence of potential evapotranspiration on the radiation term of Eq. (3) has given rise to a series of formulas based upon solar radiation measurements. These formulas eliminate the effect of the surface albedo and minimize the contribution of the aerodynamic term.

MAKKINK (1957) introduced the following equation:

$$E_p = R_s \Delta / (\Delta + \gamma) + 0.12 \text{ mm day}^{-1}. \quad (5)$$

R_s is the incoming solar radiation expressed in equivalent mm day^{-1}.

The aerodynamic term of Eq. (3) is replaced by the constant 0.12. Makkink's equation gives good results in areas like Holland where the climate is cold and wet, but is not satisfactory in arid regions. This can be explained by the fact that during the growing season in Holland, the contribution of the aerodynamic term is small, and that the water-vapor saturation deficit of the air varies little. In contrast, in arid climates the aerodynamic term is quite large and varies over a wide range.

JENSEN and HAISE (1963) proposed an equation similar to Eq. (5), but which includes the air temperature **T** in degrees C:

$$E_p = R_s (0.025 \text{ T} + 0.08) \text{ mm day}^{-1} \quad (6)$$

The air temperature indirectly introduces the contribution of the aerodynamic term into the estimate of E_p. Eq. (6) has been derived from data collected in arid regions of the western part of the United States and should yield satisfactory results in areas with similar climate.

Other empirical relationships between potential evapotranspiration and incoming solar radiation assume the form of a simple linear regression:

$$\text{E}_p = mR_s + n \text{ mm day}^{-1} \quad (7)$$

STANHILL (1961) found for alfalfa in Israel that $m = 0.72$ and $n = 0{,}87$ for weekly estimates, and $m = 0.72$ and $n = 1.04$ for monthly estimates. Other proposed values for m and n are given by TANNER (1967).

A major disadvantage of the formulas based upon solar radiation is that this statistic is not routinely recorded at standard meteorological stations. In Israel, for example, solar radiation is regularly recorded at only three sites. Solar-radiation maps of Israel drawn by Stanhill indicate that this number is inadequate if the data are to be used at locations that are not in the vicinity of the site of measurement.

Table 2. Comparison between various equations used to calculate potential evaporation using radiation data

	Eq. [1]	Eq. [3]	Eq. [5]	Eq. [6]	Eq. [7]	Mean	Coefficient of variation%
E_p mm day^{-1}	6.5	7.9	6.0	5.0	7.0	6.48	17

An example of various computations of potential evapotranspiration is summarized in Table 2. The meteorological data used are representative of an April week in the northern Negev of Israel. The crop is an irrigated alfalfa field where the canopy has an average height of 40 cm ($z_o = 4$ cm, $d = 24$ cm). Mean solar radiation is 500 cal cm^{-2}

day^{-1}; mean net radiation is 280 cal cm^{-2} day^{-1}; mean air temperature is 20°C; mean relative humidity is 50%; and mean wind speed at 200cm height is 100 km day^{-1}.

Aerodynamic Formulas

Another series of formulas estimate E_p only the aerodynamic term of Eq. (1). Most proposed formulas are of the following form:

$$E_p = c(a + \mathrm{bu})\ (e_s \text{ -}e) \tag{8}$$

Here a, b and c are empirically determined coefficients. This approach utilizes the wind run u and the water-vapor deficit the air, which are recorded at most weather stations. However a, b and c vary considerably from one site to another, and from one period of the year to another. This method is mainly used for computing evaporation from large water reservoirs.

Temperature Formulas

There is no direct relationship between evaporation and air temperature. Temperature and potential evapotranspiration are positively correlated only because they depend upon the same meteorological factors. Nevertheless, the temperature methods are widely used because temperature is the most readily available meteorological parameter.

A simple formula has been proposed by BLANEY and CRIDDLE (1950) for monthly potential evapotranspiration:

$$E_p = CDT \tag{9}$$

Here C is a „consumptive-use“ coefficient that varies with time, location, and crop; D is the ratio of daylight hours of the considered month over the yearly total of daylight hours; and T is the mean monthly air temperature. It is necessary to determine locally the value of C for each crop to obtain good results from Eq. (9).

The Thornthwaite equation (1948), which is more complex in form, does not include a crop factor:

$$E_p = 1.6\ (L/12)\ (N/30)\ (10\ \mathrm{T/I})^a \ \text{cm month}^{-1} \tag{10}$$

where L is actual day length in hours N is the number of days in the month and T is mean monthly air temperature,

$$I = \sum_1^{12} (\mathrm{T}/5)^{1.514}$$

and

$$a = (0.675\ \mathrm{I}^3 - 77.1\ \mathrm{I}^2 + 492{,}390)\ 10^{-6}$$

Eq. (10) generally underestimates the measured potential evapotranspiration in arid areas. The use of the formula is facilitated by available tables and graphical solutions (e.g., THORNTHWAITE and MATHER, 1955).

Evaporation-Pan Formulas

Potential evapotranspiration and evaporation from pans are governed by the same meteorological factors. Consequently, there exists a strong correlation between them. However, as pans and crop surfaces react differently to the meteorological conditions, it is necessary to establish experimentally the relationship between potential evapotranspiration and pan evaporation.

The relationship also depends upon the size, the shape, and the exposure of the pan. If the pan calibration is to be extrapolated to other sites, it is necessary to standardize the type of the pan and its exposure.

The U.S. Weather Bureau class A pan is the most commonly used evaporation pan. It is easy to install; also the basic investment and the maintenance costs are low (STANHILL, 1961). The class A pan has a diameter of 120 cm and a height of 25 cm, and is installed on a leveled wooden platform about 10 cm above the ground. When used in arid areas, the evaporation pan should be screened to keep out birds and other animals in quest of water. As screens intercept part of the radiation, it is necessary to construct them to well-defined standard specifications.

The relationship between E_p and pan evaporation E_o is generally obtained from linear regressions using determinations of E_p and E_o over weekly to monthly periods.

$$E_p = f E_o + s \text{ mm day}^{-1} \quad (11)$$

Values of f and s, quoted by TANNER (1967), are summarized in Table 3.

Table 3. Relation between potential evapotranspiration and pan evaporation according to Eq. (11)

Period	Time interval	Location	Crop	f	s	r[a]
All year	month	Israel	alfalfa	0.70	0.47	0.95
All year	week	Israel	alfalfa	0.75	0.36	0.77
Jan.-May	day	California	rye grass	0.67	0.45	0.94
Jan.-May	month	California	rye grass	0.79	0.08	0.99
July-Dec.	day	California	rye grass	0.77	0.03	0.90
July-Dec.	month	California	rye grass	0.76	-0.02	0.98

[a] r = coefficient of correlation.

The relation between E_p and E_o is often given as a simple ratio. This form of the relation is very convenient for computing irrigation requirements. FUCHS and STANHILL (1963) found that the water requirement of cotton after the first irrigation was equal to

$$0.69 \sum_{i=1}^{n} E_o(i)$$

where n is the number of days between two successive irrigations. The coefficient 0.69 was valid throughout the active growing season of cotton and for all regions in Israel.

Reported values of E_p/E_o range from 0.6 to 2.0 Most of the variation can be explained by pan exposure. The larger values generally obtain when the pan is in a partly shaded site or inside an irrigated field. TANNER (1967) recommends that the evaporation pan be placed within the boundaries of the irrigated field. This procedure should improve the correlation between pan evaporation and potential evapotranspiration, because the pan and the crop then have the same microclimate. However, this practice requires periodic adjustment of the height of the pan-mounting platform as the crop grows. In the case of overhead irrigation, evaporation cannot be measured at irrigation time. On the other hand, FUCHS and STANHILL (1963) have obtained good results by siting the pans in unirrigated areas with a fetch of at least 200 meters. Under these conditions, a single evaporation pan can be used for several fields and crops.

Other types of evaporation devices, such as sunken tanks and atmometers, are sometimes used. The sunken tanks are difficult to install and to maintain. Atmometers and small pan evaporimeters are generally less reliable than the class A pan. They should be proscribed because they introduce diversity where uniformity and standardization are absolutely required.

The Surface Measurement of Evapotranspiration

Two methods of estimating the evapotranspiration from land surfaces are based upon changes of the water content of the soil. First, we can periodically monitor the soil-moisture profile and by comparing two successive profiles compute the water loss from the soil. However, this method does not separate evaporation from downward drainage through the profile, the heterogeneity of the soil causes difficult sampling problems, and the minimum time lapse between two successive sets of soil-moisture measurements that yield reasonably accurate evaporation estimates is of the order of a week. Furthermore, measurements cannot include the period immediately following an irrigation because the gravitational redistribution of the water in the soil profile may require several days.

The second method is weighing lysimetry. A lysimeter is a tank filled with soil in which all the conditions of the field surrounding the tank are reproduced as exactly as possible. By monitoring the drainage and the change of weight of the tank, the evaporation can be assessed. Good lysimeters are difficult to design and the cost is always high. TANNER (1967) gives an excellent discussion of lysimetry and its problems. The main drawback of lysimeters for the irrigationist and hydrologist is that they are not mobile. Their usefulness is limited to basic studies and to the calibration of other methods of estimating evaporation.

Concluding Remarks

All the empirical methods using meteorological parameters to determine irrigation requirements need to be checked and locally calibrated for best results. The choice of a method should be governed by practical considerations. When introducing irrigation in a previously unirrigated area, one should base the choice on the kind of meteorological data that have been collected there in the past. If the method is to be used to compute the amount of water to be applied from one irrigation to the next, one should consider the time that is required for the relevant meteorological data to become available in processed form.

The various methods can be graded on the basis of the observed correlation coefficients. Accordingly, solar radiation methods offer the best results. Methods using pan evaporation are next, and they are very well suited for indicating irrigation needs when the time interval between successive irrigations is more than a week. The temperature methods are generally not very reliable unless used over periods of a month. They are mainly used as guidelines in the design of irrigation systems and for predicting long-term irrigation water requirements.

Finally, it is necessary to emphasize that the meteorological methods estimate potential evapotranspiration. The resulting recommendations imply that optimum yields are obtained when the water status in the soil permits potential evapotranspiration. Crop physiologists and economists may find that this need not be the case. Nevertheless, recommendations can also be expressed as a fraction of the potential evapotranspiration. An example of such a situation is provided by high-yielding irrigated orange groves in Israel. KALMA (1969) found that the evapotranspiration of these orange groves was only 22% of the potential evapotranspiration computed by Eq. (1). Information of this nature should be used in applying potential evapotranspiration and for optimizing the estimate of irrigation needs.

Literature

BAVEL, C. H. M. VAN: Potential evaporation: the combination concept and its experimental verification. Water Resources Res. **2,** 455–467 (1966).

BLANEY, H. F., CRIDDLE, W. D.: Determining water requirements in irrigated areas from climatological and irrigation data. USDA Soil Conserv. Serv. TP–96, 48 pp. (1950).

FUCHS, M. STANHILL, G.: The use of class A evaporation pan data to estimate irrigation water requirements of the cotton crop. Israel J. Agric. Res. **13,** 63–78 (1963).

HAGAN, R. M., HAISE, H. R., EDMINSTER, T. W.: Irrigation of Agricultural Lands. Amer. Soc. Agron., Madison, Wis., 1180 pp., 1967.

JENSEN, M. E., HAISE, H. R.: Estimating evapotranspiration from solar radiation. Amer. Soc. Civ. Eng. Proc. **89** (IR4), 15–41 (1963).

KALMA, J. D.: Some aspects of the water balance of an irrigated orange plantation. Ph. D. thesis, Hebrew University, Jerusalem 1969.

MAKKINK, G. F.: Ekzameno de la formula de Penman. Neth. J. Agr. Sci. **5,** 290–305 (1957).

PENMAN, H. L.: Natural evaporation from open water, bare soil, and grass. Roy. Soc. London, Proc. Ser. A. **193,** 120–146 (1948).

STANHILL, G.: A comparison of methods of calculating potential evapotranspiration from climatic data. Israel J. Agric. Res. 11, 159–171 (1961).

STANHILL, G.: Solar radiation in Israel. Bull. Res. Counc. Israel **11 G,** 34–41 (1962).

TANNER, C. B.: Measurement of evapotranspiration. In: Irrigation of Agricultural Lands. HAGAN, R. M., HAISE, M. R., EDMINSTER, T. W. (Eds.). Amer. Soc. Agron., Madison, Wis., pp. 534–574 (1967).

THORNTHWAITE, C. W.: An approach toward a rational classification of climate. Geog. Rev. **38,** 55–94 (1948).

THORNTHWAITE, C. W., MATHER, J. R.: The water balance. Publ. in Climatol. **8,** 1–104 (1955).

5

Water Status in Plants – Methods of Measuring

D. SHIMSHI

Introduction

The transpiration flux is envisaged as a catenary process, whereby water moves from the bulk of the soil, through the roots, stems, and leaves of plants to the atmosphere in response to an interplay of potential levels at the various point on the path, and of the resistances along the various segments of this path (VAN DEN HONERT, 1948). Although the predominant tendency has been to relate plant physiological responses to the amount of soil moisture (i.e., to the potential levels in the *soil),* it is now established that these responses are in fact related to the water status in the *plant tissues,* and it is indices of this status that should be measured in order to gain an understanding of the water relations of plants. The following is only a partial survey of methods for characterizing the water status in plants. For more complete information, two outstanding reviews on measurement of water status in plants are by SLATYER and SHMUELI (1967) and BARRS (1968).

Tissue-Water Content

The water content of plant tissue can be measured by weighing the fresh material, oven-drying to remove the water, and reweighing the dry matter. The results can be expressed on a fresh-weight basis: 100×(FW−DW)/FW (where FW = fresh weight and DW = dry weight), with values always lower than 100%, or on a dry-weight basis: 100 (FW − DW)/DW, with values often exceeding 100 percent, and even reaching 1000% in succulent tissues. Although the tendency is for the water content to decrease with increasing water stress, this index is rather unsatisfactory in view of the different proportions of water and dry matter in the various parts of the plant.

Relative Water Content (RWC)

This index measures the actual water content of the plant tissue in relation to the water content of the completely saturated tissue (WEATHERLEY, 1950). It is obtained by weighing the fresh material; allowing it to become saturated, usually by floating it on distilled water for a suitable period; reweighing it after saturation; and then oven-drying it and weighing the dry matter. The index, 100 (FW−DW)/SW−DW, where SW is the weight after saturation, is commonly referred to as the "relative turgidity", but the term "relative water content" seems more appropriate. The saturation procedures vary according to the plant material; in addition to floating leaf disks or segments on water, saturation may be achieved by dipping the petioles of leaves into water. The method requires detachment of the tissue from the plant and is therefore a destructive method.

Since leaves of mesophytes may show severe stress when RWC drops to 80%, a more convenient and sensitive index is the "relative water deficit" (RWD), which is 100−RWC. The saturation period should be long enough to approach full saturation, but short enough to prevent appreciable changes in the dry weight due to respiration or other metabolic changes. A period of 3–4 hours is common for floating disks.

Although RWC and RWD may reveal differences of water status even between various parts of the plant, they may be affected by factors that are not related to water status, such as the rigidity of the cell walls.

Beta Gauging Methods

This method gives an indirect measurement of the mass of a given area of leaf tissue. The leaf is placed between a beta-ray source and a detector, the count rate decreasing with the increase of mass/area value (Fig. 1). Over short periods, the changes in this value are mainly attributable to changes in water content of the leaf (MEDERSKI, 1961). This method is nondestructive, and it enables continuous observation of the water content of any particular leaf area. However, calibration for each type of leaf is necessary; also, the equipment is expensive and requires access to electronic workshop maintenance.

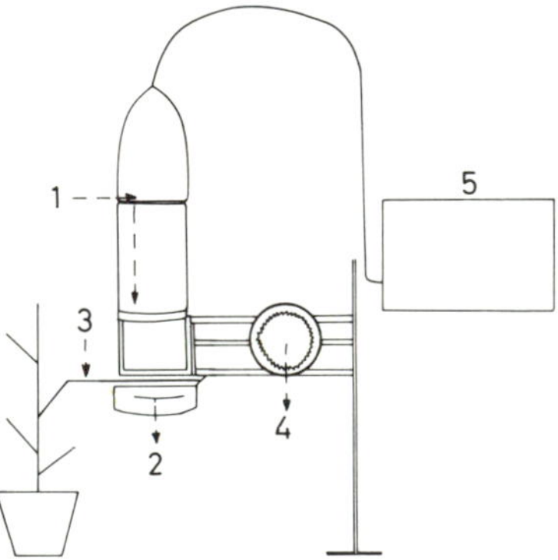

Fig. 1. Schematic description of beta-ray gauging of leaf thickness. 1−thin window G. M. detector; 2−beta-ray source; 3–leaf; 4−device for positioning source and detector; 5−scaler

Measurements of Water Potential

Since it is believed that the behavior of plant tissue is intimately related to its water potential, much work has been done to develop methods of measuring the potential, using various approaches. Following are some examples.

Contact with Graded Osmotic Solutions

These methods involve the use of the isopiestic principle. Multiple samples of tissues are brought into contact with a series of water solutions having different osmotic potentials. The equilibration process between the tissues and the solutions will cause changes in some properties of the tissue (and sometimes of the solution), except in that system where the water potential of the tissue and the osmotic potential of the solution are initially equal. Since the processes on the two sides of the equilibrium system proceed in opposite directions, it is usually enough to observe the initial direction of the processes, and to find by interpolation the null point of solution concentration equivalent to the plant-water potential. The changes observed are of several kinds:

1) The most widely used method is based on floating leaf disks on graded osmotic solutions for a certain period. By weighing the disks before and after floating (with due care to blot off the water from the disk surfaces), it is possible to determine the osmotic potential of the solution (called the isotonic solution) that caused no change in the weight of the disks.

2) When the tissue is placed in contact with a small volume of osmotic solution, the changes occurring in the concentration of the solution are large enough to be measured by various methods, such as by the refractive index. The SHARDAKOV (1938) dye method is another such method: duplicate series of osmotic solutions are prepared; one series consists of small volumes of the appropriate solutions, labeled with some osmotically inactive dye. After bringing samples of tissue into contact with the dyed solutions for a period of time, these solutions are gently injected into the corresponding undyed solutions. The dyed solution will move upward or downward, according to whether its osmotic potential has been raised or lowered during equilibration. The transition between the rising and the sinking solution roughly locates the range of tissue-water potential.

Contact with Graded Vapor Concentrations

The approach is similar to the previous methods in that it is based on the isopiestic principle. Samples of tissue are placed in vapor chambers of various vapor pressures, and the vapor pressure that causes no change in tissue weight is converted to the equivalent units of water potential. The vapor equilibration methods require the use of accurately thermostated baths, since very small temperature gradients may introduce large errors of estimation (a change of 0.1° C may cause an error of about 10 bars).

The main disadvantage inherent in both the liquid and vapor equilibration methods is the use of multiple samples of tissue for determining the isopiestic point. This may cause errors due to differences of water potential between the samples (e.g., the disks punched out of the same leaf), and it also involves tedious manipulation of multiple samples. Also, the methods are destructive.

Psychrometric Methods

These methods are based on the following principle: Samples of tissue are enclosed in a small chamber, and the system is allowed to reach a vapor pressure that is in equilibrium with the tissue-water potential. A drop of pure water suspended in this system will evaporate and cool, according to the vapor pressure in the chamber. This degree of cooling is linearly related to the vapor-pressure deficit, and thus to the water potential of the tissue sample. The temperature of the evaporating water is measured by means of thermocouples, and the electromotive force generated by the cooling is amplified and recorded.

This principle has been developed along two parallel lines. In one, the thermocouple is wetted by dipping it in pure water before operating the apparatus; this is the RICHARDS-OGATA (1958) method (Fig. 2). In the other, water is condensed on the thermocouple by cooling it by means of the Peltier effect; this is the SPANNER (1951) method.

The principal advantage of the psychrometric method is that it is a single-sample method. Also, the readings of the amplified electromotive force are linearly related to the water potential. The apparatus is calibrated by placing in the equilibration chamber disks of filter paper soaked with solutions of various known osmotic potentials.

The disadvantages of the method include the need for precise temperature control, the complexity and cost of the equipment, and the dependence on a power supply. These disadvantages make the method unsuitable, at present, for field operation. For laboratory work, the psychrometric method is probably the most precise and reliable.

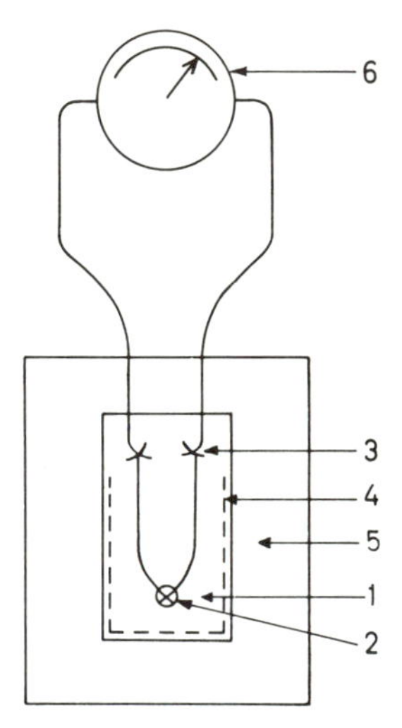

Fig. 2. Thermocouple psychrometer. 1 – equilibration chamber; 2 – thermocouple junction with water drop; 3–dry reference junctions; 4–leaf sample lining chamber wall; 5–thermostat bath; 6 – voltmeter

Pressure-Chamber Method

This method (also referred to as the "pressure bomb"; SCHOLANDER *et al.*, 1965), is based on one of the relations expressing the water status of the tissue:

$$\psi = -\pi + TP \tag{1}$$

where ψ is tissue-water potential, π is the osmotic pressure of the tissue water, and TP is the turgor pressure of the plant tissue.

A twig section (or a leaf) is enclosed in a steel pressure chamber with the cut stem (or petiole) end protruding out through an airtight opening (Fig. 3). Air pressure is gradually applied to the chamber until xylem water begins to be pressed out of the cut end of the stem or petiole. This method assumes that the water pressed out from the tissue back into the xylem moves through semipermeable membranes that do not allow solutes to move with the water. If the applied pressure at equilibrium is AP, then the total plant water potential is zero; accordingly,

$$\psi = -\pi + (TP + AP) = 0 \tag{2}$$

The applied pressure at equilibrium is therefore a measure of the water potential of the tissue. It should be noted that it is assumed that the exuded sap does not contain solutes. If it does, its osmotic potential should be determined, and its value π (*xyl*) subtracted from the equilibrium pressure. The method is now being improved and adapted to rapid field operation.

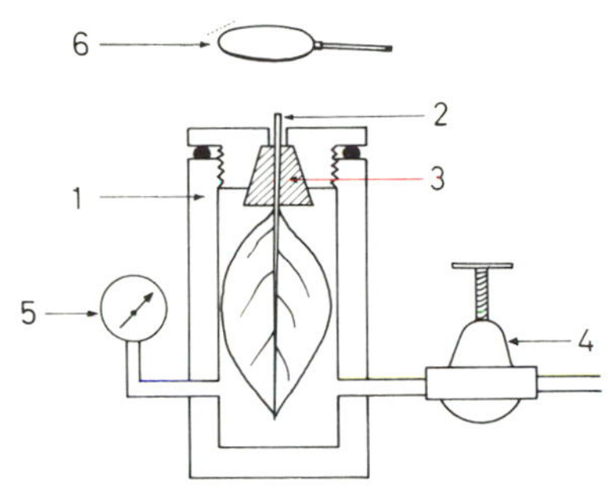

Fig. 3. Pressure chamber method. 1–pressure chamber; 2–leaf petiole protruding through 3; 3–airtight rubber bushing; 4–pressure regulator; 5–pressure gauge; 6–magnifying lens

Measurement of Concentration or Osmotic Potential of Plant Sap

As the plant develops water deficits, its degree of hydration is reduced. This dehydration manifests itself not only through changes of relative water content, but also through changes in the solute concentration of plant sap. The increase of solute concentration may be a passive result of the loss of water; quite frequently, however, water stress causes metabolic processes in the plant that convert insoluble metabolites such as starch into osmotically active solutes (sugars), thereby increasing solute concentration more than can be attributed to mere dehydration. These changes thus may serve as a reliable index of the degree of water stress. Two methods of measuring or estimating osmotic potential of plant sap are described.

Cryoscopy

The depression of the freezing point of sap is linearly related to the lowering of the osmotic potential due to the increase in solute concentration. This method usually requires fairly large volumes of sap (about 0.5 ml), and is suitable mainly for laboratory work. It usually utilizes thermistors and potentiometric recorders.

Refractometry Plus Conductimetry

This method is used to estimate the osmotic potential of drop-size samples of sap, such as are obtained in the field (SHIMSHI and LIVNE, 1967). By means of a microconductivity cell, the concentration and osmotic potential of the electrolytes in the sap are determined; the refractive index as measured with a refractometer is used to evaluate the concentration and osmotic potential of the metabolites. The sum of these two estimates of osmotic potential has been found to agree well with that obtained by means of the standard cryoscopic methods.

All methods of obtaining sap from leaf tissue are based on the disruption of the tissue, usually by freezing and thawing. This causes the resulting sap to be a mixture of various origins (vacuolar, vascular, and cell-wall sap), which does not necessarily represent the osmotic potential of vacuolar sap. It is this sap that is believed to determine the relations between water potential and metabolic processes, since the cytoplasm is probably in equilibrium with this portion of plant water.

Observation of Plasmolysis

By microscopic examination of suitable tissue (e. g., epidermal strips) that have been dipped in a series of graded osmotic solutions, it is possible to determine the osmotic potential of the solution that will cause plasmolysis in 50% of the tissue cells. This concentration is considered to be equal to that of the vacuolar solution.

Stomatal Aperture

It has long been noted that during the daylight hours the degree of stomatal opening may serve as a sensitive indicator of water stress in plants, as stomata tend to close whenever plants develop water deficits (HEATH, 1959). Stomata have microscopic dimensions (usually 10–30 μ long, and up to 15–20 μ wide) and are very sensitive to manipulation. Several methods have been devised to evaluate, either directly or indirectly, the degree of stomatal opening.

Direct Observation

Epidermal strips are quickly removed from the leaf and dipped in absolute alcohol (LLOYD, 1908). This will cause a fixation of the tissue so that stomatal aperture is "frozen" at its original dimensions; the strips thus treated may be mounted and observed with a microscope. This method is slow and tedious, and is suited only to plant species where the epidermis is easily peeled off the leaf.

Epidermal Impressions

The method is based on application of viscous fluids to the stomata-bearing epidermis, which harden or set in a short time as a result of solvent evaporation (collodion, cellulose acetate, etc.) or catalyzed polymerization (silicone rubber). When removed from the leaf surface, the hardened material bears the negative impression of the epidermal relief. By treating this relief with a transparent lacquer or varnish, a positive relief, duplicating the original leaf surface, is obtained, and may be observed with a microscope. The use of silicone rubber (type RTV 11 with a quick-setting catalyst) is very satisfactory, and cellulose acetate "prints" made from silicone rubber bear very clear outlines of the stomata. Although numerous impressions can be performed in the field, their subsequent processing and microscopic observation is still a time-consuming procedure.

Infiltration of Oily Liquids

When a drop of oily liquid (paraffin, kerosene, petroleum ether, benzene, etc.) is placed in contact with a leaf epidermis having open stomata, the liquid will quickly infiltrate into the mesophyll and a dark spot will appear on the leaf; if the stomata are closed, infiltration of the liquid will be very slow (MOLISCH, 1922). This simple method can be used to diagnose the onset of moisture stress in the plant. Each type of leaf requires a different liquid of suitable viscosity. The main disadvantage is the difficulty of expressing the process of infiltration quantitatively.

Viscous-Flow Porometers

These instruments measure the air permeability across the leaf thickness; they can be used only on plants that have stomata on both leaf surfaces (upper and lower), and where the ratio of stomatal frequency between one surface to the other is not greater than 4:1. The permeability of the leaves is determined by the degree of stomatal opening. The porometer usually consists of a rigid container in which a certain air pressure is built up, and then released to a pair of cups clamped on the leaf. The rate of pressure drop in the system is proportional to the permeability of the leaf, and is related to the degree of stomatal opening (SHIMSHI, 1967). The field operation is rapid and simple, and the observations can be interpreted on the spot.

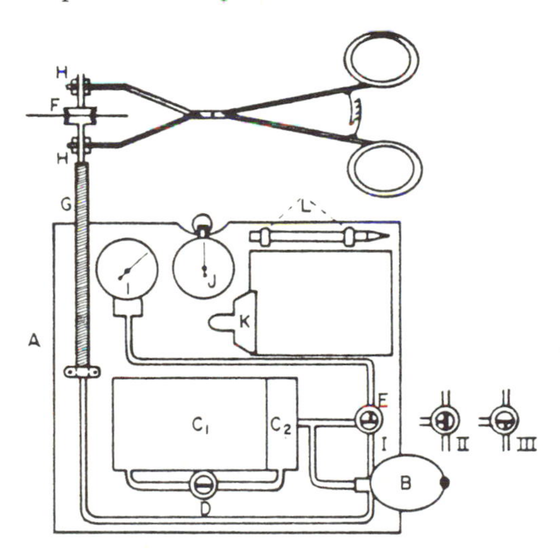

Fig. 4. The variable-sensitivity pressure-drop porometer for measuring stomatal resistance. (From SHIMSHI, 1967)

Diffusion Porometers

These porometers measure the diffusive permeability across the leaf. The rate of diffusion can be monitored through the use of gaseous radioisotopes, by means of differential diffusion of light gases such as hydrogen, or by infrared gas analysis. Because these porometers measure diffusion, their results may be useful in evaluating the diffusive resistance to transpiration.

When evaluating water status in the plant by means of stomatal measurements, it should be borne in mind that water stress is not the only factor determining stomatal behavior. Other factors are light intensity and level of leaf chlorophyll (chlorosis caused by nutrient deficiency is usually associated with a reduced ability of stomata to open). Wheat varieties were found to differ widely in their stomatal behavior under identical conditions of moisture supply. Therefore, any attempt to use stomatal opening as a diagnostic tool in the study of the water relations of plants requires a thorough preliminary investigation of the nature of correlation between water status of any particular plant species and its stomatal behavior: Whenever possible, porometer readings should be corroborated with measurement of stomatal dimensions (e.g., by silicone rubber impressions).

Measurement of Transpiration and Resistance to Vapor Transfer

Just as in the case of stomatal opening, the ability of the plant to transpire under given evaporative conditions can indicate the internal water balance of the plant. Whenever plants are well supplied with water, the rate of transpiration will depend almost entirely on the evaporative demand of the atmosphere above the plants, which, in turn, is determined by the energy balance of the system, including incoming and outgoing radiation, latent heat of vaporization, advective energy, and heat storage of the crop and the soil. This rate of water loss is defined as the potential evapotranspiration. Under conditions of water deficit the actual rate of transpiration will be smaller than the potential evapotranspiration. The reduction of water flux in the system is caused by the lower water conductivity of the dry soil of the root zone, and by the greater diffusive resistance of the closed stomata. Following are some methods of measuring transpiration rate and resistance to vapor transfer.

Weighing of Potted Plants

This method is suitable for laboratory studies, where plants are grown in containers. If the dry weight of the soil in the pot is known, and the dry-matter accretion of the plant is taken into account, then periodic weighings will serve for the determination of the rate of water loss, and its relation to soil moisture, plant relative water content, etc.

Weighing of Detached Leaves

This method is based on the assumption that when a leaf is removed from the plant it will continue, for a short time, to lose water to the atmosphere at the same rate as it did when still attached (HUBER, 1922). This assumption is, admittedly, far from correct, as several workers have shown. However, there usually is a relationship between the rate of water loss and the degree of plant hydration. Special torsion balances are employed, and the weighings are carried out immediately after detaching, and 2–5 minutes afterwards.

Soil-Moisture Data from the Field

The depletion of soil moisture in the root zone of a plant community is mainly a result of plant transpiration. The rate of this change is therefore a measure of the ability of the plant to transpire. The main disadvantage of this method is the inability to obtain reliable short-term values; the shortest time interval between successive samplings of soil is about 2–3 days, and even then the data may not be reliable, because of sampling errors. Soil-moisture data are therefore suitable for the study of the gross response of plants to moisture supply, such as in standard irrigation experiments, but their use for the study of short-term response of plants is limited.

Leaf Chambers with Hygrosensors, or Infrared Gas Analyzers

The principle underlying these devices is the measurement of the rate of vapor flux from the leaf surface, or the resistance to vapor diffusion out of the leaf. The hygrosensor chambers are clamped over the leaf surface (in a fashion similar to that of porometer cups). Immediately after clamping, the enclosed space is flushed with relatively dry air, and the rate of increase in the relative humidity in the chamber (caused by transpiration) is monitored by means of electric hygrosensors, such as LiCl resistance elements (VAN BAVEL *et al.,* 1965). This method seems quite promising, but it requires some improvements in its field portability and rapidity of operation. The infrared gas analyzer (IRGA) method measures the change in the relative humidity of air passing over a leaf area enclosed in a chamber. The method requires highly sophisticated equipment, both for controlling the enclosed environment in the leaf chamber, and for measuring the differences in vapor concentration between incoming and outgoing air. As a laboratory method, it makes possible a continuous measurement of transpiration.

Short-term Growth Analysis

Water status has a marked influence on the growth rates of various plant organs. Adequate moisture supply usually results in high growth rates since it stimulates cell division, and also promotes the enlargement of individual cells before differentiation or maturation, through the distension of cell walls by the turgor. Increasing water stress will result in reduced cell division and enlargement, and, in cases of severe dehydration, in actual shrinkage of tissues. Certain parts of the plant lend themselves to short-term (daily) measurement of their growth rate, and, thus, may serve as indicators of the internal water status.

Stem elongation of actively growing plants will usually come to a halt when water supply is curtailed. The rates of elongation are generally obtained by daily measurements of the stem length. In some plants, the youngest, rapidly growing end of the stem may have a different appearance than the more mature stem below; the length of this juvenile stem segment may indicate the conditions of water supply to the plant. Similarly, the growth of fruit diameter and trunk diameter can serve as such an indicator. It should be noted, however, that diurnal amplitudes are often superimposed over the general growth trend, since fruit or trunks may shrink somewhat during the daytime when water tension develops in the plant, whereas this tension is dissipated during the night. In such cases, these amplitudes may serve as a better index of water status than the overall growth. Changes in tree-trunk dimensions are detected with special measuring devices called dendrometers, or, more recently, with electric strain gauges, which can be connected to recording instruments.

Applications in Agricultural Research and Practice

The aforementioned methods have been used mainly as diagnostic tools for the monitoring of moisture stress in irrigation research. However, they may be applied (and in several cases have actually been applied) to commercial agricultural practice, particularly where norms for irrigation of a new crop species or in a new region have not yet been established. In such cases, rapid field methods of detecting the onset of moisture stress in the plants may be used to decide when irrigation should be applied. In a series of cotton irrigation experiments in Israel, OFIR *et al.* (1968) established the relationship between stomatal infiltration and the degree of depletion of available moisture in the main root zone. Cotton irrigation, whenever stomatal infiltration was sharply reduced, gave a higher yield of lint than cotton irrigated at a predetermined schedule (1640 vs. 1400 kg/ha), although the number of irrigations was the same (six), and the net amount of irrigation was lower (630 vs. 700 mm). Stomatal infiltration was suggested as a guide for the timing of irrigation in citrus (OPPENHEIMER and ELZE, 1941) and bananas (SHMUELI, 1953), whereas sap refractometry has been widely used in the U.S.S.R. on various crops (FILIPPOV, 1965). In fact, further development of rapid field techniques that will enable the farmer to irrigate his fields just when the crop is about to suffer from moisture stress—neither earlier nor later—may result in the avoidance of crop damage due to inadequate water supply and of waste of precious water due to unnecessary irrigation.

Literature

BARRS, H. D.: Determination of water deficits in plant tissues. In: Water Deficits and Plant Growth. KOZLOWSKI, T. T. (Ed.) p. 235–368. New York: Academic Press 1968.

FILIPPOV, L. A.: The refractometric method of diagnosing water deficits in plants. In: Water Stress in Plants. SLAVIK, B. (Ed.) p. 136–144. The Hague: W. Junk 1965.

HEATH, O. V. S.: The water relations of stomatal cells and the mechanism of stomatal movement. In: Plant Physiology – a Treatise. STEWARD, F. C. (Ed.) Vol. 2, 193–250. New York: Academic Press 1959.

HUBER, B.: Zur Methodik der Transpirationsbestimmung am Standort. Ber. Deut. Bot. Ges. **45**, 611–618, 1927.

LLOYD, F. E.: The physiology of stomata. Carnegie Inst., Washington, D.C. **82**, 1–142 (1908).

MEDERSKI, H. J.: Determination of internal water status of plants by beta ray gauging. Soil Sci. **92**, 143–146 (1961).

MOLISCH, H.: Das Offen- und Geschlossensein der Spaltöffnungen, veranschaulicht durch eine neue Methode. Z. Bot. **4**, 106–122 (1922).

OFIR, M., SHMUELI, E., MORESHET, S.: Stomatal infiltration measurements as an indicator of the water requirement and timing of irrigation for cotton. Exptl. Agric. **4**, 325–333 (1968).

OPPENHEIMER, H. R., ELZE, D. L.: Irrigation of citrus trees according to physiological indicators. Pal. J. Bot. Rehovot **4**, 20–46 (1941).

RICHARDS, L. A., OGATA, G.: A thermocouple for vapor pressure measurement in biological and soil systems at high humidity. Science **128**, 1089–1090 (1958).

SCHOLANDER, D. F., HAMMEL, H. T., BRADSTREET, E. D., HEMMINGSEN, E. A.: Sap pressure in vascular plants. Science **148**, 339 (1965).

SHARDAKOV, V. S.: Determination of sucking force of vegetal tissues by the method of small jets (in Russian). Izv. Akad. Nauk. SSSR Ser. Bil. **5**, 1297–1302 (1938).

SHIMSHI, D., LIVNE, A.: The estimation of the osmotic potential of plant sap by refractometry and conductimetry: a field method. Ann. Bot. **31**, 505–511 (1967).

SHIMSHI, D.: Some aspects of stomatal behavior as observed by means of an improved pressure-drop porometer. Israel J. Bot. **16**, 19–28 (1967).

SHMUELI, E.: Irrigation studies in the Jordan Valley. I. Physiological activity of the banana in relation to soil moisture. Bull. Res. Counc. Israel **3**, 228–247 (1953).

SLATYER, R. O., SHMUELI, E.: Measurement of internal water status and transpiration. In: Irrigation of Agricultural Lands. HAGAN, R. M., HAISE, H. R., EDMINSTER, T. W. (Eds.). Amer. Soc. Agron., Madison, Wis., pp. 337–353 (1967).

SPANNER, D. C.: The Peltier effect and its use in the measurement of suction pressure. J. Exptl. Bot. **2**, 145–168 (1951).

VAN BAVEL, C. H. M., NAKAYAMA, F. S., EHRLER, W. L.: Measuring transpiration resistance of leaves. Plant Physiol. **40**, 535–540 (1965).

VAN DEN HONERT, T. H.: Water transport in plants as a catenary process. Discuss. Faraday Soc. **3**, 146–153 (1948).

WEATHERLEY, P. E.: Studies in the water relations of the cotton plant. I. The field measurement of water deficit in leaves. New Phytol. **49**, 81–87 (1950).

VI. Salinity and Irrigation

The problems of soil salinity and of saline irrigation water are not restricted to any specific region of the world. They are well known in the humid countries of Europe (Holland, Sweden, Hungary, Russia) as well as in the arid regions of the Mediterranean bain, the Middle East and India, South America, southwestern United States, etc. In short, there is hardly a region without some problems involving salinity. However, in the arid and semiarid regions the problem is much more acute and widespread, and therefore more interest has been centered around it here and more effort has been directed to its solution. It is estimated that a third of the 12 million hectares of irrigated land in the world is affected by salinity problems.

Every soil contains a certain amount of soluble salts. However, not every soil is considered saline. It is only when soluble salt accumulation in the soil reaches a level harmful to plant growth that a salinity condition is said to have developed–which means that the plant defines the salinity of the soil. A soil saline to one type of crop might be quite suitable for another.

The development of irrigation in arid and semiarid areas requires permanent control of salinity in soils and in irrigation water. The calculation of water requirement in an irrigation system must take into consideration not only the water consumption of the irrigated crops, but also the quantity of water necessary to displace the salts from the root zone. This quantity is known as the leaching requirement. For this calculation, the irrigation specialist must know the plant tolerance to salts. From the knowledge of the salinity of the irrigation water and of crop water requirements, it is possible to predict the accumulation of salts in the root zone and to compute the amount of additional water necessary to keep the salt content in the soil sufficiently low for normal plant growth. It may be concluded that handling of salinity problems is completely integrated into the irrigation technology of arid and semiarid zones.

1

Irrigation with Saline Water

J. SHALHEVET

The principal difference between irrigated and nonirrigated agriculture, when considered in relation to persistence and permanence, arises from salinity. All irrigation water contains soluble salts, and it is only a matter of time until a salinity or alkalinity problem arises unless measures are taken to prevent it. Therefore, the control of the salinity regime in the root zone is one of the main problems of irrigation in arid and semiarid zones.

Forecasting Salt Accumulation in the Soil

In irrigation system planning it is essential to understand the dynamics of salt movement in the soil under any specific set of conditions. This is necessary in order to determine the quantities of water to be conveyed, the design of drainage systems, the use of amendments, and the selection of crops.

The two main factors that control salt accumulation are the quality and quantity of irrigation water. Additional factors of secondary importance are the hydraulic characteristics of the soil, the rainfall pattern, and the evapotranspiration *(ET)* conditions. The last is intimately connected with the quantity of irrigation water. As water is removed from the soil by evaporation and transpiration, the salt remains and accumulates at various depths. If the quantity applied *(D)* equals *ET*, then the salt accumulation will be linearly related to *D*. If the quantity applied exceeds *ET*, then some of the water will seep

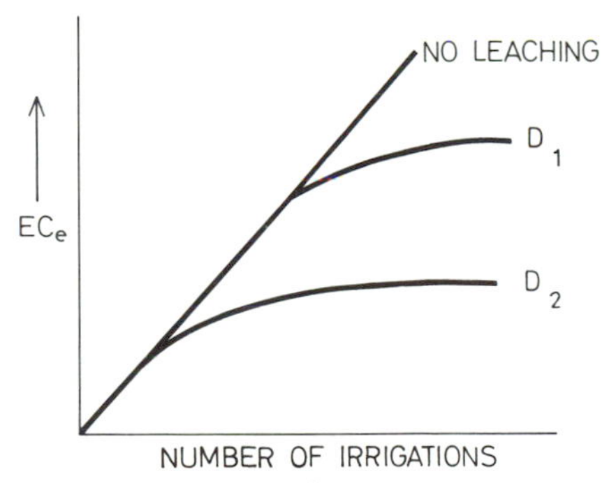

Fig. 1. Schematic relationship between salt accumulation in the soil (EC_e) and the quantity of water applied *(D)*. $(D_1 > D_2)$ when irrigation water salinity is higher than the initial soil salinity

below the root zone and an equilibrium level will be reached, depending on the quantity of excess water applied. The lower limit of the salinity of the soil solution under „field capacity" conditions will be the salinity of the irrigation water. This lower limit may be approached under very heavy leaching. Fig. 1 shows schematically the relationship between the mean electrical conductivity of the soil solution (EC_e) and the number of irrigations, for three quantities of water applied: $D = ET, D_1 > ET, D_2 > D_1$.

A few mathematical models have been proposed in the past to describe the movement of salts in the soil. All of models relate the salt concentration at any point in the soil profile to the quantity of irrigation water that passed the point, and its salt content. More rigorous, but also more complicated, models include the physical and chemical characteristics of the soil.

The simplest and most widely used approach is the leaching requirement concept based on the conservation–of–mass principle. This concept states that at steady state and where no changes in salt content occur within the soil profile, all the salts introduced into the soil profile must also leave it (U.S. Salinity Lab. Staff, 1954).

The concept was developed from the general water–and salt–balance equations and thus represents steady–state conditions. The general water–balance equation may be written as follows:

$$D_q + D_i + D_r = D_e + D_d + D_s \tag{1}$$

where

D_q is the depth of seepage water (net) entering the area; D_i is depth of irrigation water; D_r is depth of rainfall; D_e is depth of evapotranspiration water; D_d is depth of drainage water; and D_s is net depth of water stored in the soil.

For an isolated irrigated field in an arid region, we may assume $D_q = 0$,
$D_r = 0$, and $D_s = 0$, and the equation becomes:

$$D_i = D_e + D_d \tag{2}$$

Introducing the appropriate salt concentrations C into the water-balance equation, and setting $C_e = 0$ (no salt uptake by the crop), we get

$$D_i C_i = D_d C_d$$

$$\text{or} \tag{3}$$

$$\frac{D_d}{D_i} = \frac{C_i}{C_d}$$

When this ratio is used to compute the amount of leaching water D_d required to keep the salinity of the soil C_d from exceeding a specified value (usually the salt-tolerance level of the crop), the above ratio is termed the „leaching requirement (LR) equation".

In order to compute D_i from the usually available data of plant salt tolerance (C_d), salt concentration of the irrigation water (C_i), and the depth of evapotranspiration or consumptive use water (D_e), Eq. (2) may be used to transform Eq. (3) into Eq. [4]

$$D_i = D_e \left(\frac{1}{1-LR}\right) = \left(\frac{C_d}{C_d - C_i}\right) D_e \tag{4}$$

The leaching requirement should not be confused with irrigation efficiency, although they are related superficially. The former term (LR) takes into account the part of the irrigation water (D_i) that actually goes beyond rooting depth. The latter term (IE), on the other hand, considers the difference between the amount of water applied in irrigation and that which is actually used in evapotranspiration. Although some of the applied water goes beyond the depth of rooting, not all of it should be considered as leaching water.

C_d is chosen as the salt-tolerance level of the crop to be irrigated. This is an empirical value obtained by experimentation and represents the mean salinity of the soil profile at which a 50% reduction in yield may be expected, and is expressed in terms of the conductivity of the soil saturation paste extract. The electrical conductivity of the drainage water in the field (C_d), when drainage occurs in a deep soil profile, is approximately twice the conductivity of the soil saturation paste extract (EC_e); see Table 1. The use of EC_e in place of C_d is justified because the crop salt-tolerance values given in the literature in terms of EC_e refer to the mean salinity of the soil profile, when salt concentration increases with depth, and C_d is the salinity at the bottom of the root zone. Under steady-state conditions, the value of C_d would be about twice the mean salinity of the soil profile.

Table 1. The relation between field capacity, wilting point, and saturation percentage,

$$k = \frac{\text{saturation \%}}{\text{Field capacity \%}}$$

Water content by weight

Soil type	Field capacity (%)	Wilting point (%)	Saturation (%)	k
Hydromorphic clay	32	21	65	2.0
Sand	8	3	21	2.5
Silty loam	18	8	32	1.8
Sandy loam	12	6	28	2.3
Clay	30	19	62	2.0

BOWER *et al.* (1969) tested this approach using lysimeters where drainage occurred into a water table, and showed that the calculated concentration of the drainage water agreed well with experimentally determined values. In our work, where drainage occurred into a deep dry soil, the calculated values agreed with experimental results when the latter were expressed in terms of the concentration in the soil solution rather than in the soil saturation extract (SHALHEVET and REINIGER, 1964).

Rainfall is not included in the leaching requirement as presented above. In order to take rainfall into account, a weighted average of rainfall and irrigation water quality is generally computed (U.S. Salinity Lab. 1954) as follows:

$$C_{(i+r)} = \frac{C_i D_i + C_r D_r}{D_i + D_r} \tag{5}$$

This approach is reasonable when rainfall is evenly distributed throughout the year. However, when the rainy season is short, as is the case in the Mediterranean region, this approach might result in erroneous conclusions. Furthermore, soil type is not taken into account in the leaching-requirement approach. As long as rainfall is not a factor, soil type would have a minor effect at equilibrium. However, when rainfall plays a major role, soil type becomes extremely important, as is illustrated in the following example.

Let us consider two soils, a sand and a clay, both situated in a 60cm rainfall region.

$C_i = 200$ mg chloride/1

$C_r = 40$ mg chloride/1

$D_r = 36$ cm (the effective rain for leaching, from a total rainfall of 60 cm)

$D_i = 70$ cm (irrigation requirement of citrus)

$$C_{(i+r)} = \frac{200 \times 70 + 40 \times 36}{70 + 36} = 146 \text{ mg/1}$$

If we assume that 700 mg chloride/1 in the soil solution at field capacity is the tolerance limit of citrus, the leaching requirement will be

$$\frac{C_{(i+r)}}{C_d} 100 = \frac{146}{700} \times 100 = 21\%$$

For the irrigation water alone, without rainfall, we obtain

$$\frac{C_i}{C_d} 100 = \frac{200}{700} \times 100 = 27\%$$

Thus, rainfall reduced *LR* from 27 to 21%.

Now consider the sandy soil with 12% moisture by weight at field capacity, a bulk density of 1.5 g/cm^3, and containing 18 cm of water in the main root zone (100 cm depth), and the clay soil with 31% moisture at field capacity, a bulk density of 1.3 g/cm^3, and containing 40 cm of water in the main root zone. An effective rainfall of 36 cm will replace twice the stored soil moisture in the sandy soil, but it will be insufficient to replace the pore water of the clay soil. Obviously, the sandy soil will be completely leached after the winter, whereas the clay soil will be only partially leached.

If we take into account the fact that the leaching efficiency and the amount of effective rainfall (less runoff) are higher in sandy than in clay soils, it is seen that with 60 cm of rain the sandy soil will not require leaching at all, whereas the clay soil might require more leaching during the irrigation season than was calculated above. Thus, one should use the averaging procedure represented by Eq. [5] cautiously.

Another approach, based on the conservation-of-mass principle, was proposed by BRESLER (1967). By use of this model, the distribution of salts in the soil profile after each irrigation can be computed:

$$D_j \; D_j - (D_i - \sum_{k=1}^{i} E_{k,j}) \frac{C_{i,j-1} + C_{i,j}}{2} = \sum_{k=1}^{i} (C_{k,j} - C_{k,j-1}) B_k \tag{6}$$

where subscripts i and j refer to consecutive soil layers and consecutive irrigations, respectively: D is depth of irrigation; C is concentration of soil solution; E is water deficit in the soil from field capacity; and B is the moisture content of the soil layer.

In the formulation of the model the following assumptions were made:

1) Water flow is vertical at a known water content 0, and after irrigation the water content of the soil reaches field capacity.

2) The concentration of the soil solution leaving an arbitrary soil layer is the mean of the concentration in the layer before and after leaching:

$$C_d = \frac{C_{i,j} + C_{i,j-1}}{2}$$

3) No preciptation of salts in soil or salt uptake by plants occurs.

The comparison of computed and experimental results showed good agreement under many conditions. In certain extreme cases where assumption 2 did not hold, the agreement was not very good.

Generally speaking, solute movement in the soil takes place with water flow. When saline water enters a wet solute-free soil, mixing takes place at the boundary between the two miscible fluids. The degree of mixing depends on microscopic flow velocities, partition and adsorption processes, and diffusion rates. Equations were developed on the basis of chromatographic (VAN DER MOLEN, 1956) and hydrodynamic dispersion theories (DAY, 1956), to describe this flow phenomenon. The equations require the determination of experimental coefficients and thus are not very useful for prediction purposes.

The above discussion referred to the movement of non exchangeable ions such as chlorides, and under many conditions this analysis is satisfactory. However, when the irrigation water contains a high proportion of sodium, the exchange reaction between the cations must be included in the analysis (BOWER *et al.,* 1957), and its effect on the physical properties of the soil has to be taken into account (MCNEAL, 1968).

Cultural Practices in Relation to Salinity

Irrigation Method

The irrigation method may play an important, and sometimes decisive, role in the use of saline water for irrigation.

1) *Sprinkler and Flood Irrigation.* Some plants, especially stone fruits, citrus, and woody ornamentals, absorb ions through their leaves. Sprinkling on the leaves may result in excessive foliar absorption of salts, especially sodium and chloride, with consequent tip and marginal burn of the leaves and ultimate defoliation. This is particularly important when sprinkling is done intermittently, as with the hammer sprinkler, when there is a rapid concentration of the drops that remain on the leaves (EHLIG and BERNSTEIN, 1959). Keeping the foliage continuously wet during sprinkling and using low-head sprinklers might reduce foliar absorption and damage to the leaves. It should be pointed out that the relation between yield and foliar absorption in fruit trees has not yet been established.

Sprinkler irrigation provides a relatively efficient method of reducing the harmful accumulation of salt in the soil. Leaching is more efficiently done when the moisture content is low. Consequently, leaching by sprinkling at rates lower than the infiltration capacity of the soil will result in a more efficient removal of accumulated salts than will leaching by flooding. In a field experiment (NIELSEN *et al.,* 1966), flood irrigation required three times as much water as sprinkling to reduce soil salinity, by the same increment. A similar, but less marked, effect was achieved when flooding was applied intermittently rather than continuously. On the other hand, leaching by flooding is advantageous in that it takes less time than sprinkling to remove the same quantity of salt. Thus, where time is the prime consideration, continuous flooding is preferred; if it is more important to save water than time, sprinkling is recommended.

2) *Furrow Irrigation.* The salt content of the soil vary much more from place to place with furrow irrigation than with either flooding or sprinkling. It has been shown (Fig. 2) that salts accumulate at the wetting front of the advancing water. Salts may accumulate to extremely high levels at the soil surface between furrows and thereby have a strong effect on germination. The planting of row crops, such as cotton and sugar beets, should

be guided by the pattern of salt accumulation on the ridge of the irrigation furrow. A sloping bed system has been proposed in which the soil surface configuration prevents salt accumulation at the location of seed placement (BERNSTEIN and FIREMAN, 1957).

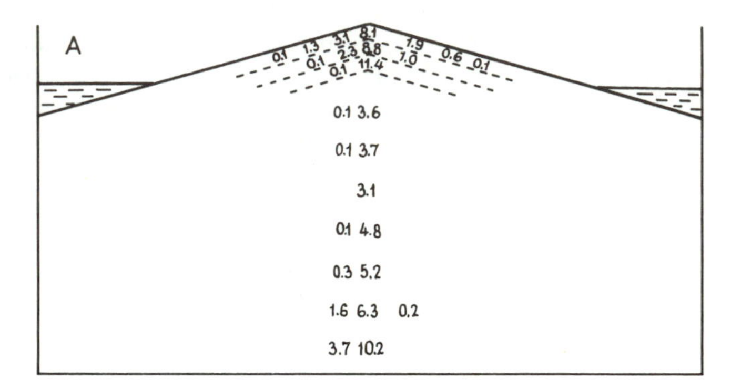

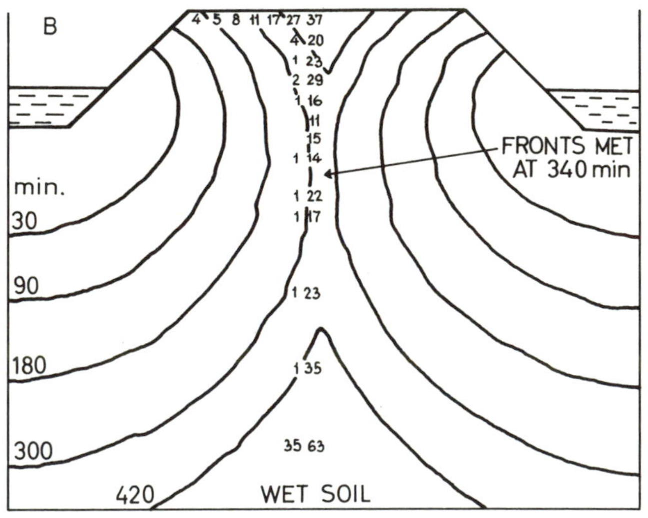

Fig. 2. Salt-distribution pattern (relative values) after furrow irrigation: A sloping bed. B normal furrow; (From BERNSTEIN and FIREMAN, 1957)

Table 2. Pepper-root distribution as a function of distance from the tricklers. (From M. SHMUELI, 1968)

Depth below trickler or distance from trickler (cm)	Roots (% of Total)
0–10	44
10–20	43
20–30	13

3) *Trickle Irrigation.* Trickle irrigation is a new method that has been used successfully with water of high salt content. Its advantages, with regard to salinity, is that it keeps the soil moisture continuously high, at least in part of the root zone, therefore maintaining

a low salt concentration level, and that it results in leaching in the zone below the tricklers. The roots of the growing plants tend to cluster in this leached zone of high moisture near the tricklers and so avoid the salt that accumulates at the wetting front (Table 2).

At the end of the growing season there is a very high salt concentration between the rows, the extent of which depends on whether the wetting zones of two adjacent tricklers overlap, as demonstrated in Fig. 3. The soil salinity pattern obtained is very similar to that with furrow irrigation, as would be expected from the similarity in water-flow patterns in the soil.

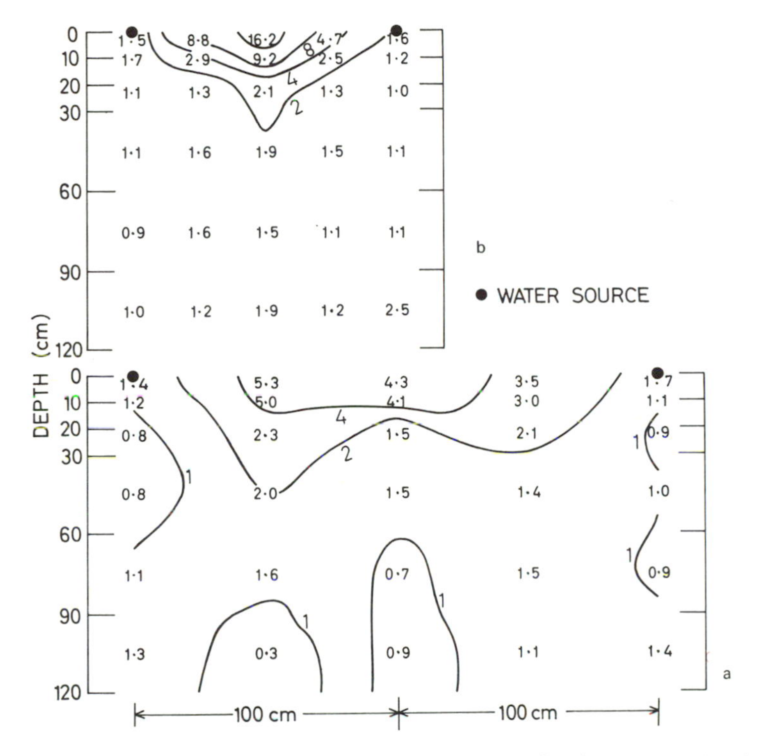

Fig. 3. Salt distribution pattern after trickle irrigation: a no overlap between wetting fronts; b with overlap. (From YARON *et al.*, 1972)

In regions of low rainfall, locally accumulated salts must be leached out by sprinkling before a new crop is planted, or the new crop (if a row crop) must be planted in the same rows as the harvested crop. Failure to do so will result in very spotty germination of the following crop, as demonstrated by wheat sown after cotton in the experiment described by YARON *et al.* (1972). This occurred even after a presowing irrigation of 100 mm was provided for the wheat.

Drainage

Drainage control is an essential element in irrigation agriculture especially when the salt content of the irrigation water is high and a certain amount of leaching is required to maintain a favorable salt balance.

When natural drainage conditions are satisfactory (e.g., soils and water-bearing aquifers are permeable or a water table exists at very great depths), no artificial drainage is required. When natural drainage conditions are limiting, or a rise of water table into the root zone may be anticipated, a provision for drainage must be included in the irrigation system design.

It is beyond the scope of this work to consider in detail the problems involved in the design of drainage systems for irrigated lands. In the following section we shall discuss in general terms only the considerations that should be taken into account when drainage problems are expected to arise as a result of leaching. In irrigated fields, drainage problems may also stem from causes other than leaching, such as canal seepage, regional hydrologic conditions, and rainfall. These will not be treated here.

There are three main factors to be considered in designing a drainage system:

1) The permissible depth to water table, either permanently or temporarily.

2) The required rate of removal of excess water.

3) The physical characteristics of the soil.

1) *Water-Table Depth.* The desired water-table depth will depend on crop and soil considerations. For irrigated areas in arid and semiarid countries, it may be safely assumed that groundwater (which generally has a high salt content) should not serve as a direct source of irrigation water (subirrigation) except in very special circumstances. Thus, as far as most annual crop plants are concerned, a water table deeper than about 100 cm during the irrigation season should be sufficient to permit normal root development and adequate aeration in most soils. For short periods immediately after an irrigation, an increase in the water table to 30 cm in depth should not be detrimental. For perennial crops, a somewhat deeper water table may be required (e.g., 150–180 cm).

During the period between irrigation seasons, the required water-table depth will depend on rainfall conditions and subsurface seepage. Where there is no seepage supply or rainfall, the water table will naturally drop rapidly to such a depth that capillary moisture movement and upward salt transport become negligible. Where seepage from outside sources exists, the depth at which the water table is maintained becomes important. This depth should be at least 1.50–1.80 m for medium-textured soils, less for sandy and clay soils, and more for silty soils (VAN DEN BERG, 1967).

Where there is winter rainfall and no seepage supply, the danger of upward movement of salts from the water table, either during the irrigation season or during the rainy season, is negligible. In this case, the water-table depth should be determined by the aeration requirements of the crop and by soil stratification.

2) *Removal of Excess Irrigation Water.* The quantities of water to be removed by drainage depend mainly on the amount of leaching required to prevent soil salinity from exceeding a specified level (the leaching requirement). The leaching-requirement concept has been discussed previously, and the following example will serve to illustrate its practical use.

Assume the salt tolerance of a crop (C_d) to be 6 mmho/cm, the conductivity of the irrigation water (C_i) to be 2 mmho/cm and the water requirement of the crop (D_c) to be 60 cm for the season, applied in 6 irrigations of 10 cm each. The depth of drainage water required in order to maintain the maximum permissible salinity at the bottom of the root zone is

$$D_d = \left(\frac{C_i}{C_d - C_i}\right) D_e = \left(\frac{2}{6-2}\right) 60 = 30 \text{ cm}$$

Thus, at each irrigation 5 cm of water has to be drained beyond the root zone and D_i becomes 15 cm instead of 10 cm. The drainage system has to be designed to carry 500 cubic meters per hectare within a few days after irrigation. Alternatively, a certain salt accumulation may be allowed, depending on the tolerance of the crop, with leaching being done every second or third irrigation. If leaching is provided every third irrigation, the capacity of the drainage system will have to be increased accordingly to carry 15 cm of water within a few days. Consequently, the choice of a leaching procedure will also depend on the cost of the drainage system.

The leaching quantities computed here are approximate, and are based on the assumption of no rainfall and steady-state conditions. In general, the estimates obtained are extravagant. Using the model proposed by BRESLER (1967), better estimates of salt accumulation may be obtained, and leaching quantities may be computed accordingly.

When there is rainfall between two irrigation seasons, the leaching due to the rain should be taken into account, as demonstrated previously. In such a case, it may sometimes be economically desirable to allow the accumulation of salts during the season, with leaching provided by the rain. Such a procedure should be used with extreme care, however, and a careful agronomic and economic analyses made in order to prevent undue damage.

3) *Physical Characteristics* of the soil. The two most important soil properties for drainage design are the hydraulic conductivity (K) and drainable porosity (p) of the soil. Given a specified quantity of water to be removed, the hydraulic conductivity will determine, to a large extent, drain spacings. The hydraulic conductivity of the soil is essentially the rate of saturated water flow in the soil under a unit gradient. The actual rate of movement of water toward drains will generally be sonsiderably lower. The saturated hydraulic conductivity is generally higher for a sandy soil than for a compacted clay soil. However, a clay soil with a good stable structure may have a hydraulic conductivity higher than a sandy soil (SHALHEVET, 1965).

A few field and laboratory methods are available for evaluating the hydraulic conductivity of the soil. Complete descriptions of the various methods, with their advantages and limitations, are given gy BOERSMA (1967) and KLUTE (1967).

The two most commonly used techniques are the auger-hole method for field determination of K under a water table, and the undisturbed core sample method for laboratory determination when there is no water table.

The determination of the mean effective hydraulic conductivity of a field to be drained is critical for proper drainage design. When it is possible to construct one or two short drainage lines in the field, prior to draining the entire field, or when a drainage system has already been constructed under similar conditions, the potential-discharge method provides the best means of determining K. This method utilizes drainage theory that relates tile discharge rates to the hydraulic head midway between drains, as expressed, for example, in Eq. (7). When an impermeable layer exists at great depths, a plot of h, vs. V gives a straight line with slope K.

The drainable porosity of the soil is the pore volume that drains out when the water table is lowered a specified distance. It depends on soil texture and structure and will generally be higher for sandy and loamy soils and lower for a compacted clay soil.

The drainable porosity affects the extent and rate of water-table fluctuations as a result of water inputs from rain or irrigation. Under steady-state conditions, where the water-table position is in equilibrium with rainfall, the drainable porosity is of minor importance. Under nonsteady-state conditions the drainable porosity of the soil will determine the height of water-table rise and the rate of its descent, and thus might have a profound influence on drain spacing.

4) *Determination of Drain Spacing.* It is assumed in this section that the drainage situation is such that a parallel system of buried drains is desirable. This is the most common situation in irrigated fields. Other situations may occur, such as seepage from unlined canals, artesian pressure, and localized springs, which will for a specific solution-an interceptor drain, pumping, etc.

Drain spacing is generally determined by the use of a drainage equation. Several such equations are available, each applicable to a different situation (SCHILFGAARDE *et al.*, 1956). They all describe the quantitative relationship between the factors governing flow of water to drains. These factors are (1) the permeability or hydraulic conductivity of the soil (K); (2) the hydraulic head (h) or the height of the water table above the drain level; (3) drain spacing (S), which, together with the hydraulic head, determines the hydraulic gradient i.e., the driving force for flow; (4) depth to an impermeable layer (D); and (5) the rate of water discharge from drains (V). In addition, under nonsteady-state conditions, the drainable porosity (p) will also govern flow, as pointed out above.

Most drainage situations are nonsteady; i.e., flow in drains will vary considerably with time, due to variability in the rate of water input by rain or irrigation. However, practically speaking, steady-state solutions to drainage design are generally adequate for most situations, and are much simpler to apply. The most frequently used solution is the one developed by HOOGHOUDT, which is based on the assumption of horizontal flow to drains (the so-called Dupoit-Forchheimmer, or DF assumption) with appropriate correction for flow convergence near the drains (LUTHIN, 1965).

$$S^2 = \frac{8K_1dh}{V} + \frac{4K_2h^2}{V} \tag{7}$$

The symbols are defined in Fig. 4.

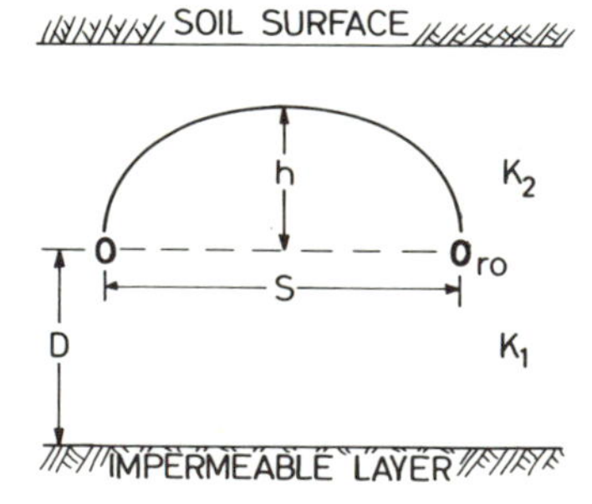

Fig. 4. A cross section of a tile-drained field

The first term on the right-hand side of Eq. (7) represents the flow below the tile-line level and the second term represents flow above the tile-line level (Fig. 4). Consequently, in a layered soil, K_1 may be taken as the hydraulic conductivity of the layer below and K_2 of the layer above the drain level. When drains are laid of on an impermeable layer ($K_1 = 0$), the first term may drop; when the impermeable layer is infinitely deep ($D \gg h$), the second term may be negligibly small.

In Eq. (7), d represents an equivalent depth to an impermeable layer and depends on drain radius r_o, drain spacing S, and the actual depth to an impermeable layer D. Values of d are given in tables or nomograms. Table 3 gives d values for various spacings and for $r_o = 50$ mm.

Table 3. Equivalent depth to an impermeable layer d, in m to be used in Hooghoudt's equation: $r_o = 50$ mm
D = depth to impermeable layers, S = drain spacing

	S (meters)					
	20	25	30	40	50	100
D (meters)						
0.5	0.47	0.47	0.48	0.49	0.49	0.49
1.0	0.83	0.86	0.88	0.91	0.92	0.96
1.5	1.08	1.15	1.20	1.26	1.30	1.40
2.0	1.25	1.35	1.43	1.55	1.62	1.79
2.5	1.38	1.52	1.63	1.80	1.89	2.16
3.0	1.46	1.63	1.76	1.97	2.12	2.49
4.0	1.56	1.70	1.96	2.26	2.48	3.06
5.0	1.61	1.87	2.09	2.45	2.73	3.54
6.0			2.17	2.58	2.92	3.94
∞	1.62	1.94	2.24	2.84	3.41	6.08

Soil Management

Some management practices may influence the efficient control of salinity and alkalinity. Uniformity of surface slope, uniformity of water application, surface mulch to reduce evaporation and runoff, and deep plowing to improve infiltration capacity are but some of the practices.

Improvement of Salt-Affected Soils

Relamation of Saline Soils

The only way known to effectively remove excess salts from the soil is by leaching. Growing salt-tolerant crops during the leaching period may sometimes help the process, but does not serve as a substitute.

On the basis of leaching theory, it can be shown that one replacement of the pore volume will reduce soil salinity by one-half, and 1.5 to 2.0 pore volume replacements of the soil solution should remove about 80% of the soil salt content. This volume corresponds approximately to an equivalent depth of water per unit depth of soil (50% pore volume), as was actually found in many field trials (Fig. 5).

Recent work indicates that leaching under unsaturated flow conditions results in more efficient displacement of the soil solution than under saturated conditions (see discussion under irrigation methods).

When natural drainage conditions are limited and an artificial drainage system is required to remove leaching water, problems with leaching efficiency may arise. Because the hydraulic gradient is steeper near the drains, a larger fraction of the leaching water will flow through the soil next to the drains.

It was shown by LUTHIN (1951) that this amount depends on drain spacing (S) and depth (Y). When $S/Y = 2.5$, then 30% of the area between drains will be leached with maximum efficiency, whereas when $L/Y = 5$, only 15% will be thus leached. As a result,

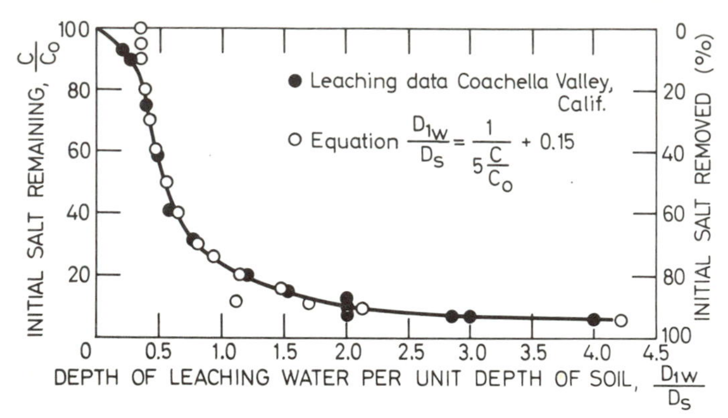

Fig. 5. Depth of water per unit depth of soil required to leach a highly saline soil. (From REEVE *et al.*, 1955)

good leaching will require very close spacing. More practical solutions may be leaching in basins and in stages, i.e., first leaching the area in the center between drains and then closer to the drains (ZASLAVSKY and LEVKOVITCH, 1967).

Reclamation of Sodic (Alkali) Soils

Saline soils that do not contain gypsum tend to become sodic during leaching, and some soils are naturally sodic. A characteristic of sodic soils is their high content of exchangeable sodium (more than 15% of the exchange capacity) and low electrolyte concentration (less than 45 meq/1 in the saturation extract). Saline soils, in comparison, contain 100–400 meq/1 (EC_e = 8–40 mmho/cm).

The high electrolyte concentration of saline soils causes flocculation of soil particles, thus producing good structure and good infiltration capacity. As the total concentration is reduced, the soil particles disperse due to the effect of sodium, and the infiltration capacity of the soil decreases drastically. The reduction of the infiltration capacity makes it difficult to continue the leaching operation unless some measure is taken to replace the excess sodium.

The most common method of replacing the exchangeable sodium ion is by adding Ca^{++} to the soil or by adding acid or acid-forming amendments (sulfur, iron sulfate, aluminum sulfate) that dissolve the lime present in the soil and bring calcium into the soil solution. The cheapest and most common source of calcium is gypsum ($CaSO_4.2H_2O$). In regions where an economical source of $CaCl_2$ is available, as near some chemical industries, this salt may be preferred because of its much higher solubility.

Assuming quantitative exchange, it would take 4.2 tons of gypsum per hectare to replace 1 meq of exchangeable sodium/100 gm to a depth of 30 cm. However, since exchange is not quantitative, the gypsum value should be multiplied by 1.25. The rate of reaction of gypsum is limited by its relatively low solubility (less than 0.2% under field conditions). Therefore, it will take more than 1200 m^3 of water to dissove one ton of high-grade gypsum or 500 mm of irrigation water to leach 1 meq of sodium to a depth of 30 cm.

Another method of reclaiming sodic soils tried recently is the successive dilution of initially high salt water (sea water). By starting with water of high electrolyte concentration, good soil permeability can be maintained. The equivalent dilution method is based on the principle that divalent cations tend to replace monovalent cations as a result of dilution of the soil solution (REEVE and BOWER, 1960).

If the dilution factor is $d = \frac{V_w}{V_s} + 1$, where V_w is the volume of saltfree water and V_s is the volume of saline water, then

$$(SAR)_{dil} \frac{\sqrt{Na^+/\,d}}{\sqrt{\frac{Ca + Mg}{2}}\sqrt{1/d}} = (SAR)_o d^{-1/2} \tag{8}$$

where $(SAR)_{dil}$ and $(SAR)_o$ refer to the diluted and the original solution, respectively.

When sea water of $(SAR)_o = 60$ is diluted by an equal volume of distilled water $(d = 2)$, its $(SAR)_{dil}$ value will change to 42 according to Eq. (8). By successive dilution, the *SAR* can be gradually reduced to less than, 10, at which point the soil is considered reclaimed.

The volume of water required for reclamation by the dilution method will be considerably larger than with normal irrigation water and gypsum, but the time needed for reclamation may be reduced by using water with a calcium content higher than in normal sea water. The divalent cations should, in any case, constitute at least 30% of the total cations in the reclamation water, if the reclamation process is to be carried out within a reasonable time.

Literature

VAN DEN BERG, C. (Ed.): Drainage systems and management. In: Intern. Source-Book on Irrigation and Drainage in Relation to Salinity and Alkalinity. FAO/UNESCO, Paris (draft edition) 1967.

BERNSTEIN, L., FIREMAN, M.: Laboratory studies on salt distribution in furrow irrigated soil with special reference to pre-emergence period. Soil Sci. **83**, 249–263 (1957).

BOERSMA, L.: Field measurement of hydraulic conductivity below and above a water table. In: Methods of Soil Analysis, Pt. I. Amer. Soc. Agron., Madison, Wis., pp. 222–252 (1967).

BOWER, C. A., GARDNER, W. R., GEERTZEN, J. O.: Dynamics of cation exchange in soil columns. Proc. Soil. Sic. Soc. Amer. **21**, 20–24 (1957).

BOWER, C. A., OGATA, G., TUCKER, J. M.: Root zone salt profiles and alfalfa growth as influenced by irrigation water salinity and leaching fraction. Agron. J. **61**, 783–787 (1969).

BRESLER, E.: A model for tracing salt distribution in the soil profile and estimating the efficient combination of water quality and quantity under varying field conditions. Soil Sci. **104**, 227–233 (1967).

DAY, P. R.: Dispersion of a moving salt water boundary advancing through saturated sand. Trans. Amer. Geophys. Union **37**, 364–369 (1956).

EHLIG, C. F., BERNSTEIN, L.: Foliar absorption of Na and Cl as factors in sprinkler irrigation. Proc. Amer. Soc. Hort. Sci. **74**, 661–670 (1959).

KLUTE, A.: Laboratory measurement of hydraulic conductivity of saturated soil. In: Methods of Soil Analysis, Pt. I. Amer. Soc. Agron., Madison, Wis., pp. 210–221 (1967).

LUTHIN, J. N.: Drainage Engineering, 250 pp. New York: Wiley 1965.

LUTHIN, J. N.: Proposed method of leaching tile drained land. Proc. Soil Sci. Amer. **15**, 63–68 (1951).

MCNEAL, B. L.: Prediction of the effect of mixed salt solutions on soil hydraulic conductivity. Proc. Soil Sci. Amer. **32**, 190–193 (1968).

VAN DER MOLEN, W. H.: The desalinization of saline soils as a column process. Soil Sci. **81**, 19–27 (1956).

NIELSEN, D. R., BIGGAR, J. W., LUTHIN, J. N.: Desalinization of soils under controlled unsaturated flow conditions. 6th Congr. Irrig. and Drain. Ques. **19**, 19.15–19.24 (1966).

REEVE, R. C., PILLSBURY, A. F., WILCOX, L. V.: Reclamation of a saline and high boron soil in the Coachella Valley of California. Hilgardia **24**, 69–91 (1955).

REEVE, R. C., BOWER, C. A.: Use of high salt waters as a flocculent and a source of divalent cations for reclaiming sodic soils. Soil Sci. **90**, 139–144 (1960).

VAN SCHILFGAARDE, J., KIRKHAM, D., FREVERT, R. K.: Physical and mathematical theories of tile and ditch drainage and their usefulness in design. Iowa Agric. Exp. Stat. Res. Bull. **436.** (1956).

SHALHEVET, J., REINIGER, P.: The development of salinity profiles upon irrigation of field crops with saline water. Israel J. Agric. Res. **14**, 187–196 (1964).

SHALHEVET, J.: Drainage of fine-textured soils – Hazorea experiment. Volcani Inst. Agric. Res., Div. of Irrigation Spec. Rep. 94 pp. (In Hebrew with English summary) (1965).

SHMUELI, M.: Trickle and sprinkler irrigation of pepper. Hebrew Univ. Faculty of Agric., Dept. of Irrig. Pamphl. **13.** Hebrew (1968).

U. S. Salinity Laboratory Staff: Diagnosis and improvement of saline and alkali soils. U. S. D. A. Handbook **60** (1954).

YARON, B., SHIMSHI, D., SHALHEVET, J.: Patterns of salt distribution under trickle irrigation. (In: Physical Aspects of Soil Water and Salt in Ecosystems. Ecological Studies **4**, 389–394. Berlin – Heidelberg – New York: Springer 1973.

ZASLAVSKY, D., LEVKOVITCH, A.: The hydraulics of leaching salts into a drainage system. Technion Res. and Develop. Foundation Pub. **37.** Hebrew (1967).

2

Crop Growth under Saline Conditions

A. MEIRI and J. SHALHEVET

Different plants have a wide range of *salinity tolerance;* sensitive plants are called *glycophytes,* and highly tolerant ones, *halophytes.* Most agricultural crops are salt-sensitive, and therefore salinity may be one of the major factors determining yields in irrigated areas, especially under arid and semiarid conditions.

Plant growth is affected by the salinity level of the soil. When irrigation water is applied by sprinkling and the foliage is wetted, further damage may result from the high level of residual salt accumulated on the leaves, and from salt absorbed by the shoot.

Adding salt to the growth medium results in (a) an increase in the concentration of the specific ions that cause the salinization; and (b) a reduction in the osmotic potential, and consequently the water potential, of the medium. These two factors may have an unfavorable effect on plant growth. If growth retardation can be correlated with the reduction in substrate water potential, it is called an *osmotic effect.* If the decrease is due to an excessive concentration of specific ions, and is more pronounced than that expected from an equivalent concentration resulting from other electrolytes, it is called a *specific ion effect.*

In general, plants that are most sensitive to total salinity, such as citrus, avocado, and deciduous trees, are also sensitive to specific ion toxicity. Therefore, for the same total ion concentration, sodium and chloride salinity is more harmful to these crops than sulfate or calcium salinity. The more tolerant crops, like many of the field and forage crops, can tolerate high salt concentrations, regardless of the source of salinity. In this case, the osmotic effect is the major salinity factor.

The osmotic effect has been demonstrated in various ways: (1) A similar reduction in growth was found in isosmotic solutions of various salts (GAUCH and WADLEIGH, 1944). If the effect was not an osmotic one, it would have to be assumed that all these salts have the same effect on plant metabolism. (2) In soils, the effects of matric potential and osmotic potential on plant growth were shown to be additive, and termed "total soil moisture stress" (TSMS) (RICHARDS and WADLEIGH, 1952). (3) Transpiration and water uptake by roots were reduced by salinity, and a drastic increase in salinity of the medium was followed by wilting. Salinity produces many symptoms similar to those caused by drought. Plants are smaller, and have a lower shoot/root ratio. The leaves are smaller and have a deep, bluish-green colour. In some cases, especially in sensitive plants such as avocado and some stone fruits, necrosis and leaf burn occur.

This evidence led to the conclusion that the salinity effect on plants is brought about by a disturbance of the plant water balance and induced conditions of "physiological drought."

The specific ion effect in inert growth media such as nutrient solutions or sand cultures may be due to direct toxicity, which is related to an excessive accumulation of the specific ion, or the formation of toxic metabolic products in the plant tissues; or interference with the nutrient ion balance of the plant. In soils, the composition of the exchange complex may have an added indirect effect as a result of the deterioration of the soil's physical conditions. When sodium predominates, these effects occur simultaneously and cannot be separated.

The dominant ions in soils are sodium, calcium, magnesium, chloride, sulfate, bicarbonate, and carbonate. Most common are the sodium and calcium chloride and sulfate types of salinity. Bicarbonate may also be injurious. Salinity problems also may be caused by high concentrations of other ions such as nitrate applied as fertilizer (YARON *et al.*, 1969), or by small increases in the concentration of microelements such as boron (WILCOX, 1960).

Mechanisms of Salt Effect on Plants

a) *The Water-Availability Mechanism.* Reduction in water availability to plants and disturbed water balance in the plant cells will result in reduced turgor pressure and a lower growth rate.

The effect of salt on the water balance of cells has been attributed to two causes—the change in water-potential gradient between the growth medium and plant cells, and the salt distribution within the plant tissue.

Water availability may be expected to decrease with a reduction in the osmotic potential of the medium if the plant behaves like an ideal osmometer; i.e., no change occurs in the internal solute content or in membrane resistance to water uptake. In this case, the reduction in osmotic gradient from the growth medium to the plant cells will decrease water availability, turgor pressure, and growth rate. However, under saline conditions there is an increase in the plant's sap concentration, and its osmotic potential is reduced. This reduction may be smaller, similar, or larger than that in the osmotic potential of the medium (BERNSTEIN and HAYWARD, 1958; GALE *et al.*, 1967). A similar reduction in osmotic potentials in the medium and the plant results in no change in the osmotic potential gradient between the growth medium and plant cells when conditions change from nonsaline to saline. Such plants are said to be *osmotically adjusted* (BERNSTEIN, 1961). It was also shown that the osmotic adjustment may result in an adjustment of total water potential and turgor pressure (SLATYER, 1961).

Recent investigations of the osmotic effect have shown that osmotic adjustment due to ion accumulation has a time lag behind the increase in the salinity of the growth medium that will exist throughout the period of substrate salinization. The magnitude of the lag depends on the salinization rate (BERNSTEIN, 1963; JANES, 1966; MEIRI and POLJAKOFF-MAYBER, 1969), plant species (GALE *et al.*, 1967) and type of salinity (STROGANOV, 1962; MEIRI *et al.*, 1971). Full adjustment may be expected only when the substrate salinity remains nearly constant. The lag in adjustment should result in growth retardation due to reduction in water availability and turgor pressure.

Growth reduction due to salinity also occurs in adjusted plants. The growth process itself results in the dilution of the vacuole sap in the growing tissues of the plant. Apparently, the adjusted plants are unable to absorb salts at a sufficiently high rate to overcome this dilution effect, and consequently plant turgor is reduced, causing a reduction in growth (GREENWAY and THOMAS, 1965; OERTLI, 1968).

Patterns of growth rate under abrupt and gradual salinization are represented schematically in Fig. 1. Abrupt salinization is followed by a severe temporary reduction in growth rate, from which there is gradual recovery as osmotic adjustment takes place. Under gradual salinization, growth retardation is proportional to the substrate salinity.

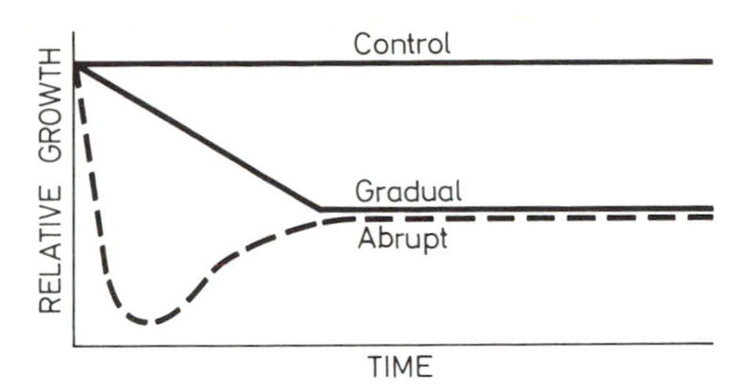

Fig. 1. Relative growth rate in response to abrupt and gradual salinization

Occasionally, higher concentrations of metabolites are found in plants grown under saline conditions (BERNSTEIN and AYERS, 1953). The metabolite accumulation results in an increase in sap concentration and in osmotic adjustment independent of ion uptake, which, in turn, will result in growth retardation, as it reduces the amount of metabolites available for growth. The induced growth retardation may incur an accumulation of primary synthesis products.

Under saline conditions, the distribution of salts within the plant cells may also result in turgor reduction and growth retardation. The main pathway from the xylem vessels to the evaporating surface is through cell walls (WEATHERLEY, 1963). If the rate of salt supply from the medium exceeds the rate of ion uptake by the cells, the salt concentration in the cell walls will build up and the potential of the cell-wall water will be reduced, causing dessication of the vacuoles and reduction in turgor pressure, and, in extreme cases, even death of the tissue (WEATHERLEY, 1965; ROBINSON, 1971). Thus, the damage will not be due to the water-potential gradient from external medium to tissue, but to the salt concentration and water-potential gradients within the tissue. This mechanism was suggested as a possible cause of salinity damage, not only by elements absorbed in large quantities, but also by boron, which is toxic in relatively small quantities (OERTLI and ROTH, 1969).

Growth retardation as a result of the osmotic effect of salinity is not necessarily followed by a similar reduction in transpiration (SHALHEVET and BERNSTEIN, 1968). The same reduction in growth rate was found for isosmotic concentrations of sulfate and chloride salinity, whereas there were larger differences in transpiration reduction caused by the different media. Under chloride salinity, transpiration rate was lower (MEIRI *et al.*, 1971). The effect of salinity in reducing water uptake and transpiration in adjusted plants has not yet been explained. It has been shown that salinity affects root (KIRKHAM *et al.*, 1969; MEIRI, unpublished data) and stomatal (GALE *et al.*, 1967) resistance to water flow. Further investigations on the response of these resistances to salinity may increase our understanding of the mechanism of the osmotic effect on water availability.

b) *The Hormone Mechanism*. The balance between root and shoot hormones changes considerably under saline conditions. Stress reduces cytokinin production in the roots and its transport to the shoot (ITAI *et al.*, 1968). The supply of cytokinin to the leaves increases transpiration (LIVNE and VAADIA, 1965), and synthesis of some proteins (VAADIA and ITAI, 1969). Therefore, a reduction in the supply of root hormones to the leaves should result in reduced transpiration and growth rate. On the other hand, water stress in the leaves results in rapid accumulation of abscissic acid, which reduces transpiration (MIZRACHI *et al.*, 1970). The balance between root and shoot hormones may result in the osmotic effect on growth retardation and transpiration suppression under saline conditions.

c) *Damage to Plant Cells and Cytoplasmic Organelles.* The accumulation of excessively high levels of ions in plant leaves results in cell death and necrosis (PEARSON, 1960; WILCOX, 1960; BERNSTEIN, 1965). It has been shown that salinity changes the structure of the chloroplasts and the mitochondria in leaves (BLUMENTAL-GOLDSMIDT and POLJAKOFF-MAYBER, 1968; GAUSMAN *et al.*, 1972). Such changes in structure of the organelles may interfer with normal metabolism and growth.

d) *Interference with Normal Metabolism.* Salinity increases respiration (NIEMAN, 1962) and reduces photosynthesis of crop plants (BOYER, 1965; GALE *et al.*, 1967), resulting in a reduction in the photosynthetic products available for growth. The increase in respiration is the result of the energy required for ion uptake. The reduction in photosynthesis is attributable to stomatal closure under saline conditions and to an increase in resistance to CO_2 entering the leaves; the other is the interference with biochemical reactions. Both of these effects were measured on whole plants.

A low osmotic potential of the leaf sap and a high concentration of electrolytes in the cells may also change enzyme structure and activity (KESSLER *et al.*, 1964). Such interference has been shown for protein (KAHANA and POLJAKOFF-MAYBER, 1968) and carbohydrate (PORATH and POLJAKOFF- MAYBER, 1968) metabolism.

Crop Response to Salinity

Much work has been done to establish quantitatively the response of various crops to total salinity and to the concentration of specific ions. Such information is necessary for the intelligent selection of crops and management systems when an irrigation water salinity problem is encountered. The subject was reviewed by HAYWARD and BERNSTEIN

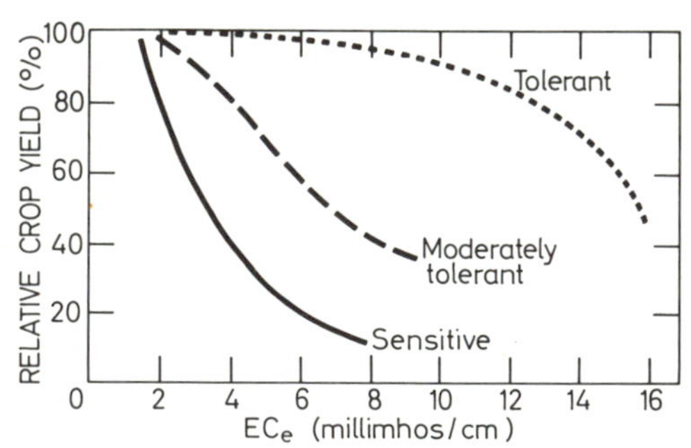

Fig. 2. Salt-tolerance curves for tolerant, moderately tolerant, and sensitive crops. (From REVEE and FIREMAN, 1967)

(1958), and the data have been organized into tolerance tables and figures (U.S. Salinity Lab., 1954; PEARSON, 1960; WILCOX, 1960; ALLISON, 1964; BERNSTEIN, 1964, 1965; DE FORGE, 1970).

The effect of salinity on commercial yields depends on the plant species and on the

part of the plant that is harvested. There are plants such as barley, wheat, and cotton in which vegetative growth is reduced markedly while full yields of seed are produced. Rice, on the other hand, may show normal vegetative growth without producing any grain (HAYWARD and BERNSTEIN, 1958). The effect on most fruits and vegetables is expressed by a reduction in both number and size of fruits. The nutritional value of forage crops may be reduced. Rhodes grass, for example, which is a most tolerant crop, may accumulate salt to a very high level, which causes cattle diseases. However, salinity may improve the quality of some crops such as melons, carrots, and sugar beets by causing an increase in sugar content.

a) *Osmotic Effect.* The general response curve to salinity shows a constant high yield, or sometimes a slight yield increase at a "low" salinity level, a relatively rapid decrease at "medium" levels, and a slow decrease at "high" levels, until no yield is obtained. The salinity level in the transition zone is characteristically different for various crops (Fig. 2).

Yield reduction is gradual and there is no clear-cut threshold of salinity tolerance. Therefore, crop tolerance was compared at arbitrary values of yield decrements as a percentage of the yield under nonsaline conditions (and when all other growth requirements are supplied at optimal levels). In Holland, a 25% yield decrement is used (VAN DEN BERG, 1950), and in North Africa a 20% yield decrement is used (GRILLOT, 1954). In the American literature, the 50% yield decrement is most common (U.S. Salinity Lab., 1954). Recently, a most useful graphic presentation was adopted of the complete response curve relating yield to soil salinity (Fig. 3) (BERNSTEIN, 1964, DE FORGE, 1970).

b) *Specific Ion Effects.* Chloride, sodium, and boron toxicity are the most common cause of specific ion effect. The toxic levels of these ions differ considerably between plants, and are highest for chloride and lowest for boron.

Different criteria have been used to compare crop response to these ions. For chloride, its concentration in the soil solution was used (BERNSTEIN, 1965); for sodium, the exchangeable sodium percentage (ESP) was employed (PEARSON, 1960); and for boron, its concentration in the irrigation water utilized (U.S. Salinity Lab., 1954; WILCOX, 1960).

Chloride (Cl^-) is known to exist in a very wide range of concentrations in soils and plants. It does not interfere with the uptake of nutrient elements.

The typical symptom of chloride injury is leaf burn starting from the leaf margins. The critical levels of chloride in the leaves are not well defined. Higher concentrations have sometimes been found in undamaged than in damaged leaves. Most sensitive to chloride are fruit trees, such as stone fruits, almonds, avocado, citrus, and grapes. For them, the critical concentration in plant tissue is in the range of 0.5–1.8%, on a dry-weight basis. Tolerant plants may accumulate up to 4% Cl^- with no apparent injury (BERNSTEIN and HAYWARD, 1958).

Fruit trees, which are specifically sensitive to both chloride and sodium, and at the same time are also sensitive to total salinity, show large differences in tolerance according to variety and rootstock. Here, differences are emphasized with respect to specific ions and seem to be a result of the great differences in the rates of ion uptake by the different rootstocks. The specific injury symptoms appear before any effect of total salt concentration is observed. For instance, a chloride level of 10 meq/l in a saturated paste extract is considered toxic to sensitive citrus rootstocks (causing a 10% reduction in yield). However, the same rootstock can tolerate a higher total salinity if it is not due to chloride salts. Therefore, classification of fruit crops with respect to specific salinity according to

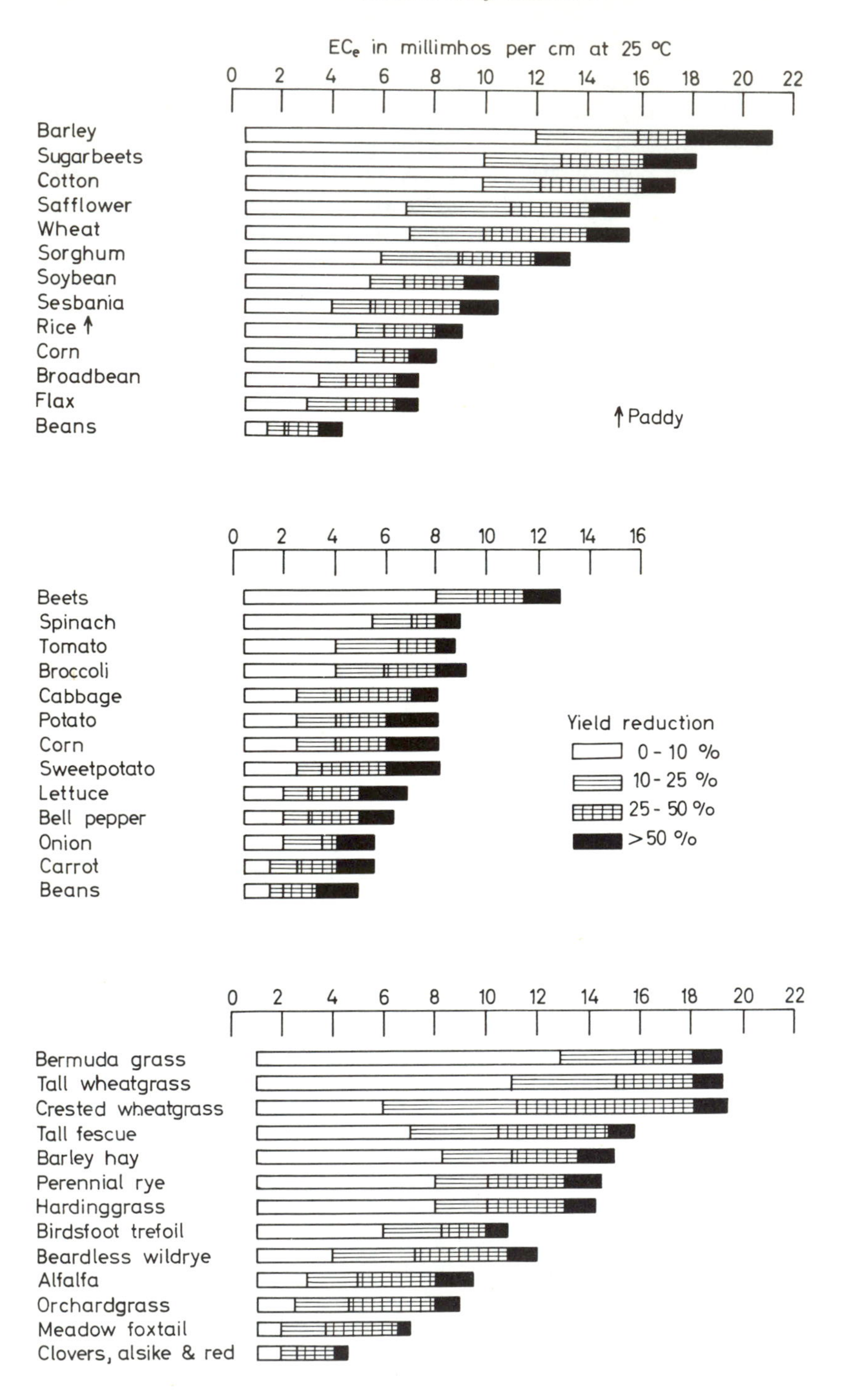

Fig. 3. Salt tolerance of field vegetables and forage crops. (From BERNSTEIN, 1964)

varieties and rootstocks is of major importance. Such a tolerance classification was presented by BERNSTEIN for chloride (Table 1).

Sodium (Na^+) is one of the more difficult and problematic ions. It may be toxic, especially to fruit trees. In sensitive plants, a sodium level of 0.05% of dry weight may

Table 1. Hazardous chloride levels in saturation extracts for fruit crop varieties and rootstocks (From BERNSTEIN, 1965)

Crop	Variety or Rootstock	Chloride (meq/l in saturation extract)
Citrus rootstocks	Rungpur lime, Cleopatra mandarin	25
	Rough lemon, Tangelo, Sour orange	15
	Sweet orange, Citrange	10
Stone fruits rootstocks	Mariana	25
	Lovel, Shalil	10
	Yunnan	7
Avocado rootstocks	West Indian	8
	Mexican	5
Grape varieties	Thompson seedless, Perlette	25
	Cardinal, Black Rose	10
Strawberry		5–8

Table 2. Tolerance of various crops to exchangeable sodium percentage. (From PEARSON, 1960)

Tolerance to ESP and range at which affected	Crop	Growth response under field conditions
Extremely sensitive (ESP = 2–10)	Deciduous fruits Nuts Citrus Avocado	Sodium toxicity symptoms even at low ESP values
Sensitive (ESP = 10–20)	Beans	Stunted growth at low ESP values even though the physical condition of the soil may be good
Moderately tolerant (ESP = 20–40)	Clover Oats Tall fescue Rice Dallisgrass	Stunted growth due to both nutritional factors and adverse soil conditions
Tolerant (ESP = 40–60)	Wheat Cotton Alfalfa Barley Tomatoes Beets	Stunted growth usually due to adverse physical conditions of soil
Most tolerant (ESP = more than 60)	Crested and Fairway wheatgrass Tall wheatgrass Rhodes grass	Stunted growth usually due to adverse physical conditions of soil

result in toxic symptoms. It may disrupt the nutritional balance of plants by causing calcium deficiency, and it may adversely affect the physical conditions of the soil and thus indirectly cause poor plant growth through lack of aeration, poor water transmission, and physical impedance of root development and seedling emergence. PEARSON (1960) presented a classification of crop tolerance to ESP and indicated the major reason for growth retardation (see Table 2). The most sensitive fruit trees suffer from direct sodium toxicity at low ESP values that cannot interfere with the physical properties of the soil. Crops with high tolerance suffer from nutritional imbalance and adverse physical conditions of the soil.

Increasing soil salinity reduces the physical effects of ESP (see chapt. 4), but then growth may be restricted by the effect of total salinity.

Boron (B^{+3}). This ion is present in most irrigation water and saline soils. It is toxic to most plants at a concentration only slightly in excess of that required for optimum growth. Small quantities of boron absorbed by the roots are accumulated in the leaves, and values above 250 ppm (0.025%) are in the toxic range and result in typical leaf burn. Boron is very difficult to leach out of the soil, and therefore the use of irrigation water containing too high concentration of this ion should be avoided. Limiting levels of boron in irrigation water for various crops are shown in Table 3.

Table 3. Limits of boron in irrigation water for crops of different degrees of boron tolerance (From WILCOX, 1960)

Tolerant	Semitolerant	Sensitive
4.0 ppm of boron	2.0 ppm of boron	1,0 ppm of boron
Athel *(Tamarix aphylla)*	Sunflower (native)	Pecan
Asparagus	Potato	Walnut (black and Persian, or English)
Palm *(Phoenix canariensis)*	Cotton (Acala and Pima)	Jerusalem artichoke
Date palm *(P. dactylifera)*	Tomato	Navy bean
Sugar beet	Sweetpea	American elm
Mangel	Radish	Plum
Garden beet	Field pea	Pear
Alfalfa	Ragged-robin rose	Apple
Gladiolus	Olive	Grape (Sultanina and Malaga)
Broadbean	Barley	Kadota fig
Onion	Wheat	Persimmon
Turnip	Corn	Cherry
Cabbage	Milo	Peach
Lettuce	Oat	Apricot
Carrot	Zinnia	Thornless blackberry
	Pumpkin	Orange
	Bell pepper	Avocado
	Sweet potato	Grapefruit
	Lima bean	Lemon
2.0 ppm of boron	1.0 ppm of boron	0.3 ppm of boron

Sulfate ($SO_4^{=}$) toxicity has been shown to occur in citrus. A 1% SO_4 concentration in the leaves is considered a toxic level. Most irrigation waters do not contain sufficient sulfate to cause injury. However, $SO_4^{=}$ promotes the uptake of sodium and restricts uptake of calcium. Thus, depending on substrate composition, $SO_4^{=}$ might cause Ca^{++} deficiency or Na^{+} toxicity.

Bicarbonate (HCO_3^{-}) ion may be injurious to some plants by causing lime-induced chlorosis. However, its effect is more important in precipitating calcium in the soil, thus causing an increase in the exchangeable sodium percentage and a deterioration of soil structure.

Magnesium (Mg^{++}) is known to affect plant growth mainly by reducing calcium uptake and causing calcium deficiency. When high concentrations of magnesium occur together with high concentrations of calcium, there is no specific ion effect of Mg^{++}.

Factors Affecting Plant Response to Salinity

Salinity tolerance levels were obtained under a variety of conditions. Despite the variability, there is general agreement regarding the relative tolerance of different crops. Table 4 shows the salinity at which a 10, 20, and 50% reduction in yield occurs for a few crops and locations. Obviously, there are significant differences among the various sources. Consequently, in order to be able to use the tolerance values reported in Fig. 3 and Tables 1, 2, and 3, the reasons responsible for the differences must be evaluated. The major factors that can alter the results are discussed below.

Table 4. Soil salinity level (mean EC_e in mmho/cm) that results in the indicated reduction in yield of marketable products (data from various sources). (From SHALHEVET, 1971)

Crop	Tomatoes				Alfalfa			Barley			Corn			
Source	1	2	3	4	1	3	5	1	6	7	1	3	8	9
Yield reduction, %														
10	4.1	4.1	3.8	2.1	3.0	5.0	1.5	12	7	11	2.5	3.1	2.0	2.5
20	5.3	6.5	4.3	3.0	4.5	6.0	3.5	15	12	15	3.6	4.0	4.0	4.0
50	8.0		5.9	5.0	8.0		8.5	17			6.0	6.0	8.0	7.8

Source: 1) BERNSTEIN (1964).
2) SHALHEVET and YARON (1973)
3) DE FORGES (1970).
4) BIERHUIZEN (1969).
5) SHALHEVET and BERNSTEIN (1968).
6) VAN BIEKON *et al.* (1953).
7) NOURI *et al.* (1970 a).
8) NOURI *et al.* (1970 b).
9) KADDAH and GHOWAD (1964).

a) *Heterogeneity of Salinity in Space and Time.* The salinity regime of a plant is a combination of the stage of growth when salinization occurs, the rate of salinization, the absolute salinity levels, the length of time of exposure to salinity, and, if leaching is carried out, the periods of low salinity level. Under field conditions, uniformity with depth and time is the exception, and therefore it is not always simple to use in the field information obtained under uniform experimental conditions.

Salt content at any depth usually varies during the growth period due to salt accumulation or leaching. Moisture conditions in the field may play an important role. The ratio of saturated-paste to field-capacity moisture contents may vary between 1.7 and 2.5 for different soils (see p. 265). Thus, data reported on the basis of saturated-paste moisture content when translated to field conditions may result in differences by a factor of two, depending on the moisture characteristics of the soil. The drier the soil, the more concentrated the soil solution will become, and the greater the effect salinity will have on plant growth. Consequently, the longer the irrigation interval, the greater the effect of salinity.

If the period of plant exposure to salinity is extended, growth retardation is more severe. For many plants, the response is directly related to the period of exposure to salinity (LUNIN *et al.*, 1961; MEIRI and POLJAKOFF-MAYBER, 1970). But it may also depend on specifically sensitive growth stages, as was shown for rice (PEARSON and BERNSTEIN, 1959).

Regarding salt distribution with depth, different results were reported. For corn, it was shown that if relatively small parts of the root zone were nonsaline, the plants might develop as though the total root zone were not saline (LUNIN and GALLATIN, 1965; BINGHAM and GARBER, 1970). However, for alfalfa, groundnuts, barley, and beans, the effective value was the mean salinity of the root zone (SHALHEVET and BERNSTEIN, 1968; SHALHEVET *et al.*, 1969; BOWER *et al.*, 1969; KIRKHAM *et al.*, 1969). MEIRI and SHALHEVET (1972) have shown for peppers that the mean salinity is also the effective value for different salinity regimes, as achieved by variations in salt distribution with time and space (Fig. 4).

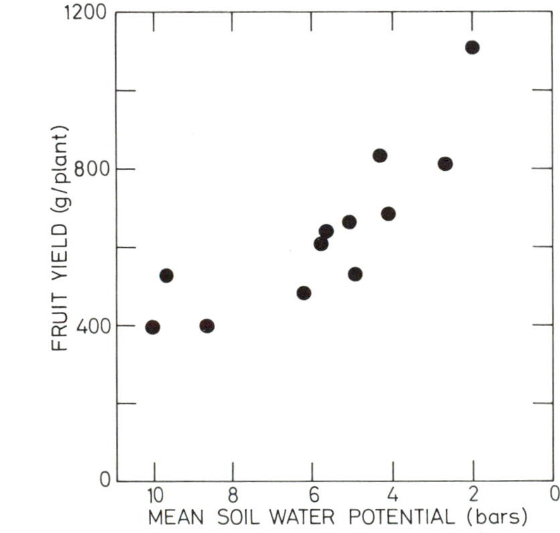

Fig. 4. Fruit yield of pepper plants related to mean soil-water potential for nonuniform salt distribution over depth and time, as achieved by irrigation with water of different salinity levels and with different leaching regimes. (From MEIRI and SHALHEVET, 1973)

The problem of integrating the entire salinity regime is very complicated, because of the many plant and environment interactions that exist, and thus a great deal of clarification is still required. At this stage, the mean salinity value seems to be the best choice.

The above discussion pertains to salinity build-up due to irrigation with saline water after the plants are established. It was shown (WADLEIGH *et al.*, 1947) that root development is limited by soil salinity, and thus, under preexisting conditions of nonuniformity, roots will develop only in the less saline portions of the soil, and the mean salinity may not be used to represent the effective salinity. This is important with regard to trickle and furrow irrigation (see p. 268–269).

Spotty growth in saline fields leads to the popular belief that plants are most sensitive to salinity during germination. Investigations of the salinity effect during the first stages of plant growth have shown that in most cases germinating seedlings have a tolerance similar to that of mature plants (BERNSTEIN and HAYWARD, 1958). However, there may be higher tolerance, as in corn, or lower, as in beets, and usually young seedlings are more sensitive to salt. Thus, the major reasons for a poor stand in saline fields is the accumulation of salt in the surface layer of the soil by the evaporation process and the high salinity to which the germinating seeds and the roots of young seedlings are exposed. A small water application sufficient to leach the surface layer before sowing or at the initial stages of growth should improve the crop stand.

Unlike most field crops, which are more salt-resistant at the mature stage, fruit trees generally become more sensitive as they grow older.

b) *Soil Chemical and Physical Conditions.* The most important soil factors that interact with salinity are fertility, structure, aeration, and moisture content—in addition to the chemical composition of the exchange complex (ESP), which was discussed previously.

Generally speaking, the relative effect of salinity will be stronger under favorable conditions than under limiting conditions of any of the variables. Thus, at high levels of fertility (LUNIN *et al.*, 1964), aeration (AUBERTIN *et al.*, 1968), or moisture (WADLEIGH and AYERS, 1945), salinity reduces yield more than at low levels. Consequently, salinity response curves obtained under low-productivity conditions will show apparent tolerances that do not really exist. The absolute yield level, not only the relative level, is also of importance when applying tolerance data.

c) *Climatic Conditions.* A comparison of plant response to salinity under various climatic conditions has shown stronger effects under high temperatures and low air humidities than the converse (MAGISTAD *et al.*, 1943; LUNT *et al.*, 1960). Conditions that increase water demand and transpiration should accentuate the osmotic effect. Likewise, the increased transpiration rates may result in higher ion concentrations near the roots and the specific ion effect may also be stronger.

Practically speaking, results obtained from one climatic region should not be applied indiscriminately to another.

Conclusions

Salinity and its specific components may have a variety of effects on plant development. In addition, salinity influences different plant species and varieties of the same species in various ways. The physiological nature of plant response to salinity, as well as the nature of plant tolerance, still require intensive study. However, the existing information can be exploited for productive agriculture under saline conditions if proper judgement is used in the selection of crops according to their tolerance to total salinity and to specific ions. For fruit trees, the tolerance of both the rootstock and the commercial variety should be of concern. The irrigation method should prevent the accumulation of excessively high salt concentrations in the root zone at any growth stage. For plants susceptible to leaf injury by direct contact with saline water, overhead sprinkler irrigation with saline water should be avoided. If the plants have special sensitive stages of development, leaching at these times may help to reduce the salinity effect considerably.

Literature

ALLISON, L. E.: Salinity in relation to irrigation. Adv. Agron. **16**, 139–180 (1964).

AUBERTIN, G. M., RICKMAN, R. W., LETEY, J.: Differential salt-oxygen levels influence plant growth. Agron. J. **60**, 345–349 (1968).

VAN BEEKON, C. W. C., VAN DEN BERG, C., DE BOOR, TH. A., VAN DEN MOLEN, W. H., VERHOVEN, B., WESTERHOF, J. S., ZUUR, A. J.: Reclaiming land flooded with salt water. Netherland J. Agric. Sci. **1**, 153–163 (1953).

VAN DEN BERG, C.: The influence of salt in the soil on the yield of agricultural crops. 4th Int. Cong. Soil Sci., Trans. vol. **1**, 411–413 (1950).

BERNSTEIN, L.: Osmotic adjustment of plants to saline media. I. Steady state. Amer. J. Bot. **48**, 909–918 (1961).

BERNSTEIN, L.: Osmotic adjustment of plants to saline media. II. Dynamic phase. Amer. J. Bot. **50**, 360–370 (1963).

BERNSTEIN, L.: Salt tolerance of plants. USDA Inform. Bull. no **238**, 23 pp. (1964).

BERNSTEIN, L.: Salt tolerance of fruit crops. USDA Inform. Bull. no. **292**, 8 pp. (1964).

BERNSTEIN, L., AYERS, A. D.: Salt tolerance of five varieties of carrots. Proc. Amer. Soc. Hort. **61**, 360–366 (1953).

BERNSTEIN, L., HAYWARD, H. E.: Physiology of salt tolerance. Ann. Rev. Plant Physiol. **9**, 25–46 (1958).

BINGHAM, F. T., GARBER, M. J.: Zonal salinization of the root system with sodium chloride and boron in relation to growth and water uptake of corn plants. Soil Sci. Soc. Amer. Proc. **34**, 122–126 (1970).

BIERHUIZEN, J. F.: Water quality and yield depression. Inst. Land Water Manag. Res. Tech. Bull. **61**, 163–173 (1969).

BLUMENTHAL-GOLDSCHMIDT, S., POLJAKOFF-MAYBER, A.: Effect of substrate salinity on growth and on submicroscopic structure of leaf cells of *Atriplex halimus* L. Aust. J. Bot. **16**, 469–478 (1968).

BOYER, J. S.: Effects of osmotic water stress on metabolic rates of cotton plants with open stomata. Plant Physiol. **40**, 229–234 (1965).

BOWER, C. A., OGATA, G., TUCKER, J. M.: Root zone salt profiles and alfalfa growth as influenced by irrigation water salinity and leaching fraction. Agron. J. **61**, 783–785 (1969).

DE FORGES, J. M.: Research on the utilization of saline water for irrigation in Tunisia. Nature and Resources **VI**, 2–6 (1970).

GALE, J., KOHL, H. C., HAGAN, R. M.: Changes in the water balance and photosynthesis of onion, bean and cotton plants under saline conditions. Physiol. Plant. **20**, 408–420 (1967).

GAUCH, H. G., WADLEIGH, C. H.: Effects of high salt concentration on growth of bean plants. Bot. Gaz. **105**, 373–387 (1944).

GAUSMAN, W., BAUR, P. S., Jr., PORTERFIELD, M. P., CARDENS, R.: Effects of salt treatments of cotton plants *(Gossypium hirsutum L.)* on leaf mesophyll cell microstructure. Agron. J. **64**, 133–136 (1972).

GREENWAY, H.: Plant response to saline substrates. VII. Growth and ion uptake throughout plant development in two varieties of *Hordeum vulgare*. Aust. J. Biol. Sci. **18**, 763–769 (1965).

GREENWAY, H., THOMAS, D. A.: Plant response to saline substrates. V. Chloride regulation in the individual organs of *Hordeum vulgare* during treatment with sodium chloride. Aust. J. Biol. Sci. **18**, 505–524 (1965).

GRILLOT, G.: The biological and agricultural problems presented by plants tolerant of saline or brackish water and the employment of such water for irrigation. In: Utilization of Saline Water. UNESCO, Paris, pp. 9–35 (1954).

HAYWARD, H. E., BERNSTEIN, L.: Plant-growth relationships on salt-affected soils. Bot. Rev. **24**, 584–635 (1958).

ITAI, C., RICHMOND, A. E., VAADIA, Y.: The role of root cytokinins during water and salinity stress. Israel J. Bot. **17**, 187–195 (1968).

JANES, E. E.: Adjustment mechanisms of plants subjected to varied osmotic pressure of nutrient solutions. Soil. Sci. **101**, 180–188 (1966).

KADDAH, M. T., GHOWAIL, S. I.: Salinity effects on the growth of corn at different stages of development. Agron. J. **56**, 214–216 (1964).

KAHANE, I., POLJAKOFF-MAYBER, A.: Effect of substrate salinity on the ability for protein synthesis in pea roots. Plant Physiol. **43**, 1115–1119 (1968).

KESSLER, B., ENGELBERG, N., CHEN, D., GREENSPAN, H.: Studies on physiological and biochemical problems of stress in higher plants. Volcani Inst. Agric. Res. Spec. Bull. **64** (1964).

KIRKHAM, M. B., GARDNER, W. R., GERLOFF, G. C.: Leaf water potential of differentially salinized plants. Plant Physiol. **44**, 1378–1382 (1969).

LIVNE, A., VAADIA, Y.: Stimulation of transpiration rate in barley leaves by kinetin and gibberellic acid. Physiol. Plant. **18**, 658–664 (1965).

LUNIN, J., GALLATIN, M. H.: Zonal salinization of the root system in relation to plant growth. Soil Sci. Soc. Amer. Proc. **29**, 608–612 (1965).

LUNIN, J., GALLATIN, M. H., BATCHELDER, A. R.: Effect of stage of growth at time of salinization on the growth and chemical composition of beans. II. Salinization in one irrigation compared with gradual salinization. Soil. Sci. **92**, 194–201 (1961).

LUNIN, J., GALLATIN, M. H., BATCHELDER, A. R.: Interactive effect of soil fertility and salinity on the growth and composition of beans. Proc. Amer. Soc. Hort. Sci. **85**, 350–360 (1964).

LUNT, O. R. OERTLI, J. J., KOHL, H. C.: Influence of environmental conditions ns the salinity tolerance of several plant species. 7th Int. Cong. Soil Sci. Trans. vol. **1**, 560–570 (1960).

MAGISTAD, O. C., AYERS, A. D., WADLEIGH, C. H., GAUCH, H. G.: Effect of salt concentration, kind of salt and climate on plant growth in sand cultures. Plant Physiol. **18**, 151–166 (1943).

MEIRI, A., POLJAKOFF-MAYBER, A.: Effect of variations in substrate salinity on the water balance and ionic composition of bean leaves. Israel J. Bot. **18**, 99–112 (1969).

MEIRI, A., POLJAKOFF-MAYBER, A.: Effect of various salinity regimes on growth, leaf expansion and transpiration rate of bean plants. Soil Sci. **109**, 26–34 (1970).

MEIRI, A., KAMBUROFF, J., POLJAKOFF-MAYBER, A.: Response of bean plants to sodium chloride and sodium sulphate salinization. Ann. Bot. **35**, 837–847 (1971).

MEIRI, A., SHALHEVET, J.: Pepper plant response to irrigation water quality and timing of leaching. In: Physical Aspects of Soil Water and Salts in Ecosystems. HADAS, A. *et al.* (Eds.) Ecological Studies **4**. 421–431 Berlin–Heidelberg–New York: Springer 1973.

MIZRAHI, Y., BLUMENFELD, A., RICHMOND, A. E.: Abscisic acid and transpiration in leaves in relation to osmotic root stress. Plant Physiol. **46**, 169–171 (1970).

NIEMAN, R. H.: Some effects of NaCl on growth, photosynthesis, and respiration of twelve crop plants. Bot. Gaz. **123**, 279–285 (1962).

NOURI, A. K. H., JAMES, V. D., KNUDSEN, D., OLSON, R. A.: Influence of soil salinity on production of dry matter and uptake and distribution of nutrients in barley and corn. I. Barley (*Hordeum vulgare* L.). Agron. J. **62**, 43–45 (1970).

NOURI, A. K. H., JAMES, V. D., KNUDSEN, D., OLSON, R. A.: Influence of soil salinity on production of dry matter and uptake and distribution of nutrients in barley and corn. II. Corn (*Zea mays* L.). Agron. J. **62**, 46–48 (1970).

OERTLI, J. J.: Effects of external salt concentrations on water relations in plants. 6. Effects of the external osmotic water potential on solute requirements, salt transport kinetics and growth rates of leaves. Soil Sci. **105**, 302–310 (1968).

OERTLI, J. J., ROTH, J. A.: Boron nutrition of sugar beet, cotton and soybean. Agron. J. **61**, 191–195 (1969).

PEARSON, G. A.: Tolerance of crops to exchangeable sodium. USDA Inform. Bull. **216**, 4 pp. (1960).

PEARSON, G. A., BERNSTEIN, L.: Salinity effects at several growth stages of rice. Agron. J. **51**, 654–657 (1959).

PORATH, E., POLJAKOFF-MAYBER, A.: The effect of salinity in the growth medium on cabohydrate metabolism in pea root tips. Plant Cell Physiol. **9**, 195–203 (1968).

REEVE, R. C., FIREMAN, M.: Salt problems in relation to irrigation. In: Irrigation of Agricultural Lands. HAGAN, R. M., HAISE, H. R., EDMINSTER, T. W. (Eds.) Amer. Soc. Agron., Madison, Wis. pp. 988–1008, 1967.

RICHARDS, L. A., WADLEIGH, C. H.: Soil water and plant growth. In: Soil Physical Conditions and Plant Growth. B. T. SHAW (Ed.), pp. 73–251. New York: Academic Press 1952.

ROBINSON, J. B.: Salinity and the whole plant. In: Salinity and Water Use. A National Symp. on Hydrology. TALSMA, T., PHILIP, J. R. (Eds.), pp. 193–206. Aust. Acad. Sci. London: MacMillan 1971.

SHALHEVET, J.: The use of saline water for irrigation. In: Dialogue in Development. 2nd World

Congress of Engineers and Architects, Jerusalem. GERSHON, D. (Ed.): Jerusalem Academic Press, pp. 437–442, 1971.

SHALHEVET, J., BERNSTEIN, L.: Effects of vertically heterogeneous soil salinity on plant growth and water uptake. Soil Sci. **106,** 85–93 (1968).

SHALHEVET, J., REININGER, P., SHIMSHI, D.: Peanut response to uniform and non-uniform soil salinity. Agron. J. **61,** 384–387 (1969).

SHALHEVET, J., YARON, B.: Effect of soil water salinity on tomato growth. Plant and Soil **23,** (1973).

SLATYER, R. O.: Effects of several osmotic substrates on the water relationships of tomato. Aust. J. Biol. Sci. **14,** 519–540 (1961).

STROGONOV, B. P.: Physiological basis of salt tolerance of plants. Israel Program for Scientific Translations, Jersulam, 1962 (translated from Russian, 1964).

U.S. Salinity Lab.: Diagnosis and Improvement of Saline and Alkali Soils. USDA Handbook 60, 160 pp. (1954).

VAADIA, Y., ITAI, C.: Interrelationship of growth with reference to the distribution of growth substances. In: Root Growth. WHITTINGTON, W. J. (Ed.), pp. 65–79. London: Butterworths 1969.

WADLEIGH, C. H., AYERS, A. D.: Growth and biochemical composition of bean plants as conditioned by soil moisture tension and salt concentration Plant Physiol. **20,** 106–132 (1945).

WADLEIGH, C. H., GAUCH, H. G., STRONG, D. G.: Root penetration and moisture extraction in saline soil by crop plants. Soil Sci. **63,** 341–349 (1947).

WEATHERLEY, P. E.: The pathway of water movement across the root cortex and the leaf mesophyll of transpiring plants. In: The Water Relations of Plants. 3rd Symp. Brit. Ecol. Soc., pp. 85–100, Oxford: Blackwell 1963.

WEATHERLEY, P. E.: The state and movement of water in the leaf. In: The State and Movement of Water in Living Organisms. Symp. Soc. Exp. Biol. **19,** 157–184 (1965).

WILCOX, L. V.: Boron injury to plants. USDA Inform. Bull. **211,** 7 pp. (1960).

YARON, B., ZIESLIN, N., HALEVY, A. H.: Response of Baccara roses to saline irrigation. J. Amer. Soc. Hort. Sci. *94,* 481–484 (1969).

3

Evaluation of Salinity in Soils and Plants

A. MEIRI and R. LEVY

Three factors influence crop growth and yield under saline and alkali conditions: (1) the water potential of the growth medium (the osmotic effect); (2) the accumulation of toxic concentrations of specific ions or imbalance of nutrients in the plant (the specific ionic effect); and (3) the physical deterioration of the soil because of the accumulation of high concentrations of exchangeable sodium (the alkali effect). Plant response to salinity is investigated by measuring the parameters connected with one or all of the aforementioned effects. The choice of the parameter should take into account the specific problem to be solved, and the interrelationships that exist between soil characteristics, salt composition, and plant species.

Defining Soil Salinity Status

Two distinctly different approaches are commonly used for defining the salinity status of a given soil. In the first one the salinity of the soil is expressed as a percentage of salt in the soil on a dry-weight basis (g salt/100 g dry soil) (ASTAPOV, 1958). Since plants respond to the concentration of salt in the soil solution, this expression can be meaningless unless the moisture characteristics of the soil are known. In the second approach, the electrical conductivity (EC_e in millimho/cm) of the soil's saturation extract is used as a measure of the salinity status (U.S. Salinity Lab., 1954). The advantages of this approach are several:

1) The saturation extract is a function of the moisture characteristics of the soil. The moisture saturation percentage (SP) of most medium- and heavy-textured soils was found to be twice that of field capacity.

2) The electrical conductivity of the soil solution is proportional to its salt concentration. The soil solution is a mixture of salts, and on the average $C = EC_e \times 10$ at the higher concentrations and $C = EC_e \times 12$ a the lower consentrations (where EC_e = electrical conductivity of the soil's saturation extract at 25° C and C = the ion concentration, meq/l).

3) The electrical conductivity of the saturation extract is proportional to the osmotic pressure (OP) of the solution, and in many cases it is the latter that determines plant response. [The U.S. Salinity Lab. (1954) proposed the empirical relationship $OP = 0.36 \times EC_e$.]

For special experimental conditions it is more convenient to prepare soil extracts by using different soil-water ratios (on a weight basis); e.g., 1:1 or 1:5. These techniques may be used when repeated sampling is done on the same soil in order to determine changes in salinity due to a given treatment. The limitations of the diluted extracts are: (a) by adding more water to the soil than is necessary to reach saturation, larger amounts of the slightly soluble $CaSO_4$ and $CaCO_3$ are dissolved; (b) by diluting the soil solution and changing the ratio of the soluble salts, the ratio of the exchangeable ions also changes.

A soil is considered saline if the electrical conductivity of the saturated extract exceeds 4 mmho/cm, and it is considered alkali if more than 15% of the total exchange capacity of the soil is saturated by sodium (U.S. Salinity Lab.,1954; REEVE and FIREMAN, 1967). A criterion for evaluating the alkalinity status of a given soil from data obtained from the analysis of the saturation extract was proposed by the U.S. Salinity Lab. (1954):

$$ESR = k\ SAR$$

where $ESR = \dfrac{ES}{CEC-ES}$; $SAR = \dfrac{Na^+}{\sqrt{(Ca^{++}+Mg^{++})/2}}$

here, ES is exchangeable sodium meq/100 g soil; CEC is the cation exchange capacity, meq/100 g soil; ESR is the exchangeable sodium ratio; SAR is the sodium adsorption ratio; Na^+, Ca^{++}, Mg^{++} are the concentration of cations in the saturation extract, meq/l; and k is a constant.

The facts that make SAR a good indicator of the alkali status of a soil should be emphasized: The concentration of the ions in the soil extract are readily determined, and ESR is a linear function of SAR.

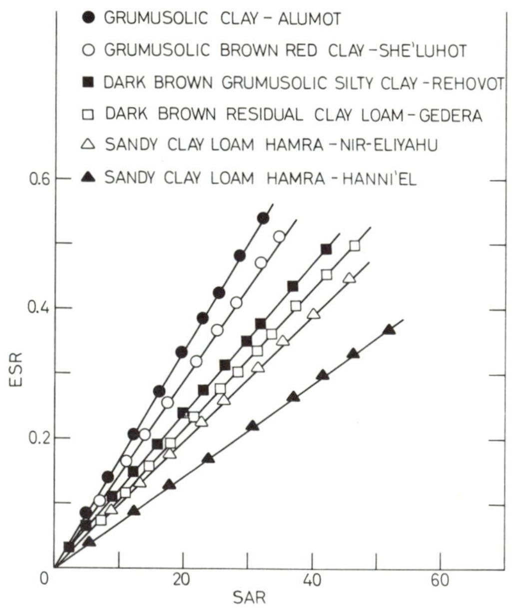

Fig. 1. Relationship between exchangeable sodium ratio (ESR) and sodium adsorption ratio (SAR) in six Israeli soils

The ESR–SAR relationship has been found to be good for a large number of soils (U.S. Salinity Lab., 1954). However, if a better estimation of the exchangeable sodium status is required, the constant k of the relationship should be determined experimentally for each soil (LEVY and MOR, 1965). When a soil is reclaimed, it is sufficient to follow the changes that take place in the soil extract and to calculate the changes in the soil complex by use of the ESR = k SAR relationship and the experimentally determined constant k. In Fig. 1, the ESR–SAR relationships of six different soils of Israel are plotted, and it is seen that k has different values, varying from 0.0072 to 0.0169.

Assessing the Salt Tolerance of Crops

The absolute yield obtained under saline conditions is of great economic significance. However, as yield integrates all the genetic and environmental effects acting upon the plant, the absolute yield under given conditions may not be obtained under other conditions. Experimental results, therefore, cannot be projected to commercial fields, and a comparison cannot be made between crops on the basis of absolute yields from isolated experimental plots. As a consequence, the most meaningfull criterion is relative yield obtained under saline conditions compared with nonsaline conditions, with all other factors being kept equal.

There are four principal methods for assessing plant response to salinity, and the choice depends on the accuracy required and on the purpose of the test.

a) *Salinity Survey.* A salinity survey obtains data from soils and crops over large areas of variable salinity. Because the data vary considerably, this method may provide results of a general nature and its accuracy is poor. However, such a method was used successfully in Holland after the flood of 1952, and in Israel after the introduction of Lake Tiberias water into the National Water Carrier.

b) *Field Experiments.* Field experiments are performed on plots, about 500 m^2 in area or 5 × 5 trees per plot. Irrigation is by the commonly used method. In addition to the different salinity levels of the water, treatments may include leaching intensity, irrigation frequency, irrigation method, and cultural practices.

c) *Artificially Salinized Field Plots.* This is the most widely used method of determining salt tolerance. The plots may be 2×2 m to 5×5 m in area, or, if the interest is in tree fruits, each plot may contain one tree. To avoid any horizontal variation in salt distribution, the plots are carefully leveled; and to avoid vertical variations in salt distribution, excess amounts of water are applied in each irrigation. The tolerance limits obtained under these conditions are for uniform saline profiles. Care must be given to guarantee proper fertilizer practice, especially concerning nitrogen, as the large water applications may result in the leaching of nutrients. Treatments may range from very low to very high salinity levels, depending on the crop sensitivity. The salinized water is usually applied to the soil after the crops are established, and the salinity is increased gradually to the desired level. If the increase in salinity is abrupt, the crop may be severely injured.

d) *Pot Experiments.* Pot experiments are performed in order to obtain information on certain problems, such as the effects of specific ions, the effects of well-defined salt distribution in the root zone, the effect of environmental conditions, etc. Soil, sand, or nutrient solutions can be used as the growing medium, and the experiments are carried out in greenhouses or growth chambers. Different types of pots can be used according to the problems being investigated. If soil is the growing medium, the salt content of the soil, the salt distribution within the profile, and the moisture must all be controlled. This can be achieved by drainage or leaching while suction is applied to the bottom of the pots

(YARON and MOKADY, 1962). When plant response to nonuniform salinity is under investigation, special pots and techniques are adopted. For example, in studying the effect of salt distribution in space, the split-root technique is used (SHALHEVET and BERNSTEIN, 1967), and for testing the variation in salinity with time, a streaming system is used (MEIRI and POLJAKOFF-MAYBER, 1970).

Methods for Soil and Plant Analysis

Measurements may be performed *in situ* or on samples brought to the laboratory. *In situ* measurements can be repeated at the same point, and so may reduce the number of sampling points required to obtain significant information about the effects of any treatment. The *in situ* methods for nondestructive determinations may save labor and time if their performance involves less work than that required for sample removal, preparation of the samples for analysis, and making the analysis.

Parameters to be measured are soil- and plant-water contents and the various components of their water potential. Under saline conditions, special attention should be given to the osmotic component of the water potential. Measurements are carried out *in situ*, or on soil samples, to determine soil-water status. Soil extracts are analyzed for ion concentration, electrical conductivity, and osmotic potential. Plant tissue is analyzed for water status and chemical composition. Extracted plant sap is used to determine the osmotic potential of the sap, and the concentration of the soluble organic and inorganic constituents.

As osmotic potential is difficult to determine and requires expensive equipment; other parameters correlated with OP are usually determined. For soil extracts, the electrical conductivity is converted to osmotic potential (see p. 291). For the sap of plant leaves, electrical conductivity plus refractive-index values may be converted to osmotic potential (SHIMSHI and LIVNE, 1967).

Soil

Methods of determining soil-water content and status are described previously.

Soil Sampling. A good sampling program is important for obtaining representative data. Sampling intensity will depend on the heterogeneity of the soil and the accuracy of the values required. Saline soils are usually extremely variable in their salt content. Factors such as climate, microtopography, soil structure and texture, and agricultural management influence the salt distribution. Preliminary information on the heterogeneity of the soil will help in deciding what sampling pattern to follow. The soil samples are either removed according to the different soil genetical horizons, or are taken in predetermined increments, usually every 30 cm, to the depth of rooting or to the depth of the water table. The soil samples are transferred quickly to the laboratory and spread out to air-dry. They are then passed through a 2-mm screen. If average values of parameters from a field or a plot are sufficient, much work can be saved by using composite samples.

In order to obtain a composite sample, samples of similar characteristics and of the same weight are selected and carefully mixed. The air-dried soil can be stored for very long periods of time.

Extraction of Water-Soluble Constituents from the Soil Samples. The concentration of the water-soluble constituents in the soil solution increases with decreasing soil-moisture content. The concentration values of agricultural significance are those at a moisture content below field capacity. However, extraction of a soil solution at field-moisture content is difficult, and so the saturated-paste-moisture content has been adopted as the most useful one. The method of preparation of a saturated soil paste is described in Handbook No. 60 (U.S. Salinity Lab., 1954).

Extraction of Soluble Constituents in the Field. In the field, the soluble constituents are extracted by the use of porous ceramic cups (REEVE and DOERING, 1965). The cups are connected to rigid tubes and buried at the desired depth in the soil. There should be a tight contact between the sampling tube and cup and the soil, so that the sample taken will truly represent the soil solution. Suction is applied to the tube *via* a vacuum pump or evacuated bottle. The cup must be wet and have a higher air-entry value than the suction applied. The system must be closed tightly to prevent any air leaks. The rate of soil solution accumulation in the cup will depend on the suction gradient from the soil to the cup and on the soil's hydraulic conductivity. When extration is complete, the suction is released and atmospheric pressure pushes the solution from the cup into a collection vial (Fig. 2).

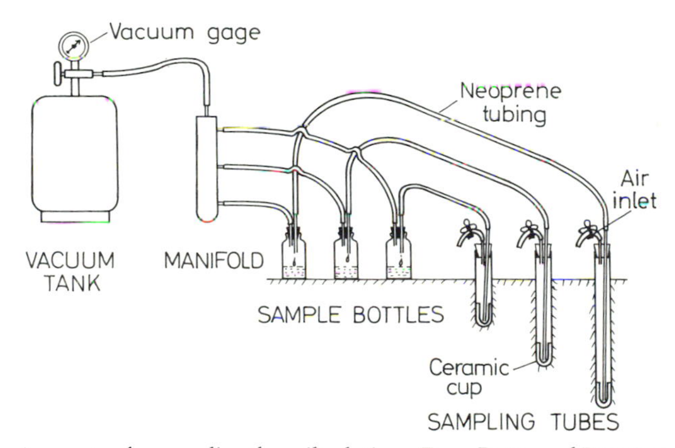

Fig. 2. Apparatus for sampling the soil solution. (From REEVE and DOERING, 1965)

The time during which sampling can be done after an irrigation is limited. Sufficient time is required for equilibrium to be reached by the soil solution before suction is applied. Also, sampling cannot be overly delayed because soil hydraulic conductivity decreases and water retention forces increase as the soil dries out. Practically speaking, the suction should be applied two or three days after irrigation, and for a period of several hours.

The porous material of the cup should have the smallest possible electric charge. Charged material may alter the composition of the collected solution because an exchange process between the cup and the solution may take place, and also because it may have a sieving effect on the entering solution by ion exclusion.

Sensors for in situ Measurements of Soil Solution Conductivity. The drawbacks of the *in situ* measurements are that conductivity varies with solute concentration, with soil-moisture content, and with soil characteristics; mainly, the cations exchange capacity.

Salinity sensors are constructed of two electrodes and a thermistor (RICHARDS, 1966) inserted into a small unit of fine porous material. These are buried in the soil at the desired depth, with electric wires leading to the soil surface. The hole in the soil around the leads should be carefully filled to reduce any interference in water flow in the soil. Conductivity bridges can be used for the conductivity and temperature measurements made on the solution within the sensor. The air-entry value of the porous material is high, so that the moisture content of the material remains constant throughout the range of soil-moisture contents, and any variations in electrical conductivity are a result of variations in concentrations of the sensor solution. Concentration equilibrium between solutions in the unit and in the soil is achieved through diffusion. The units are thin, and have high porosity and good contact with the soil, so equilibrium time is reduced. In aqueous solutions, about two hours are required to reach 90 % of equilibrium (Fig. 3).

The porous material should have a minimum electric charge to reduce interference in measurements due to ion exchange or ion sieving effect. The cell constant of every sensor should be determined by use of standard solutions.

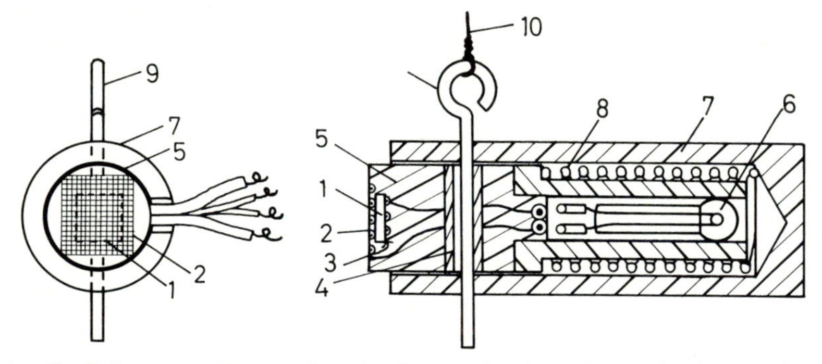

Fig. 3. A soil salinity sensor. Front and sectional views showing: (1) ceramic plate, (2) front screen electrode, (3) back electrode, (4) nylon tubing, (5) epoxy block, (6) thermistor, (7) outer Lucite case, (8) spring for holding the sensitive element against soil, (9) release pin, and (10) pull wire. (From RICHARDS, 1966)

Plant. Methods of determining plant-water content and plant-water status are described in a previous Chapter.

Plant Sampling. Leaves, stems, and roots vary markedly in their composition. Since this composition changes with age, it is advisable to collect and analyze leaves and stems of similar age. Usually, fully developed leaves that do not show visual symptoms of salinity injury are chosen for analysis. In order to correlate results obtained from plant analysis with soil parameters, in a heterogeneous area plant and soil samples should be collected separately to represent different parts of the area.

Surface contamination of the plant material should be removed by brushing, rapid rinsing with a weak detergent solution and distilled water, and blotting gently between sheets of filter paper. If analysis will be performed on the whole plant material it should be dried rapidly in a forced-draft oven at 60°C and ground. Determination of the percentage of water in the fresh material is important for evaluating the results on a dry- or fresh-weight basis.

If analysis will be performed on the sap, weighed fresh samples are frozen rapidly, using dry ice, liquid air, or some other cooling mixture. If a cooling mixture is used, the sample should be closed tightly to prevent contamination. The frozen samples are stored in a freezer. In order to obtain the sap, the samples should be pressed at a pressure of

10,000 to 15,000 psi. The freezing destroys all biological membranes so that filtration of the vacuole sap by the tissue during extraction is minimized, and the extract is similar in composition to the plant sap. Sometimes centrifugation is required to separate the sap from solid particles. The sap is diluted to a proper volume and analyzed for its constituents.

Extraction of Inorganic Constituents. For chemical analysis of inorganic constituents of the dry matter, the organic matter should be destroyed. This can be achieved by dry-ashing or by wet digestion. If only some specific elements are to be determined, extraction of the material will be sufficient. Selection of the method to be used depends on the information required and on its expense.

a) *Dry-ashing.* Dry, ground plant material is ashed in an oven at about 550° C. The ash is dissolved in acid, usually hydrochloric acid, and the solution is diluted to a desired volume. The dry-ashing method can be used for the determination of calcium, magnesium, potassium, and sodium, and other microelements; for elements such as chloride, boron, and sulfur, the method should be modified to prevent volatilization losses. If sulfur and phosphorus are to be determined, the dry matter should be mixed with magnesium nitrate before the dry-ashing. If chloride or boron is to be determined, the dry plant material should be mixed with calcium oxide before the ashing to make it basic.

b) *Wet Digestion.* The dry plant material is digested in a heated concentrated acid, and after the material has dissolved it is transferred to a volumetric flask and diluted. It is easier to dissolve the plant material by wet digestion than by the dry-ashing method, and the contents of calcium, magnesium, sodium, and potassium in the solution can be determined.

c) *Ion Extraction.* The method is easy to carry out and is very useful for ions that are not constituents of organic components. It is used only if these ions are of interest. Sodium and potassium, which can accumulate as inorganic salts due to the growth of the plant under saline conditions, can be extracted by shaking the ground plant material for about 30 minutes in 1*N* ammonium acetate. The extract can be directly analyzed for the two ions.

Chloride can be extracted by shaking the ground plant material for a short period in dilute nitric acid or allowing it to stay overnight in contact with the weak acid. The extract can be directly analyzed for chloride.

Analytical Methods

Once a solution extract is available, the analytical methods for the determination of inorganic ions are the same, for both soil and plant extracts. Of main importance in evaluating the salinity status of soils and plants are sodium, potassium, calcium, magnesium, chloride, bicarbonate, and sulfate. The analytical method chosen for the determination of each of the above-mentioned elements will depend on the presence of the other ions found in the extract that may interfere with the analysis.

There are several books that can be consulted for detailed analytical methods for the determination of the above ions in irrigation water, in soil-water extracts, on the exchange complex, and in plant extracts (U.S. Salinity Lab., 1954; CHAPMAN and PRATT, 1961; JACKSON, 1964; BLACK, 1965).

A laboratory equipped with an emission flame photometer, an absorption flame photometer, and a chloridometer can analyze, quickly and accurately, most of the important elements in salinity investigations. The emission flame photometer is very good for the determination of sodium and potassium, although for calcium and magnesium it is not so accurate. The absorption flame photometer is more suitable for the determination of calcium and magnesium than for sodium and potassium. Calcium and magnesium can be successfully determined by titration with EDTA, but only in the absence of organic compounds such as ammonium acetate, which is present when exchangeable cations are extracted. Plant extracts obtained by wet digestion are also unsuitable for titration with EDTA because of the appreciable amounts of acid present. For the determination of exchangeable calcium and magnesium extracted with acetate and acid plant extracts, the absorption flame photometer is indispensable. The determination of chlorine is by titration with $AgNO_3$. It can be performed either by use of a chloridometer (potentiometric titration), or by titration with $AgNO_3$ with an indicator. When small amounts of the extract are available or the concentration of the chlorine is very low, titrations with an indicator end point is hard to perform. Bicarbonate and sulfates are analyzed by volumetric and gravimetric methods, respectively. In an investigation of a salinity problem their determination is usually of secondary importance.

The cation-exchange capacity of soils is determined by saturating the soil with a given cation by use of a 0.1 or 1.0 *N* solution of a salt of this cation. The excess salt is washed, and the cation extracted by an exchange reaction with another salt. The exchangeable cations are extracted with a 0.1 or 1.0*N* solution of a given salt. The choice of the saturating and extracting salts will depend on the facilities available for the determination of the saturating cation, and on the possibility of using an extracting salt with the least interference in the determination of the extracted cations. The saturating salt most widely employed for the determination of the cation-exchange capacity of soils is Na acetate, because the determination of Na is easy to perform when a flame photometer is available. The most widely used extracting salt is NH_4Ac, because the NH_4 cation and the Ac anion do not interfere in the determination of Na, K, Mg, and Ca if they are determined by flame photometry methods.

Literature

ASTAPOV, C. B.: Reclamation of Soils (in Russian). Government Editions of Agricultural Literature, Moscow (1958).

BLACK, C. A. (Ed.): Methods of Soil Analysis, Pt. 2. Amer. Soc. Agron., Madison, Wis. (1965).

CHAPMAN, H. D., PRATT, P. F.: Methods of analysis for soils, plants and waters. Univ. of Calif., Riverside (1961).

JACKSON, M. L.: Soil Chemical Analysis. Englewood Cliffs, N. J.: Prentice-Hall 1964.

LEVY, R., MOR, E.: Soluble and exchangeable cation ratios in some soils of Israel. J. Soil Sci. **16**, 290–295 (1965).

MEIRI, A., POLJAKOFF-MAYBER, A.: Effect of various salinity regimes on growth, leaf expansion and transpiration rate of bean plants. Soil Sci. **109**, 26–34 (1970).

REEVE, R. C., DOERING, E. J.: Sampling the soil solution for salinity appraisal. Soil Sci. **99**, 339–344 (1965).

REEVE, R. C., FIREMAN, M.: Salt problems in relation to irrigation. In: Irrigation of Agricultural Lands. HAGAN, R. M., HAISE, M. R., EDMINSTER, T. W. (Eds.): Amer. Soc. Agron., Madison, Wis. pp. 988–1008 (1967).

RICHARDS, L. A.: A soil salinity sensor of improved design. Soil Sci. Soc. Amer. Proc. **30**, 333–337 (1966).

SHALHEVET, J., BERSTENIN, L.: Effect of vertically heterogeneous soil salinity on plant growth and water uptake. Soil Sci. **106**, 85–93 (1967).

SHIMSHI, D., LIVNE, A.: The estimation of the osmotic potential of plant sap by refractometry and conductimetry: a field method. Ann. Bot. **31**, 505–511 (1967).

U.S. Salinity Lab. Staff: Diagnosis and Improvement of Saline and Alkali Soils. USDA Handbook **60** (1954).

YARON, B., MOKADY, R.: A technique for conducting pot experiments with saline water. Plant and Soil **17**, 392–398 (1962).

VII. Irrigation Technology

The purpose of any irrigation system is to convey water from a source to the field and to deliver it to the root zone of the crop. A well-designed and -operated system will perform this task while meeting three general requirements: assurance of maximum economic return to the farmer; minimal loss of water during conveyance and application; and maintenance of long-term productivity of the land through prevention of erosion, soil-structure degradation, soil salinization, and raising of the groundwater table.

Many factors interact in determining the extent to which these three requirements are met, and therefore no ready-made "recipe" can be prescribed for the choice of irrigation method and for the design of a particular system. These factors include topography; soil depth, texture, and structure; climate; crop characteristics; size and type of water source; quality of water; depth and quality of groundwater; the relative cost of irrigation equipment; land preparation and labor; the cost of credit; and the availability and skill of farm labor.

The responsibility of the designer extends beyond the provision of an efficiently functioning mechanical system. Not only must soil and cropping factors be considered during the design, recommendations should be given to the farmer on how best to operate his system, and enough flexibility should be incorporated in the design to accommodate changes in cropping patterns that are liable to occur in the future.

The principal methods used to apply water to the root zone may conveniently be divided into the following main types: (a) airborne or sprinkler irrigation techniques; (b) irrigation based on gravity flow; (c) trickle irrigation; and (d) special techniques.

By the end of this chapter, the reader will find examples of practical ways for planning irrigation at the farm level by using suitable irrigation techniques and considering all the agronomical and environmental aspects.

1

Sprinkler Irrigation

E. RAWITZ

General Considerations

Before discussing various irrigation methods it is necessary to define a number of parameters common to all methods. For the purposes of practical planning, the total amount of water withdrawn from a source for irrigation is defined as the *gross application amount* (W_a) and the amount of water actually added to the root zone (possibly arbitrarily defined) of the crop is the *net application amount* (W_r). The difference between the gross and net application is defined as "losses". These losses may be divided into water lost in transit from the source to the field to be irrigated (due to leakage and evaporation) called *conveyance loss;* and water lost through evaporation, deep percolation, surface runoff, and wind drift as *field loss* (i.e., loss with respect to a particular irrigation application and area of land). Conveyance losses concern the civil engineer more than the agronomically oriented irrigation engineer, and will not be discussed in further detail here. The *"field irrigation efficiency"* (*Ea*) (or the water application efficiency) is the percentage of the water delivered at the edge of the field that is actually added to the root zone and is available for evapotranspiration ($E_a = W_r / W_a \times 100$). The field loss is controllable to some extent by proper design and operation of an irrigation system. For example, the shorter the time during which there is free water on or above the field, the smaller will be the *evaporation loss,* and this is only one of several reasons why a fast water application is desirable, provided it is consistent with the infiltration capacity of the soil. *Deep percolation* may be caused not only by gross overirriagation, but also by nonuniform distribution of the correct amount of water – with some parts of the field getting less and others much more than the desired amount, part of which is lost by deep percolation. Surface runoff is most often found in improperly designed or operated flood and furrow irrigation systems, but may also result from sprinkling at a rate higher than the infiltration capacity of the soil. *Wind drift* results when sprinklers are operated under conditions improper for sprinkling, and is aggravated when they are operated at too high a water pressure. As a result, the water jet breaks up into very fine drops (aerosol formation), which take much longer to fall to the ground and are thus more likely to be transported laterally out of the area to be irrigated.

Evaporation loss from wet soil and from wetted foliage is a function of the irrigation frequency, and is independent of the size of a single application. To illustrate this, it may be assumed that foliage will intercept the maximum possible amount of water within a very short period, and that this amount will be "lost" through evaporation after irrigation is terminated[1].

---

1 Evaporation from wet foliage and from wet soil is not necessarily a true loss since it uses up energy that would otherwise have been available for normal evapotranspiration. However, according to the above definition, it is technically a water loss.

It may also be assumed that a certain constant percentage of the water stored in the top layer of soil (e.g. 30 cm) will rapidly evaporate after each irrigation and will not be available for transpiration. As the amount of water lost this way is constant for each irrigation, the more frequent the irrigations, the more water will be lost during a season, thus reducing the application efficiency. Therefore, from the point of view of water application efficiency, it is desirable to decrease application frequency and increase the single-irrigation dosage, provided this is consistent with maintaining soil-moisture availability at a satisfactory level.

Classification and Types of Sprinkler Irrigation Systems

All sprinkler systems have certain basic components: a source of water under pressure, a system of pipelines to convey the water from the source to the point of delivery, and some arrangement of nozzles or orifices through which the water is distributed over the land. Additional components may be used for starting and ending the flow of water, for regulating the water pressure, for draining part or all of the conveyance network after irrigation, for measuring flow and pressure of water, and for "changing the set" – i. e., either moving the sprinkler line to a new area, or shutting off one line and activating another. Many of these secondary components range from very simple devices to highly sophisticated automatic equipment. It is beyond the scope of this chapter to discuss all of them in detail, and emphasis will be put on those systems most commonly used in agriculture.

Sprinkler systems are classified both according to type of conveyance system and type of discharge device. Conveyance systems are divided as follows:

Permanent or Fixed Conveyance Systems

All distribution pipelines are permanently in place. In some systems, sprinkler heads and risers are moved for each "set"; in others, they, too, remain fixed. The pipes are usually buried in a trench to facilitate traffic and to protect the pipes. Permanent systems have the highest initial cost and the lowest labor cost. They are particularly well adapted for automation and for irrigation of perennial crops such as orchards.

Semipermanent (Semiportable) Conveyance Systems

Similar to the permanent systems, the large-diameter pipelines comprising the mains and submains are usually permanent and buried in the ground. Portable laterals, typically of aluminum or plastic, are attached to risers equipped with valves that are part of the mains system. Semipermanent systems are the most common type in intensive irrigated agriculture. Various devices for moving entire laterals at a time instead of one or two lengths of pipe by hand have lowered labor costs and have improved the efficiency of these systems.

Fully Portable Conveyance Systems

Here all the system components are portable. This type of system requires a trailer-mounted pump with access to the source of water at several locations. Fully portable systems have the lowest initial cost but the highest running costs of the three types of systems. Labor costs for moving the network are considerable; due to the constant moving, depreciation rate of the equipment is higher; and pumping with portable gasoline–or diesel powered pumps is more expensive than pumping with electric power

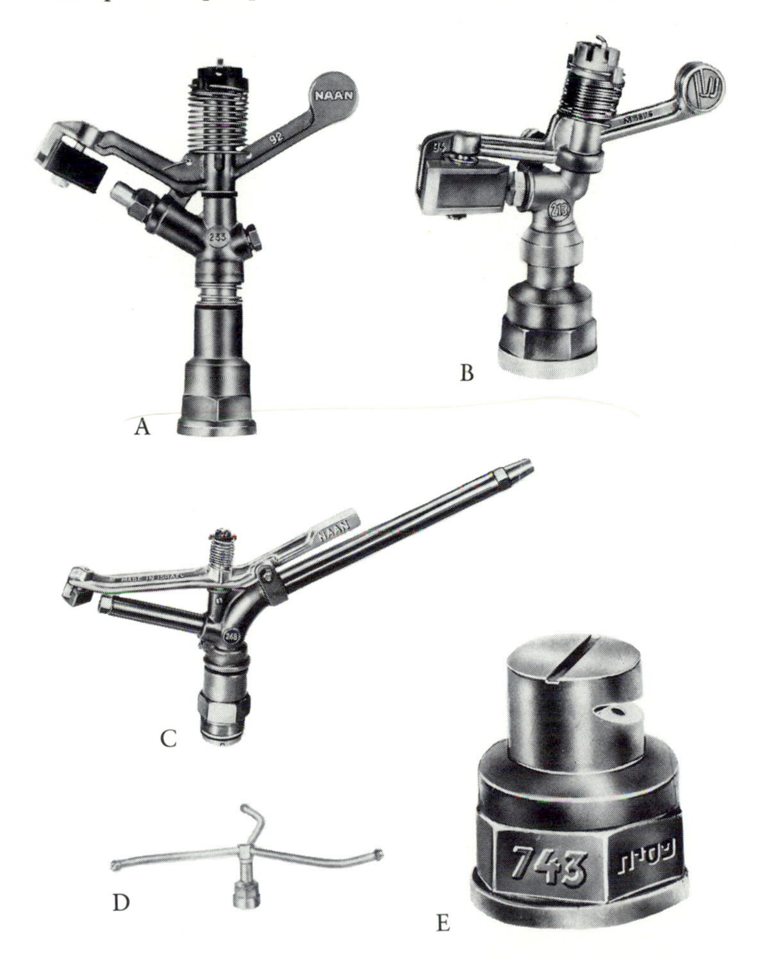

Fig. 1. Various sprinkler heads used in Israel: A general-purpose slowly rotating; B under-tree slowly rotating, single nozzle; C giant sprinkler, slowly totating, three nozzles; D fast-turning "Whirlybird"; E static head. (Courtesy of Na'an Metal Works, Ltd.)

or large stationary diesels. Fully portable systems are, of course, the most flexible in operation. They are thus best adapted for supplementary irrigation, where perhaps only one or two irrigations are applied per year.

The discharge devices used in sprinkler systems are divided into three general classes:

Static Sprinklers

Static sprinklers have a small range and suffer from poor distribution, especially in windy conditions. They are generally used only for special purposes, such as in private gardens, as fog sprinklers in nurseries and greenhouses, and for aiding in the establisment of young trees.

Nozzle Lines

Nozzle lines are of two types. They may be static lines, usually of aluminum pipes 2–4 in. in diameter, with several rows of holes drilled into them radially to distribute the water at different angles. They operate at relatively low pressure, the width of wetted

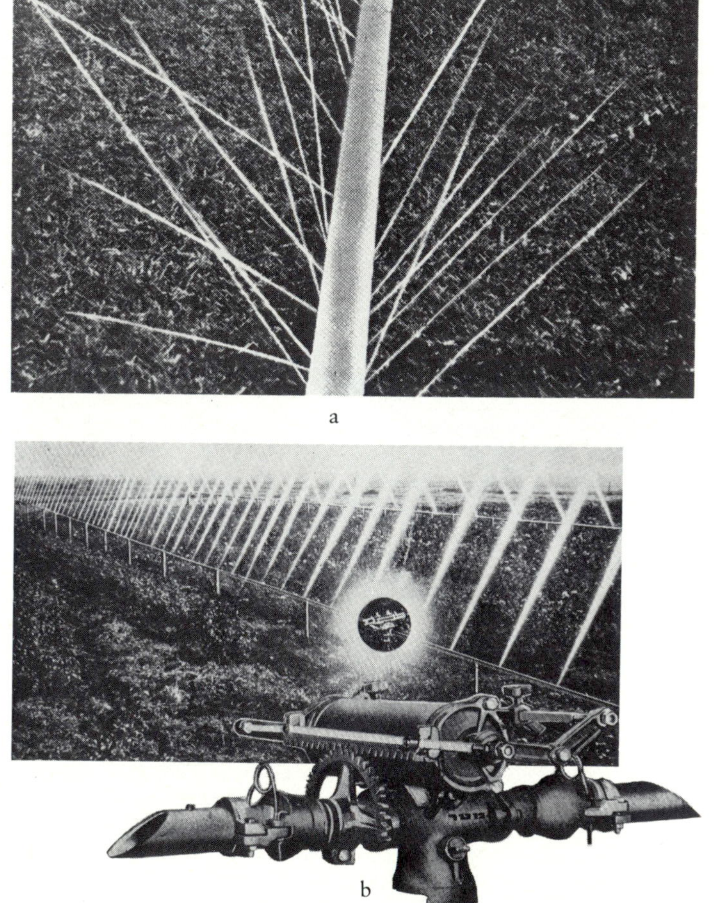

Fig. 2. a Static nozzle line operating in pasture. b Oscillating nozzle line on permanent post supports. Insert: hydraulic piston motor with rack and pinion provides power to oscillate line

strip being a function of the pressure. The application intensity (per unit area) is usually independent of pressure within the recommended limits. They are available with application rates of 1 in. (25 mm) per hour, or 2 in. (50 mm) per hour. The high application rate limits this equipment to soils with relatively high intake rates. The holes tend to clog unless filtered water is used, and they eventually become enlarged. Static nozzle lines are in limited use in agriculture, mostly as under-tree sprinklers in orchards and with low-growing crops.

The second kind of nozzle line is the oscillating nozzle line. This consits, as a rule, of galvanized iron pipe, 1 in. or 1 1/2 in. diameter, resting on supports 50 to 100 cm high. Brass nozzles are tapped into the pipe, and the oscillation is provided by a hydraulic motor whose piston is driven by the irrigation flow. These nozzle lines are permanent, in contrast to the first type. They operate at medium pressures, and their distribution pattern is very sensitive to wind. Initial cost is high and operating cost low, as in all permanent systems. Oscillating nozzle lines interfere with the movement of farm machinery and are not in use with field crops. They are commonly utilized in nurseries of various kinds.

Rotating Sprinklers

Rapidly and slowly rotating sprinklers are in use. The rapidly turning sprinklers ("whirlybird") have two to four radial arms with nozzles at their ends, and operate at relatively low pressure. Because of various limitations, these sprinklers are used mainly in private gardens. The slowly rotating sprinklers are the most important type in use in agriculture. They make a complete revolution in 30–120 sec., driven by the jet of water from one nozzle impinging on a spring-loaded impact arm. Some types have a counterweighted arm without a spring. These sprinklers are equipped with from one nozzle for small, low-pressure models, to 4 or 5 nozzles in giant sprinklers. 1 and 2 nozzle sprinklers are the most common. Special types are available for use with sediment-laden water, for frost protection, for upside-down operation in greenhouses, for part-circle operation, and for landscape garden use. They are reliable, and easy to dismantle for cleaning and replacement of worn parts.

Giant Sprinkler Machines

A few words should be devoted to a special self-contained, integrated, self-propelled sprinkler system, sometimes called a giant sprinkler machine. At least three types have been produced in the U.S. and in the U.S.S.R.: a large-diameter (3–4 in.) part-circle high-

Fig. 3. Giant boom sprinkler irrigation 1/2 to 2 ha

pressure nozzle mounted on and powered by a crawler tractor moving along an open ditch and pumping from it; a cantilevered system of pipes with static or slowly rotating sprinklers mounted along it at intervals, carried on a crawler tractor moving along an open ditch; and a slowly rotating giant sprinkler consisting of a framework of pipes with many regular, slowly rotating sprinkler heads, connected to widely spaced permanent risers. These systems may irrigate up to one hectare per set. All of them are in very limited use because of technical problems and their high cost.

Accessories such as valves, pressure regulators, water meters, and automatic shut-off and starting devices my be used with any of the systems. Table 1 summarizes the various combinations of sprinkler systems and their components as they are commonly used.

Choosing the Type of System

The first step in designing a particular sprinkler system is to select the type of distribution system and discharge device. The most common choice for permanent, irrigated farming is a semiportable system using medium-pressure slowly rotating sprinklers mounted on portable aluminum laterals. Where the system is to be used for supplementary irrigation, a fully portable system may be preferable.

Mains System

The choice of power unit and pump and the designing of the mains system is best assigned to a hydraulic engineer hired for this purpose. The choice of pipeline material (e.g., steel, asbestos-cement, or plastic) is often affected as much by considerations of availability and price as by technical considerations. It should be pointed out, however, that a common problem involves the hitting by farm machinery of the 2 or 3 in. risers suplying

Table 1. Common combinations of sprinkler system components

Component	Type of System		
	Permanent	Semiportable	Portable
Source	Stationary pump, electric or diesel powered, regional major pipeline	Same as for permanent	Mobile pump powered by integral gasoline or diesel engine or tractor PTO
Mains	Underground steel, asbestos-cement, or concrete pipe	Same as for permanent, may be above ground	Aluminum or plastic pipes on surface connected with quick couplers or dressers
Laterals	Surface or underground steel, iron, or asbestos-cement pipe; plastic, copper, or aluminum tubes on surface	Surface aluminum pipes with quick couplers, plastic tubes, copper tubing, sheet-steel tubes	Same as for semiportable
Risers	Galvanized iron, aluminum, or plastic pipe with supports where needed; same for all three types of systems		
Sprinklers	Static sprinklers, oscillating nozzle lines, slow- or fast-rotating sprinkler heads	Same as for permanent, except that oscillating nozzle lines are avoided, and static nozzle lines are used; giant sprinkler machines.	

the laterals. In order to avoid severe damage to the main lines, and expensive and often time-consuming repairs, a safety device should be incorperated that will cause the riser, not the main line, to be damaged by the impact. The use of a plastic riser, or the inclusion of a weak section in the riser, solves this problem adequately.

Laterals

Laterals can be chosen only after a decision has been made on how these will be moved. Aluminum laterals of 2–4 in. in diameter are the most common, and plastic laterals up to 1 1/2 in. in diameter are also popular.

Aluminum laterals are more adaptable to varying conditions. If the lines are to be hand-moved in individual sections of 20 or 40 ft (6 or 12 m), a quick coupler is chosen that can be engaged easily when the pipe is held at its midpoint during assembly of the line. If the laterals are to be moved by towing in a direction along the longitudinal axis of the pipe, the coupling must be able to withstand tension. If the lateral is to be rolled perpendicular to its long axis, the couplings must be able to withstand twisting action. These types of line-moving systems are shown in Fig. 4. Laterals that are moved without dismantling should be equipped with a drain valve to drain the line automatically as soon as the water pressure is shut off.

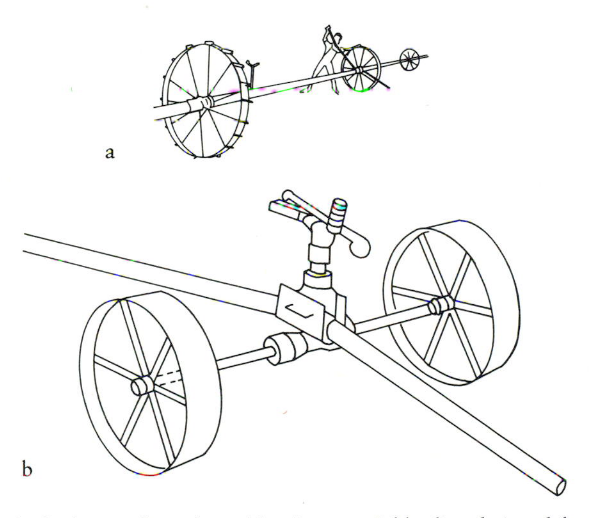

Fig. 4. Two types of wheel-move laterals: a Aluminum sprinkler line designed for rolling perpendicular to pipe axis. b Plastic lateral designed for towing along axis of tubing

Plastic laterals are specifically designed for moving by towing, and a small sled or a pair of wheels is mounted under each sprinkler. Flexible plastic tubing is used, and special fittings are available for connecting the sprinklers, etc. Plastic laterals are usually shorter than aluminum laterals, and have proven particularly useful for under–tree sprinkling of orchards. Such laterals have the advantage of being flexible and light, enabling the irrigator to pull them toward the riser on the main line immediately after shutting off the water without having to walk over the muddy soil just irrigated. Quick transfer of the laterals in this manner saves equipment and labor while avoiding soil compaction. It is not practical to mount sprinklers on risers higher than about 2 ft (60 cm) on any lateral that is moved without dismantling.

Sprinkler Heads

There are many brands of sprinklers available on the world market, and each manufacturer offers a large variety of models. Suppliers usually have catalogs that help the customer to choose a sprinkler, and also give information on application rates, and special features and limitations, as well as design criteria for choosing the proper lateral diameter for each type of sprinkler. The main points to be considered in choosing a sprinkler are as follows:

a) *Application Rate.* The application rate must be such that it will be possible to apply irrigations of the desired magnitude without exceeding the infiltration capacity of the soil. On the other hand, the highest allowable application rate is desirable since this represents optimal utilization of the equipment.

b) *Operating Pressure.* Each sprinkler operates properly only within a certain range of pressures, and the power source and distribution system must be designed to meet this demand.

c) *Sprinkler Spacing.* Recommended spacings are given in the catalogs, and sprinkler spacing must be coordinated with the spacing of laterals and risers from the mains. The wider the permissible spacing, the lower the initial cost of the system and the cost of moving laterals.

d) *Drop Size.* Large, coarse drops may damage certain delicate crops, and may cause damage to soil structure if bare soil is exposed. On the other hand, very small droplets are easily carried away by wind. Most sprinklers do not approach either extreme, and drop size is not usually specified in catalogs.

e) *Uniformity of Water Distribution and Pattern Stability.* Most sprinkler manufacturers adhere to the generally accepted standard that each sprinkler have a Christiansen uniformity coefficient of 84% or better. (This will be discussed in more detail in the next section.) Since uniformity of water distribution is affected by wind, catalogs generally specify the range of wind speeds within which the sprinkler can be operated, and what changes in spacing should be made if operation at higher wind speeds is unavoidable. Generally speaking, pattern stability is better with larger drop size and closer sprinkler spacing. Better uniformity is often obtained if sprinklers are placed in a triangular rather than a square or rectangular pattern in the field.

Design Recommendations

The irrigation network is designed "upward" from the single sprinkler in the individual field with its laterals, to blocks and their submains, to entire farms and groups of farms with their mains system. First, the highest expected irrigation frequency (i.e., shortest interval) must be decided upon. Second, the maximum or the desired water application rate is determined, as well as the number of hours per day that the field will be irrigated. Before these and the following decisions can be made, it is necessary for the planner to have topographic and soil maps of the area, as well as information on the planned land use and crop rotation, estimated maximum daily consumptive use of water, the estimated infiltration capacity and water-holding capacity of the soils, and the size, type, and quality of water source available. Taking into account the time required for moving laterals, as well as the time during each irrigation cycle when no irrigation is carried out (reserve time for cultivations, sprayings, down-time of equipment), the sprinkler head can be chosen and the number of laterals required is determined. To obtain best uniformity, laterals should run perpendicular, rather than parallel, to the direction of prevailing wind. Also, lateral diameter should be chosen so that there is no more than a 10% difference in discharge between the first and last sprinkler due to pressure differences. In order to achieve this, the pressure difference between the first and last sprinkler must be less than 20%. It is good practice, therefore, to lay the laterals on the contour or on a downslope, but not sloping upward. "Tapered" laterals are sometimes used, e.g. part 3 in. and part 2 in. pipes.

The mains and submains systems should be planned for the "worst case," i.e., maximum demand. Where more than one lateral is used per field, they should initially be placed as far apart as possible to minimize pressure losses in the submain, but in the "worst case" position of each lateral in its own sector. This will be demonstrated in the following example.

Where conditions allow this, it is ideal to plan the farm so as to have square or rectangular fields of roughly equal size. If all laterals are the same length, they can be more easily transferred for use in different fields. A symmetrical system of mains and submains is the most economical one to install and operate. When the mains system is very long it may happen that, in order to provide adequate pressure at the far end, pressures near the head of the line are too high to operate sprinklers efficiently. In such cases pressure reducers must be installed in the upstream submains, which add to the cost of the system. In practice, compromises are often necessary due to conflicting conditions: topography vs. prevailing wind direction, high-pressure small-diameter mains with high pumping costs to provide pressure vs. large-diameter mains with higher initial cost but lower pumping costs, etc.

Field Tests

Although most sprinklers conform to the standard of uniformity of distribution, it sometimes is necessary to perform a field test. The method generally used is that of Christiansen, and the reader should refer to the source for details. One or more sprinklers are operated in the field under known conditions of water pressure and discharge, wind speed and direction, and sprinkler spacing. The sprinkler head should be at least 30 cm above ground level. The irrigation water is caught in a numer of receptacles, which may range from proper rain gauges to simple tin cans placed on a square grid on the test area. The cans should be deep enough to eliminate splash, and no more than about 15 cm above the ground surface. Sophisticated test systems may feature cans sunken in the ground so that their top edge is flush with the soil, or they may be placed on a nonsplashing surface (grass, plastic screening, etc.).

The cans should be no more than 2 meters apart in the grid, and the time of run should be long enough to catch an appreciable amount of wate A one-hour run is usually adequate. The water from each can is measured with a graduate cylinder to the nearest ml, and the amount is recorded either on data sheets or directly on a map of the area. If the test area was covered by a complete sprinkler pattern (sprinklers were operated in all the positions from which water is contributed to the area in a field pattern), analysis can proceed. If fewer sprinklers were used for the test, their individual patterns are assumed to be representative, and the observed values are superimposed on the map according to the spacing of sprinklers to be investigated. This technique also makes it possible to use one set of data to calculate the distribution pattern obtainable with various sprinkler spacings. If the distribution pattern was distorted by wind, care must be taken to place each sample quadrant in the proper position on the map; this is illustrated in Fig. 5. The contributions from each sprinkler overlapping at each point are then added up to obtain the total catch at that point, and the anlysis proceeds.

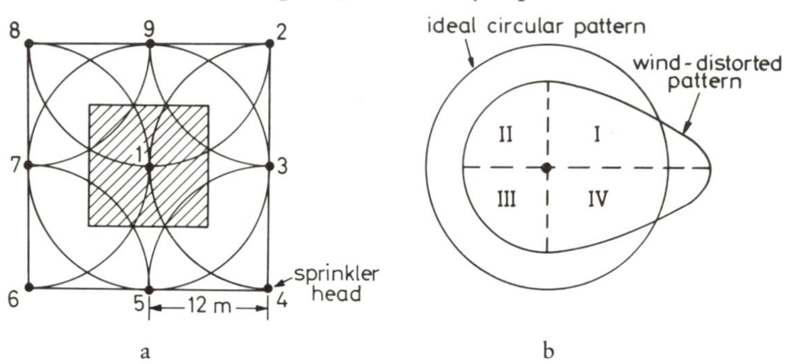

Fig. 5. a Plan of field sprinkler test installation, showing locations of nine sprinklers necessary for complete, overlapping coverage of the test area (cross-hatched square). b Wind-distorted sprinkler pattern divided into quadrants for use in constructing synthetic patterns on paper

The coefficient of uniformity C_u as defined by CHRISTIANSEN is calculated according to the equation:

$$C_u = 100(1.0 - \Sigma|x| \quad mn)$$

where $|x|$ is the deviation of individual observations from the mean value m, and n is the number of observations. Comparison between the above method and more sophisticated statistical procedures has shown that for the purpose of sprinkler performance evaluation there is nothing to be gained from use of more complicated procedures. A uniformity coefficient of 84% is generally accepted as the minimum requirement for a sprinkler operating under test conditions without wind distortion. Visual examination of the precipitation map, or of profiles drawn from it, is useful for diagnosis of sprinkler patterns with unsatisfactory coefficients.

A Sample Sprinkler System Design

Let us assume the following rather simple and ideal situation.

Area to be irrigated: 25 ha of level land with shape as shown in Fig. 6.

Cropping pattern: 6.25 ha of orchard, 12.5 ha of field crops in rotation (to include corn, cotton, sugar beets, tomatoes, alfalfa), and 6.25 ha of permanent pasture.

Soils: Loams and silt loams with a minimum final infiltration rate of 15 mm/hr, fairly uniform, deep profile, no salinity or drainage problems. Field capacity about 24% by weight; permanent wilting percentage about 15%, estimated bulk density, 1.3.

Water supply: the entire area is underlain by a high-yielding aquifer of good-quality water, 30 m thick, at a depth of 70 m. A well can be sunk at any chosen point, and electric power is available for pumping.

Special conditions: Nights are still; daytime windy hours generally from 10:00–17:00.

The detailed design procedure will entail the following steps:

Location of pump and main lines

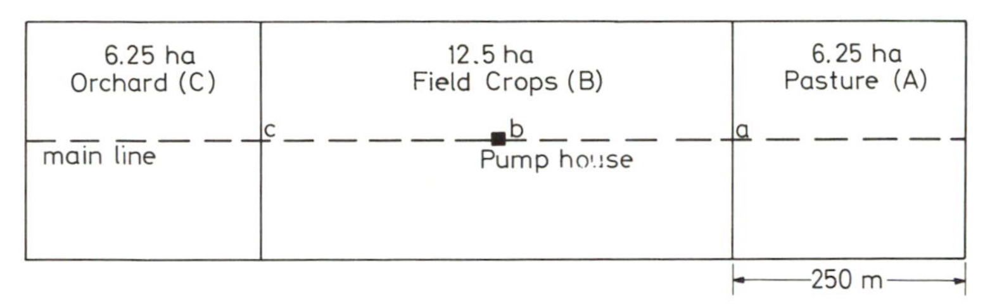

Fig. 6. Schematic map of area for which a sprinkler system is to be planned

For each separate cropping area:

Irrigation frequency and size of single application

Choice of sprinkler, its discharge and spacing

Number of laterals required

Diameter of laterals and of main line

Location of Pump and Main Line

The shape of the area makes this choice obvious. The well will be drilled in the geometrical center of the field, and the main line will bisect the field along its long axis.

Design for Pasture Area (Area A)

a) *Irrigation Frequency and Size of Single Application:* Pasture grasses usually have shallow root systems, and we shall assume an active root zone of 60 cm depth. We shall further assume a maximum daily consumptive use (evapotranspiration rate) of 7 mm, and that we do not wish to exhaust more than 50% of the total soil moisture in storage in order to maintain rapid vegetative growth. Total "available" water storage in calculated as follows:

$$d = \frac{DS_a\,(P_w)_{FC} - (P_w)_{PWP}}{10}$$

where d is the numer of surface millimeters of water available in the root zone; D is the depth of the root zone, cm; S_a is the bulk density of the soil; P_w is the soil moisture content at field capacity (FC) and permanent wilting percentage (PWP), respectively.

Using our assumed data:

$$d = \frac{(60)\,(1.3)\,(24-15)}{10} = 70.2 \text{ mm}$$

Irrigation will therefore have to be applied after 35 mm of water has been used up, or every five days at the maximum-use rate. Assuming that some of the irrigation may have to be applied in the daytime under windy conditions, we shall plan for an application efficiency of only 70%. The size of a single application will therefore be 35 mm/0.7, or 50 mm.

b) *Choice of Sprinkler:* A manufacturer's catalog may tell us, for example, that for use under windy conditions, with a square or rectangular pattern, a 50% overlap of the sprinkler pattern diameter should be allowed to obtain satisfactory distribution unifor-

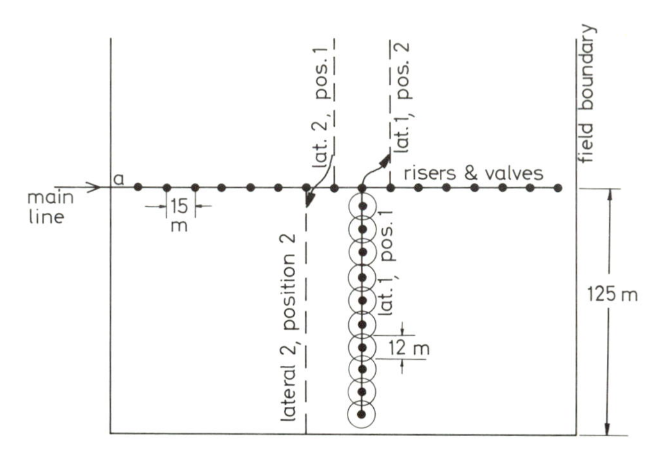

Fig. 7. Design of pasture field indicating lateral design and transport scheme

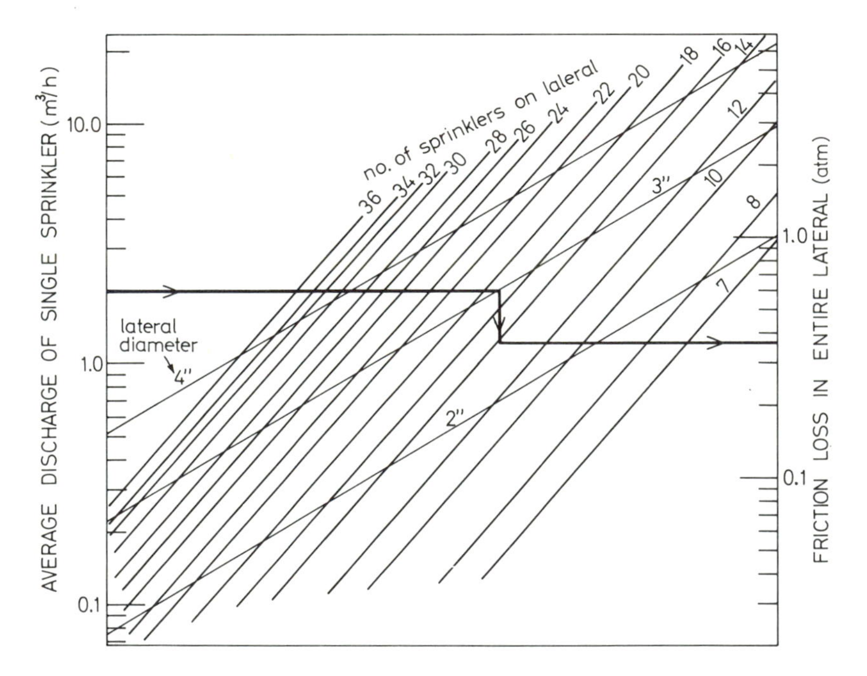

Fig. 8. Nomogram for calculating friction loss in sprinkler lateral. Heavy line with arrows indicates sample computation. To calculate loss in a 3-inch lateral carrying 14 sprinklers with a discharge of 2 m^3/hr, follow line to lateral diameter (3 in.). Move up or down to number of sprinklers on lateral. Move horizontally to right ordinate to find total friction loss of 0.35 m. Nomogram is designed for sprinklers spaced every 12 m along the lateral. For any other spacing multiply friction loss obtained by $S/12$ where S is the spacing used in meters. (Modified after CHRISTIANSEN)

mity. A sprinkler is selected that has a rate of application of, say, 14.2 mm/hr at 2.5 atm and a 12-by 15-m spacing, when the nozzle diameters chosen are 5.5 and 4.2 mm. Pattern diameter is 31 m at 2.5 atm, giving the desired 50% overlap in both directions. The sprinkler is very flexible, as it will operate at all pressures between 1.5 and and 4.0 atm, with application rates ranging from 3.5 to 45 mm/hr at various spacings, pressures, and nozzle sizes.

At the design pressure of 2.5 atm, the time for applying 50 mm will be 50 mm/14.2 mm/hr = 3.5 hrs. This will allow two settings per shift per line.

c) *Number of Laterals Needed:* 32 "sets", or lateral positions, are needed to cover the field. If we assume a 15 hr irrigation day, each lateral can over 4 sets per day, and thus two laterals will be required in this field. This allows coverage of the entire field in four days. Fences will be necessary to manage the grazing intensity and to keep the livestock

Table 2. Values of the factor F by which the friction loss in pipe must be multiplied to obtain the actual loss in a line with multiple outlets. (After CHRISTIANSEN)

Number of outlets	$m = 1.85$	$m = 1.90$	$m = 2.00$
1	1.0	1.0	1.0
2	0.630	0.634	0.625
3	0.535	0.538	0.518
4	0.486	0.480	0.469
5	0.457	0.451	0.440
6	0.435	0.433	0.421
7	0.425	0.419	0.408
8	0.415	0.410	0.398
9	0.409	0.402	0.391
10	0.402	0.396	0.385
11	0.397	0.392	0.380
12	0.394	0.388	0.376
13	0.391	0.384	0.373
14	0.387	0.381	0.370
15	0.384	0.379	0.367
16	0.382	0.377	0.365
17	0.380	0.375	0.363
18	0.379	0.373	0.361
19	0.377	0.372	0.360
20	0.376	0.370	0.359
22	0.374	0.368	0.357
24	0.372	0.366	0.355
26	0.370	0.364	0.353
28	0.369	0.363	0.351
30	0.368	0.362	0.350
35	0.365	0.359	0.347
40	0.364	0.357	0.345
50	0.361	0.355	0.343
100	0.356	0.350	0.338
∞	0.351	0.345	0.333

from trampling recently wetted soil. Laterals will be moved back and forth across the main at each set change, being towed in the longitudinal direction.

d) *Diameter of Lateral:* The net length of the lateral is 114 m carrying ten sprinklers for a total discharge of 25.5 m^3/hr. Allowable head loss in the lateral due ot friction is 20 percent of 2.5 atm, or 5 m.

A numbers of aids are available for calculating friction losses in both mains and laterals in the form of nomograms, slide rules, and tables, which can often be obtained without charge from equipment manufacturers. These are most often based either on Scobey's or Williams and Hazen's formulas. Since the discharge of a multiple-outlet pipe decreases with distance, it is necessary either to use a specialy prepared nomogram, or to calculate what the head loss H_t would be if the entire flow went through the pipe to the end, and then multiply this value by a factor that is a function of the number of outlets on the lateral. Some typical values of the factor are given in Table 2, which is based on the work of CHRISTIANSEN. Fig. 8 is a nomogram for direct calculation of head loss in sprinkler laterals. To calculate lateral diameter, by trial and error:

Let us examine the suitability of a 3 in. diameter lateral for the pasture area. Assuming a roughness coefficient of $C = 140$ (Williams and Hazen formula), we find that friction loss, $h_f = 3.8$ % or meters per hundred if the entire discharge of 25.5 m^3/hr were to flow to the end of the line. The multiple-outlet factor F for ten sprinklers can be taken as 0.4. The total head loss in the multiple-outlet lateral $H_f = 3.8$ 114/100) (0.4) = = 1.76 m. Since the pressure loss along the line is less than 20% of the nominal pressure recommended for the sprinkler chosen, 3 in. diameter lateral is adequate.

e) *Desgin of Main Section for Field A:* Assume buried steel pipes with 2" risers and valves, having a roughness coefficient corresponding to 15-year old pipe, C = 100. In this and following examples, only the acceptable solution for the worst case is given. Sometimes recalculations were done when the first estimate proved unacceptable, and this is the procedure usually followed in practice. This trial-and-error procedure is simple and, after some experience, it is easy to judge what combination of pipe diameter to choose. Use 200 m of 4 in. line and 50 m of 6 in. line. Worst case is when both laterals meet at the center of the field, 125 m of line carrying 51 m^3/hr (Fig. 6).

6 in. line: $h_f = 0.81\%$; for 50 m $H_f = 0.40$ m
4 in. line: $h_f = 5.8\%$; for 75 m $H_f = 4.35$ m

Total head loss in main section $H_f = 4.75$ m

Pressure needed at point *a* on map:

Average pressure of sprinklers	25.00 m
50% of lateral loss at head of line	0.90 m
Head loss in 50 cm high riser	0.50 m
Local loss in 50 cm high riser	1.00 m
Local losses in tee-joints, assumed	0.60 m
Friction loss in main line	4.75 m
Pressure needed at point *a*	32.75 m

Design of Orchard (Area C)

a) *Irrigation Frequency and Size of Single Application:*
Assuming:

Depth of root system = 1.2 m
Maximum consumptive use rate = 4 mm/day

Allowable moisture extraction = 60% of maximum storage
Application efficiency = 75%

Then:

Preirrigation water deficit in root zone = 84 mm
Size of single application = 112 mm

b) *Choice of Sprinkler:* The trees are spaced 5 m apart in the orchard, and the crowns are relatively high. The plan is for under-tree irrigation during daytime hours, as wind velocity inside the orchard is low. It is possible to irrigate alternate rows with water

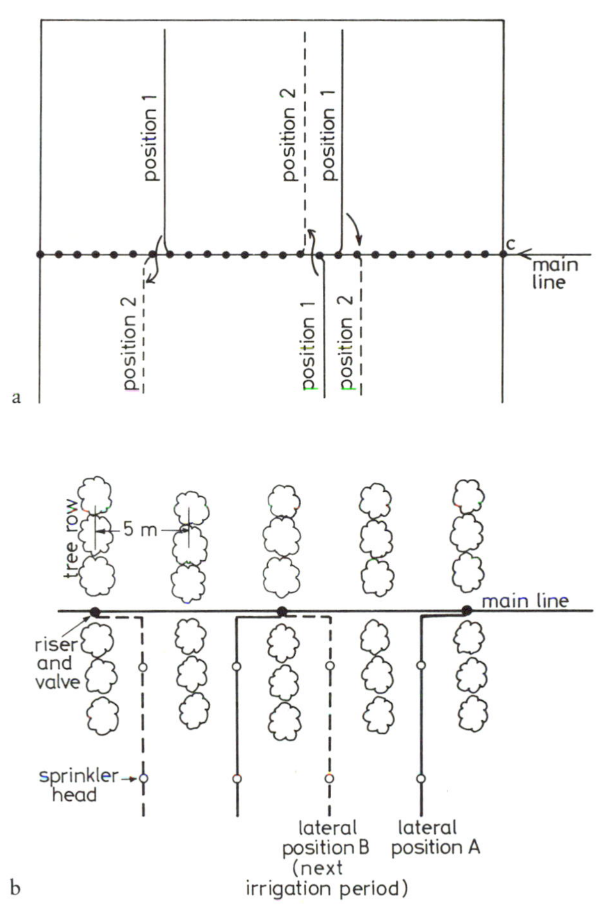

Fig. 9. a Schematic layout of orchard plot irrigation design indicating transport of laterals. b Detail showing valve placement in relation to tree rows, and scheme of alternating lateral placement every second irrigation

reaching into the unirrigated row, and the set should be alternated by one row every second irrigation. The sprinkler chosen from the catalog has a single 4.0-mm nozzle, and is specially designed as an under-tree sprinkler. It has a discharge of 0.76 m^3/hr at 2.5 atm, a radius of coverage of 8 m, and an application rate of 12.7 mm/hr at a spacing of 6 by 10 m. Time per lateral set will be 8.8 hours, and the maximum irrigation interval is 84/4, or 21 days. To allow for spraying, etc., the cycle should be finished in about 16 days.

c) *Number of Laterals Needed:* Along the 250 m of main there will be 25 valves spaced 10 m apart, or 50 lateral sets. Using three laterals, the irrigation will take 17 days.

d) *Diameter of Lateral:* Number of sprinklers per lateral = 20
Discharge of entire lateral = 20 (0.76) = 15.2 m^3/hr
Allowable friction loss along lateral = 5 m

Using the same technique as for area *A*, we find a head loss for the line without multiple outlets equal to 11.4 m in a 2 in. line, $F = 0.37$, and a total head loss in the lateral of 4.2 m.

e) *Design of Main Line:* The maximum, worst-case loss in a 4 in. steel main line, as shown in Fig. 7, is 4.59 m.

Pressure needed at point *c* on map:

Average sprinkler pressure	25.00 m
Head loss in 30 cm riser	0.30 m
50% of lateral loss	2.10 m
Local loss in valve	1.00 m
Local losses in tee-joints	0.60 m
Maximum friction loss in main line	4.60 m
Pressure needed at point *c*	33.60 m

Design of Field Crops Area (Area B)

a) *Irrigation Frequency and Size of Single Application:*
Assuming:
Depth of root system = 1.2 m
Maximum consumptive use rate = 6 mm/day
Allowable moisture extraction = 65%
Application efficiency = 70%
Then:
Preirrigation water deficit in root zone = 90 mm
Size of single application = 130 mm

b) *Choice of Sprinkler:* The same model sprinkler will be used as in the pasture, but with 4.6 × 3.8-mm nozzles. At a pressure of 2.5 atm, this sprinkler has a discharge of 1.90 m^3/hr and a 15 m radius of coverage. At a spacing of 12 × 18 m it gives an application rate of 8.8 mm/hr. As field crops often require high risers, hand moving will be employed, and night irrigation will be practiced. Since laterals should not be moved while the bare soil (between rows) is muddy, only one lateral move per day is planned. Set time per lateral will be 130 mm/8.8 mm/hr, or 15 hours. The maximum interval is 15 days, and the cycle time for design purposes will be about ten days.

c) *Number of Laterals Needed:* Since the block of land will be split in two and probably planted in different crops, and since the hydraulics of the main lines differ, one-half of the field is the basis for this design. There will be 14 valves 18 m apart, requiring 28 sets of one lateral. Using three laterals, we obtain a 9 1/2-day irrigation cycle.

d) *Diameter of Lateral:* Number of sprinklers per lateral = 10
Dischare of entire lateral = 19 m^3/hr
Allowable friction loss along lateral = 5 m

In a 3 in. lateral, the head loss if the entire discharge passed through the pipe would be 2.5 m, $F = 0.4$, and the actual loss in the lateral is 1.0 m. The loss in a 2 in. lateral would be an excessive 8 m. A tapered lateral could be used, but it was decided to use a 3 in. lateral throughout.

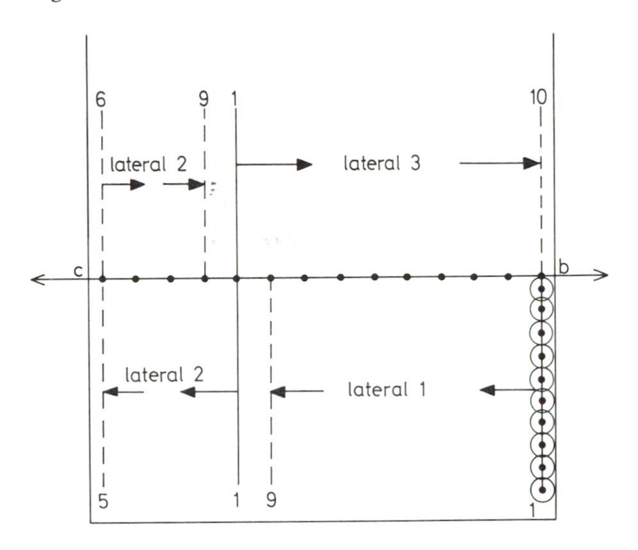

Fig. 10. Layout of one-half of area *B* showing sequence of lateral moves

e) *Design of Main Line:* Since the orchard is irrigated during the daytime, and the pasture as well as the field crops area are irrigated at night, different discharges will be required in the two sections of main opposite each other. The side toward the orchard will be designed first. The worst case is shown in Fig. 10. After some trial and error it was found that the first 144 m of main should be 6 in. in diameter, and the rest 4 in. Then, when the orchard is being irrigated, the pressure required at the pumphouse is 39.64 m, and when the field crops are being irrigated in the daytime, it is 41.9 m. This correspondence is close enough, since some reserve should always be designed into a system.

If only 6 in. main is used in field *B* on the pasture side, and both field *B* and the pasture are irrigated simultaneously, the pressure required at the pumphouse is 37.4 m for the worst case. This, too, is considered to be close enough to balance the pressure difference by use of a valve. By using a short section of 4 in. main in field *B*, exact balancing could be easily obtained. Risers 1.5 m high were considered for this field for use with tall crops.

This brings us to the end of the detailed design of each section of the farm. Irrigation frequency and size have been determined, appropriate sprinklers chosen, the number of laterals calculated in view of the length of the irrigation cycle, and, finally, hydraulic calculations made to determine the diameters of both the lateral and main lines. However, number of other factors also must be considered in order to arrive at the complete solution of our design problem, and this follows below.

Transport of Laterals

Both in the pasture and in the orchard, lines will be towed across the main line along the direction of the lateral. This can be done by using either a small sled or a set of wheels under every other sprinkler (every 12 m). Tractor-towing would be best for the rather heavy laterals. Hand-moving is planned for field *B* because towing is difficult in row crops, especially with tall risers.

Valve Placement in Orchard

Valves will be placed in line with the tree rows to allow irrigation of alternate rows at alternate irrigations, and also to keep the interrow area free of obstacles to traffic.

Capacity and Power Required at Pump

During the daytime, there is a demand for 46 m^3/hr to irrigate the orchard, at night the needed discharge is 165 m^3/hr. Once the design and operating conditions laid down at the beginning are accepted, there is not much that can be done to correct the unbalanced demand for water. During the daytime hours the pump will work at low efficiency. Pump horsepower is determined as follows:
1 hp = 75 kg-m/sec
Weight of 1 m^3 of water is 1000 kg
Pumping lift in well is about 85 m, pressure head needed is 42 m, total head reguired is about 130 m.

$$\text{Water hp} = \frac{(130\text{ m})\,(165{,}000\text{ kg})}{\text{hr}} \cdot \frac{1\text{ hr}}{3{,}600\text{ sec}} \cdot \frac{1/\text{sec}}{75\text{ kg-m}} = 80\text{ hp}$$

By water horsepower it is meant that 80 hp is the useful power actually needed to move the water. However, pumps do not operate at 100% efficiency, and at best usually around 60%, so that in our case the motor output should be about 130 hp.

Scheduling of Operations and Instructions to Farmer

The night-irrigated pasture has a net irrigation time of 14 hr/day.

Wheel-move of laterals is fast, and a 15 hr/day should be adequate. Sprinkler nozzles were chosen for the field crops so that time per set will also be 15 hr. There the laterals must be moved during the day following irrigation. The 9-hr set time in the orchard in combination with the 15-hour night irrigation allows for 24 hr operation, which is the most efficient utilization of equipment. In practice, it might be decided to sacrifice some application efficiency in favor of lower initial cost and more efficient equipment utilization by irrigating part of the field crops in the daytime to balance the demand. This would be carried out between the hours 17:00 and 08:00, and daytime irrigation during the hours 08:00 and 17:00. Laterals should be set initially as shown in the figures accompanying the design of each field.

Evaluation of Sprinkler Irrigation

Advantages

a) Fields with steep slopes or irregular topography, as well as with irregular shapes, can be irrigated with relative ease. In most cases land leveling is not necessary, thus avoiding the removal of the topsoil.

b) The amount and rate of water application are easily controlled, making it possible to avoid large losses due to surface runoff and deep percolation. This ability to control the water application is of particular advantage where small applications are needed because of coarse soil texture, shallow soil, or soil underlain by hardpans, claypans, or a high water table, or for germination and establishment of young crops.

c) Sprinkler systems can utilize a small, continuous supply of water more efficiently than gravity methods of water application.

d) Sprinkler irrigation systems do not require land for canals, turnout structures, head ditches, and borders. This saves about 5% of the land area, facilitates the movement of farm machinery, decreases conveyance losses, and decreases problems of ditch maintenance and weed concentrations around the irrigation structures.

e) It is easier to train new farmers in the operation of sprinkler systems than in the application of surface (gravity) irrigation methods. Also, the results of operator mistakes are usually less disastrous than can be the case with surface systems.

Disadvantages

a) Under certain conditions the initial cost of a sprinkler system may appreciably exceed that of gravity distribution and application.

b) Sprinklers require a minimum water pressure of about 2 atm (30 psi) to operate satisfactorily, and in most cases this pressure must be provided by a pump even when a water source is available that could be used for gravity distribution with a surface system. This increases both the initial and operating costs.

c) The cost moving sprinkler laterals and their high rate of depreciation may lead to higher operating costs than those of a surface system. This is particularly true in the case of tall-growing crops such as cotton, corn, and sorghum.

d) The uniformity of water distribution is very sensitive to wind. Under windy conditions the application efficiency decreases due to losses by deep percolation (uneven distribution), evaporation, and wind drift.

e) Some crops are adversely affected when their foliage is wetted. Citrus is known to suffer from the absorption of salts from irrigation water through the leaves, and a number of bacterial and fungal plant diseases are much more severe under sprinkler irrigation than under surface water application.

It is often claimed that irrigation can be carried out with fewer losses by sprinkling than by flood or furrow irrigation. This is not necessarily so, and well-designed surface irrigation systems can be as economical and as efficient as sprinkler systems. However, in practice losses generally prove to be smaller with sprinkler systems, owing to various technical as well as human factors.

Literature

CHRISTIANSEN, J. E.: Irrigation by Sprinkling. Calif. AES Bull. **670,** (1942).
CRIDDLE, W. D., DAVIS, S., PAIR, C. H., SHOCKLEY, D.: Methods for evaluating irrigation systems. USDA-SCS Handbook **82,** (1956).
KING, H. W.: Handbook of Hydraulics (3rd ed.), 617 pp. New York and London: McGraw-Hill 1939.
PAIR, C. H.: Sprinkler Irrigation. USDA Leaflet **476,** (1966).
SCHWALEN, H. C., FROST, K. R.: Sprinkler Irrigation. Univ. of Arizona, AES Bull. A-**24,** (1965).
WOODWARD, G. O. (Ed.): Sprinkler Irrigation (2nd ed.). Sprinkler Irrigation Asso., Washington, D. C. (1959).

2

Gravity Irrigation

E. Rawitz

The most important feature of gravity irrigation is that the ultimate distribution of water over the land surface is controlled by the land surface itself and not by some mechanical device. The manner in which this distribution takes place can be modified by changing the condition of the land surface (e.g., land leveling to smooth the surface or change the slope, making of furrows to direct the flow). However, since our ability to change the face of the earth is llimited, design of surface systems essentially involves the proper choice of length and direction of run[1] and of stream size consistent with local conditions of soil and topography.

Classification of Gravity Systems

Elements of the Gravity System

All gravity systems consist basically of a source of water, a distribution and conveyance network, and a land surface arranged so as to direct the flow of water. Distribution systems may consist of pipelines or open ditches, the pipelines usually operating at low pressure. Turnout structures[2] and pressure-regulating devices therefore are essential parts of the distribution system.

Types of Systems and Equipment

Surface irrigation systems are classified according to how the irrigation stream is controlled by the land surface. Wild-flooding, border-check, and furrow irrigation are the three types commonly used.

a) *Wild Flooding.* Wild flooding is the least efficient and least common system. Temporary ditches are opened either along or perpendiculat to contour lines, and are made to overflow at a number of points. Water progresses uncontrolled down the slope, and any excess is intercepted by the next ditch (see Fig. 1). This method is used as a temporary system for irrigating pastures, hay, and small-grain crops. Water and labor efficiency are likely to be low, even under favorable conditions. Its use is warranted only by very special conditions.

b) *Border Checks.* This method is used for irrigating all crops (with the possible

---

1 Length of run is the downslope distance from the point where water is discharged onto the land to the end of the plot dash the furrow or border check.

2 Devices for controlled discharge of water from a pipe or canal onto the land, often also incorporating flow-measuring and energy-dissipation devices.

exception of row crops) where large planar surfaces of land are available and slopes are moderate. Checks are bounded by low borders or levees made of earth, and the cross-slope (perpendicular to the direction of water flow) must be zero. Relatively large flows

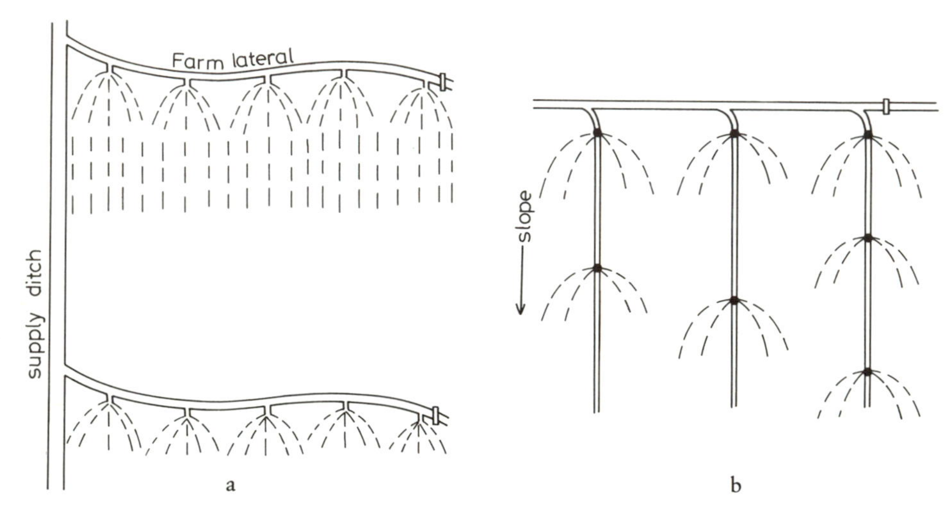

Fig. 1. Wild flooding from contour ditches: a on steep slopes with contour ditches; b on almost flat slope, with ditches following the slope

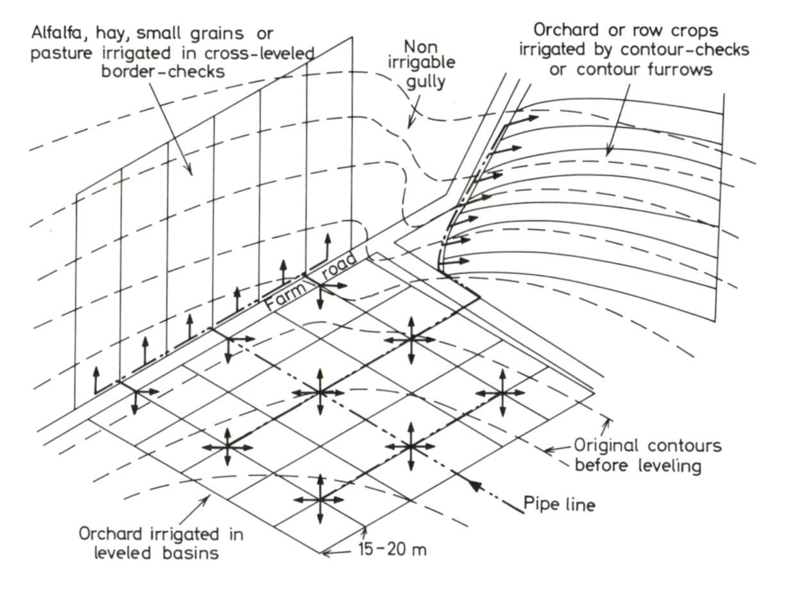

Fig. 2. Schematic gravity irrigation farm layout showing use of basins, leveled checks, and contour checks or furrows

of water are required. The width of a check is a function of the infiltration capacity of the soil, the head of water available (size of stream), and the cross-slope of the land. The permissible length of a check is a function of the infiltration capacity, head of water, surface roughness, and average slope of the land. The border width should be a multiple of the width of the widest agricultural machinery to be used in the field (e.g., mower, combine, cultivator). Border checks are usually permanent in location, although contour checks may have their levees rebuilt every year. The distribution system therefore is usually permanent also. Border checks may be subdivided into rectangular, sloping border checks, irregularly shaped contour checks, and flat, square, or nearly square basins (Fig. 2).

c) *Furrows*. Furrows are used mainly for the irrigation of row crops on land having up to moderately steep slopes. Corrugations are a variation of this method and are really small furrows, that sometimes are used to irrigate land not in row crops, and to guide the water where the cross-slope is not perfectly flat (as inside border or contour checks). In contrast to furrow irrigation, "overtopping" of the ridges by the flowing water is permissible in corrugations. Furrows are usually rebuilt every year, and are served by what may be called semipermanent distribution systems consisting of permanent mains and portable or temporary "laterals." This will be explained in greater detail further on.

Equipment and Installations

The irrigation water is distributed either in open systems or by pipelines.

a) *Hydraulics*. The principles of distribution system layout are the same as for sprinkler installations, and the hydraulics of the pipelines identical.

For both technical and economic reasons, it is common practice to use open pipe systems whose components cannot withstand high pressures. It therefore becomes important to consider the topography so that the line does not depart too much from the line of hydraulic gradient, as this requires high standpipes. This is illustrated in Fig. 3.

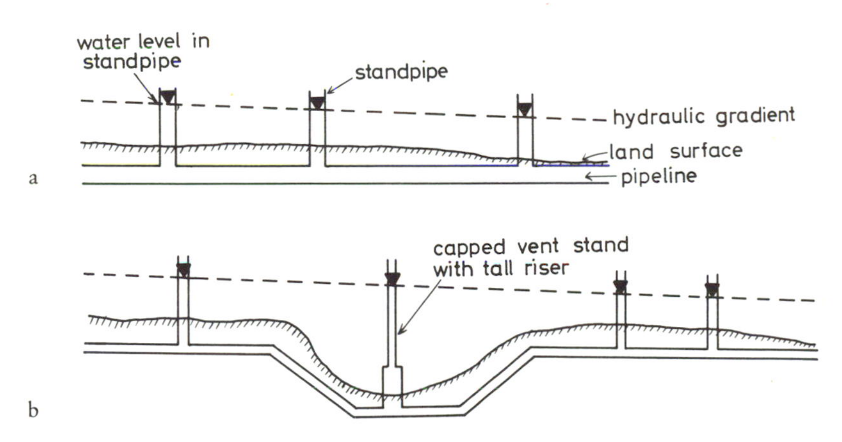

Fig. 3. a Low-pressure conveyance pipeline with standpipes; b showing problems arising from irregular topography requiring use of high risers

Open ditches must follow contour lines and provide the proper hydraulic gradient for water conveyance. Unlike pipelines, it is not possible to increase the capacity of ditches appreciably by simply adding "a little more pressure." Although the hydraulics is straightforward, the construction of low-pressure concrete pipelines and of lined ditches requires detailed knowledge of techniques and devices that are beyond the scope of this chapter. Some pertinent references are provideds, but it is advisable to leave the design and supervision of construction to an experienced engineer.

b) *Open Systems.* Open systems include unlined earth ditches; ditches lined with concrete, asphalt-cement, or plastic; and metal and wood flumes. Since some small delivery head is required, the ditches are often built on "ditch pads," which are low embankments of earth raised over the general level of the land into which the ditches are placed. Unlined ditches are a poor choice, except as a temporary expedient. They have large seepage losses that not only waste water but cause soil water-dogging and spread of weeds, and result in high maintenance costs. Ditches take up as much as 5% of the land surface, impede traffic and thus require bridges, and are subject to water loss by evaporation. It is therefore common to use a mains system of pipes, with ditches only as laterals. Water can be distributed from the ditch to the field by means of permanent wood, metal, or concrete turnout structures or gates, by portable siphons, by tubes or "spiles" running through the ditch bank, and even by simply breaching the earth bank of the ditch. The degree of control over the irrigation stream decreases in the order that the above devices are listed. The water level in the ditch must be adjusted for each field, and for this purpose overflow structures are placed in the ditch. Again, these may be permanent, adjustable weirs, canvas "flags" with wood or metal frame, or earth thrown in the ditch to form a temporary dam (Fig. 4).

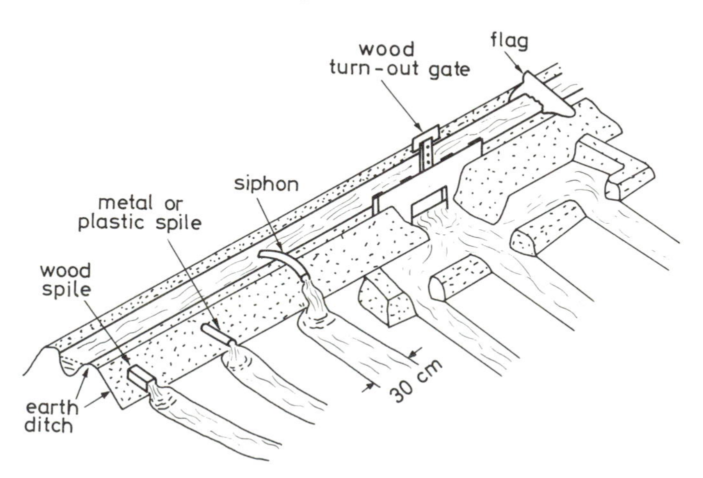

Fig. 4. Different ways of turning out water from a farm ditch into irrigation furrows

Water-measuring devices such as measuring weirs, flumes, calibrated control sections, and orifice plates are often incorporated into permanent turnout structures. Automatic metering gates can be set for a certain discharge, which will be maintained by the self-adjusting gate. Open-ditch distribution systems are not practical where slopes are steeper than about 0.5% along the route of the ditch, or where topography is too rolling.

c) *Pipelines.* Pipelines for gravity irrigation must be able to deliver large discharges at low pressure, and low-pressure, large-diameter pipes are commonly used in open systems. Such a system is characterized by a number of open standpipes that regulate the pressure along the line. The standpipes generally are up to 3 m high, although in certain instances they may be as tall as 10 m. Concrete pipe, 12–48 in. in diameter, buried underground, is commonly used. Risers are installed at the required intervals; they are usually made of 6–10 in. concrete pipe with a metal cover and valve (Fig. 5).

d) *Pressure System.* One exception to the type of system described above is a pressure

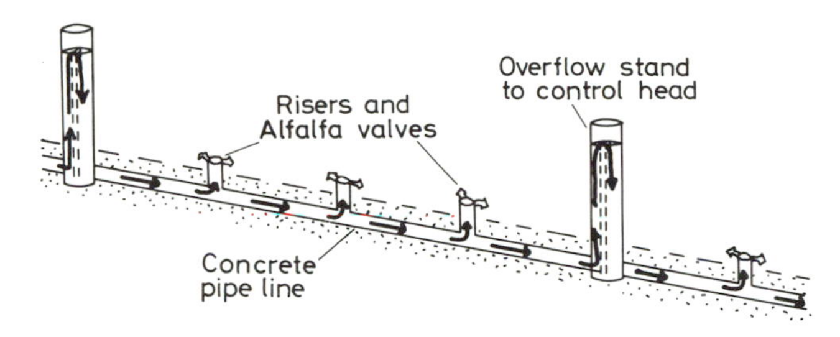

Fig. 5. Low-pressure concrete distribution lateral showing turnout risers and overflow-type standpipes for pressure regulation

Fig. 6. Gated pipe used for furrow irrigation

system designed exclusively for irrigating furrows with gated aluminum pipes. Although gated plastic pipes operate at low pressure and have large diameters, and can be supplied either from ditches or from large-diameter pipes, aluminum gated-pipe laterals usually have a smaller diameter (4–8 in.) and are supplied with water under greater pressure, similar to sprinkler laterals. Such systems can thus be supplied from a pressure system, and in fact are compatible with sprinkler installations (Fig. 6).

Principles of Gravity Irrigation

An ideal system would apply exactly the required amount of water to the field with perfect uniformity, with no deep percolation or runoff losses, by means of a network that is economical to install and operate. This ideal is unattainable in practice, because of the inherent variablity of soil and crop conditions and because of the hydraulic problems involved.

Since there is no difference in principle between border and furrow irrigation, the basics will be described together. A shallow sheet of water spreads and advances over the land from either a point or a line source. Where slope and surface roughness conditions resemble open-channel flow (slope larger than about 0.5%, relatively smooth surface), the water surface will be roughly parallel to the land surface. When slope is flat and impedance to flow high, the water will have to "build up its own slope" in order to sustain advance of the stream, and the water surface will assume a slope greater than that of the soil surface. However, only on soils with a very low infiltration capacity can the advance of water over the land be even approximately described according to the hydraulics of open-channel flow.

As a rule, the situation is complicated by the fact that as the sheet of water advances over the land, part of the water infiltrates into the soil. However, the infiltration capacity is not a constant, but decreases with time. As a larger area of land is covered by the advancing stream, a relatively larger part of the input discharge is required for infiltration and less of it is available for further advance. Thus, advance rate tends to decrease with the distance of the advance front from the source. On the other hand, as the infiltration capacity of the soil that was wetted earliest decreases (in the upstream part of the furrow or check), relatively more water becomes available for advance and for infiltration downstream. In practice, the first phenomenon generally dominates, and advance rate decreases with time. From this qualitative description alone, it is clear that the hydraulics of surface irrigation is extremely complex, even for ideal cases.

A mathematical description of the advance of water over a strip of land would have to incorporate two processes: (1) open-channel flow, described by Manning's or similar equations, having the general form:

$$V = cR^m S^p / n$$

where V is flow velocity: R is hydraulic radius (cross-sectional area of flow divided by the wetted perimeter); S is slope of the water surface; n is roughness coefficient; and c, m, p are constants and (2) the infiltration function, having the general form:

$$I = at + st^n$$

where I is cumulative infiltration; t is time; and a, s, n are constants.

Both equations are empirical, and their coefficients have to be experimentally determined for each set of conditions, or estimated on the basis of past experience. Once the design engineer either has made the empirical tests or has acquired enough experience, he can probably apply this knowledge directly to the design of the system without recourse to the equations. Knowledge of the principles involved does help, however, in the application of judgment to the results of the empirical tests.

Practical Approaches to Design

Although all types of surface irrigation methods are similar in principle, when it comes to the details of an actual design problem, one must remember that there are appreciable differences not only between borders and furrows, but between different types of borders. To discuss these it is necessary to examine what happens as the water is initially applied to the land, and also what happens at the end of the irrigation.

If a layer of water of uniform depth were infiltrating through a perfectly level land surface, water depth would decrease with time, and the water would eventually disappear from the entire land surface simultaneously and instantaneously. On the other hand, if a water layer of uniform depth were flowing over a sloping land surface when the input stream as cut off, both infiltration and downhill flow would continue, thus exposing bare land at the upstream end while the lower portions were still covered with water. We can collect data on the advance of water at the beginning of irrigation and, conditions permitting, on the recession of water at the end. When plotted on a graph (Fig. 7), it is readily seen that the vertical distance between the advance and recession curves tells the time that water was available for infiltration at each point along the length of the wetted strip. This is called the intake opportunity time (IOT). Two very important points immediately become obvious: (1) Assuming uniform infiltration properties in the field, uniform application of water is obtained if the advance and recession curves are parallel; if they are not parallel, the graph is a good diagnostic tool telling us which end of the strip received too much water, and which end too little. (2) Provided the two curves are parallel, it is not important, from the point of view of uniformity, what shape or slope these curves have.

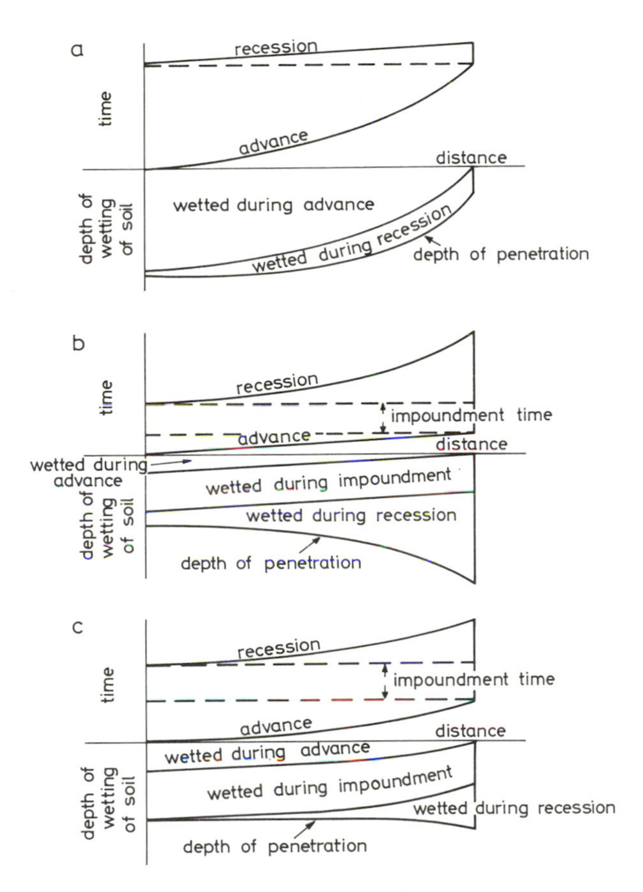

Fig. 7. How advance and recession of water determine uniformity of water distribution. (Upper half of each diagram depicts relation between time and distance of advance or recession, lower half represents the corresponding wetting pattern of the soil.). a Advance too slow, upstream end got more water than lower end; b Advance too fast, lower end got overirrigation; c Advance and recession lines parallel, resulting in nearly ideal water distribution

In practice, the slope of the field should not depart by much from the natural land slope, and it is difficult to alter the infiltration properties of the soil. Since slope and infiltration capacity are the two factors determining the recession curve, the design problem often boils down to choosing the input stream that will produce an advance curve parallel to the recession curve. When the direction of flow departs from the maximum land slope, as in contour checks or furrows, both slope and input stream must be chosen.

Three generalized models are available for design purposes, and each will be presented in detail.

The Instantaneous Recession Model

Infiltration is assumed to cease simultaneously over the entire wetted area. In order to obtain a uniform application, the initial wetting must also take place instantaneously. In practice this is approximated by turning the maximum possible input stream onto the land. One example is basin irrigation, usually of orchards, where the basins are approximately square, and level. High and solid levees are necessary to impound the water. Usually the entire volume of the irrigation is applied as fast as possible and the input stream diverted to the next basin. The role of intake rate in uniformity of distribution is minimized.

Due to the partial wetting of the land surface, the small amount of water impounded, and the steeper slope necessary to move water through relatively rough and long furrows, recession in furrows is often very nearly instantaneous over most of the length, with impoundment taking place only very near the downstream end of the furrow if it happens to be blocked. Since it is not possible to apply large streams to furrows without overtopping, it is difficult to obtain unform application in furrows, especially if they are long or if infiltration capacity is high. Uniformity is enhanced by steeper slope, lower infiltration capacity, and shorter runs. A good "rule of thumb" for this model is that advance time should be kept to less than 25% of the total irrigation time for satisfactory uniformity. An unsatisfactory irrigation of land approximating this model may be seen in Fig. 7, case (a).

The Balanced Advance-Recession Model

This model is comon to fairly long border checks with an appreciable slope and fairly high roughness coefficient. The irrigation stream is adjusted so as to give an advance curve as nearly parallel as possible to the recession curve. Length of run and irrigation stream must be adjusted so that the amount of water applied during the time required for the stream to reach the far end of the check does not exceed the smallest irrigation application to be given; see Fig. 7, case (c). If this is not accomplished, it will be necessary to continue applying water until the advance front reaches the end of the field. Thus there will be need to overirrigate, resulting in deep percolation losses at the upper end of the check. If the irrigation stream reaches the far end of the check before the design application is given, the irrigation time must be extended. If no adjustments are made for this, runoff of "tailwater" will result. This problem can be solved in three ways. Sometimes the tailwater is collected in a ditch and rerouted for reuse elsewhere. The irrigation stream can be decreased or "cut back" so that input exactly balances infiltration. Since infiltration is not constant in many cases, this can only be done approximately, and small runoff losses may still occur. It is not realistic to demand of the farmer that he be in continuous attendance to constantly readjust the input stream.

Combination Impoundment and Balanced Advance-Recession Model

The third way to assure sufficiently large applications where the advancing stream reaches the far end of the check before this is accomplished is to impound water at the lower end. Since the check does have some slope, there are obvious limitations as to maximum allowable slope, maximum height of levee that will hold without breaching, and maximum length of check over which water can be impounded. It is generally not practical to build levees higher than about 30 cm. One common arrangement of this model is a variation of the instantaneous-recession model. Since truly instantaneous application of water to a long strip is impossible from the narrow end, there will be an advance curve even for the largest practical input stream. The system is so designed that the total fall in elevation over the length of the check exactly equals the depth of the design application, and that the time required for the water to advance to the far end is just enough to allow the minimum design application to infiltrate at the upstream end of the check. The water is then cut off and recession begins immediately, with the water in the check very quickly assuming a level surface profile. Thus the upper end of the check receives its application entirely during the advance stage, and the lower end entirely during the recession stage, with intermediate conditions prevailing along the rest of the check. If it is desired to apply more than the design application, this is possible without affecting the uniformity of distribution, provided, of course, that the impounding levees are high enough and strong enough.

Design Recommendations

Although no general recipe can be given, experience has shown what limits of slope and length of run must be observed in order to achieve reasonable uniformity and control of water and to avoid accelerated soil erosion. The information given here is meant to be a general guideline, and should not be applied unquestioningly.

Slope. The finer-extured the soil, the less the danger of erosion, and the steeper the permissible slope. Likewise, a sodded pasture is better protected against erosion than a bare furrow. As a rule, the slope along the direction of flow in a check should not exceed about 0.5%, and the maximum allowable furrow slope on fine-textured soil is about 2%.

In addition to the danger of erosion occurring as a result of controlled flow, land with a cross-slope is very vulnerable to overtopping of border levees or furrow ridges. Overtopping results in uncontrolled downhill flow capable of causing great damage to the land and wasting large amounts of water. In order to minimize possible damage, it is good practice to begin irrigation with the highest-lying furrows or borders and progress down the cross-slope. If a breakthrough should occur, water will flow onto as-yet-unwetted soil, which, having an initially high infiltration capacity, may well absorb all of the uncontrolled flow. This not only minimizes waste of water but prevents a destructive chain reaction of breakthroughs that would occur if one took place above a recently irrigated part of the field. The maximum cross-slope for furrows is about 10%; the greater the furrow slope, the greater will be the allowable cross-slope. For basins or border checks, the maximum cross-slope is not defined as such, but experience has shown that the maximum difference in elevation between neighboring checks should not exceed about 6 cm to avoid danger of overtopping. The permissible width of check or basin is therefore a function of the slope.

Input Stream for Advance. With border checks the problem is more often the provision of a minimal satisfactory stream than the need to limit the stream to prevent damage. In vegetated borders there is usually little danger of erosion, and in many cases uniformity of application could be improved by a larger stream, but this may not be available. The ability of the levees to contain the water may also be a limitation. The common range of input streams is between 10 and 50 $m^3/hr/m$ width of check, depending on infiltration capacity of the soil, slope, and vegetative cover.

Furrows are much more sensitive to erosion, and because of the variety of depths, widths, and cross-sectional shapes, a rather wider range of input streams may vary between 1 and 15 $m^3/hr/furrow$.

Length of Run. It is advantageous to have the longest possible length of run consistent with uniformity of distribution. Permissible length of run is a function of maximum permissible noneroding stream, soil infiltration capacity, roughness, and slope. With the exception of basin irrigation, it is not worthwhile to have runs shorter than about 60 m; if such short runs appear necessary, it is an indication of very high infiltration capacity, and a switch to sprinkling should be considered. On heavy, sticky soils with a low infiltration capacity borders and furrows may be as long as 350 m. Run lengths between 100 and 200 m are common and acceptable.

General Remarks. To recapitulate, the design problem of surface irrigation is to determine the most efficient combination of slope, length of run, and size of stream for a given set of conditions defined by the infiltration function of the soil, surface roughness, susceptibility to erosion, and size of irrigation to be applied. To date, no fundamental equation has been developed to solve even part of this problem, and recourse must be had to empirical methods. Where this is possible, the best empirical method is still a field trial of an actual check or furrow. Fortunately, adjustments usually can be made in an existing installation by changing the input stream or the length or run to compensate for errors of prediction during the planning process. In order to ensure the ability to make such changes, the original design should be so constructed that none of the parameters are close to the extreme recommended values.

Field Tests

Procedure. Much can be learned from tests carried out on existing installations if there are such in the vicinity of a new project. If there are none, and if a typical area can be found with water available, a simple set of field tests will enable the engineer to arrive at a satisfactory design. The procedure is similar for borders and furrows, except that it may not be practical to test borders with different slopes, or even full-width borders. In furrow tests, at least one furrow on each side of the test furrow should be irrigated as a buffer zone in order to simulate normal conditions of lateral infiltration. Provision must be made for measuring the input stream and for controlling it at any desired discharge. The longitudinal profile of the test plot is surveyed, and distances are marked along the plot with stakes. If it is desired to measure the intake rate as well, provision must also be made for outflow of water from the downstream end of the border or furrow, and its measurement. For each slope several plots will be necessary in order to test several sizes of input stream. Water is turned into the check or furrow, and discharge is adjusted to the desired value. The time of water entry is noted, as well as the time at which the advance front reaches each stake. When the stream reaches the lower end, it may be either "cut back" (input stream decreased) by estimate, or runoff may be allowed and measured. In this case, a portion of the average infiltration capacity curve of the plot will be obtained. When the design application has been given, the water is shut off, and the recession of water is observed in the same way as the advance. The data are then plotted and diagnosed, as shown in Fig. 7. The information can be confirmed by utilizing appropriate sampling methods to check the depth of wetting along the length of the plot.

If the number of test borders is insufficient, the available data may be analyzed and extrapolated by fitting to one of the empirical formulas for predicting advance and recession.

Index of Uniformity. A numerical index of uniformity of water distribution developed by Rawitz is analogous to Christiansen's index of uniformity for sprinkler irrigation. It is based on a mathematical function expressing the integral of moments of points, representing intake opportunity time or depth of wetting at any point along the furrow or border check, about the line representing uniform IOT or depth of wetting.

The coefficient of distribution uniformity is defined as follows:

$$C_u = 100\left[1 - \frac{I^{1/2}}{m}\right]$$

and

$$I = I' - m^2$$

where C_u is expressed as a percentage; I' is the integral of second moments of all observations around the line of uniform IOT; m is the integral of first moments of all observations around the line of uniform IOT; and I is an integral of the difference between individual y-values and m^2.

A simple numerical method has been developed for the calculation of C_u and the following example demonstrates its use.

Table 1. Sample calculation of the uniformity of distribution index of a furrow irrigation

1 x (meters)	2 Y IOT[a] (minutes)	3 Avg. of Y's	4 Y^2	5 Avg. of Y^2's
0	110	117.5	12,100	13,863
15	125	137.5	15,625	19,062
30	150	157.5	22,500	24,863
45	165	167.5	27,225	28,062
60	170	177.6	28,900	31,563
75	185	190.0	34,225	36,125
90	195	217.5	38,025	47,812
105	240		57,600	

[a] Intake opportunity time.

A table (1) is constructed and in its first two columns are entered the distances (x-values) from the head of the plot at which advance and recession times were observed, and the IOT or depth of wetting (y-values). In column 3 are entered the averages of each successive pair of y-values. Column 4 contains the square of each y-value, and column 5 is for the averages of each successive pair of y^2-values. From this tubulation one can easily compute m, which is the arithmetic average of all the values in column 3. I' is the average of all values in column 5. Having obtained these two values, the rest of the calculation is a matter of four simple additional steps.

$$m = \frac{(117.5 + 137.5 + \ldots . 217.5)}{7} = 166.428$$

$$m^2 = 27{,}698$$

$$I' = 28{,}764$$

$$I = I - m^2$$

$$I = 1066$$

$$I^{1/2} = 32.65$$

$$UI = \frac{I^{1/2}}{m} = \frac{32.65}{166.43} = 0.196$$

Coefficient of uniformity = 1.000−0.196 = 0.804 = 80,4%

Land Grading

This is the basic and most important operation in the construction of a surface irrigation system. The work represents a large part of the initial investment, and defects are difficult to remedy once the system is operating. The steps involved are as follows:

1) Approximate choice of field shape and irrigation direction from existing maps or inspection of the "lay of the land."

2) Staking – the placement of wood stakes on a grid, typically in a square pattern with a 20-by 20-m spacing.

3) Survey and drawing of contour map. Notation is shown in Fig. 8.

4) Selection of areas for separate grading; this is extremely important, for proper choise will keep depths of cut and fill to a minimum, saving both money and valuable topsoil.

5) Planning of water distribution and drainage system. It is desirable to plan square or rectangular fields, without wedges between them; of course, the distribution lateral (pipe or ditch) must run along the high side of each field. Provision also should be made for removal of excess rainwater and possible runoff water from irrigation mishaps.

6) Land-grading calculations. Cut and fill should be kept to a minimum, since excessive cutting is expensive and removes the topsoil, and deep fills have a tendency to consolidate with time, causing undesirable low spots in the field. Several methods are used for grading calculations, depending on the situation. When the land is nearly plane, the entire field is brought to the same slope and the least-squares method is used. When the topography is steeper and the farm layout such that there will be an appreciable cross-slope perpendicular to the direction of irrigation, it is preferable to divide the field into several "steps," each at a different level, and to grade each strip separately. This method is also applicable

to contour checks or furrows where some grading is necessary. For such cases the average profile method is best.

All methods of grading calculations yield both the original and the predetermined desired slope and land elevation, or "grade" after leveling with equal volumes of cut

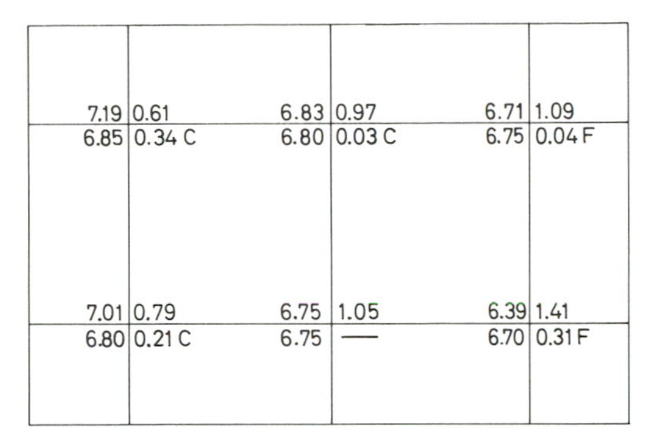

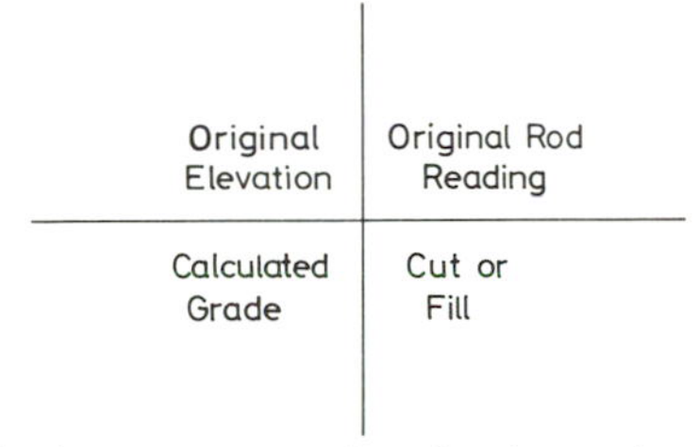

Fig. 8. Sample of survey notes and grading instructions on a grid map

and fill, since all involve the rotation of the line or plane of average land profile around the centroid of the field to achieve the desired slope. However, experience has shown that for various reasons more cut than fill must be allowed in order to complete a grading job. The excess of cut over fill must be determined by trial and error in the field during grading, the first estimate being made on the basis of local experience and general knowledge. When planned cuts are small, of the order of 15 cm or less, as much as 100% excess of cut over fill is allowed. When the required depth of cut exceeds about 20 cm, and volume of cut is of the order of 1000 m^3/ha, the adequate excess of cut over fill volume is generally between 20% and 45%.

The calculated cut or fill and grade are entered at the grid intersections on the contour map, as shown in Fig. 8. For ease of supervision in the field, cuts are marked in red and fill in blue. The same color code is then used in marking the stakes in the field, to guide the equipment operator during grading. One system of marking in common use shows the desired final grade plus a constant height added to it on each stake. In case of cuts, the depth of cut is indicated by the length of the mark, and grade is at the bottom of the mark. Fills also are indicated by length of mark, but grade is at the top of the mark, as shown in Fig. 9. After the first rough grading is completed, the area is surveyed again, and corrected estimates are made, if necessary, for excess of cut over fill.

After rough leveling it may be advisable to chisel (or subsoil) the compacted land; final smoothing then is done with a landplane.

The levees are built last, and the ratio between their base and height should be as large as possible. A good design for border checks is about 2 m wide at the base and 25 to 40 cm high. The levees should not be constructed by travel along the axis of the levee with a disk-ridger or an A-frame, since the result is a levee with too narrow a base. Also, it leaves a furrow running along each edge of the levee base, which would cause an excess of irrigation water to run at the edge of the border check. The proper way to construct the levees is to run a bucket- or dump-scraper or a road grader perpendiculat to the levee axis, shaving a uniformly deep layer of soil from the border-check surface and depositing it on the axis of the levee. The gentle slopes of the levee allow them to be seeded, thereby avoiding concentrations of weeds and waste of land area, as well as permitting crosstraffic of farm implements without damage to the levee.

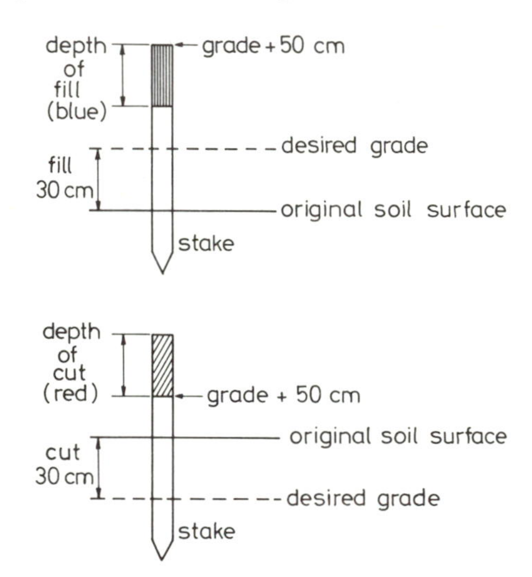

Fig. 9. Method of marking stakes at grid intersections with instructions for equipment operator

Some fill areas may consolidate or subside during the first few years of use, and it is often necessary to repair the grading job after some time. It is thus inadvisable to plant perennial crops on freshly graded land. After these repairs, only an occasional touch-up landplaning may be needed, and the depreciation rate of leveled borders is very slow.

Evaluation of Gravity Irrigation

Finally, it cannot be overemphasized that the operation of a gravity system requires more understanding attention than a sprinkler system. Although the engineer may design a system that is capable of being operated efficiently, it is the farmer who has to live with it. If the farmer has not received adequate instructions, or if the system is not convenient to operate, the design will be a failure. The farmer often thinks in terms of immediate problems, and will stress short-term low costs and convenience at the expense of efficient use of water and land. The advantages and disadvantages of gravity irrigation may be summarized as follows:

Advantages

a) Where the land is naturally smooth and nearly level, very little initial investment may be required, and depreciation rate will generally be very slow.

b) No pressure is required for the actual distribution of water over the land. Where available, surface sources can be diverted directly onto the fields.

c) Labor costs are minimal when systems are well designed.

d) Operation is not affected by wind conditions.

Disadvantages

a) Removal of topsoil during leveling may impair soil fertility and productivity for many years.

b) Large but intermittent discharges of water are required.

c) Even relatively small flaws in design or operation of the system may have drastic results, since control of the irrigation stream becomes difficult. This results in a sharp rise in labor costs, unavoidable waste of large amounts of water, and soil erosion.

d) It may be difficult or impossible to make relatively minor modifications in the system in order to improve performance, because such changes affect the entire system and may require a complete redesign of the distribution network.

Literature

Anonymous: Irrigation on western farms. USDI Bur. of Reclamation and USDA-SCS Agric. Infor. Bull. **199,** (1959).

Criddle, W. D., Davis, S., Pair, C. H., Shockley, D.: Methods for evaluating irrigation systems. USDA-SCS Handbook **82,** (1956).

Israelsen, D. W.: Irrigation Principles and Practices (2nd ed.). New York: Wiley 1950.

Jensen, M. E. (Ed.): Hydraulics of surface irrigation. USDA-ARS Public. 41–43 (1960).

Johnston, C. N.: Farm irrigation structures. Calif. AES Circ. **362,** (1945).

Marr, J. C.: Grading land for surface irrigation. Calif. AES Circ. **438,** (1954).

Phelan, J. T., Criddle, W. D.: Surface Irrigation Methods. In Water. USDA Yearbook of Agriculture, pp. 258–266 (1955).

Pillsbury, A. F.: Concrete pipe for irrigation. Calif. AES Circ. **418,** (1952).

Rawitz, E.: Surface irrigation research in Israel. Int'l Comm. on Irrig. and Drainage, 5th Congr., vol. R. 1, Question 16, pp 1–12 (1963).

Willardson, L. S., Bishop, A. A.: Analysis of surface irrigation application efficiency. J. Irrig. and Drainage Div., Proc. Amer. Soc. Civil Eng., vol. 93, no. IR2, Proc. Paper 5267, pp. 21–36 (1967).

3

Trickle Irrigation

J. HELLER and E. BRESLER

The conventional irrigation methods described previously have certain advantages, as well as limitations, when one considers their technical, economical, and crop-producing values. The trickle-irrigation method was developed for specific conditions of an intensive irrigated agriculture. Some of the technical and agronomical objectives in selecting the optimal irrigation method for such conditions are listed below.

1) The possibility of obtaining high average values (over time) of soil-water content, or low values of suction, without causing soil aeration problems.

2) Minimizing water-content fluctuations during the irrigation cycle.

3) Avoiding destruction of the soil-surface structure and the development of surface crust.

4) Restricting water supply only to those parts of the soil where water uptake by the root system is the most efficient. Selective wetting of the soil surface has additional beneficial results, such as reducing water evaporation, limiting the growth of weeds, decreasing the need for weed control, and enabling more convenient pest control.

5) Minimizing the salinity hazard to plants by (a) displacing the salts beyond the efficient root volume, (b) lowering the salt concentration by maintaining a high soil-water content, and (c) avoiding the burning of leaves and damage due to salt accumulation on the surfaces of leaves in contact with irrigation water. A dry foliage may retard the development of leaf diseases that require humidity and does not necessitate the removal of plant-protecting chemicals from the leaves by washing.

6) Optimizing the nutritional balance of the root zone by directly supplying nutrients to the most efficient part of the root zone.

7) Saving water by (a) minimizing evaporation from the soil surface, (b) reducing runoff in low permeable or crusted soil, (c) contour cultivation on slopy hills, and (d) preventing water loss beyond the borders of the irrigated field by wind convection.

In addition, the optimal irrigation method should rely on a relatively low operational pressure and a small pipe diameter, and should operate 24 hours a day, including windy hours, with a minimum of manpower.

Trickle irrigation has been developed with these objectives in mind.

Principles of the Trickle Irrigation Method

Trickle irrigation is based on the discharge of small amounts of water from small-diameter orifices in plastic tubings located on or immediately below the soil surface. The pressure required for the operation of a trickle system may be lower than that required for sprinkler irrigation. A low discharge is achieved through the use of tiny outlets and low-pressure heads in the supply line. For a given discharge, decreasing the size of the orifice below a certain minimum may cause clogging and require higher pressure in the supply line. On the other hand, increasing the size of the orifice to achieve a given discharge may require a lower operating pressure, which, in turn, may adversely affect the uniformity of the discharge along the line, owing to differences in elevation.

The inner diameter of the pipe (D) and the discharge of the trickler (Q) determine the Reynolds number and the water-flow regime, according to the equation,

$$R_E = \frac{4Q}{\pi . D . \nu} \tag{1}$$

where ν = kinematic viscosity (L^2T^{-1}).

For a given discharge, the diameter of the pipe must be larger in laminar flow than in turbulent flow. Head loss (h_L) in the trickler depends on the flow regime. In laminar flow, it is given by

$$h_L = \frac{L . \nu . 4Q}{g . \pi . D^4} \tag{2}$$

where L is length of the path flow and g is acceleration due to gravity.

In turbulent flow, the head losses are

$$h_L = f \frac{8LQ^2}{\pi^2 . g . D^5} \tag{3}$$

where f is friction factor.

From these equations one can see that the relationship between discharge and head loss is linear in laminar flow and parabolic in turbulent flow. For given discharge differences along the trickler lateral, the use of laminar flow tricklers require larger-diameter pipes or shorter runs than the use of turbulent flow tricklers.

The rate of discharge and the diameter and length of the pipe, as well as convenience in operating the system, are determined by the size of the orifice and the operational pressure. The discharge from an orifice depends on the number of openings along the line, the amount of water to be applied, frequency of irrigation required by the particular crop, and soil and climatic conditions. The application of small amounts of water, at intervals or through continuous irrigation, results in a low rate of discharge. This makes it possible to use smaller-diameter pipes and thereby reduce the cost of investment. The spacing between the openings along the line, as well as the distance between lines, depend on the response of the crop to water. These, together with the physical characteristics of the soil, the discharge of the openings, and the total amount of water to be applied, affect the wetting pattern of the soil and the spacing between the trickle sources (see p. 344).

Different trickle systems are characterized by different types of nozzles, such as perforations in the plastic pipe, nozzles forced into the tubing wall, tricklers designed as pressure regulators that reduce the existing pressure in the tubing by means of a long spiral or straight path through which the water flows, or a multiexit orifice that first reduces the pressure in the supply line by means of a tiny perforation or by a spiral-flow path and then directs the water into secondary openings connected to fine tubings. The intention in designing these various types of tricklers is to obtain a low discharge with as large an opening as possible. Tricklers having a spiral-flow path can yield a low discharge with an opening of about 1 mm in diameter and more (and are thus less prone to clogging) and operat at a pressure of about 1 atm. Tricklers with a short flow path have orifice diameters below 1 mm, are of a simple design, and are relatively cheap. However, they also clog easily. The various types of nozzles that are presently in use are illustrated in Fig. 1. Their respective characteristics are summarized in Table 1.

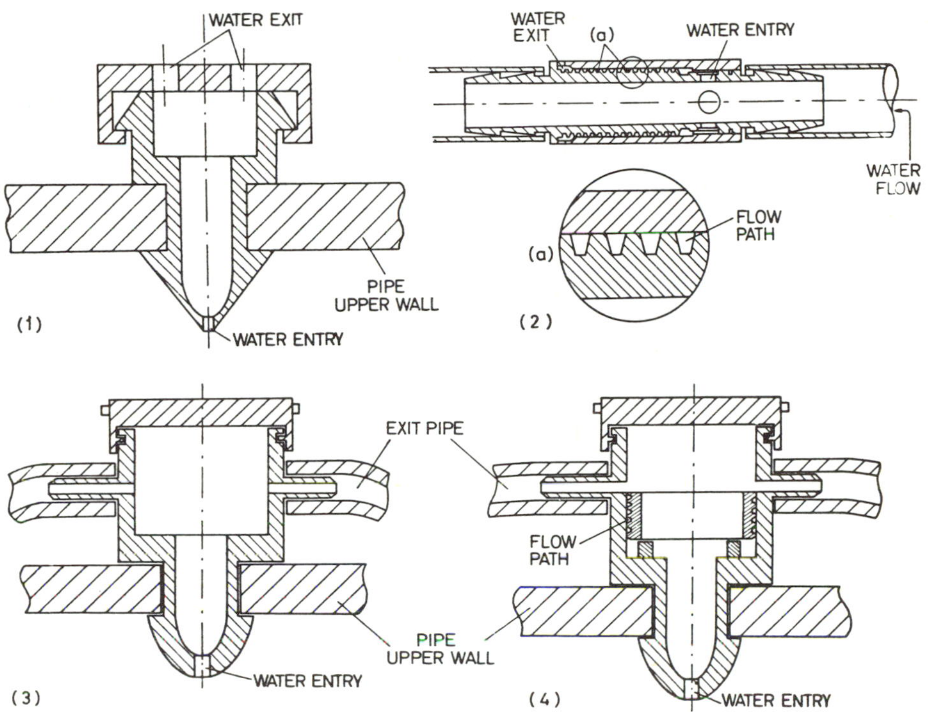

Fig. 1. Various types of nozzles. (After KARMELI, 1971)

Table 1. Common qualities of drippers. (After KARMELI, 1971)

Type of dripper qualities	Orifice (1)	Length path flow (2)	Multi exit orifice (3)	Multi exit long path flow (4)
Flow cross section (clogging problems)	Very narrow – (serious clogging possibilities)	Narrow (clogging possible)	Narrow (clogging possible)	Wide (clogging only in low-quality water)
Low discharges	Almost impossible	Most possible	A minimum of 4 l/h can be provided per exit	Dependent on number of exits
Q (h) relationships	$Q = f(\sqrt{h})$	$Q = f(h)$	$Q = f(\sqrt{h})$	$Q = f(\sqrt{h})$
Average quality of manufacturing capability	Inferior	Very good	Inferior	Good
Water distribution uniformity (over soil surface)	Nonuniform	Nonuniform	Uniform, dependent on number of exits	Uniform, dependent on number of exits
Relative total coast of system	High	Highest	Low	Low

In selecting a trickle nozzle, one must consider the cleanliness and composition of the water; the magnitude and uniformity of the discharge; cost of investment, depreciation, and convenience in operating the system; soil-water characteristics; and crop response.

Soil-water Regime during Trickle Infiltration

Trickle irrigation is relatively new and, research concerning its effect on the soil-water regime has barely started. Recently, the problem of infiltration under trickle irrigation was investigated by BRANDT *et al.* (1971) and by BRESLER *et al.* (1971). The present discussion therefore is limited to the soil-water regime during infiltration.

The system to be considered in this section consists of a small screwshaped plastic tube, fitted into an outer case. Water enters through an orifice, travels the length of the spiral path, which reduces its pressure, and discharges as a trickle at a predetermined rate. The irrigation trickle nozzle is placed directly on the soil surface, so that the area across which infiltration takes place is very small compared with the total soil surface. As a result, one has a case of three-dimensional, transient infiltration of water into the soil. This differs from the usual one-dimensional case of flood or sprinkler infiltration, where the area across which water enters the soil is assumed to be that of the entire soil surface.

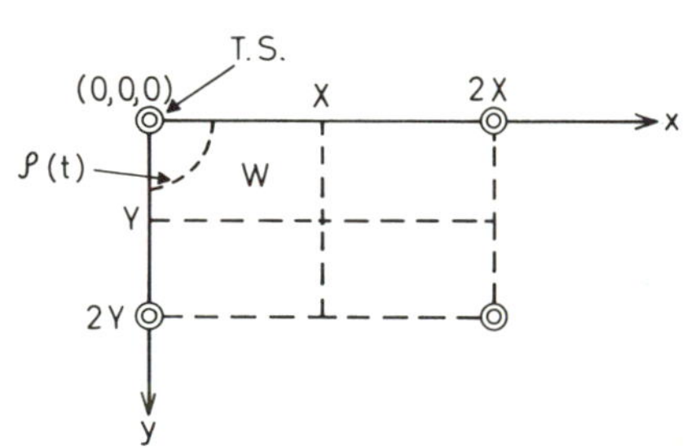

Fig. 2. Schematic representation of trickle-irrigated field. (After BRANDT *et al.*, 1971)

Consider a field that is irrigated by a set of trickle sources, spaced at regular intervals, $2X$ and $2Y$, as shown in Fig. 2. Due to the symmetry of the pattern, one can subdivide the entire field into identical volume elements, W, of length X, width Y, and depth Z, where the latter always remains below the wetting front. Here, each volume element acts as an independent unit in the sense that there is no flow from one element to another. Therefore, in order to describe the soil-water regime in the entire field, it is sufficient to analyze the status of the water in a single element, W. This, of course, is true only for the interior part of the field that is not too close to the margins.

The effect of any irrigation method on the soil-water regime depends primarily on the conditions prevailing at soil-surface boundary. In the case of trickle irrigation, these conditions may be defined by the trickle discharge (Q) measured as the amount of water per unit time, by the horizontal area across which infiltration takes place (A), and by the rate of evaporation at the soil surface (E). The last factor becomes important only when the potential evaporation is extremely high and the saturated hydraulic conductivity of the soil is very low.

It was observed that, in general, a radial area of ponded water develops in the vicinity of the trickle source. This area is initially very small, but its radius becomes larger with time. Since the ponded body of water is usually very thin, the effect of water storage at the soil surface can safely be neglected. This means that water from the trickle source is able to infiltrate into the soil, or evaporate into the air, instantaneously. Obviously, the soil-water content immediately beneath the ponded area, $A = \pi\varrho^2$, is always equal to the water content at saturation, θ_s. This saturated area is a function of time and is the only place where water can infiltrate into the soil from the surface. Thus, the rate at which water enters the soil across this area is equal to the rate of trickle discharge, minus the rate of evaporation. The size of this saturated water entry zone, which is an increasing function of time, is therefore an important factor in determining the soil-water regime during trickle irrigation. Neglecting evaporation (E), the size of this area depends on the rate of trickle discharge (Q) as well as the saturated hydraulic conductivity (K_{sat}) of the soil. Obviously, the area (A) of this water-entry zone is always smaller than the ratio between trickle discharge and the saturated hydraulic conductivity of the soil ($A < Q/K_{sat}$ or $\varrho < \sqrt{Q/\pi K_{sat}}$). If evaporation is too significant to be

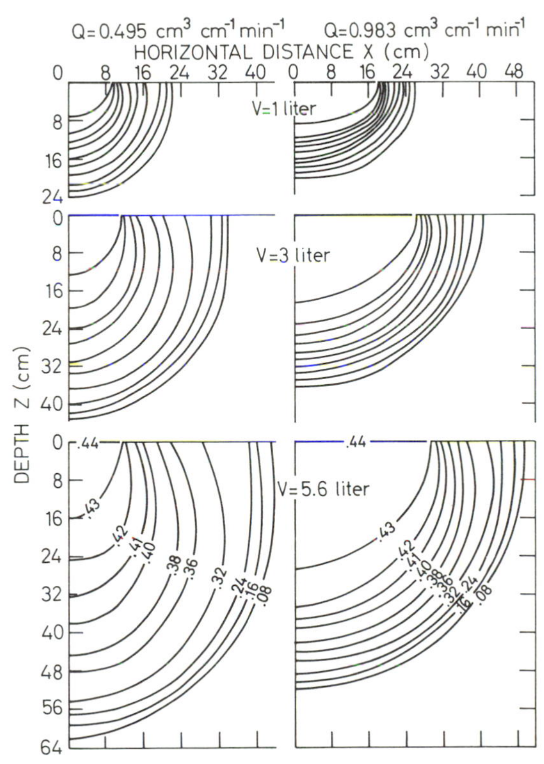

Fig. 3. Water-content field as a function of cumulative infiltration (V) for two cases of discharge (Q). The number labeling the curves indicate water content (θ). (After BRANDT *et al.*, 1971)

neglected, then, of course, $A < (Q-E)/K_{sat}$. Thus, for a praticular soil with given hydraulic conductivity and evaporative conditions, the size of the water-entry zone, as well as the soil-water regime in the entire irrigated field, is related to the discharge rate of the trickle sources.

To illustrate the water-content distribution, one horizontal dimension, x, and the vertical dimension z are considered. Examples of water-content distribution for two cases of trickle discharge (Q, expressed in terms of discharge per unit length) are given in Fig. 3, which shows how the water content changes with time and position during trickle infiltration. The illustration also shows the effect of trickle discharge on the water content. The saturated water-entry area (A), and the rate at which it changes with time, are shown by the particular line of saturated water content at the surface, where $\theta_{sat} = 0.44$. In general, the area increases as the rate of trickle discharge increases. In the vicinity of the source ($x = 0$, $z = 0$ in Fig. 3), the moisture gradients increase when the rate of discharge decreases. These gradients can be calculated from the distances between lines of equal water content. This condition reverses as the wetting front is approached. The overall shape of the wetted zone also depends on the trickle discharge. The vertical component of the wetted zone becomes larger and the horizontal component narrower as the rate of discharge decreases.

In the case of infiltration into a dry soil, a distinct boundary exists between the wetted zone and the dry zone; this is known as the wetting front. This wetting front is important in trickle infiltration because it indicates the boundaries of the irrigated soil volume. Fig. 4 shows the location of the wetting front, as a function of the space coordinates (x, z)

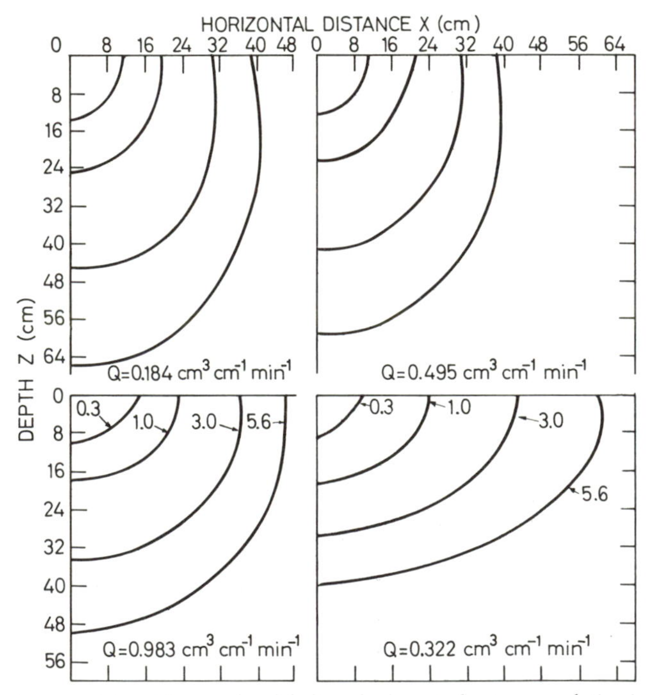

Fig. 4. Wetting front position. The number labeling the lines indicates cumulative infiltration water (in liters). (After BRESLER *et al.*, 1971)

and the total amount of infiltrated water, for 4 different trickle discharges. The total amount of infiltrated water (in liter) is indicated on the wetting front lines appearing in the figure. The trickle discharge, in terms of rate per unit length Q, is also indicated.

The data presented in Figs. 4 and 3 clearly demonstrate that the rate of trickle discharge has a marked effect on the shape of the wetted soil zone. Increasing this rate results in an increase in the horizontal wetted area and a decrease in the wetted soil depth. This is probably due to a change in the size of the water-entry saturated zone for each rate of discharge (Fig. 3), the latter zone becomes larger as the trickle rate increases.

The possibility of controlling soil-water content, suction, fluxes, and the wetted volume of the soil by regulating the trickle discharge (Figs. 3 and 4), is of practical interest in problems such as those discussed in the introductory section of this chapter, and in the design of field irrigation systems.

Description of the Operational System

Every trickle system includes a control unit through which all the water passes. This unit consists of a valve, generally up to 2 in. in diameter, connected to the water source, and an automatic flow rate (or discharge) meter. In addition, it includes a nonreturn valve (to prevent contamination of the main water grid by the fertilizer solutions), a vacuum breaker, a water filter, and a manometer with a 2 or 3-way tap to measure the pressure before and after the filter. The water can be cleaned or filtered by means of a screen, or by means of gravel filters provided with a flashing arrangement. Usually, a fertilizer tank (25–100 l) is installed near the control unit to mix the fertilizers with the water and discharge the solution into the water stream before it enters the filter. A typical trickle-control unit is shown schematically in Fig. 5.

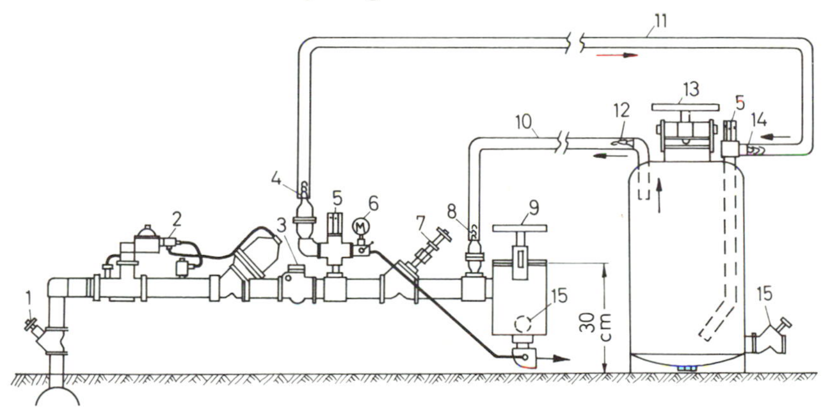

Fig. 5. Trickle irrigation control system unit. (1) Connection to main water supply. (2) Automatic metering valve. (3) Non return valve. (4) Entry of water to fertilizer tank. (5) Vacuum valve. (6) Pressure gauge with 3-way valve. (7) Pressure-reducing valve. (8) Exit of water from fertilizer tank. (9) Water filter. (10) Tubing for diluted fertilizer. (11) Tubing for water to fertilizer tank. (12) Exit of diluted fertilizer from the tank. (13) Fertilizer tank. (14) Entry of water to tank. (15) Flushing valve

The control unit is usually located on the highest place in the field and, in order to gain energy, the pipelines are placed parallel to the slope. One or more main lines, made of polyethylene, are connected to the control unit. Their diameter depends on the shape and size of the field, and the maximum flow rate required. The main line is often divided into secondary branches that connect to the lines with the tricklers. The main and secondary branches are frequently fitted with flow or pressure regulators. A device that can be opened and closed to flush and clean the system is connected to the end of each line. The diameter of the main line is usually between 25 and 50 mm, and that of each secondary branch is 10–20 mm. Their length varies between 25 and 100 m.

The system is operated by first adjusting the automatic water meter to the required amount of water, and then irrigating the entire plot as one unit, using the control unit. The customary spacing between nozzles along the line is 60–200 cm, depending on crop needs, soil-water characteristics, climatic conditions, and discharge from the trickle source. In row crops, lines are usually placed in each row. In orchards, the number of trickle lines per row of trees varies from 1 to 2, depending on the distance between the rows, the age of the trees, the root system distribution, etc. The energy head in the branches is about 10–12 m of water. Maintenance of the systems requires periodic cleaning of the filters at the control unit, and checking the proper operation of the tricklers. For this reason, it is preferable to lay the tricklers on the soil surface and not to place them underground. The extend of clogging depends on the cleanliness and composition of the water, the rate of flow, the aperture size, and the duration of irrigation. The usual discharge from each nozzle is 4–10 l/hr in orchards and 1–2 l/hr in vegetables. If the rate of discharge decreases with time due to the depositions of carbonates on the walls of the tubing, passing an acid of low concentration through the system may often help dissolve the carbonates. Standard maintenance includes cleaning the filter, periodically checking the tricklers, and changing or cleaning clogged tricklers.

Design of Trickle-Irrigation Systems

An accepted criterion in the design of trickle-irrigation systems is to allow the discharge to fluctuate within a range of up to 10% over the irrigated field. The permissible head loss for such a change in discharge depends on the type of trickler and the relation between discharge and pressure. Therefore, pipe diameter depends on trickle discharge, pipe length, and slope of the field along the line. Pipe diameter is selected with the aid of a nomogram that gives the relative head loss for different pipe diameters as a function of discharge (Fig. 6).

The diameter and length of the line are established by determining the head loss in a pipe without nozzles, based on discharge data (Fig. 6), and multiplying this by the appropriate factor (F) for a pipe with nozzles, as was done in the case of sprinkler lines (see p. 315). In practice, the correction factor for a perforated pipe is often taken as $F = 0.4$.

The head loss along a pipe with tricklers depends on the form and structure of the nozzles. The discharge coefficient c_q of the nozzle is a function of the form and structure of the orifice. With a low coefficient, the head losses are smaller, and therefore the F-value can be lower. The correction factor F must be selected in accordance with the particular type of nozzle used.

KARMELI and BEN SHAUL, (1970) developed an equation similar to that proposed by CHRISTIANSEN (1942) for designing a trickle-irrigation line, assuming a constant distance between the nozzles and a pressure drop between the openings equal to that in a pipe of similar diameter without openings. The general equation for the trickle discharges along the irrigated line is based on the Darcy-Weisbach Eq. (4), and the orifice discharge Eq. (5):

$$h_L = f\frac{L\,v^2}{D\,2g} \tag{4}$$

where h_L is head loss, m; f is friction factor; L is pipe length, m; D is internal diameter of the pipe, m; g is acceleration due to gravity, m/sec^2; and v is average water velocity, m/sec.

Tests of a polyethylene pipe without openings have shown that the relationship between the friction factor f and the Reynolds number Re is given by

$$f = \frac{0.291}{R_e 0.246} \qquad \text{for } R_e > 2000$$

and, of course,

$$f = \frac{64}{R_e} \qquad \text{for } R_e < 2000$$

According to the orifice discharge equation,

$$q = c_q A \sqrt{2gh/0.2778} \tag{5}$$

where q is discharge of the orifice, l/hr; c_q is discharge coefficient of the orifice; A is area of the orifice, mm^2, and h is total head at the orifice, meters.

The coefficient 0.2778 is a correction unit factor. The typical value for c_q for the tested nozzle was 0.88.

In order to calculate the discharge of any trickler along the irrigated line, the pressure head and the discharge at the first opening (q_1) must be known. The discharges along the line are then calculated from Eq. (6), which is based on Eqs. (4) and (5) when the slope of the field is also considered.

$$q_j = \left[q_{j-1} - E\left(\frac{1 - \sum_{i=1}^{j-1} q_i}{Q} \right) \pm \frac{SZ}{100} \right]^{1/2} \qquad j = 2,3,\ldots,N \tag{6}$$

$$E = \frac{\beta \times S \times Q^2 \times q_1{}^2}{h_1 \times D^{\alpha+d}}$$

where q_1 is discharge of the first nozzle, l/hr; q_j is discharge of the j–th nozzle; S is space between nozzles, m; Q is flow rate into the line, l/hr; D is internal diameter of the pipe, mm; h_1 is pressure head at the first nozzle, m; z is slope of the soil surface, (%); d is the nozzle diameter and α and β are empirical constants.

The values of β and α depend on relations between Reynolds number and the friction-loss factor. It was found that when the Reynolds number was smaller than 2000, α was 1.754 and β was 0.4758. When $Re > 2000$, $\alpha = 1.0$ and $\beta = 1.183$.

From Eq. (6) it is possible (a) to calculate the minimum pipe diameter required when the distance between the nozzles is constant, or (b) to change the size of the nozzle and the diameters of different pipe sections, in order to obtain the derived system.

To illustrate the design of a trickle-irrigation system, let us consider a citrus grove with the follwing specifications:

Area of the grove	0.8568 ha
Planting slope along the rows	1.5%
Tree spacing:	4 m within the row, 6 m between the rows
Length of a trickle irrigation line (L)	68 m
Distance between trickles on the line	1 m
Number of trickle lines per row of trees	2 (each 1 m from trunk)
Type of trickler	"Netafim" (2, Fig., table 1)
Discharge of trickler in an operating head	2 l/hr. (0.002 m^3/hr)
Head at the trickler	10 m
Total number of tricklers	2856
Maximum flow rate	5.71 m^3/hr
Water requirement of the crop	4 mm/day
Frequency of irrigation	3 days
Amount of water per set	102.8 m^3
Duration of irrigation	18 hr
Maximum variation in discharge of the system	10%

A map of the trickle irrigation layout in this grove is given in Fig. 7. The pipe diameters, the discharge rates, and total heads are also given in the figure.

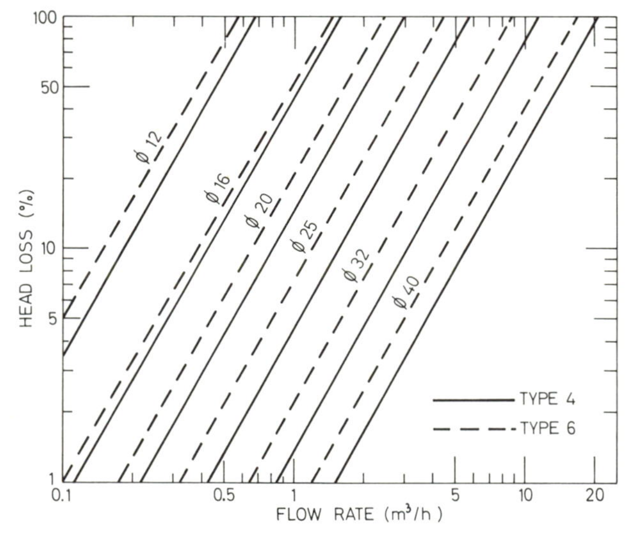

Fig. 6. Monogram giving the relative head loss at different pipe diameters. (After Israeli Institute of Water Appliances)

From the above design specifications, it is seen that the maximum permissible head differences in the entire trickle system is 1 m (according to the 10% criterion) when the operating pressure of a trickler discharging at a rate of 2 l/h is 10 m. In order to achieve the desired discharge, the maximum permissible head loss along the supply line must not exceed 100 cm of water. This means that the head differences between A and C in Fig. 7 should not exceed 1 m.

Calculation of head losses along the trickler line: The total discharge of a trickle line is

Q = 68 tricklers per line × 0.002 m³/h per trickler = 0.136 m³/hr. The head loss with this discharge in a thin polyethylene tube 12 mm in diameter (grade 4) is 5.6% the nomogram in Fig. 6). The value of F for 66 openings and an exponential value of $m = 1.85$[1] is 0.359. The head losses along the line are thus

$$\frac{0.359 \times 5.6 \times 68}{100} = 1.37 \text{ m}$$

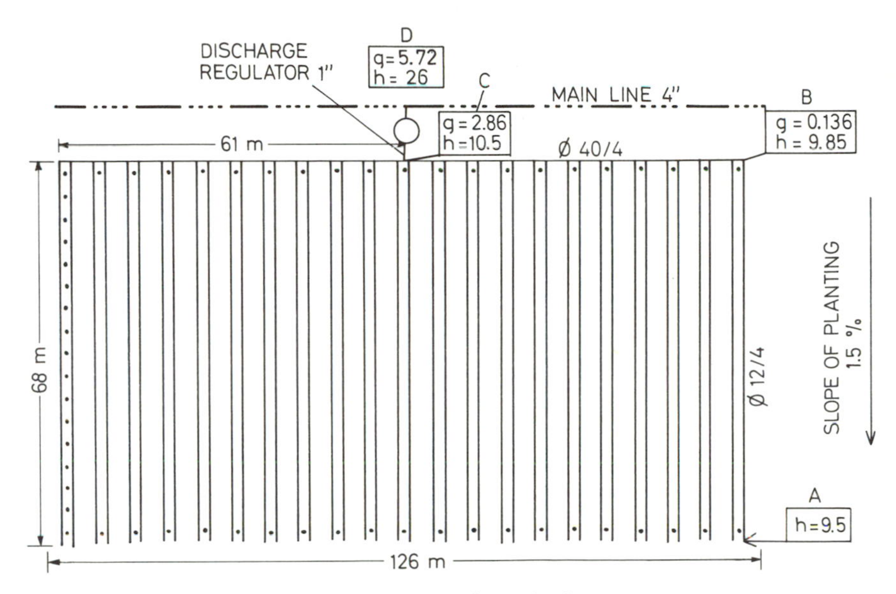

Fig. 7. Trickle irrigation layout in citrus grove

Head gain due to elevation along a line 68 m long is

$$\frac{1.5 \times 68}{100} = 1.02 \text{ m}$$

Therefore, the total head loss along the trickle line is 1.37−1.02 = 0.35 m. The head at point B is 9.5 + 0.35 = 9.85 m

Head losses in the rigid supply line: Owing to symmetry, it is sufficient to perform the calculation for half on the plot. Therefore, the length of the supply line is 61 m, and the discharge of the 21 lines is 0.136 × 21 = 2.86 m³/hr. The head loss in a blind line 40 mm in diameter (grade 4) is 2.86% (Fig. 6). If the value of F is taken as $F = 0.373$, then the head loss along the supply line is

$$\frac{0.373 \times 2.86 \times 61}{100} = 0.65 \text{ m}$$

and the total head requirement at point C is 9.85 + 0.65 = 10.5 m. Head loss in the irrigated field is 0.65 + 0.35 = 1 m which is the maximum permissible loss, as mentioned

---

1 The exponent m (CHRISTIANSEN, 1942) is smaller for polyethylene tubing, and so the F-values must be somewhat higher than those given by CHRISTIANSEN for aluminium pipes.

earlier. Thus, in order to maintain at each trickler a total head of 10 ± 0.5 m, a head of 10.5 m must be kept at C, which results in a head of 9.5 at *A* (Fig. 7).

Head losses at the control system unit:		
Discharge regulator	=	8.00 m
Riser	=	0.30
Filter	=	4.00
Automatic metering valve	=	1.00
Fertilizer tank	=	2.00
Total		15.30 m

It follows, therefore, that the minimum pressure head required at point *D* is 10.5 + 15.3 ≈ 26 m.

An irrigation cycle of 3 days, and an irrigation duration of about 18 hours, demand opening the water in the field every 54–55 hours. The water can be opened manually or automatically, using a system of sequentially operating valves.

Current prices for a permanent trickle-irrigation system depend on the spacing and discharge of the nozzles employed. It should be mentioned that irrigating an orchard with one line of tricklers per row of trees was found to be the cheapest of all the solid-set irrigation systems, as far as investment cost is concerned. However, the effect of such partial wetting on crop performance and economic yield is not yet known.

Some Problems Concerning Practical Use of the Trickle Irrigation Method

Operational difficulties may sometimes arise from the clogging of tricklers. This results in nonuniform discharge and wetting of the soil, which, in turn, may cause water deficits and salinity problems. Overcoming these difficulties involves using an efficient filtering system and periodically cleaning or replacing the tricklers, and may also involve extra expenses and manpower when a larger number of tricklers or lines per unit area are needed. Relatively expensive equipment is required for crops that are sensitive to the water distribution in the soil, such as spread sowing or narrow row crops. In arid regions where saline water must be used, there is a tendency for salt accumulation near the margins of the wetted zone, and especially on the soil surface. This salt must be leached before the next crop is planted. If there is not sufficient rainfall for the leaching process, then leaching must be accomplished by sprinkler or flood irrigation, both of which raise the cost considerably. Another solution is to place the tricklers much closer together, which is also expensive.

In cases where there are no salinity problems, the wetted volume of water should be kept within the borders of the efficient root zone. In order to prevent losses of water beyond the efficient root zone, one must determine the proper trickler discharge, the duration of irrigation (amount of water applied) for given soil-water characteristics, the space distribution of soil-water content prior the each irrigation, the crop root distribution, and the pattern of water uptake by the roots. These factors determine the dynamics of the soil-water system, and therefore the volume of soil to be wetted, and they can be established once the appropriate soil and plant parameters are known. Determination of the proper trickler spacing and discharge under given soil, crop, and climatic conditions requires knowledge of crop response to the size and form of the wetted soil volume and to the soil-water content distribution and fluctuation within this volume.

Literature

BRANDT, A., BRESLER, E., DINER, N., BEN ASHER, I., HELLER, J., GOLDBERG, D.: Infiltration from a trickle source. I. Mathematical Models. Soil Sci. Soc. Amer. Proc. **35**, 675–683 (1971).

BRESLER, E., HELLER, J., DINER, N., BEN ASHER, I., BRANDT, A., GOLDBERG, D.: Infiltration from a trickle source. II. Experimental data and theoretical predictions. Soil Sci. Soc. Amer. Proc. **35**, 683-689 (1971).

CHRISTIANSON, J. E.: Irrigation by sprinkling. Calif. AES Bull. **670** (1942).

KARMELI, D., BEN SHAUL, S.: Methods of trickle irrigation (in Hebrew). Faculty of Agric. Eng. Technion, Bull **47** (1970).

KARMELI, D.: Hydraulic aspects of various drippers. Israeli Tricklers Congr., July 1971.

4

Special Irrigation Methods and Accessory Devices

E. Rawitz

A fairly large number of special methods of irrigation have been proposed from time to time. Some of these methods, which solve specific and unusual problems, are briefly presented here.

Subirrigation

There are a number of areas in the world where a very permeable soil (often organic or peat soils) in the root-zone depth is underlain by an impermeable layer, and often also by a high water table. In this method, the root zone is wetted by raising the water table, after which it is lowered again until the next irrigation is required. Sometimes the water table is adjusted as to allow capillary rise of groundwater into the root zone. Technically this is a very efficient method, but it is flawed by the fact that permanent and intensive irrigation is generally practiced in fairly dry climates where groundwater invariably contains at least some amounts of dissolved salts. These salts are brought to the root zone with the irrigation water, but remain behind when the water evaporates or is transpired by plants. Although this also happens with other irrigation methods, subirrigation has two major drawbacks in this respect: First, shallow groundwater generally has the highest salt content of all available water sources. Second, where water is applied from above, there exists the possibility of leaching out accumulated salts, which is not the case with subirrigation. Two large areas of organic soils that were once irrigated this way are the Sacramento River delta in northern California, and the Huleh Valley of northern Israel. Both of these areas are now irrigated by sprinkling. However, subirrigation may have limited use under special conditions, such as supplementary irrigation during a drought in more humid climates.

The advent of cheap plastic tubing led to the suggestion that a network of small-diameter plastic tubes be buried within the root zone in soils with a normal profile, as a sort of underground trickle-irrigation system. The main claims made for this idea were the ability to irrigate continuously, thus increasing soil-moisture availability and therefore yields, and water savings of up to 30% due to decreased evaporation. However, point sources of water placed deep enough to decrease evaporation substantially would be too deep for many crops and soils, whereas adequate water supply by subirrigation would be unsatisfactory because of salt accumulation and excess deep percolation. The delivery devices were plagued by root penetration, and the network was subject to damage by agricultural machinery and by rodents. Nevertheless, the suggestion received a good deal of publicity, possibly because it was both new and commercially attractive, and because some sponsors applied more enthusiasm than analysis to the problem.

Autoirrigators

A number of devices have been suggested that will automatically apply water when a certain threshold condition is reached. Although conventional irrigation systems can certainly be automated, autoirrigators usually contain the sensor and the applicator in one small device. In one such arrangement a small, vented container containing a sponge is buried under a plant. When the sponge dries and contracts, it releases a valve and opens the flow of water; as the sponge wets it expands again and shuts the water off. Such devices have not proved practical.

Automated Irrigation Systems

Labor makes up a large part of the cost of irrigation, and operator convenience is another important consideration when the purchase of an irrigation system is being contemplated. A well-designed system that requires the farmer to go to the field to shut off water and move a lateral in the middle of the night just won't operated efficiently in practice. In addition to the various aids in transporting laterals that were discussed earlier, a number of possibilities exist for automatic shutoff and changing of set.

Where the farm is supplied from its own well, perhaps the simplest device is an electric timer that shuts the pump motor off at a set time. If application rates are known, this is a very good solution. The same system may be used for turning the pump on, although it is advisable to supervise and check a system as water is turned on to correct possible leaks and other malfunctions.

If water is drawn from a regional supply pipeline, mechanically actuated shutoff valves are available for sprinkler laterals and other small-diameter pipes. These operate either on the timer principle, or, better yet, are actuated by an attached, integrating water meter after a predetermined volume of water has passed through the meter. Water meters can be made to close or open electric circuits by closing contacts or actuating micro-switches, opening the way to the use of solenoid valves, relays, and electric motors that can be used to open or close control devices on even the largest pipelines or canals. Thus the use of automation for performing only one operation, either opening or closing the flow of water, is relatively simple, but also limited. It would obviously be desirable to automate the operation of "changing the set," i.e., stop irrigation of one area and divert the water to another one.

Since automatic moving of laterals has not been developed, such automation has been possible so far only for permanent systems, or where at least all the laterals to be used during the period of automated move are installed and ready to operate. A number of systems actuated by timers or pressure or water-level sensing devices have been constructed for this purpose. Because they are both complicated and expensive, they have not been put into widespread use.

All the automated schemes described so far depend on the possibly arbitrary decision of the operator as to how much water should be applied, and when. It is also possible, however, to apply the criterion of soil-moisture availability to automated irrigation, and this has been carried out in practice in a number of installations. The simplest example is the irrigation of potted plants or greenhouse crops. Commercially made scales are available on which a pot is placed, and the scale adjusted so that irrigation is started and stopped when the pot reaches a certain minimum and maximum weight, respectively. The scale includes a device that actuates the water-supply valve. Although this system is too expensive for individual pots except for research purposes, one pot and scale could be used as a pilot for many pots, or even for plants growing on benches in beds, provided the soil-moisture extraction rate of the indicator plant can be assumed to be representative of all the other plants.

A more sophisticated and more flexible scheme is to use either a tensiometer or electrical resistance unit as indicator of soil-moisture status. Both devices can be made compatible with electric controls by the use of the appropriate interface (mercury switches, relays, electronic decision-making circuits of the "go-no go" type). One sensor may be used to initiate and terminate irrigation, or two sensors can be used to improve sensitivity and flexibility. For example, a tensiometer at the 50-cm depth may be used to start irrigation when a certain tension is reached, and another tensiometer or a resistance block at the bottom of the root zone (e.g., at 100 cm) may be used to terminate irrigation when the irrigation water has reached that depth. Such systems of course require both a permanent installation of delivery devices and access to the water source upon demand. This is most often the case in small installations such as greenhouses.

Literature

HAGOOD, M. A.: Automated irrigation is here. Agric. Engng. 53, 16–18 (1972).

HAISE, H. R., KRUSE, E. G., ERIE, L.: Automating surface irrigation. Agric. Engng. 50, 212–216 (1969)

RENFRO, G. M.: Applying water under the surface of the ground. In: Water. USDA Yearbook of Agriculture, pp. 273–278 (1955).

WILDER, L. D., BUSH, C. D.: Push button control for water systems management. Agric. Engng. 51, 408–409 (1970).

VIII. Crop Water Requirement

The irrigation requirement determined for a specific crop is not universally applicable to a variety of environmental conditions. Even within a given arid or semiarid zone, the variations in climatic conditions are so great that differences in evapotranspiration are appreciable. This chapter is concerned with the irrigation of a wide range of crops that can generally be cultivated in arid and semiarid regions having a long, dry summer season. In contrast to the other chapters, which are of a more general nature, this one contains information based largely on experience gained in Israel.

The general considerations concerning the effect of soil-moisture availability and the use of plant-water status indicators for determining the water requirements of different crops, together with some practical views on irrigation scheduling, are included in the section entitled "Prediction of Irrigation Needs."

The data presented for different crops serve to illustrate the approach used in determining irrigation requirements under different experimental conditions, and examples are given for each main group of irrigated crops. Field crops are exemplified by wheat, cotton, and peanuts; forage crops by corn and alfalfa; vegetables by potatoes and tomatoes; and fruit trees by citrus and apples. For some of the crops, a comparison is made between the irrigation requirements in Israel and in other countries. However, the discussion is generally related to those conditions prevailing in Israel.

Considerable research has been conducted in Israel in order to achieve maximum water-use efficiency in the irrigation of various crops. According to the experience gained in the determination of water requirements, it is clear that irrigation practice must be based on local conditions. The use of an approach similar to that which proved successful in Israel is recommended.

1

Prediction of Irrigation Needs

H. BIELORAI

Criteria most suitable for scheduling irrigations vary from one situation to another. Where water is scarce or expensive, irrigations should be planned to maximize crop production per unit of applied water; where good land is scarcer than water, irrigation should be planned to obtain maximum production (or income) per unit of planted area.

Several other factors must be taken into consideration when planning an irrigation schedule for a certain area or a single farm. For example, irrigation schedules may be modified in order to control low penetration in heavy soils; groundwater table by shallow wetting; or leaching salts accumulated in the soil profile (by applying heavy irrigations). The interaction between soil nutrients, interrow spacing, and water use must also be considered. Additional problems that may affect irrigation planning include available irrigation water resources, the need or supplemental irrigation in dry farming, and the delivery and distribution of the irrigation water on a regional and single-farm basis.

In all cases, the criteria for irrigation should be selected to achieve favorable crop yields and efficient use of water, and careful consideration must be given to the various factors during the planning stage.

Various indicators are used to assess irrigation needs, including soil-moisture measurements, plant indicators, and climatological indices (see Chap. V).

Soil-Water Availability

Two important concepts resulted from early work relating soil water to plant response. The first was the *field capacity* concept (ISRAELSEN and WEST, 1922; VEIHMEYER and HENDRICKSON, 1927), which became identified with the upper limit of available water in soil. The second, the *wilting coefficient* (BRIGGS and SHANTZ, 1911), led to the permanent wilting percentage (HENDRICKSON and VEIHMEYER, 1929) that gained wide acceptance as the lower limit of available water. The utility of these two concepts is in their universal application in determining practical irrigation schedules.

The availability of soil water to plants is of particular importance for irrigated agriculture in arid regions. For some decades the problem connected with soil-water availability has been the subject of diverse opinion between opposing schools. VEIHMEYER and his co-workers claimed that soil water is equally available over the range between field capacity and permanent wilting point. On the other hand, RICHARDS and WADLEIGH (1952) made a strong case for the concept of *decreasing availability* as the soil-water content decreased from the upper to the lower limit.

A third approach is represented by a group of irrigation scientists who suggested the existence of a "critical soil-moisture level" (in the available soil-moisture range). Below this critical level, a significant decrease in yield is noted. In fact, each of these three concepts may apply under certain conditions. The three concepts are presented schematically in Fig. 1.

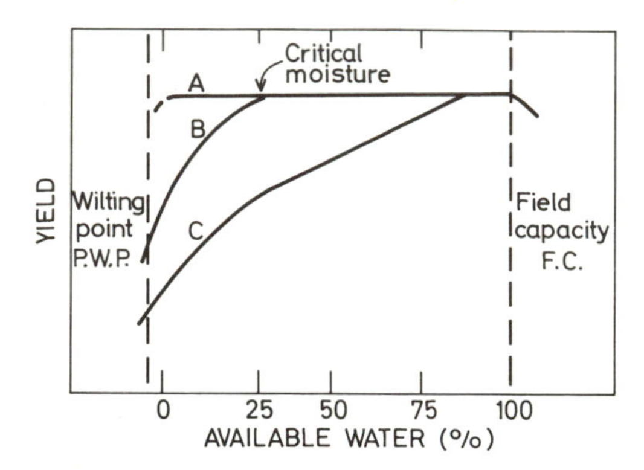

Fig. 1. Three conceptual relationships between available water and yield. (Adapted from "Water-Soil-Plant Rleations." Calif. Agric. Vol. **11**, 1957)

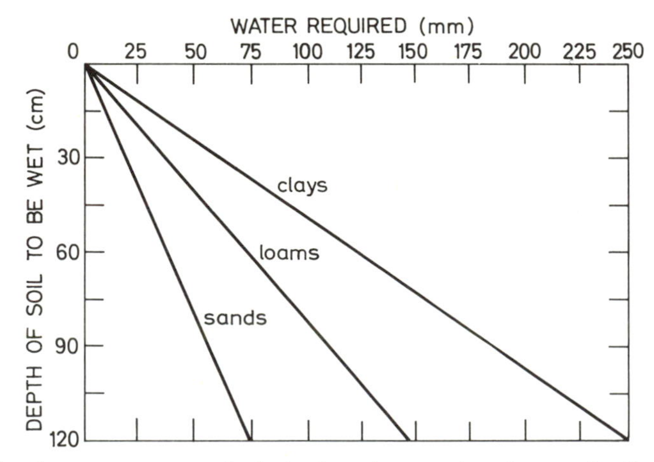

Fig. 2. Relation between water required, depth to be wetted, and type of soil. (Adapted from "Water," The Yearbook of Agriculture 1955, USDA)

Curve *A* represents the concept of equal soil-moisture availability, in the range between field capacity (FC) and permanent wilting point (PWP). Curve *B* describes the existence of a critical soil moisture within the available soil-moisture range. Curve *C* represents the decrease in yield as soil moisture decreases gradually. As a first approximation in establishing an irrigation schedule in different soil types under prevailing climatological conditions, the available soil moisture between FC and PWP is taken into consideration, through the design schedule will not, as a rule, cause the extraction of all the "available moisture" between irrigations.

By means of the nomogram presented in Fig. 2, the amount of water required to wet a certain root zone can be obtained for three types of soil: sand, loam, and clay. In Fig. 3 the approximate intervals (for the same types of soil) between irrigations can be determined taking into consideration soil texture, depth of roots, and water-storage capacity.

Fig. 2 shows that in order to replensih the soil water in the main root zone, 0–90 cm, an irrigation of 50 mm of water should be applied in a sandy soil, in a loamy soil, 100 mm, and in a clay soil, 175 mm of irrigation water is needed. On the other hand, because of the

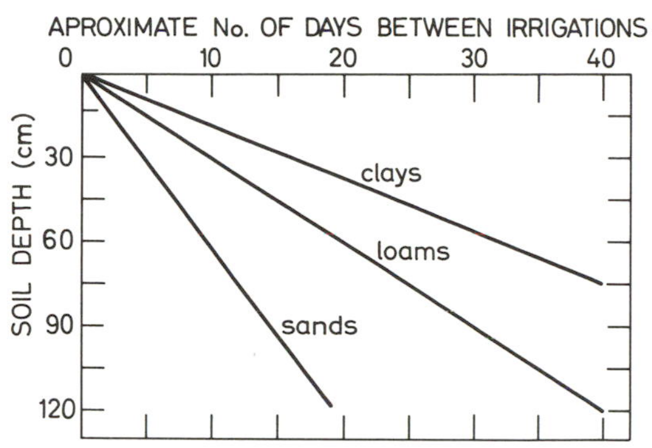

Fig. 3. The relation between actual rate of soil-moisture depletion and irrigation frequency. (Adapted from "Water," The Yearbook of Agriculture 1955, USDA)

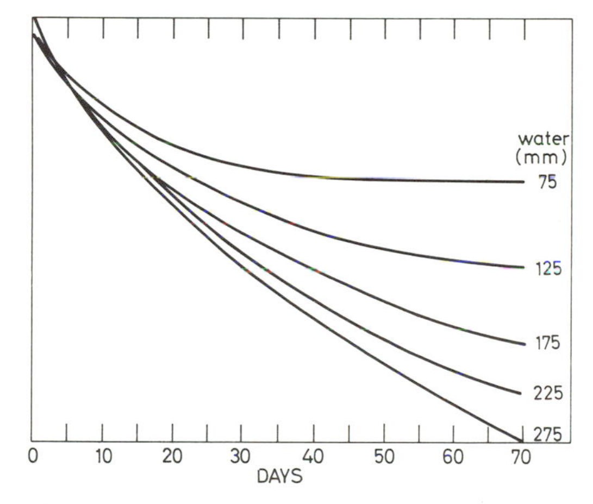

Fig. 4. Irrigation interval as influenced by soil texture and water storage in the root zone (assuming a water use of 4.0 mm/day). (Adapted from "Water," The Yearbook of Agriculture 1955, USDA)

greater storage capacity of the clay soils, the intervals between irrigation, taking into account the actual depletion from the main root zone, are considerably greater than those of sandy soils. Average irrigation frequencies in clay soil are 30 days and in sandy soils about ten days (Fig. 3).

Immediately after a soil is wetted, evaporation and transpiration by plants begin to reduce its water content. As the soil dries, the rate of evapotranspiration diminishes. In all soils, evapotranspiration initially takes place at almost the maximum rate, but by the time about 25 mm of water has been removed, the depletion rates for different soils begin to differ.

Fig. 4 shows the actual rate of soil-water depletion from soils holding different amounts of water in the root zone, assuming a constant potential evapotranspiration rate of 4 mm/day.

When about 75% of the available water is depleted, the plants begin to suffer from water stress and the yield decreases. Supposing the rate of evapotranspiration to be constant (a simplifying assumption), then two thirds of the moisture in a sandy soil will be depleted within 10 days or less, but about 30 days will be required under the same conditions in a fine-textured soil (the amount of water stored in the root zone of a fine-textured soil is considerably greater).

The computations presented in these nomograms can be compared with actual soil-moisture determinations. The percentage of soil water is expressed on a dry-weight basis. To convert this value of soil-water content to a volume basis (in order to know the amount of water required to replenish the deficit from field capacity in the soil profile), the bulk density of the soil must be known. The soil-water measurements should be made to the depth of the entire root zone. The amount of water to be applied per single irrigation is calculated by means of the following equation:

$$eh = \frac{Pw \cdot DbZ}{10}$$

where eh is the amount of water, mm; Db is the bulk density, g/cm^3; and Z is the depth of the soil profile to be wetted, cm. The soil-water content may be determined by any one of several methods.

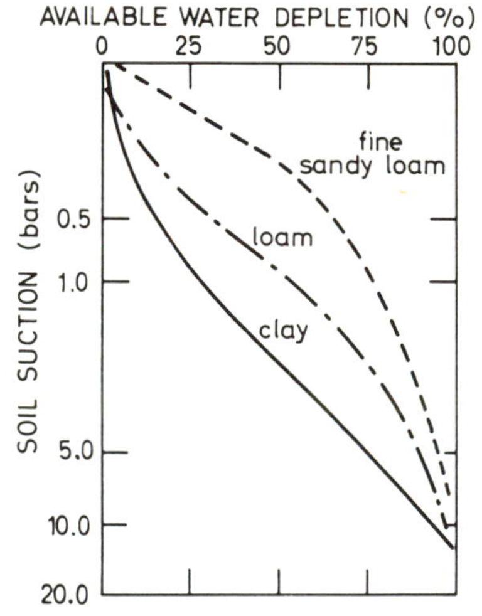

Fig. 5. Relation between percent available water depletion and soil-water potential (bars). (Adapted from "Irrigation of Agricultural Lands." Agronomy Monograph, 11, Amer. Soc. Agron.)

Irrigation schedules based on soil-water content, water use for a given soil, or climatic conditions at a given experimental site have limited application. The use of soil-water suction or potential to schedule irrigation overcomes much of the difficulty in applying results from one area to another, if the relation of crop response to soil-water availability as expressed by soil-moisture potential is known. Since plant response to irrigation is better correlated with water potential or suction than with soil-water content, measurement of soil-water potential or suction provides a useful tool for scheduling irrigation.

Figure 5 presents water-retention curves for several soils plotted in terms of percent available water removed (RICHARDS and MARSH, 1961). It can be seen that the soil water suction at 50% available water depletion in a clay soil is considerably higher than in loam and sandy loam soils. This means that the transfer of irrigation needs from one location to another by means of suction data is valid only if the same type of soil is involved in both places.

In Table 1 are presented the soil-water potential values at which water should be applied for maximum yield of selected crops grown in deep, well-drained and fertilized soils.

Table 1. Maximum yield obtained at specific soil-water potential for different crops

Crop	Soil suction[a] (bars)	Reference
	Vegetative crops	
Alfalfa (Forage)	1.50	TAYLOR, Utah, 1959
Alfalfa (seeds)	2 –8	TAYLOR, Utah, 1959
Lettuce	0.40–0.60	MARSH, California, 1961
Sweet corn	0.50–1.00	VITTUM *et al.*, New York, 1963
Corn	1.0 –2.0	BIELORAI and RUBIN, Israel, 1957
	Root crops	
Sugar beets	0.40–0.60	TAYLOR, Utah, 1963
Sugar beets	1.0 –2.0	BIELORAI and RUBIN, Israel, 1957
Potatoes	0.30–0.50	TAYLOR *et al.*, Utah, 1956
Potatoes	0.50–1.00	SHIMSHI and RUBIN, Israel, 1962
	Fruit crops	
Oranges	0.2 –1.00	STOLZY *et al.*, Riverside, 1963
Oranges	1.00	MANTELL and GOELL, Israel, 1970
Grapefruit	0.50–1.20	BIELORAI and LEVI, Israel, 1971
Bananas	0.30–1.50	SHMUELI, Israel, 1953

[a] Soil-water potential can be determined by tensiometers or electrical resistance units.

In deciding the water deficit that is permissible prior to irrigation, primary consideration must be given to plant response, such as plant root and shoot characteristics, and critical growth stages of a certain crop–e.g., jointing, flowering, and milk stage in corn; jointing, heading, and milk stage in sorghum; flowering and boll set in cotton. The data collected during irrigation experiments, including soil-moisture depletion patterns and yield response to various soil-moisture treatments, provide the necessary information to determine the optimal irrigation requirement, timing of irrigation, size of a single irrigation, irrigation efficiency, and seasonal water use.

It is now widely recognized that plant growth is directly related to the water balance in plant tissue. This balance is determined by the rate of water uptake and water loss (transpiration).

Plant-water Indicators and Measurements

There are several methods of appraising water need by the plant.

Plant Color. A practical guide for irrigation requirement of some crops is color change. Leaves of beans *(Phaseolus vulgaris)*, cotton *(Gossypium hirsutum)*, and peanuts *(Arachis hypogaea)* become dark green as the available soil water is depleted and moisture stress begins to affect the plant.

Plant Movement or Elongation. Corn and sorghum leaves curl or change their angle of orientation when soil moisture decreases.

The data in Table 2 present the effect of soil-water potential in the main root zone on corn leaf elongation. The daily rate of leaf elongation decreases considerably when soil-water potential reaches 1.4 atm, and indicates the need for irrigation.

Table 2. Daily elongation rate of corn leaves at different soil-water potentials, corn irrigation experiment at GILAT, 1956 (BIELORAI *et al.*)

Age of plants (days)	Number of days after irrigation	Soil-water potential (bars) in the root zone (0–60 cm)	Leaf elongation rate (cm/day)
50–59	0–9	0.2–0.4	5.9
	20–29	1.4–5.3	3.1

These measurements must be carried out by skilled personnel, and are thus impractical for routine use by farmers.

Fruit Growth. FURR and TAYLOR (1938) demonstrated a positive correlation between enlargement of lemon fruits and availability of water. OPPENHEIMER and ELZE(1941) recommended irrigation of orange trees whenever the daily increment in fruit circumference falls below 0.2–0.3 mm during the summer. Similar observations have been made for apples. However, fruit growth is also influenced by rapidly fluctuating conditions of humidity and temperature that accelerate water loss by the fruit and interfere with soil-moisture measurements.

Stem and Trunk Growth. Stem and trunk growth are affected by water stress, as demonstrated by VAADIA and CAJIMATIS (1961) for grapes, by CLEMENTS *et al.* (1952) for sugar cane, by MARANI and HORVITZ (1963) for cotton, and by BIELORAI and LEVY (1971) for citrus.

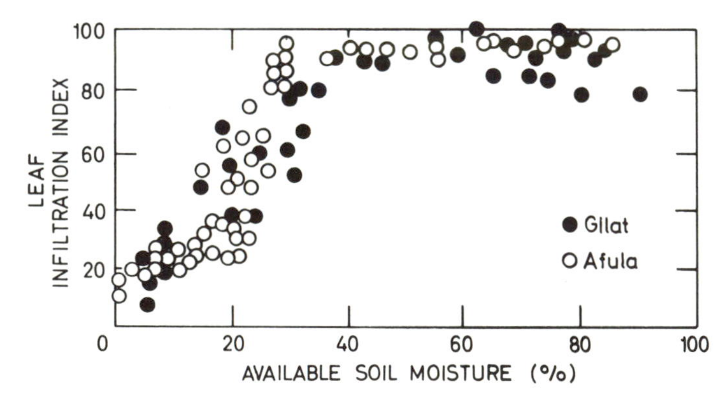

Fig. 6. Relation between stomatal aperture and available soil moisture (leaf infiltration index). (Adapted from OPHIR *et al.*, Exp. Agric. 4, 325–333 (1968))

The development of recording dendrometers for measurement of trunk growth gives some hope for using the method in determining irrigation schedules, although climatological fluctuations interfere with finding precise correlations between trunk growth and soil-water status (see Chap. V, Contr. 4).

Stomatal Movement. Several workers have proved the existence of a close relationship between stomatal aperture and the availability of soil moisture. The data presented in Figs. 6 and 7 show that the need to irrigate cotton is indicated whenever a sharp decrease of stomatal aperture is observed, and that this occurs when the available moisture in the mean root zone drops to 20–30%. When irrigating a field, information is required on both the timing of irrigation and the amount of water needed. It seems that the combination of physiological measurements (stomatal aperture indices) and meteorological indices can provide a fair estimation of both the water requirement and the timing of irrigation.

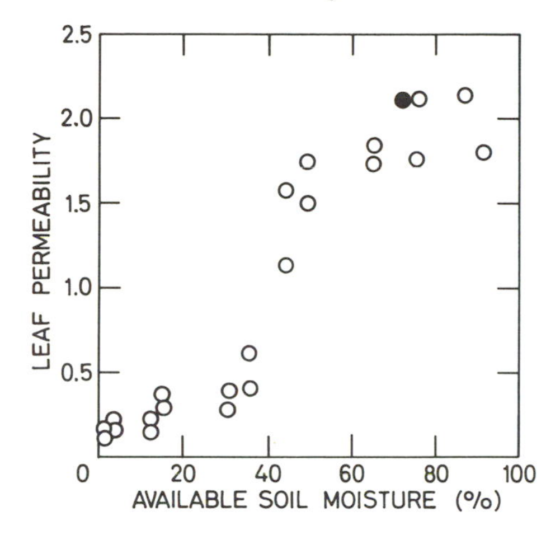

Fig. 7. Relation between stomatal aperture (expressed by leaf permeability) and available soil moisture (porometry). (Adapted from OHPIR *et al.*, Ex. Agric. 4, 325–333 (1968))

Rooting Habits of Different Crops

Although plant roots absorb water freely from soils at field capacity, rooting habits vary according to species. The main root zone is concentrated in the upper 0–90 cm and about 80–90% of the water is extracted from the upper layers. Alfalfa is known as a deep-rooted crop, reaching 300 cm in depth. Onion has shallow roots (30–40 cm), and potatoes have a sparse root system. Roots of fruit trees extend down to 150 cm. Shallow soils restrict root growth, and most of the roots concentrate in the upper layer.

Young plants that only partly cover the ground have very small root systems. Irrigation prior to seeding will provide a certain moisture reserve for later root development. Plant roots will grow deeper in soils that are kept moist but not wet; lack of aeration in wet soils will hinder root development. The growth of roots ans shoots is retarded when moisture is deficient at any stage.

Interaction of Fertility and Soil Water

The effects of frequency of irrigation and amount of water applied are affected by the fertility of soils. Unless enough nutrients are available to the plants to produce maximum growth, there is no advantage to maintaining available soil water at low tension. The most desirable frequency of irrigation can have maximum benefit on the crop only if a supply of readily available nitrogen is present in the soil. An example is presented soil in Fig. 8 of the relation between consumptive use and kernel yield of the corn at different nitrogen levels.

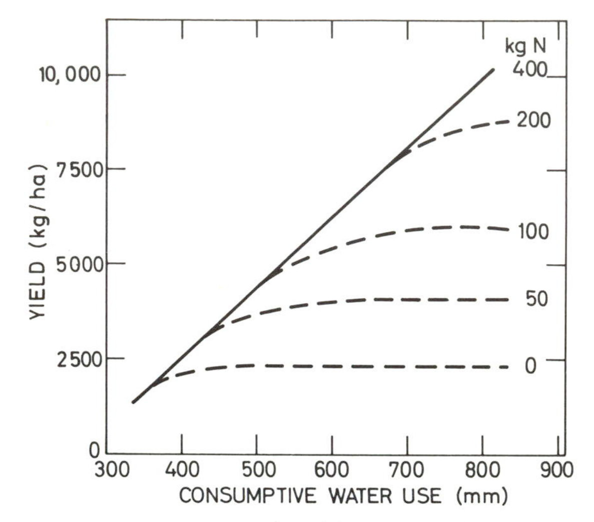

Fig. 8. Relation between consumptive water use and kernel yield of corn at five nitrogen fertilizer levels. (Adapted from Shimshi, Ph.D thesis submitted to the Hebrew University, Jerusalem, 1967)

Practical Considerations in Scheduling Irrigation

a) Instruments and other means of indicating the irrigation need are most useful in water-deficient areas where water is expensive and where high-value crops are grown. Crops with established root systems (such as orchards) are suitable for the use of tensiometers and neutron scattering. Where qualified technical personnel is available, plant indicators can be used.

b) It is not always possible for a farmer to apply water at the precise time indicated, and therefore maximum flexibility in scheduling should be allowed in order to facilitate the irrigation of a large area. To avoid crop damage, the schedule should be planned in advance.

c) The proper irrigation method should be chosen in order to improve irrigation efficiency, and for water conservation. In areas where water is plentiful and is cheap in comparison with other production costs, there will be little incentive for a farmer to extend irrigation intervals or to improve irrigation efficiency, and he should be aware of the dangers of excessive irrigation. The efficiency of water use can be increased by reducing evaporation losses, mulching, and weed control, taking into account fertilizer requirements and using suitable irrigation equipment. The irrigation schedule is frequently influenced by labor costs, which may necessitate extension of the intervals between water applications. It is desirable that the final irrigation schedule be based on results of irrigation experiments.

The prediction of irrigation needs requires an anlysis of prevailing conditions, soil factors, crop characteristics, and microclimatic factors affecting evaporation and transpiration. Also to be considered are the plant, extent of ground cover, root depth, effect of water stress on yield, soil profile, infiltration, and drainage properties of the soil.

In any country or region where irrigation water is limited, the question of its efficient use arises. How should the water be distributed on the relatively abundant land area? How should priorities be assigned to the crops? Similar questions are encountered by every farm and farmer. Should the farm have part of its fields irrigated, leaving other parts unirrigated? Should the irrigation water be applied to a certain crop and the other crops grown with supplemental irrigation only?

The proper decision should be made on the basis of input and output relations in irrigation—namely, on the basis of economic return. In modern agriculture the problem of crop rotation is of minor importance, owing to the use of fertilizers, pesticides, and insecticides. This enables one to grow a crop continuously, and, as stated previously, the decision should be made on the basis of available water and economical return. The farmer can calculate the profitability of irrigation at a number of points of time, such as before planting the crop or at various stages of growth. The economically optimal irrigation intensity will also depend on whether land, water, and other resources are limited or unlimited. In this respect, alternative situations prevail:

Situation A—Water limited, land unlimited, other resources unlimited. In this situation optimal irrigation schedule is that in which the average net income is based on m^3 of water.

Situation B—Land limited, water limited, other resources unlimited. In this situation it pays to transfer water from a field or crop with low net income to a field or crop with high net output.

Wise use of the criteria discussed will result in a successful irrigation program.

Literature

Bielorai, H., Rubin, J.: A study of the irrigation requirement and consumptive water use of hybrid corn in the northern Negev. Spec. Bull. **6**, pp. 8–15. Agric. Res. Station, Rehovot, Israel (1957).

Bielorai, H., Rubin, J.: A study of the irrigation requirements and consumptive water use by sugar beets in the northern Negev. Spec. Bull. **6**, pp. 20–24. Agric. Res. Station, Rehovot, Israel (1957).

Bielorai, H., Levy, J.: Irrigation regimes in a semi-arid area and their effects on grapefruit yield, water use and soil salinity. Israel J. Agric. Res. **21**, 3–12 (1971).

Briggs, L. Y., Shantz, H. L.: A wax seal method for determining the lower limit of available soil moisture. Bot. Gaz. **53**, 229–235 (1911).

Clements, H. F., Shigeura, C., Acamine, E. K.: Factors affecting the growth of sugarcane. Univ. Hawaii Agr. Exp. Sta. Tech. Bull. **18**, 90 pp. (1952).

Furr, J. R., Taylor, S. A.: The growth of lemon fruits in relation to the moisture content of the soil. USDA Tech. Bull. **640**, pp. 1–71 (1938).

Hendrickson, A. H., Veihmeyer, F. J.: Irrigation experiments with peaches in California. Calif. Agr. Exp. Sta. Bull. **479** (1929).

Israelsen, O. W. West, F. L.: Water holding capacity of irrigated soils. Utah Agr. Exp. Sta. Bull. **183** (1922).

Mantell, A., Goel, A.: The response of Shamouti orange trees to irrigation frequency and depth of wetting. Prelim. Rep. 709, The Volcani Inst. Agric. Res., Bet-Dagan, Israel (1970).

Stolzy, L. H., Taylor, O. C., Garber, M. J., Lombard, P. B.: Previous treatments as factors in subsequent irrigation level studies in orange production. Amer. Hort. Soc. Proc. **82**, 199–203 (1963).

Taylor, S. A.: Use of mean soil moisture tension to evaluate the effect of soil moisture on crop yields. Soil Sci. **74**, 217–226 (1952).

TAYLOR, S. A., HADDOCK, Y. L.: Soil moisture availability related to power required to remove water. Soil Sci. Soc. Amer. Proc. **20**, 284–288 (1956).

TAYLOR, S. A., HADDOCK, Y. L., PEDERSEN, M. W.: Alfalfa irrigation for maximum seed production. Agrion. J. **51**, 357–360 (1959).

VAADIA, Y., KASIMATIS, A. N.: Vineyard irrigation trials. Amer. J. Enol. Viticult. **12**, 88–98 (1961).

VITTUM, M. T., ALDERFER, R. B., JANES, B. E., REYNOLDS, C. W., STRUCHTENMEYER, R. A.: Soil-plant-water relations as a basis for irrigation. New-York Agric. Exp. Sta. (Geneva) Bull. **800**, 66 pp. (1963).

2

Irrigation of Field Crops

D. SHIMSHI, H. BIELORAI, and A. MANTELL

The successful production of field crops in arid regions is entirely dependent on the use of a sound irrigation program based on a knowledge of the effect of soil-moisture stress on the plant at different stages of its development. This is particularly critical when the crop is grown during that period of the year when there is no rainfall, or when rainfall anticipated during the growing season is not forthcoming.

The plants included in this chapter–wheat, cotton, and peanuts–represent the three most important cereal, fiber, and annual legume crops, respectively, grown in Israel. During the past decade much effort has been devoted to the investigation of their response to irrigation practice.

The Irrigation of Wheat (D. SHIMSHI)

General Considerations

Wheat is not usually considered a typical irrigated crop, although in some countries (notably Pakistan) it is grown almost exclusively under irrigation. Over much of the temperate zones wheat is grown under conditions where rainfall is sufficient for satisfactory production.

In Israel, as in many countries with a Mediterranean-type climate, spring-type wheat is grown during the mild, rainy winter, maturing at the beginning of the dry summer. Thus the life cycle is generally completed under adequate water supply. However, as one moves toward the more arid parts of the country (south or east), the amount and reliability of the winter rains diminish to a point that wheat growing becomes a risky enterprise; failure due to drought is a common occurrence in the arid regions of Israel (Negev and Bet She'an Valley), and occasionally drought conditions extend over the more humid parts of the country, resulting in widespread reduction of wheat yields.

Much work has been carried out on the water requirements of wheat, especially in the U. S. However, a large part of this effort relates to the relation between wheat yield and the moisture reserve in the root zone at the beginning of the growing season. These studies are quite inapplicable to Mediterranean conditions, where wheat is planted in the autumn on a completely dry soil, and the very initiation of its life cycle depends on the early winter rains.

The practice of irrigation of wheat is aimed to supplement the natural rainfall, so that both rainfall and irrigation ensure a favorable moisture regime in the soil conducive to high yield. Since the rainfall pattern varies (sometimes drastically) from year to year in arid regions, it follows that the optimum irrigation practices will differ accordingly. It is quite conceivable that in a relatively moist year even a single irrigation may be superfluous, whereas in a dry year, two or three irrigations may be required.

The yield of wheat does not depend solely on water supply. Other factors of production interact with its reponse to water, such as soil fertility, nutrient supply, and the yield potential of any particular wheat variety.

In Israel, the interest in wheat irrigation has lately increased because of several reasons. The standard tall variety (F.A. 8193), which has been the principal variety grown, has a limited yield potential because improved moisture and nutrient supply cause lodging (the falling over of the stems); with the introduction and breeding of dwarf and semidwarf varieties, the response has increased, sometimes doubling the yields. Irrigation networks, originally installed for typical irrigated crops such as cotton and vegetables, have become available for wheat production as a part of the crop rotation. The decrease in wheat surpluses in the U. S. and the necessity of buying wheat at world market prices has encouraged the trend toward self-sufficiency in wheat in Israel; since most of the arable land in the rainy regions is already devoted to other crops, this proposed increase in production will come mainly from the drier regions, through the application of supplemental irrigation.

Example of a Wheat Irrigation Experiment

A series of irrigation experiments has been conducted since 1954 at Gilat in the Negev region. At first the experiments were carried out with the standard tall variety (F. A. 8193), but lately a dwarf variety (N. 46) has also been tested. During nine seasons of experimentation both rainfall quantity and distribution have varied widely, and the results therefore represent the response of wheat to a combination of several irrigation practices and several rainfall patterns. Following is an example of such an experiment.

Wheat of the two varieties (F.A. 8193 and N. 46) was planted at the end of November. Some of the treatments were irrigated immediately after planting, and some at later stages. Since rain fell at about the time of the early irrigation, germination took place simultaneously in both the irrigated and unirrigated plots. The rainfall pattern was characterized by heavy rains during early winter (November–January), with a dry period during February and March, and some rain in April. The total winter rainfall was 318 mm. The irrigation treatments and the yields are presented in Table 1.

Table 1. Wheat irrigation experiments at Gilat, 1967–1968

Treatment no.	Irrigations date	Amount (mm)	Total (mm)	Grain yields (kg/ha) F.A. 8193	N. 46
1	–	–	–	1,840	2,620
2	8/12	86	86	3,360	4,820
3	8/12	86	224	4,200	6,190
	17/3	138			
4	25/2	90	90	2,790	3,840
5	25/2	90	200	3,470	4,700
	9/4	110			
6	13/3	151	151	3,600	4,630

The soil-moisture regime in the root zone (down to 1.50 m) was followed by means of a neutron probe. The soil-moisture data served for the determination of the amounts of the individual irrigations and also for the computation of evapotranspiration. Although the soil-moisture regime strongly affected the rate of water use, there was no difference in evapotranspiration (for any given irrigation treatment) between the tall and the dwarf variety. This result points out an interesting fact: The higher yields obtained with the dwarf variety did not necessitate a greater supply of water; and, conversely, the lower vegetative stature of the dwarf variety did not transpire less than that of the tall variety.

Timing of Irrigation

Some of the conclusions drawn from such experiments over several years are that among the different timings of irrigation, the early irrigation (applied immediately after planting) usually is the most efficient, resulting in the highest marginal response, even in relatively rainy years, and even in years with early rains. The marginal response ranged between 4 kg/ha/mm in a relatively rainy year, to 24–25 kg/ha/mm in dry years, where it spelled the difference between a complete failure (320 kg/ha) and a satisfactory yield (3910 kg/ha). (At present prices of wheat and irrigation water, a response of about 3 kg/ha/mm is the break-even point.) Once such an irrigation of 150 mm is applied on a dry soil, the moisture reserve built up in the soil (usually to a depth of 70–90 cm) is sufficient to carry the wheat plants for a period of about 2 months, even with little rainfall; by that time sufficient rainfall usually occurs so that the moisture supply is assured at least until heading time (beginning of March). Much of the beneficial effect of the early irrigation may be attributed to the fact that it causes early germination and development, so that the plants enter the coldest month of winter (January) when they are robust and well developed; in years when the early rains are delayed, late-germinating plants remain stunted during the winter, with subsequent reduction in performance until maturation. Over most of the experiment years early irrigation accounted for at least 2/3 of the response amplitude between the unirrigated treatment and the maximum-yield treatment.

The efficiency of later irrigation varies according to the rainfall pattern. Sometimes a late irrigation applied during a dry spell at the end of winter may be almost as effective as an early irrigation, provided a heavy rain immediately following this irrigation does not make it unnecessary. Other times late irrigations may cause only a small increase in total yield, but may considerably improve the quality of grain (weight of 1000 kernels), which may be just as important from the economic aspect.

It has been repeatedly stressed in the literature that certain stages of wheat development, notably booting and heading, are more sensitive to moisture stress. This was not the case in the present series of experiments. It was found that moisture stress, even during early stages (tillering and stem elongation), can cause damage that can only be partially remedied by a subsequent, favorable moisture regime. On the other hand, even at the soft-dough stage, after heading, a drought may cause spikelet abortion and grain shriveling, whereas a late irrigation may increase kernel weight.

Although wheat may develop roots to a depth of 2 m in deeply wetted soils, about 80% of the moisture extraction, under conditions of satisfactory yields, comes from the 0–100-cm layer. In Israel wheat is grown on medium- to fine-textured soils, which have a capacity of 120–150 mm available moisture in this 1-m layer. The critical threshold of available moisture reserve below which wheat suffers from water stress has not been sharply defined, but it appears that this threshold occurs at an available moisture deficit of about 90–110 mm in the upper 1 m; that is, after the utilization of $^{2}/_{3}$ to $^{3}/_{4}$ of the available moisture. However, under conditions of small, frequent rains, wheat may grow well even though the *average* moisture of the upper meter is quite low.

Under conditions where moisture does not limit wheat development, the daily evapotranspiration is approximately as follows (in mm/day): December, 1.5; January, 2.0; February, 3.0; March, 3.0; April, 2.5; half of May (maturation), 1.0. Prolonged drought and low soil moisture will result in much lower rates. Excessive irrigation or rainfall, especially under high evaporative conditions, will result in higher rates. The seasonal evapotranspiration under a well-balanced moisture regime (without drought or excessive wetness) is thus about 370–400 mm, probably irrespective of yield potential of the wheat varieties. The effect of nutrient supply on evapotranspiration in wheat is not yet clear.

Practical Considerations

In essence, the supplemental irrigation of wheat involves a series of decisions by the farmer, based on the certainty of events (rains) occurring prior to the decision, and on the uncertain expectation, or probability, of events that might occur after the decision. The element of probability requires a long-term analysis of the expected results of any irrigation schedule superimposed on the variable rainfall pattern of any particular location. A computer model for this analysis is now being developed on the basis of this series of irrigation experiments. The development of this model proceeds along the following stages:

1) Establishing a function describing the water use of wheat as determined by soil moisture and evaporative demand.

2) Determining the soil-moisture fluctuations caused by water use, rainfall, and irrigation.

3) Establishing a response function relating plant yield to an integrated value of soil moisture throughout the growing season.

4) Predicting the expected yield (or income) resulting from the application of several irrigation practices during the years for which there are rainfall records at a particular location.

5) Analyzing (by economic criteria) the long-term probability of success or failure of any particular irrigation schedule at the location.

Such a model may extend the conclusions drawn from the experiments to a wide area, under conditions different from those of the experimental site, and may assist in formulating a clear policy for planning the allocation of water for wheat irrigation.

It should again be emphasized that irrigation is but one of the production factors, and that attempts to rely solely on it as a means of obtaining high yields of wheat in arid regions are bound to cause disappointment, unless other factors, such as nutrient supply, improved varieties, disease and pest control, etc., are properly applied.

Literature

ROBINS, J. S., MUSICK, J. T., FINFROCK, D. C., RHOADES, H. F.: Irrigation of principal crops: grain and field crops. In: Irrigation of Agricultural Lands. HAGAN, R. M., HAISE, H. R., EDMINSTER, T. W. (Eds.). Amer. Soc. Agron., Madison, Wis. (1967).

SHIMSHI, D., GAIRON, S., RUBIN, J.: Experiments on winter cereals in the Northern Negev (in Hebrew). Nat. Univ. Inst. Agr. Prelim. Rep. **419** (1963).
SHIMSHI, D.: Supplemental irrigation of wheat 1967–68 (in Hebrew). Progress Rep., Div. of Irrigation, Volcani Inst. Agric. Res. (mimeo.) (1968).

The Irrigation of Cotton (H. BIELORAI)

General Considerations

Cotton is grown in areas that differ widely in temperature, elevation, rainfall, soils and length of growing season. The two most important species in the cotton industry are *Gossypium hirsutum* (upland cotton) and *Gossypium barbadense* (long-fiber cotton).

Upland cotton (Acala varieties) is grown in the cotton belts of Africa (except Egypt, where the Pima varieties belonging to *G. barbadense* prevail), North America, and South America (mainly in Brazil and Peru). These types do not grow in Asia (U. S. S. R.), China, and India; the main species grown in this part of the word are *G. herbareum* and *G. arborum.*

Cotton is grown in medium and heavy soils (sandy loam, loam and clay). In the United States and Minor and Central Asia alluvial soils or chernozems are used. The planting date is determined by the soil temperature, which should be no lower than 17–18° C. Therefore, the growing period begins in April and ends in October.

Most of the cotton is grown under irrigation. The actual quantity of water used by the plant during the growing season can be expected to vary between 750 mm and 1200 mm. Though the consumptive use may differ in different areas (depending on climatological factors), the water use during the growing season follows similar trends in all locations. Evapotranspiration is low early in the season when leaf area is small, root systems shallow, and temperatures low. The root development of an irrigated cotton plant is presented schematically in Fig. 1.

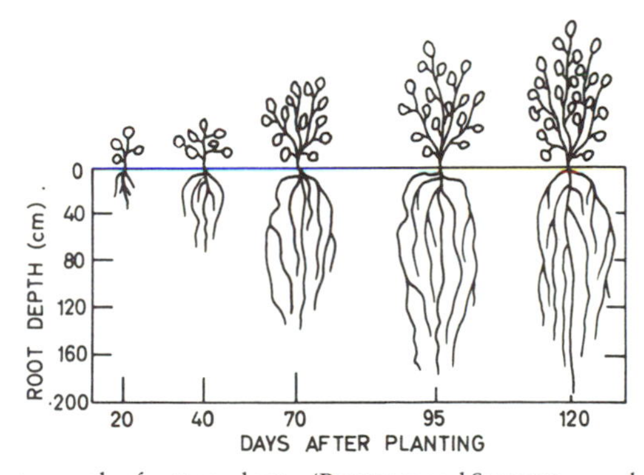

Fig. 1. Root growth of cotton plants. (BIELORAI and SHIMSHI, unpublished data)

ERIE (1963) reported that the mean consumptive use of water by cotton in Arizona for the 9-year period 1954–1962 varied from about 700 mm to 1250 mm. Experiments conducted in the Hula Valley in Israel by LEVIN and SHMUELI in 1962–1963 showed that for maximum production 5–6 irrigations were required, in addition to soil moisture stored from winter rains. Total consumptive use was 670 mm. In humid regions (such as Alabama with 700-mm rainfall), the need for supplemental water depends on the duration and frequency of summer drought. The evapotranspiration reaches peak values in the middle of the summer when leaf area is maximum, root systems are deep, temperatures are high, and humidity is relatively low. Consumptive use declines late in the season, after boll setting and when plants are mature.

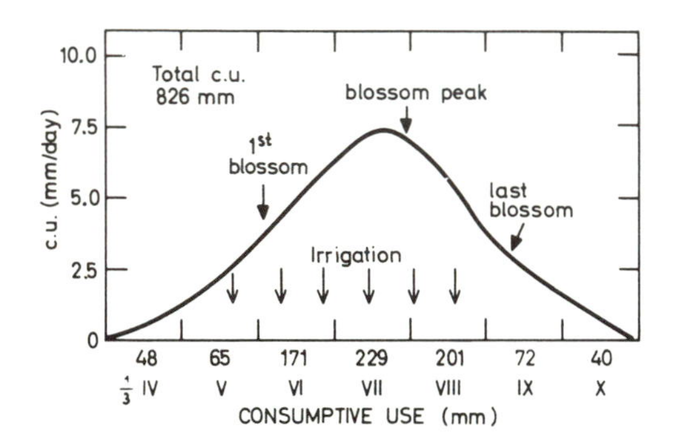

Fig. 2. Evapotranspiration rates of irrigated cotton at Gilat, Israel. (BIELORAI and SHIMSHI, 1963)

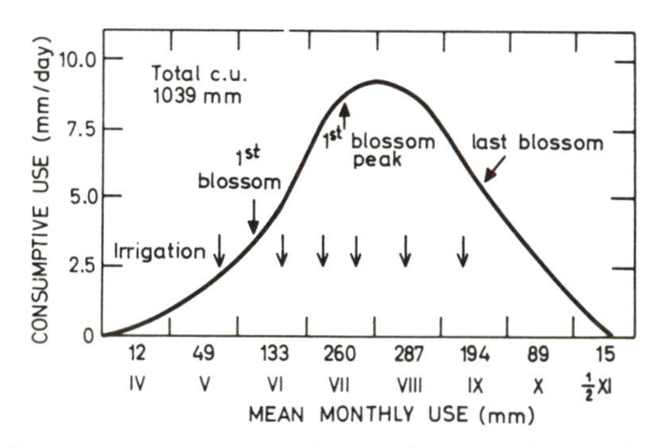

Fig. 3. Evapotranspiration rates of irrigated cotton at Tempe, Ariz. (ERIE, 1963)

Figs. 2 and 3 describe the sequence of water use during the growing season on the Arizona mesa and at Gilat in the northern Negev of Israel. It can be seen that the general shapes of the two curves presenting the sequence of seasonal water use in both locations are similar, though the rates of evapotranspiration are different, depending on the climatic conditions (temperature, relative humidity, and others). Also, the irrigation season in Israel is shorter than in Arizona. Therefore, the seasonal consumptive water use is 800 mm in Israel and above 1000 mm in Arizona. The peak monthly water use in Gilat, Israel, is in July and in Arizona is July–August.

Timing of Irrigations

In many cotton-producing regions of the world, winter rainfall is not sufficient to wet the soil to the desired depth, nor to leach salts from the potential root zone. In such areas it is recommended that the soil be wet to a depth of at least 180 cm with a preplanting irrigation. This irrigation leaches salts (if present), provides moisture for seedling emergence and early growth, and stores moisture in the root zone. The moisture reserve in the lower layers will be available for use later in the growing season when deficiencies occur in the upper soil layers.

Early-season Irrigation. Numerous research workers have shown a high positive correlation between available soil water and total vegetative growth, provided that other factors (fertility, light, and temperature) are not limiting. They suggested that vegetative growth and moisture supply are directly related to the formation of fruiting branches and to the final seed cotton yield. Experiments carried out in Israel show that the first irrigation can be delayed until the beginning of flowering, if the depletion of the available water in the main root zone does not exceed 50%. This means, in practice, that the first irrigation can be delayed until 60–65 days after planting. Contrary to findings of other workers, Israel irrigationists claim that excessive vegetative growth may cause boll loss due to shedding.

Midseason Irrigation. The timing of irrigation is critical during the main flowering and fruiting period, which is also the time of maximum water consumption. Roots have attained near-maximum depth, leaf area is greatest, and air temperatures are highest. Experiments in Yuma, Ariz., suggested that during this period irrigations at 14- to 21-day intervals are advisable. Experiments in Israel by BIELORAI and SHIMSHI (1963) showed that midseason irrigation should be applied when about 65–70% of available moisture in the main root zone is extracted.

Late-Season Irrigation. Consumptive use declines late in the season as a result of limited new growth, cooler weather, and maturation of the crop. Moisture deficits at this time are not harmful, and may even have a beneficial effect by retarding vegetative growth without causing measurable decreases in seed or lint yields. On the basis of these observations, experiments carried out by BIELORAI and SHIMSHI and others showed that it is possible to shorten the irrigation season, applying the last irrigation in the middle of August, as opposed to the previously accepted practice of applying the last irrigation in the middle of September.

In areas characterized by long, hot growing seasons and warm fall weather (Arizona, California, and Bet She'an Valley in Israel), growers may obtain an appreciable boll set in late August and September, and so late irrigations are applied.

Relation of Soil Moisture to Boll and Fiber Properties. The more important fiber properties–length, strength, and fineness–are primarily controlled by genetic factors and to a lesser extent by soil moisture and fertilizers. Results regarding the effect of the soil-moisture regime on fiber properties are controversial. Boll size is affected by the moisture status, and, as previously mentioned, the midseason moisture regime will determine the size and weight of the bolls.

Efficiency of Water Use

It has been emphasized previously that total vegetative growth of the cotton plant is largely proportional to the availability of soil moisture. Controlled applications of irrigation water are the tools by which vegetative growth can be encouraged or suppressed. A clear distinction should be drawn, however, between the relation of moisture to *growth* and the efficiency of water use in *production of marketable yield.*

Water-use efficiency: This is defined as the weight of lint produced per unit of water applied. Experiments carried out in the western Negev during three consecutive years (1964–1966) can serve as an example of scheduling an irrigation program for cotton, where water-use efficiency was the main goal. Results are presented in Table 2.

Table 2. Effect of irrigation frequency on water use and yield of cotton. (From AMIR and BIELORAI)[a]

Treatment	date of irrigation after planting (days)	Consumptive use (mm)	Water applied (mm)	Yield (kg/ha) seed cotton	lint	Relative yield (%)	Water-use efficiency, (kg cotton/mm water)
1	–	237	—	1,880	760	39	0.79
2	70	348	110	2,770	1,120	58	0.80
3	70, 92	445	208	3,830	1,500	80	0.86
4	70, 92, 104	535	208	4,750	1,870	100	0.89
5	67, 88, 109, 125	610	373	4,800	1,860	100	0.79

[a] Date of planting April 20.
There was a 10% variation in seed-cotton yield between years. The results indicated that maximum yield was achieved with three irrigations per season.

The results of a 3-year cotton irrigation experiment conducted in the western Negev of Israel indicated that the highest water-use efficiency (0.89 kg cotton/mm water) was obtained with three irrigations (treatment 4), whereas in the other treatments the value was about 0.80.

The additional irrigation in treatment 5 did not increase yield or water-use efficiency; nor was there a significant increase in the number of bolls, although there was a slight increase in weight. In one of the three years there was a decrease in yield in this treatment.

Table 3. Irrigation effects on quality of cotton yields (treatments refer to those of Table 2)

Treatment	Boll weight (g)	Number of bolls/m^2
1	7.2	26
2	7.2	37
3	7.6	52
4	8.2	54
5	8.5	55

No significant effect on lint quality was recorded

Methods Used to Schedule Irrigation of Cotton

Most cotton growers prefer to irrigate by the so-called calendar method when prior experience for a given area is available. Irrigation schedules often rely on irrigation experiments in which the response of cotton yield to different moisture regimes is tested. The results thus obtained help in working out a proper irrigation schedule, taking into account the critical stages of plant development (flowering and boll set). To estimate the irrigation water requirement of the cotton crop, evaporation data from Class A pan may be used (FUCHS and STANHILL, 1963), providing the relation between actual water use and potential evapotranspiration is known. To establish efficient schedules, use should be made of information on crop-rooting depth, water-retaining capacity of the soil, and the evapotranspiration rates for the different stages of the growing season. Visible plant-water symptoms, such as wilting or stem elongation in the flowering stage, can also be useful in some situations. Irrigation should be applied when stem elongation ceases.

As a general rule, the first irrigation before flowering can be postponed until 50% of the available water is depleted. In the critical periods (flowering and fruiting) irrigation should be continued until all the bolls have set that are expected to mature. The time required for a boll to mature after setting is about 40 days. Late irrigations in the fall usually cause undesirable vegetative growth.

To derive maximum benefit from his irrigation investment, the farmer should use good seeds of well-adapted varieties, fertilize properly, control weeds and insects, and employ other improved management practices.

Literature

AMIR, J., BIELORAI, H.: The influence of various soil moisture regimes on the yield and quality of cotton in arid zones. J. Agric. Sci. **73**, 425–429 (1969).

BIELORAI, H., SHIMSHI, D.: The influence of the depth of wetting and the shortening of the irrigation season on the water consumption and yields of irrigated cotton. Isreal. J. Agric. Res. **13**, 55–62 (1963).

ERIE, L. J.: Irrigation management for optimum cotton production. Cotton Gin and Oil Mill Press **64**, 30–32 (1963).

FUCHS, M., STANHILL, G.: The use of Class A evaporation pan data to estimate the irrigation water requirement of the cotton crop. Israel J. Agric. Res. **13**, 63–78 (1963).

The Irrigation of Peanuts (A. MANTELL)

General Considerations

Peanuts or groundnuts *(Arachis hypogaea)* are one of the leading agricultural crops of the world for the production of oil and plant protein. World production reached 15.0 million tons in 1968 from a total area of 17.6 million hectares, with yields of 0.6–3.0 ton/ha. One-third of the yield was in India. The four other most important producers are mainland China, Nigeria, Senegal, and the United States. Actually, the crop is widely grown on all the continents wherever there are favorable climatic conditions—relatively high temperatures (above 22° C) and considerable sunlight. Development seems to be unaffected by day length. In Israel, approximately 3.000 ha were planted in 1968. Average yields are 3.5–4.0 ton/ha, although successful growers have obtained up to 6 ton/ha. A recently developed local variety, Shulamit, has a potential yield exceeding 7 ton/ha.

High yields are obtained on deep, well-drained, light sandy soils. Such soils encourage early flowering, easy penetration of pegs, and rapid growth of the pods, and also facilitate removal of the pods at harvest. In finer-textured soils that tend to form surface crusts when wetted there may be a problem of peg entry into the soil, resulting in lower yields.

In many peanut-producing areas rainfall occurs during the growing season, and in times of subnormal precipitation supplementary irrigation may occasionally be used to replenish depleted water reserves in the soil. However, in arid regions, the crop can succeed only with a full irrigation schedule, and in order to determine this, one must be acquainted with the plant's growing habits and its response to various soil-moisture conditions. There have been few such investigations throughout the world, and irrigation research with regard to peanuts has not kept abreast with that for other plants.

Plant Development and Response to Soil Water

The total growing season from planting to harvest is 125–155 days, depending on variety. The plant has a long flowering period, extending from about 5 weeks after planting until the end of the growing season. Height of flowering occurs between 60 and 90 days of age. Pods developing from flowers that appear after this time do not reach full maturity. About 65 days normally elapse from flowering to pod ripening.

The plant develops a taproot system with numerous lateral branches. Growth in depth is rapid, and can reach 180 cm in light, sandy soil. The main root zone is located in the 0–90-cm soil layer. Root activity in the soil profile can be characterized by the amounts of water extracted from different depths. The contribution of each soil layer to the total consumptive use during July and August, when the root system is already fully developed and at the peak of its activity, is illustrated in Fig. 4. It is seen that the greatest extraction occurs in the upper 90 cm of the soil profile, and also that frequency of irrigation has a marked effect on the pattern of extraction.

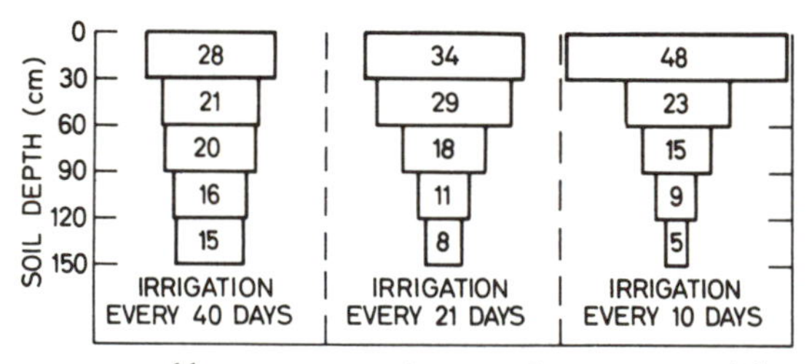

Fig. 4. Amount of water extracted by peanut roots (expressed as a percent of the total water removed) from various soil depths during July–August. (MANTELL and GOLDIN, 1964)

Three main stages of development are recognized in the peanut plant: (1) from planting until the beginning of flowering; (2) the flowering period, and (3) from beginning of pod growth until harvest.

Various investigations have been conducted to determine the effect of soil-water stress on growth and yield when it is imposed at one or more of the above-mentioned developmental periods. It has been found that for maximal yields, there must be an adequate supply of soil moisture during flowering and pod development. A study in the Soviet Union has shown that the yield of plants well supplied with water only during flowering was equal to that of plants well supplied with water at all stages of growth. Also, at flowering the plants were particularly sensitive to a soil-water shortage. There were indications that yield may not be affected by wide variations in soil water between germination and flowering. Other workers have pointed out that toward the end of the growing season there is less danger of damage by slight water stress. In Formosa, the most critical period regarding soil-moisture deficiency was found to be that from peak flowering to early fruiting.

When and How Much to Irrigate

The planning of a suitable irrigation program must obviously be based on the above-mentioned considerations. Other factors such as soil type and moisture reserves in the root zone should also be taken into account.

If rainfall has not already wetted the potential root zone to a depth of 120–150 cm before the growing season, a preplanting irrigation should be given to make up the deficit and thus guarantee an ample supply of moisture in the lower soil layers for satisfactory root development.

Immediately after planting, a light irrigation is given to aid germination. A second, light irrigation may be useful to help seedling emergence in soils that form a surface crust.

In medium- to fine-textured soils, the first irrigation after emergence is generally applied 35–45 days after seeding when about 40% of the available water is left in the root zone. With coarse-textured soils having a lower water-holding capacity, it may be necessary to irrigate somewhat earlier.

The succeeding irrigations are given at intervals of about 14–20 days, depending on soil type (in very sandy soils, intervals of only seven days between applications may be necessary). The level of available water in the upper 90 cm of soil should be maintained above 65%. Since the peanut plant is especially sensitive to aeration and excess water, irrigations should not be more frequent than necessary, or excessive in amount.

The final irrigation in sandy soils is generally given 8–10 days prior to the anticipated harvest date, and in finer-textured soils from 14–30 days. The exact date and amount depend on harvest date, soil physical conditions, amount of available water extracted in the different soil layers, and degree of pod ripening. Excess water in the soil during the ripening stage of the pods may cause rotting, discolored shells, and detachment of the pods from the plants during harvest.

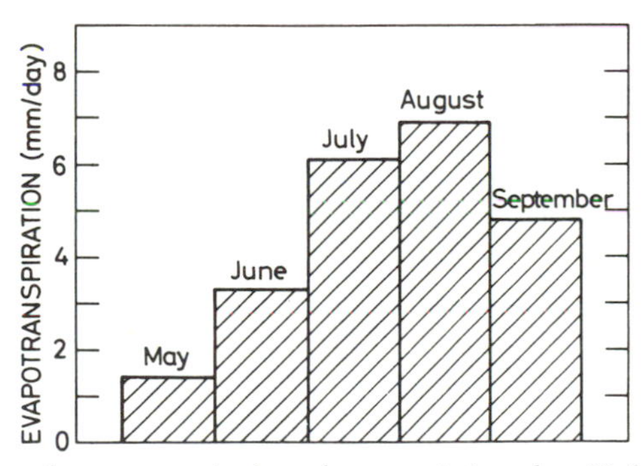

Fig. 5. Average daily rate of evapotranspiration of peanuts irrigated at 30-day intervals. (MANTELL and GOLDIN, 1964)

For best growth and yield of peanuts in Oklahoma (MATLOCK *et al.*, 1961), a seasonal application of 625 mm is required. In Israel, MANTELL and GOLDIN (1964) investigated the response of peanuts in the central coastal plain to five irrigation frequencies (10, 14, 21, 30, and 40 days) during two seasons on fine-textured sandy clay and clay loam soils. Optimal pod yield was obtained with five irrigations to a depth of 90 cm during the growing season, at a frequency of 30 days and with a seasonal water application of approximately 550 mm. Indications were obtained that it may suffice to replenish, at each irrigation, the deficit from field capacity only in the 0–60-cm layer, provided the entire root zone is at field capacity at the beginning of the growing season. The evapotranspiration rate of the optimal treatment (Fig. 5) reached a peak of almost 7 mm/day in August during the period of flowering and pod development.

In the arid northern Negev of Israel, the optimal yield of peanuts growing in a loess soil was obtained with a total water application of about 650 mm after planting (BIELORAI *et al.*, 1962). This amount was given in seven irrigations at two-week intervals, and was calculated to replenish the deficit from field capacity to a depth of 60 cm.

Effect of Sprinkling Intensity

The sensitivity of peanut plants to soil aeration, and the difficulty encountered by the pegs in penetrating surface crusts, prompted some studies in Israel to determine the possible beneficial effect of applying water at rates lower than those conventionally used. In the same experiment mentioned above, MANTELL and GOLDIN compared intensities of 6 and 12 mm/hr at different frequencies of irrigation. Although the effect on yield of unshelled pods was not consistent, there did seem to be some advantage to the low intensity. No differences were reflected, however, in shelling percentage, 1000-seed weight, or quality of the shells.

A similar study by BIELORAI *et al.* (1962) in a loess soil using the same intensities as above found no significant differences in yield or quality of the crop. However, GORNAT and GOLDBERG (1967) tested three intensities–3, 12, and 25 mm/hr–on plants growing in containers filled with sand, sandy loam, and clay soil. On the sand, sprinkler intensity had no significant effect on yield; but on the sandy loam, the yield obtained with the lowest sprinkler intensity was significantly higher than that with the other two intensities. The yield obtained on the clay soil with the highest application rate was lower than that obtained with the other two.

Irrigation of Inoculated Peanuts

The inoculation of soil with root-nodule bacteria has been found to greatly increase peanut yields. It has been claimed that the moisture content in the upper soil layer must be maintained at a relatively high level in order to facilitate nodulation and to guarantee favorable bacterial activity. However, SHIMSHI *et al.* (1967) irrigated peanuts at 7-, 14-, and 21-day intervals, and showed that the repeated drying of the upper soil layers during irrigation cycles did not inhibit nodulation. The placement of inoculum at a depth of 12 cm resulted in poorer nodulation than at a depth of 3–4 cm, despite the higher moisture content of the soil at the greater depth.

Literature

BIELORAI, H., MANTELL, A., RAWITZ, E., REISS, A., GOLDIN, E.: Irrigation of peanuts (in Hebrew). In: Research and Experiments in Peanuts. Israel Ministry of Agric. Ext. Authority Publ. **45**, 45–47, (1962).

GORNAT, B., GOLDBERG, D.: Effect of sprinkler irrigation intensity on the flowering and yield of peanuts grown in three different soils. Israel J. Agric. Res. **17**, 187–191 (1967).

MANTELL, A., GOLDIN, E.: The influence of irrigation frequency and intensity on the yield and quality of peanuts *(Arachis hypogaea)*. Israel J. Agric. Res. **14**, 203–210 (1964).

MATLOCK, R. S., GARTON, J. E., STONE, J. F.: Peanut irrigation studies in Oklahoma, 1956–1959. Bull. Okla. Agric. Expt. Sta. B-**580**. 19 pp. (1961).

SHIMSHI, D., SCHIFFMAN, J., KOST, Y., BIELORAI, H., ALPER, Y.: Effect of soil moisture regime on nodulation of inoculated peanuts. Agron. J. **59**, 397–400 (1967).

3

Irrigation of Vegetable Crops

D. SHIMSHI

Vegetable crops are widely grown under irrigation in arid and semiarid regions. Since these regions have a relatively mild winter climate, a great part of the vegetable production is aimed at export of out-of-season vegetables to cooler countries. In addition, several vegetables are grown on large areas for industrial processing, either as canned or frozen products.

Despite the wide areas of vegetables under irrigation, data available on irrigation requirements are rather sketchy and incomplete. The main reason for this is the great variety of truck crops, some of which are of relatively minor importance. Since each type of vegetable has its own specific irrigation requirements, it is not surprising that research on this subject has been comprehensive only for the more important crops such as tomatoes, cucumbers, and potatoes (the last have been often classified as field crop rather than as a truck crop). The irrigation of truck crops has been the subject of several reviews, of which that of VITTUM and FLOCKER (1967) is the most recent. It is interesting to note that much of the work mentioned in this review was concerned with the supplemental irrigation of vegetables in relatively cool or humid regions in the U.S. (New York, Connecticut, Wisconsin, Florida, Maryland). This underscores the fact that truck crops are usually more sensitive to water shortage than other crops and may require irrigation even in these regions. This sensitivity is probably attributable to two main reasons: (1) Most vegetable crops are shallow-rooted, although there are several exceptions, such as tomatoes, watermelons, artichokes, and asparagus. (2) The marketable product is usually the fresh fruit or tubers, or the vegetative portions of the plant; unlike dry-matter products (grain, fiber, sugar) the yield of these organs is more sensitive to water deficit.

Vegetables are important crops in Israel. Some are grown during the rainless summer under full irrigation; some are grown during the relatively rainy winter, but even they require some irrigation; and some are grown the year around, often under plastic cover or in greenhouses during the winter.

Although practically all the vegetables in Israel are irrigated to some extent, relatively little research has been done on the irrigation requirement of vegetables.

The Irrigation of Potatoes

Potatoes are grown in Israel during two main seasons: spring (planted in February, harvested in June) and autumn (planted in August, harvested in December). In both seasons irrigation is necessary. Spring potatoes are usually planted on a soil that has been wetted by the winter rains, but soon after germination the warm, dry season sets in, and irrigation is required to bring the crop to tuber production. Potatoes are rather shallow-rooted, most of the active roots being confined to a depth of about 60 cm; in addition, relatively frequent irrigations are required as the season advances into early summer in order to cool the soil around the tubers, since high soil temperatures usually result in *poorly filled* tubers. Experiments carried out by RUBIN and SHIMSHI (1962) in the arid Negev region indicated that spring potatoes should be irrigated at intervals of seven to ten days, the number of irrigations being ten to seven, respectively, and the total water consumption being about 450–500 mm. Unpublished work on autumn potatoes (SHIMSHI) indicates that such an irrigation frequency should also be practiced from germination until the beginning of the winter rains. A recent tendency, however, has been for a greater frequency of irrigation (every 3–4 days); to this end, the areas planted to potatoes are often equipped with a permanent and stationary sprinkler network. Such practice has resulted in somewhat higher yields, but difficulties are encountered with pest and disease control, since spraying equipment cannot be used when the soil is excessively wet.

The Irrigation of Tomatoes

The irrigation of tomatoes has been the subject of several investigations (AHARONI *et al.*, 1967, 1969). Since tomatoes for industrial processing are of the single-harvest type, their irrigation is of special interest; the practice has a marked influence not only on yield, but also on the pattern of fruit maturation, and on the processing quality (color, dry-matter content, acidity) of the fruit. In general, it has been found that tomatoes germinating on a deeply wetted soil utilize relatively little water until the beginning of flowering (age 45 days). From then on, they should be irrigated until about one month before harvest (age 80–90 days); within this period, the numer of irrigations varies according to the locality. In the humid coastal areas of Israel, two irrigations may be sufficient, and in the drier inland regions, four or five may be necessary. Applying water after the age of 90 days usually results in rotting of the early-maturing fruits. Too frequent irrigation cause slow maturation, so that at harvest time a part of the fruit is still immature while another part is deteriorating because of overripening. Water deficit during the main period of flowering (45–70 days) may cause excessive shedding of young fruit.

Although vegetables are usually irrigated by sprinkling (as are most crops in Israel), it appears that they respond favorably to the newly introduced method of trickle irrigation. In the extremely arid region of the Arava, where water is somewhat saline, trickle irrigation is practically the only suitable method of vegetable irrigation. It enables the maintenance of a continously moist root zone, and it obviates the wetting of the leaves, which results in leaf scorching (SHMUELI and GOLDBERG). Moreover, beneficial effects of trickle irrigation on tomatoes have also been obtained under less severe conditions. It has been suggested that the process results in lower incidence of fungal diseases, since the microclimate of the leaves wetted by sprinkling is more favorable to the pathogens, and the sprayed chemicals are not washed off the leaves by trickle irrigation. The yield of canning tomatoes in the Negev region was 68.0 ton/ha with trickle irrigation compared with 56.0 ton/ha with sprinkling, the amount and frequency of irrigation being equal under both methods (SHIMSHI, 1968). Furthermore, since trickle irrigation usually involves a partial wetting of the rooting volume, this method may require less water. These aspects are now under study.

Literature

AHARONI, A., ANGELCHIK, M., WEISS, M.: Irrigation practices of tomatoes for processing (in Hebrew). Israel Min. Agr. Extension Service (mimeo) (1967).

AHARONI, A., SAGI, S., YANAI, E., CAHANI, Y.: Irrigation practices of tomatoes for processing (in Hebrew). Israel Min. Agr. Extension Service (mimeo) (1969).

RUBIN, J., SHIMSHI, D.: Irrigation requirements of spring potatoes in the northern Negev (in Hebrew). Nat. Univ. Inst. Agr. Rep. 100, Bet-Dagan, Israel (1962).

SHIMSHI, D.: Experiments on trickle-irrigation of tomatoes for processing (in Hebrew). Progress Rep. Div. of Irrigation, Volcani Inst. Agric. Res. (mimeo) (1968).

VITTUM, M. T., FLOCKER, W. J.: Vegetable crops. In: Irrigation of Agricultural Lands. HAGAN, R. M., HAISE, H. R., EDMINSTER, T. W. (Eds.). Amer. Soc. Agron., Madision, Wis., pp. 674–685 (1967).

4

Irrigation of Forage Crops

A. MANTELL and H. BIELORAI

The effect of soil water on quality and yield of various forage crops has been the subject of a considerable amount of research in the past. The optimum irrigation schedule for a given crop depends on many factors, including rooting habits of the plant, water-holding capacity of the soil, peak water use during the growing season, and climatic factors such as the presence or absence of rainfall. The maintenance of a favorable soil-moisture regime is particularly critical in arid and semiarid regions when most or all of the growing season coincides with the rainless period of the year.

The two principal forage crops grown in Israel, alfalfa and corn, are very different from each other insofar as water consumption is concerned. The former is a permanently established crop with a consequently high annual water requirement, whereas the latter is an annual, short-lived crop requiring a relatively moderate seasonal water application.

Of all the forage plants, alfalfa probably has received the greatest amount of attention regarding irrigation research. The water relations of corn, a major cereal crop, have also been thoroughly investigated, and the findings are generally applicable when this plant is grown for forage rather than for its grain.

The principles of irrigation of these two crops are discussed in the following.

The Irrigation of Forage Corn (A. MANTELL)

Corn is a high-energy forage product, and probably the best silage crop in regions where water is available for irrigation. When the entire plant is used for fodder, as in silage, it excels all other forage crops in the average yield of dry matter and of digestible nutrients per unit area, even slightly surpassing alfalfa in this respect. Although corn is one of the three most important cereal crops in the world (after rice and wheat), as forage it is of major importance primarily in Europe and America. In Israel, the amount of forage corn grown has been steadily increasing in recent years. In 1967, a total area of 1000 ha was planted to corn, of which about 85% was produced for forage and silage. A normal yield of green fodder is about 70 ton/ha, although maximum yields of up to 110 ton/ha have been obtained. When used for this purpose, the plants are cut at the age of 80–90 days for varieties with a long growing season, and at 50–65 days for those with a short season. For silage, the crop is harvested after the milk stage.

Corn is a warm-weather plant, requiring considerable moisture and warmth from the time of planting to the end of the flowering period. It is raised at all latitudes in the temperate and tropical regions of the world where the growing season is sufficiently long and the temperatures are suitable. The optimum temperature range for growth is 22–30° C. Night temperature should be considerably cooler, but above 10° C during most of the growing season. In general, the plant ceases to grow at temperatures below 10° C.

The most desirable soil is considered to be deep, medium-textured, high in organic matter, well-drained, high in water-holding capacity, and containing all the essential nutrients required by the growing plant. Soil moisture is supplied to the crop either naturally by rainfall, or artificially by irrigation in arid and semiarid zones. The fertility level of the field has a marked influence on plant response to irrigation. At low fertility levels, drought effects are not as pronounced as they are when the plants have received high nitrogen applications. If nutrients are not available at all times and in sufficient quantity, little will be gained by maintaining a high moisture supply in the soil by frequent irrigations. However, the more fertile the soil, the greater the benefits to be expected from the application of irrigation water.

Methods of Irrigation

Two methods of water application are in general use–furrows and sprinkling. Furrow irrigation is suitable on almost level land or on slopes of less than 0.5%. Distribution of water to the field is accomplished by gravity from canals and ditches, or by a system of pipelines. Various techniques and devices (i.e., siphons, gated pipes, etc.) are used to direct the water into the furrows.

Sprinkler irrigation is commonly used on deep soils with a high water-intake rate, for rolling topography, on steep slopes, or where leveling would remove too much fertile soil. In Israel, almost all corn is irrigated by this method. The disadvantages are the high initial cost of installation, uneven water distribution under windy conditions, and the necessity of mounting the sprinkler heads on high risers. Moving sprinkler irrigation pipes when the corn has reached its full height (up to 3 m) is a difficult chore when done by hand, and the labor cost is high. Various techniques have been developed to use tractors for moving the lines.

Sprinkling is preferable to furrows in saline areas because of the leaching effect. With furrow irrigation salts are carried by the water into the ridges and accumulate there as the water evaporates from the soil surface. With sprinkling, the movement of water and salts is downward. On the other hand, excessive downward movement of water will leach soluble nutrients from the root zone. Irrigation applications by sprinkling thus must be controlled to avoid the leaching of elements essential for plant growth.

Corn is classified as being relatively sensitive to soluble salts. Although its sensitivity varies somewhat depending on variety and stage of growth, it is not well suited to saline or alkali soil conditions, or to irrigation with water of high salt content.

Root Development

Under optimal growing conditions the corn plants will develop a very extensive root system. In well-drained loamy soils, roots have been found to attain a lateral spread of up to 240 cm and a vertical penetration of at least180 cm. On emergence, maximal root length is about 15–20 cm and the horizontal spread about 10–15 cm. At a plant height of 20 cm, numerous roots extend beyond a depth of 60 cm, but only the upper 30 cm is thoroughly explored. Root growth progressively increases until about silking, by which time the upper 120 cm of soil is thoroughly explored by roots, and some water extraction can also be detected in the 120–180-cm zone (Fig. 1).

Needless to say, root development greatly affects the supply of water that the plant can exploit. Poor exploration of soil by a shallow and underdeveloped root system will result in wilting at a higher average soil-water content than if root development was satisfactory.

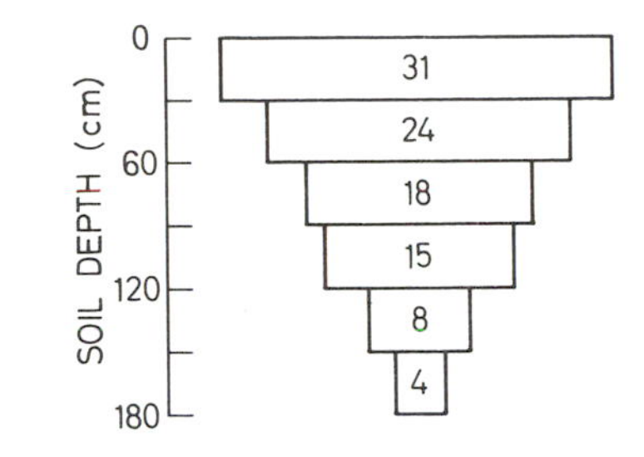

Fig. 1. Amount of water extracted by roots of corn plants at the dough stage. The plants, grown in a clay-loam soil, received a total of 5 irrigations after planting. The values represent percentages of the total water removed to a depth of 180 cm. (MANTELL, unpublished data)

Evapotranspiration by Corn

Water use varies according to the stage of plant development, with maximal consumption occurring between tasseling and the hard-dough stage of grain production. Then, as physiological activity declines and temperatures drop in late summer and early fall, water use gradually decreases.

The average daily rate of water usage seldom exceeds 2–3 mm until the corn is about 20–30 cm tall. It then increases to 6–8 mm from silking to the soft-dough stage. Under conditions of high evaporative demand and when plants are well developed, the daily rate of evapotranspiration may rise to as much as 10 mm.

The correlation of measured evapotranspiration with water loss from a standard evaporation pan can be used as a guide in determining the magnitude of water application. DOSS *et al.* (1952) in Alabama reported that the ratio of evapotranspiration by corn to pan evaporation varied from 0.38 at emergence to 1.12 during the early dough stage, declining thereafter to 0.95 at grain maturity. Others have also reported that during the period of rapid plant development and with sufficient soil moisture present, the rate of water use may substantially exceed that measured in an evaporation pan.

When and How Much to Irrigate

In regions with a limited water supply, and perhaps no rainfall during the growing season, the problem arises of when to apply the available water and in what quantity in order to achieve maximal returns. Numerous investigations in the past have shown that soil-moisture stress between tasseling and silking is more harmful to yield than at any-other growth stage, and moisture stress during the vegetative stage of growth results in reduced plant size, accompanied by lower production of dry matter (DENMEAD and SHAW, 1960; ROBINS and DOMINGO, 1953).

The marked effect of irrigation frequency on plant development is illustrated in

Fig. 2. Although the soil profile had been wetted to a depth of about 120 cm before planting, it is seen that the infrequently irrigated plants did not have sufficient available water to maintain a rapid growth rate throughout the season.

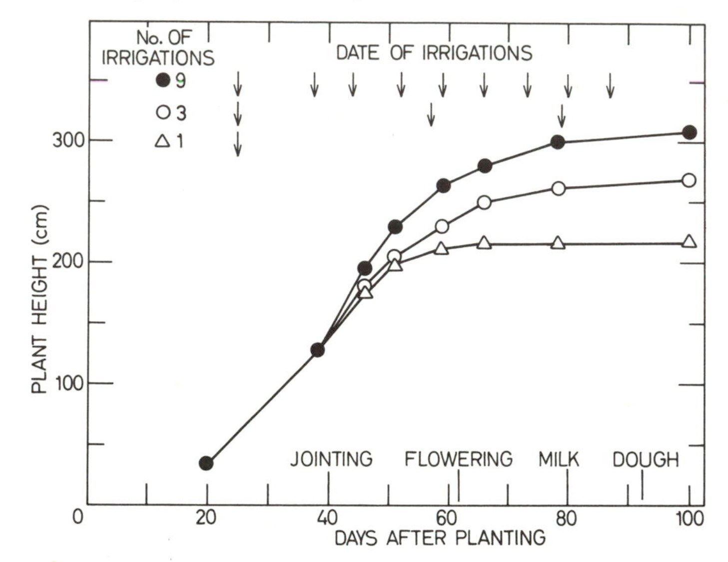

Fig. 2. Effect of irrigation frequency on height of corn plants that received 1, 3 and 9 water applications after planting. The tallest plants were irrigated weekly; the shortest ones only at 25 days of age; and the intermediate ones at 25 days, at beginning of flowering and at milk stage. (MANTELL, unpublished data)

In studying the effect of withholding water at various stages of plant development, RUSSELL and DANIELSON (1956) found that depletion of the soil water to wilting point for only 1–2 days during tasseling or silking resulted in a grain yield reduction of up to 22 %, whereas 6–8 days of wilting at this stage reduced yields by about 50 %. Later irrigations did not prevent the occurrence of these losses. Following maturity (hard-dough stage), depletion of available water had no effect on yield.

It seems that from emergence to tasseling it is safe to extract up to 60% of the available water in the main root zone (0–90 cm), and even a greater amount as the plant approaches maturity. However, from tasseling to silking, the depletion of available water should not be allowed to exceed 40–50%.

A 3-year study was conducted in Israel by SHMUELI and LESHEM (1967) in which forage corn was subjected to water stress during different growth periods. The authors concluded the following:

A. *On Highly Fertile Soil.* Dry-matter yields of up to 18 ton/ha can be achieved with a seasonal water application (excluding preplanting) of 200–300 mm given in 2–3 irrigations between 32 and 66 days after planting.

B. *On Soil of Average Fertility.* 1) A similarly high yield can be reached only if the first irrigation is applied at an early stage of plant growth, followed by three additional irrigations applied until the end of the flowering period (60–65 days).

2) With 250–300 mm of water given in 3 irrigations between 32 and 66 days, a dry-matter yield of about 15.5 ton/ha can be achieved.

3) The same yield as above can be realized with 2 irrigations and 200–250 mm of water, applied at 32–35 days and 63–65 days after planting.

4) Postponement of irrigation beyond 40 days from planting, in a regime of 2 irrigations, resulted in a significant reduction of yield compared with treatments receiving 3 or even 2 irrigations, with the first applied at 32–36 days.

When irrigation was given close to the end of the flowering period, there was a sufficient reserve of water in the soil to produce a maximal yield without the necessity of applying more water at a later date. It was found that a regime of 4 irrigations, with the first early in the growing season (during the 4th week of growth) and the last near the end of flowering, is the best schedule for growing forage corn in Israel. Using such a regime, a seasonal water application of about 300 mm is sufficient. With programs in which 2–3 irrigations are given, it is possible to realize a certain saving in water and labor, but in fields of only average fertility this is liable to be accompanied by a reduced yield.

A preplanting irrigation should be given to wet the potential root zone (about 0–120 cm) and guarantee an adequate water supply during the early stages of plant growth. If planting takes place shortly after the end of the rainy season, only a small deficit may exist (for example, less than 100 mm), and in this case the preplanting irrigation may be dispensed with and the deficient water applied together with that given for germination.

The magnitude of water application during the season should be calculated to replenish the deficit from field capacity in the main root zone. The last irrigation is given about 7–10 days before cutting the crop.

Literature

DENMEAD, O. T., SHAW, R. H.: The effects of soil moisture stress at different stages of growth on the development and yield of corn. Agron. J. **52**, 272–274 (1960).

DOSS, B. D., BENNETT, O. L., ASHLEY, D. A.: Evapotranspiration by irrigated corn. Agron. J. **54**, 497–498 (1960).

ROBINS, J. S., DOMINGO, C. E.: Some effects of severe soil moisture deficits at specific growth stages in corn. Agron. J. **45**, 618–621 (1953).

RUSSELL, M. B., DANIELSON, R. E.: Time and depth patterns of water use by corn. Agron. J. **48**, 163–165 (1956).

SHMUELI, E., LESHEM, Y.: The response of forage corn to four irrigation regimes. Volcani Inst. of Agric. Res. Pamphlet 116. 40 pp. (Hebrew, with English summary) (1967).

The Irrigation of Alfalfa (H. BIELORAI)

Importance and Establishment

Forage is the principal as well as the cheapest feed for livestock. Alfalfa is considered the most important sown forage because of its quality, high production, and longevity.

According to the available statistics, there are nearly 2 million hectares of irrigated alfalfa in the eastern U.S. in addition to 100,000 irrigated hectares producing alfalfa seed. Irrigated alfalfa is usually sown on highly productive crop land and dry-matter yields average over 7 tons per hectare. Producing areas other than the United States are the Argentine, the Mediterranean region, and some European countries.

Alfalfa *(Medicago sativa)* is usually planted in the fall or spring, when loss of moisture from the soil surface is slow. Fall plantings should be early enough to obtain adequate root growth to survive the winter. Spring sowings are frequently more satisfactory in moderate northern climates.

Preirrigating the soil during seedbed preparation is usually advisable. In those regions where rainfall is not sufficient, it is especially important to fill the potential root zone to field capacity.

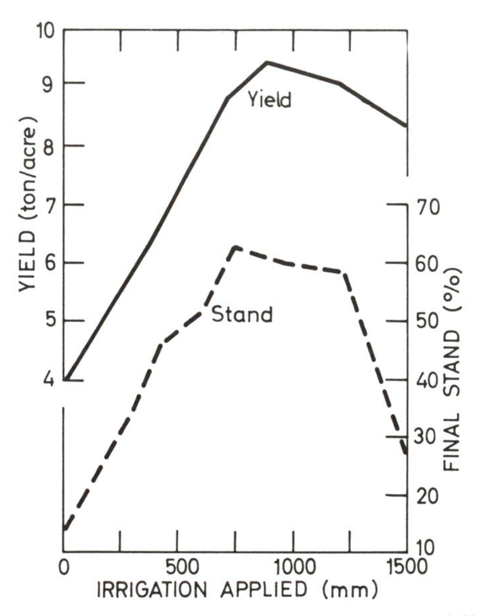

Fig. 3. Relation between final stand and alfalfa yield in response to different amounts of annual irrigation (optimum stand contained 6 plants/m^2). (From BECKETT and ROBERTSON, 1917)

Alfalfa succeeds on a variety of soil types. Deep well-drained loams with a high capacity for absorption and storage of water are preferable. High yields are obtained in arid zones when the crop is irrigated. Irrigation affects yield, plant survival, and final stand. An optimal stand contains 6 plants per m^2. The final stand (percentage of the optimal stand) in relation to the amount of seasonal water applied is presented in Fig. 3.

On the basis of the yields and plant survival (final stand) described in Fig. 3, it can be concluded that an application of about 800 mm water per season maintains a good stand and assures a 10-ton/acre yield.

Irrigation after Planting

Irrigation after plant emergence is unnecessary if the seedbed preparation and climatic conditions are satisfactory. However, some growers prefer to plant in a dry seedbed, especially in sandy soils. Alfalfa planted in a dry soil or an arid region should be irrigated immediately. Additional irrigations may be necessary to soften any crust formation so that seedlings can emerge. Frequent irrigations also may be required in sandy soils.

The main development of the alfalfa plant during the first few months is in the root system. Young seedlings are shallow-rooted. The intervals between irrigations are increased as roots penetrate deeper into the soil. If salt accumulation is noted, excess water is added during the winter to leach it out. When the crop is irrigated by flooding, alfalfa plants are injured if water stands on a field for too long a period (e.g., 24–48 hours), especially in a hot climate.

The growing season of alfalfa differs from one region to another. Its consumptive use is determined by the irrigation season and number of cuttings. The relationship between length of growing season in days and irrigation water needs in different locations is presented in Fig. 4.

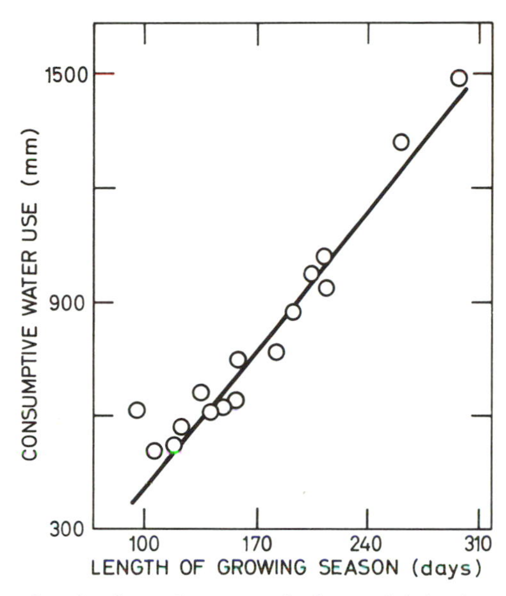

Fig. 4. Relation between length of growing season in days and irrigation water needs of alfalfa. (From BLANEY and CRIDDLE, 1962)

A very close relationship is shown between the number of days alfalfa grows and its water needs. The realationship is linear and in general agrees with practice in growing alfalfa under irrigation.

Effect of Irrigation on Yield

Irrigation of alfalfa often receives a lower priority than irrigation of other crops because alfalfa will maintain itself over a wide range of moisture conditions. However, since the foliage of the plant is the part harvested, rapid growth should be promoted. It is considered good practice to irrigate immediately after each cutting to supply sufficient moisture for rapid growth of new stems and leaves.

If the amount of available water in the root zone is low, an additional irrigation may be advisable between cuttings. This practice has some benefit in arid zones, although it requires extra labor.

In an arid zone, the yield declines during the hot summer months. A common practice in Israel is to withhold irrigation from June to August, and to grow the alfalfa for seed during this period. In cooler climates, alfalfa production remains substantially unchanged over a wide range of irrigation practices. The effect of three irrigation treatments on dry

Table 1. Effect of irrigation treatment on alfalfa production in an experiment conducted at Gan Shmuel in Israel, 1968. (After BIELORAI *et al.*)

Treatment	Water applied (mm)	Consumptive water use (mm)	Dry-matter yield per cutting (kg/ha)								Total Yield (kg/ha)
			Date of cutting April	May	June	July	Aug. 5	Aug. 29	Sept.	Oct.	
Wet	900	944	2,390	2,470	1,780	1,540	1,390	1,180	1,510	1,980	14,240
Medium	720	849	2,440	2,450	1,670	1,550	1,280	960	1,230	1,450	13,030
Seed	745	892	2,320	2,400	–	–	–	910	1,280	1,200	8,110 450 kg seed[a]

[a] This treatment produced 8,110 kg dry matter and 450 kg seed/ha.

matter and seed production in an alfalfa irrigation experiment carried out in Israel is shown in Table 1. In all the treatments, irrigation was given after cutting.

The results show that reducing the seasonal water application by 20% resulted in an 8.5% decrease in yield. Withholding irrigation from June to August caused a 40% loss in yield of dry matter for forage, but as alfalfa seeds have a fairly high value, this loss was offset by the 450 kg/ha of seed produced. The advantage of this treatment is that water was saved during the summer months when demand is at a peak, and the saved water could be diverted for example, to cotton production. The peak water use of cotton is in June, July, and the beginning of August.

In those treatments where irrigation was withheld or where the amount of water applied was less than that depleted, more water was utilized by the alfalfa crop from the deeper soil layers (where water was stored from rainfall). Taylor obtained maximum alfalfa *seed yield* by maintaining a mean annual integrated soil water stress of 2–8 bars. A comparison between the consumptive water use of alfalfa in New Mexico and in Israel is presented in Fig. 5.

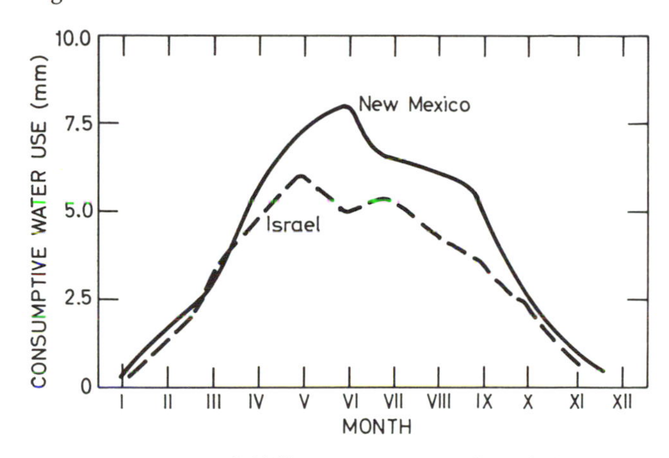

Fig. 5. Consumptive water use of alfalfa in New Mexico and Israel. (From BIELORAI, 1969)

The general pattern of the consumptive water use of alfalfa in both countries is similar, however, the monthly evapotranspiration rates in Israel are lower than those in New Mexico, and so is the annual consumptive water use. The yield in New Mexico (about 20 ton/ha) was higher than in Israel due to the longer growing season. The yields from the irrigation experiment in Israel included eight cuttings from April to October. Penman reported an increase in yield due to fertility without a change in the water requirement.

Conclusion

When and how much to irrigate depend on the rate at which alfalfa uses soil moisture and on the water-holding capacity of the root zone. Daily use should be considered when planning an irrigation schedule since demand is low in the spring and rises in midsummer (see Fig. 4), depending on growing conditions and location.

The amount of available water stored in the root zone should also be taken into account. Alfalfa is a deep-rooted crop and is able to extract water from deep layers.

A survey concerning irrigation schedules indicated that most alfalfa growers apply water after each cutting. Overirrigation should be avoided where drainage and soil permeability are poor.

Best growth can be expected when depletion of available moisture in the root zone does not exceed 60 %. It has to be remembered, however, that variable texture and depht of soil mean different capacities for holding available water. Water-use efficiency is higher when the amount of water transpired and evaporated is low relative to the dry matter produced.

Literature

BECKETT, S. H., ROBERTSON, R. D.: The economical irrigation of alfalfa in Sacramento Valley. Calif. Agr. Exp. Sta. Bull. **280** (1917).

BIELORAI, H., DOVRAT, A., COHEN, J., SAGEE, S.: Irrigation experiments on alfalfa for meal and seed production (in Hebrew). Hassadeh **49,** 935–938 (1969).

BLANEY, H. F., CRIDDLE, W. D.: Determining consumptive use and irrigation water requirement. USDA Tech. Bull. **1275,** 69 pp. (1962).

HANSON, E. G.: Influence of irrigation practices on alfalfa yield and consumptive use. New Mexico Agr. Exp. Sta. Bull. **514,** 20 pp. (1967).

PENMAN, H. L.: Natural evaporation from open water, bare soil and grass. Proc. Roy. Soc. (London) **193,** 120–145 (1948).

TAYLOR, S. A., HADDOCK, J. L., PETERSEN, M. W.: Alfalfa irrigation for maximum seed production. Agron. J. **51,** 357–360 (1959).

5

Irrigation of Fruit Trees

H. Bielorai, I. Levin, and R. Assaf

Defining Irrigation Requirements of Trees

Specific Fruit-Tree Requirements

Many factors must be considered in irrigation management of fruit crops, some of which are different from those involved in field-crop irrigation. The principal points of concern are (1) the importance of the reproductive stage in tree growth and fruit production; and (2) the cumulative response of fruit trees to moisture regime on a long-term basis–namely, the effect of precipitation or irrigation practices from one year to another. The total amount of water needed by fruit trees will vary with the climatic conditions under which they are grown, with their age and size, and with the depth and type of the soil and rooting habits of the trees.

Fruit trees may be classified in two categories: (1) deciduous fruit frees, which are indigenous to humid areas but grow well in subtropical zones; and (2) evergreen trees, which are apparently indigenous to the subarid and arid subtropical zones. The length of the irrigation season and the timing of irrigation depend on the rate of use of water by the tree, its stage of growth, bloom and fruit setting, development of the fruit on the tree, and time of harvesting.

Citrus Growth (H. Bielorai)

Citrus is a typical mesophyte that grows in subtropical areas. The trees are characterized by moderately thick, dark evergreen leaves, with some drought- and heat-resisting characteristics. Citrus trees are grown in all types of subtropical and tropical climates, and water requirements differ widely.

World Citrus Production

The increasing importance of citrus is reflected not only in the rising figures of production and export–about 16,700,000 metric tons–but also is due to its nutritional value (vitamin C). The most important citrus-producing countries are the U.S.A. (California, Florida, Arizona), Spain, Italy, Israel, Greece, Cyprus, Morocco, Algeria, the Congo, Australia, Brazil, Agentine, and Japan.

The main regions of cultivation are situated at the edge of the tropics between 20 and 40° north and south latitude. The minimum and maximum temperatures for citrus growth are 10 °C and 35 °C, respectively, with an optimum range of 20 to 30 °C.

The average temperature influences the time required for fruit development. For example, Marsh grapefruit requires about 12 months from fowering until ripening in Arizona and about nine or ten months in Gilat (Israel).

Fig. 1 shows the seasonal fruit growth of two citrus species. Whereas grapefruit gradually increases and reaches its maximum growth at the end of the growing season, the Valencia orange grows rapidly and reaches its maximum size earlier.

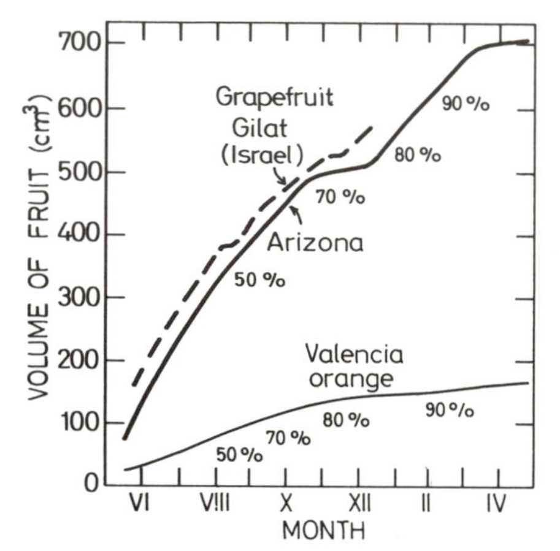

Fig. 1. Seasonal growth of grapefruit and Valencia orange in percentage of the final size of the fruit. (Adapted from HILGEMANN, 1954)

Soil Requirements

The physical condition of the soil (structure and texture) is of decisive importance for citrus growing. The main areas of cultivation are on deep, well-drained sands, sandy loams, loams, and clay loams. Loose sands of sandy loams with good physical properties are preferred. Heavy soils with a high clay content must be avoided because of their impeded drainage and insufficient aeration.

Water Requirements

As stated previously, citrus cultivation extends from humid regions to desert zones. Being evergreen, citrus trees require a sufficient supply of water throughout the entire year. Table 1 indicates the irrigation practice for the various citrus-producing countries of the world.

The diagrams in Fig. 2 present the response of fruit growth (Valencia oranges) to two different irrigation regimes—wet and dry. The upper diagram shows the increase in growth (fruit volume); the lower diagram presents the soil-water pattern in the wet and dry irrigation treatment. It can be seen that water deficiency hinders growth. Applying irrigation after a relatively long dry period somewhat compensates growth; nevertheless, the fruit remains smaller than that undergoing wet treatment.

The curves presented in Fig. 3 show the seasonal sequence of water use in two grapefruit groves, in Tempe, Ariz., and in the northern Negev of Israel. The shapes of the two curves are rather similar, indicating low evapotranspiration rates in the spring and winter and relatively high rates in the summer—June to August. However, the evapotranspiration rates and seasonal water use in Arizona are considerably higher, indicating more extreme climatic conditions than in the northern Negev of Israel. These conditions will alter the irrigation practice in both locations. The evapotranspiration rate in Arizona is close to

Table 1. Irrigation practice in different countries

No irrigation	Summer irrigation plus winter rain	Year-long irrigation
Florida	California	Texas
China	Spain	Arizona
Japan	Italy	South Africa
Brazil	Algeria	
West Indies	South Australia	
Nepal	Chile	
Uruguay	Israel	
Paraguay		

The consumptive use of citrus plantations varies from 1200 mm in Arizona to about 800 mm in Israel.

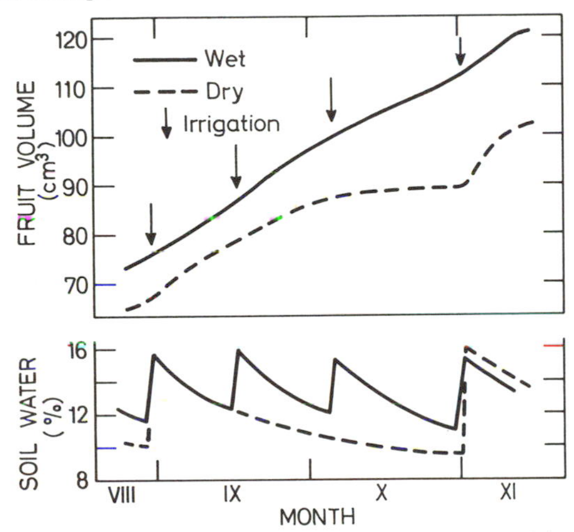

Fig. 2. Relation between Valencia fruit growth and irrigation treatment. (Adapted from HILGEMANN, 1954)

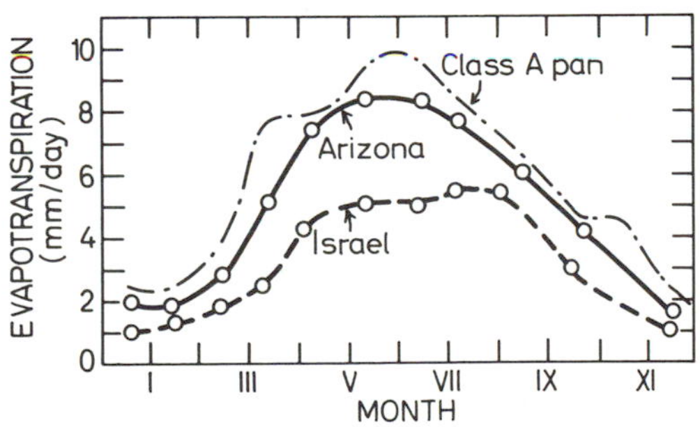

Fig. 3. Evapotranspiration rates of an irrigated grapefruit grove in Israel and Arizona. (From BIELORAI, 1966, and HILGEMANN, 1954)

potential evapotranspiration. The dotted line represents the potential evapotranspiration in the northern Negev of Israel.

Water applied to a citrus grove is used by the trees, evaporates from the soil surface, and percolates below the root zone. By measuring soil-water changes it is possible to obtain a pattern of moisture extraction from the various soil layers. And estimating water losses due to evaporation and percolation, the actual water use of citrus trees can be calculated.

The amount of water extracted from different soil depths by citrus trees (expressed as a percentage of the total water consumption) is shown in Table 2. It is seen that similar results were obtained for Valencia oranges and grapefruit on sour-orange rootstock. Most of the water is extracted from the upper three soil layers (0–90 cm). The depth of irrigation affected the water extraction in the upper layers, increasing the water losses due to evaporation from the bare soil (see last column of table).

Table 2. Extraction of soil-water percent by Valencia orange and grapefruit trees

Soil Depth (cm)	Valencia	Grapefruit	
		Entire root zone irrigated	Only upper root zone irrigated
0– 30	33	38.7	43.0
30– 60	23	26.8	33.0
60– 90	16	19.6	18.4
90–120	13	11.1	3.8
120–150	8	2.0	1.2
150–180	7	2.0	0.6

Root Development

Available information shows that root development is largely dependent on the rootstock and the characteristics of the soil profile, and is modified by the soil-water regime. Sweet-orange rootstock extends to greater depths than does the sour-orange rootstock.

It has been reported by certain investigators that citrus trees have shallow root systems. Our observations, based on root excavations and using P^{32} as a radioactive tracer, gave evidence that in deep soils the main root zone is located in the 0–90-cm layer. However, the entire root zone reaches a depth of 150–180 cm.

Studies of water extraction from the soil profile support these observations. About 85–90% of the total water is extracted from the main root zone and only about 10–15% from the deeper layers. Other data showed that citrus roots are shallow and most of the water is extracted from the upper layer. The apparent contradictions in the above observations may be explained by the specific characteristics of different rootstocks and their response to soil types and soil water.

In conclusion, the relations between soil water, root development, and yield can be simply defined as follows:

1. Deficient soil water–large root volume–low yield.
2. Adequate soil water–moderate root volume–high yield
3. Excessive soil water–poor aeration–small root volume–low yield

Flowering and Abscission of Young Fruit

The flowering period is the most critical in citrus production. Heavy blossoming can be induced by the proper control of water application: withholding irrigation causes induction of blossoming. This technique is employed with lemon trees in Italy. After blossoming, there occures an abscission of the weaker fruit, known as "June drop". According to HUBERTY(1954), maintaining a high soil-water content reduces this drop. The importance of ample water in the soil in reducing fruit drop has been emphasized by several authors. However, other workers (such as Furr) have reported that in desert areas fruit drop is not affected by irrigation.

Effect of Irrigation on Yield

HILGEMANN (1954) reports that Valencia fruit size and trunk growth were affected by drought. In Florida, KOO (1955) reported that irrigation during the spring when one-third of the available water was depleted increased production of grapefruit.

Irrigation experiments were conducted during a six-year period at the Gilat Experimental Farm in Israel to investigate the influence of various soil-moisture regimes on the water requirement, yield, and salt accumulation in the soil of a grapefruit grove.

The results presented in Fig. 4 indicate that citrus response to various irrigation regimes is cumulative, especially in young groves when trees are still developing and have not yet reached a stable level of productivity.

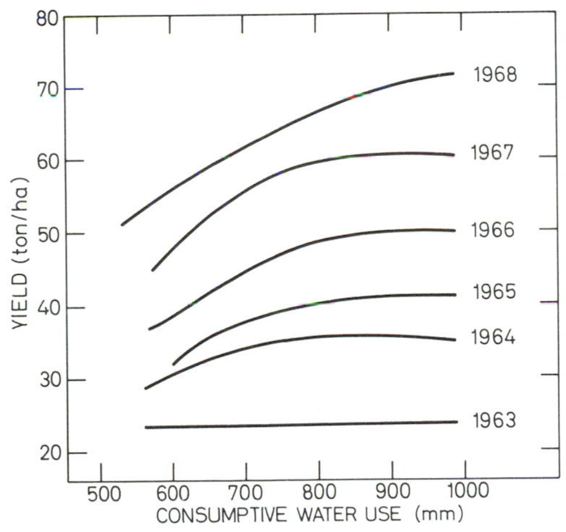

Fig. 4. Relation between consumptive water use and yield of irrigated grapefruit. (From BIELORAI and LEVY, 1971)

The results show that in the first year a uniform yield was obtained in all the irrigation treatments, but in the second year there was a decline in yield in the driest treatment. In 1968, a 28% decrease was observed in the dry treatment in comparison with the wettest one. The dry treatment was irrigated every 40 days and the wet treatment every 18 days. The treatments affected both consumptive use and the evapotranspiration rate. The consumptive use of the dry treatment was 550 mm and of the wet treatment about 900 mm.

Fig. 4 illustrates the stepwise increase of yield with time in all treatments and the effect of the driest water regime, which resulted in a lower yield in comparison with the more frequent irrigation treatments. Other results were as follows: The annual growth increment of the rootstock and of the canopy in the dry treatment was smaller than in the other treatments. Leaf flush cycles in the dry treatment were fewer than in the other treatments. Almost continuous leaf flushing in the wet treatment was observed.

Fruit size and quality were tested at the end of the irrigation season. The dry treatment produced fruit with a higher sugar content. However, rind thickness was greater in the dry treatment. There is some evidence that ripening of the fruit can be accelerated by the use of certain irrigation treatments, which produce a more favorable sucrose :acid ratio. This is important for early export marketing, resulting in a higher income for the grower.

The results of irrigation experiments with Shamouti oranges at Newe Yarak in Israel (conducted by HELLER and GOELL) on sour-orange rootstocks are presented in Table 3.

Table 3. Effect of irrigation intensity on yield of orange trees in fine-textured soils (HELLER and GOELL)

Number of irrigations	Irrigation interval (days)	Seasonal water application (mm)	Total yield 1963–1966 (kg/tree)
4	42	460	143
5–6	30	570	169
8–9	21	720	164

The purpose of the study was to examine the problem of irrigating citrus in fine-textured soils and to evaluate the relation between aeration and frequency of water application.

There was no yield difference between irrigating every 21 or every 30 days. However, the dry treatment caused a yield reduction of about 12%. The average annual rainfall was around 500 mm.

Another study with Shamouti oranges on sour-oranges rootstock was carried out in a sandy-loam soil at Bet-Dagan in Israel (conducted by MANTELL and GOELL), and a summary of the results is given in Table 4.

Table 4. Effect of irrigation intensity on yield of Shamouti oranges in sandy loam soils (MANTELL and GOELL)

Number of irrigation	Irrigation interval (days)	Seasonal water application (mm)	Yield (kg/tree)
5	40	524	119
7	30	700	123
9	21	763	127
13	14	842	110

The results show that in this mature orchard (about 35 years old) there was no difference in yield response to the various treatments. However, about 20–25 percent of the water applied can be saved by using the longer interval between irrigations.

Practical Irrigation Programs

Since climatic conditions differ each year and soils in citrus areas vary in their capacities to hold water, it is not possible to give a precise schedule for irrigation. However, average sandy-loam soils should be irrigated after about 65–70% of the available water is removed. Calculations based on water usage have shown that in a single irrigation 80–100 mm of irrigation water should be applied. In a sandy soil, the irrigation frequency should be about 14–18 days. In sandy loam or clay, the irrigation interval could be extended to 24–30 days. An example of an irrigation schedule is presented below.

Irrigation dates in Arizona

Date	Interval (days)	Date	Interval (days)	Date	Interval (days)
January 2		June 5	20	August 14	18
March 15	72	June 22	17	September 13	17
April 25	40	July 9	17	September 28	25
May 16	21	July 26	17	November 1	33

Average interval = 27 days
Range in intervals = 17–72 days

The highest irrigation frequencies are practiced from June to November, about 17–29 days.

Similar schedules are recommended in Israel where no irrigation is generally applied in the winter.

The irrigation intervals in midseason (June–July–August) should be shortened to about 14–20 days. A high moisture level should be maintained in the soil profile during the period of early blossom and fruit set. Experience during the past years has shown that tensiometers can be used as a guide for irrigation. Irrigation should be applied when the soil-moisture tension reaches 0.45–1.0 bar in the main root zone.

Increase in fruit size or trunk growth (measured daily by recording dendrometers) can be used as a guide for irrigation. However, the use of plant indicators demands skilled technical personnel. (The effect of temporary weather changes should be taken into account when interpreting these measurements.)

Methods for Irrigating Citrus Groves

Three methods of applying water to citrus groves are in wide use: border check, furrow, and sprinkler.

In border-check irrigation, small dikes are built along contour lines, and the area between the dikes is flooded. This method is used where conditions are suitable. The land as a whole must be fairly level and areas within a flood basin must be level for uniform wetting.

In furrow irrigation the water is run in small furrows between the tree rows. Furrows are usually spaced 3 to 4 feet apart, with closer spacing on sandy soils. Care is necessary to obtain even distribution of water.

Sprinkler irrigation of citrus groves is increasing greatly. Sprinklers have the advantage over furrows of more uniform and complete wetting, including the surface soil. Sprinkler irrigation can be used widely on uneven soils, mainly as under-tree sprinkling. Recently a new method, known as trickle irrigation, was introduced and is now in general use in young citrus groves in Israel.

In conclusion, the timing of irrigation of citrus trees depends on many factors, such as amount of available water, soil depth, rate of use of water by the trees, climatic conditions, time of year, stage of development of the fruit and trees, water accessibility and quality, cultural operations, and irrigation methods. Specific time and quantity of irrigation applications are dictated by the rate of removal and the quantity of soil water that may be removed from the root zone without adversely affecting yield and fruit quality.

Literature

BIELORAI, H., LEVY, J.: Irrigation regimes in a semi-arid area and their effect on grapefruit yield, water use and soil salinity. Israel J. Agric. Res. **21** (1), 3–12 (1971).

HELLER, J., GOELL, A.: Two methods of irrigating citrus grove with water of different salinity levels Haklauth be Isreal 3, 1–4 (in Hebrew) (1967)

HILGEMANN, R. H., VAN HORN, C. W.: Citrus growing in Arizona. Univ. of Arizona Bull. **258** (1954).

HUBERTY, M. R., and RICHARDS, S. J.: Irrigation tests with oranges California Agr. **8** (10) 8–15 1954

KOO, R. C. Y., SITES, J. W.: Results of research and response of citrus to supplemental irrigation. Proc. Soil Sci. Soc. Fla. **15**, 180–190 (1955).

MENDEL, K.: Orange leaf transpiration under orchard conditons. Soil moisture content decreasing. Pal. J. Bot. Rehovot **5**, 59–85 (1941).

MENDEL, K.: Orange leaf transpiration under orchard conditons. Prolonged soil drought and the influence of stocks. Pal. J. Bot. Rehovot **8**, 45–53 (1951).

OPPENHEIMER, H. R., MENDEL, K.: Orange leaf transpiration under orchard conditons. Soil moisture content high. Pal. J. Bot. **2**, 172–250 (1934).

WEBBER, J. J., BATCHELOR, L. D.: I. History of botany and breeding. II. Production of the crop (1968).

Deciduous Fruit (I. LEVIN and R. ASSAF)

The extreme southern border of the region suitable for the cultivation of deciduous fruit falls in the semiarid area. Consequently, the special climatic requirements for each fruit determine the types and varieties that can be grown in the different locations. An important limiting factor for this area is the lack of water during the dry summer season, even in districts with heavy winter rains. Therefore, the culture of deciduous trees depends on irrigation. Certain considerations pertaining to the cultivation of deciduous fruit trees under the semiarid condition of Israel are presented below.

The wide range of climatic requirements by deciduous fruits has made it possible to cultivate a variety of types throughout the country. The most important types being grown are pome fruits (apples, pears); stone fruits (peaches, plums, apricots) and nuts (walnuts, pecans, pistachis).

The most suitable varieties are selected for the each district in accordance with the climatic conditions. In Israel's hilly region with its rainy winter and cold climate, those fruit requiring frost to break dormancy are cultivated. In the warmer districts, as in the coastal plain, the interior valleys, and the Negev, varieties are grown that demand less frost, and oil sprays are used in the winter to help break the dormancy.

Water Requirements

Deciduous fruit trees require a relatively deep soil profile (at least 90 cm) as a reservoir for water and nutrients, and also as anchor for the tree roots. In those districts where the quantity of winter rainfall is insufficient to wet the soil profile, it is customary to irrigate in the winter also. This is the case, for example, in the Negev. Summer irrigation alone is required in the other regions. The water requirement is a function primarily of the fruit type and variety. Where the growing season of the fruit is longer, and the date of ripening is later in the season, the total water requirement is greater.

The following annual quantities of water are generally considered necessary for the different fruit crops: apples, pears–900 mm; plums, peaches, nuts–800 mm; apricots–600 mm.

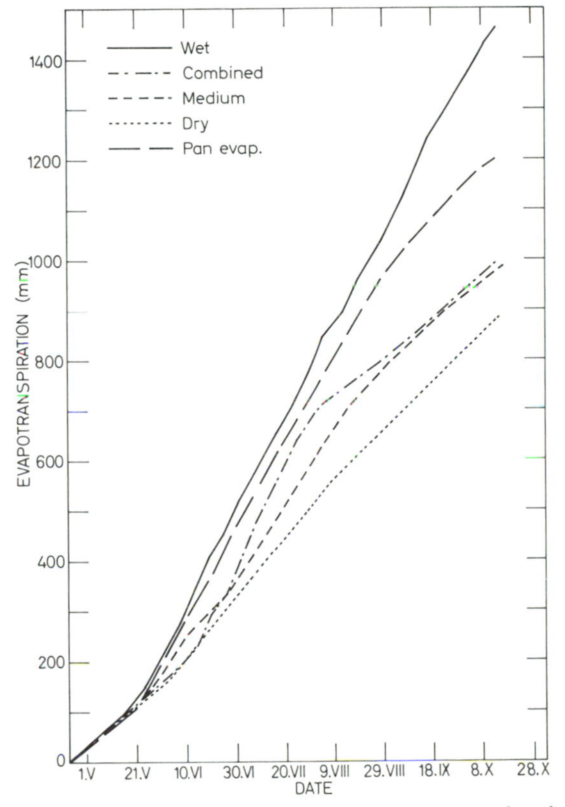

Fig. 5. Effect of the water regime on the evapotranspiration of apple trees

Adaptation of the proper amount of water and frequency of irrigation to the specific requirements of the tree under a certain set of climatic and soil conditions will determine the fate of the orchard–its yield and quality–more than any other factor controllable by the grower.

The experimental result from an apple orchard at Hulata (ASSAF *et al.*, 1968; LEVIN *et al.*, 1969) show that the total evapotranspiration is significantly dependent on the soil-moisture regime and on the quantitiy of water available to the tree (Fig. 5). In the wet treatment, the total evapotranspiration was 25% greater than the amount of water evaporated in a class A pan. This high rate of evapotranspiration was attributable to the frequent irrigations applied weekly throughout the season to well-developed trees and to a permanent vegetative ground cover between the trees. The evapotranspiration in the dry treatment was only 60% of the pan evaporation, and in the medium and combined treatments it was about 70%. The different slopes of the cumulative evapotranspiration curves in Fig. 5 show that the daily rate of evaporation varied between treatments during the growing period.

In order to determine the optimum irrigation regime that will produce the best results in the orchard, one must know the effect of soil moisture at different depths of the profile. The water extracted from each soil layer, to a depth of 180 cm, in the apple orchard experiment is shown in Fig. 6. In all the treatments the two upper layers contributed the greatest part of the evapotranspiration, from 54.5% in the dry treatment to 77% in the wet treatment. As more water was applied, the contribution of the upper layers increased and there was relatively less extraction from the deeper parts of the profile.

The effect of amount of water applied on fruit yield is presented in Table 5.

Table 5. Effect of irrigation on apple yield (HULATA, 1969)

Water applied (mm)	No. of irrigations	Average yield (kg/du)	Fruit size distribution Large	Medium	Small
630	4	6854	32	25	43
840	11	8832	44	21	35
830	12	9108	54	24	22
1285	23	8832	46	25	29

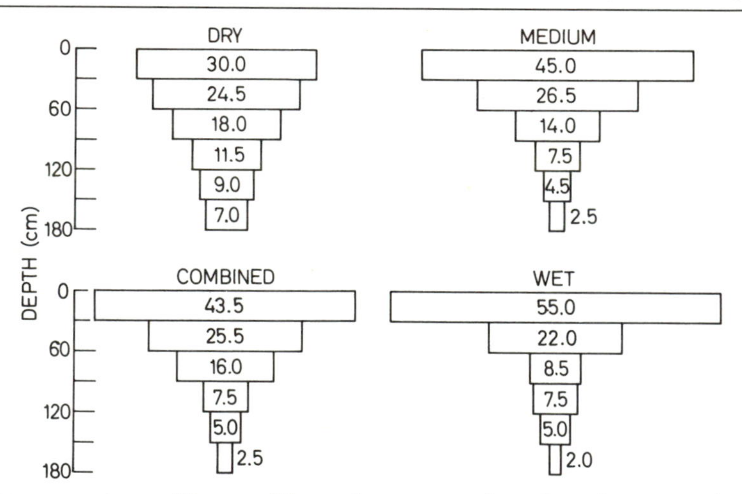

Fig. 6. Water extraction at different soil layers (in percentage from the entire root zone) as affected by the water regime

The wet treatment produced a greater yield and a higher percentage of large fruit than did the medium and dry treatments. The percentage of preharvest fruit drop was higher in the wet treatments. The total soluble solids (TSS) and turgor pressure of the fruit at harvest were greater in the dry treatments. Consequently, the keeping quality of fruit that received the wet treatments was lower. The percentage of large fruit was higher in the wet treatments.

The combined treatment, which was dry at the beginning and end of the season and wet during the main fruit-growing period, and which received a smaller water application, produced maximum yield and fruit growth similar to that in the wet treatment, but with better keeping quality.

Results of daily measurements of radial trunk growth made during the season with dendrometers are shown in Fig. 7. The vegetative growth of the trees and the trend in trunk growth were directly related to the seasonal water application.

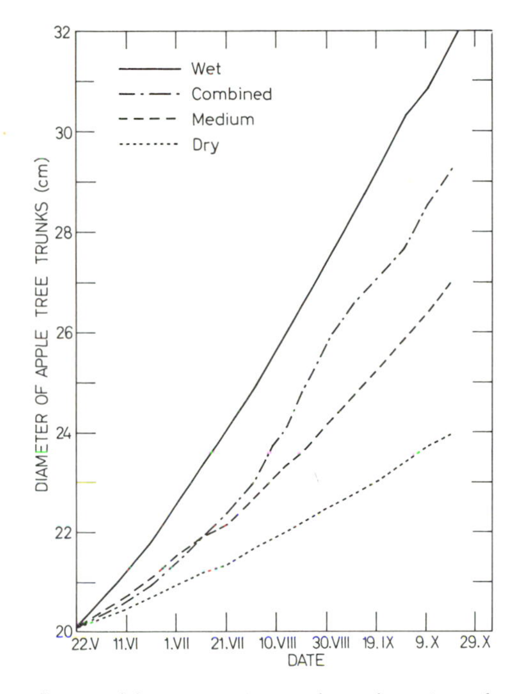

Fig. 7. Influence of the water regime on the apple tree's trunk growth

Measurements of the internal water potential of the trees (SCHOLANDER *et al.*, 1965), and of stomatal resistance and transpiration (BRAVDO *et al.*, 1972), and dendrometer measurements of the daily contractions of tree trunks showed that these factors are greatly affected by the soil-moisture status. For the present, however, these measurements are useful only as indicators of the water status of the tree, and are not practical for use by the grower in scheduling irrigations.

As a result of experiments similar to that described above, it is possible to select an irrigation regime that will produce the best yields. The program should be based on maintaining a high moisture level in the soil during the main season of fruit growth, and a relatively dry soil-moisture regime at the beginning and end of the season. A wet regime is defined as one in which the available water level does not drop below 50%. A dry regime is one in which irrigations are given when the soil moisture approaches wilting percentage.

Since the main root zone is located in the 0–60-cm soil layer, the above-mentioned moisture levels are to be maintained in this part of the profile, and the amount of water applied should replenish the deficit from field capacity in this layer. During the warm period the 60–120-cm layer must be replenished. Following is an optimum irrigation schedule for apples in Israel.

Month	Irrigations	Estimated water application (mm)	Preirrigation available water in 0–60-cm layer %
May	1–2 irrigations, according to end of rainy séason	80–160	Wilting percentage
Beginning of June to beginning of August	8–9 weekly irrigations, including one deep irrigation to 150 cm	600–700	50
Beginning of August to beginning of rains	2–3 irrigations	200–300	Wilting percentage.

The estimated total seasonal water application will be about 1000 mm, given in 12 irrigations.

For fruit trees other than apple, one must take into consideration the main periods of fruits growth in order to decide when to irrigate more frequently. For example, in the case of stone fruits, the critical period occurs after hardening of the stone. The season of frequent irrigation will be of longer duration as the variety ripens later. This accounts for the difference in seasonal water application between early and late varieties of the same fruit type. To illustrate this point, the following table presents the recommended irrigation schedule for two commonly grown peach varieties: (1) early, end-of-May harvest; (2) medium, beginning-of-July harvest.

Month	Irrigation	Schedule
	Variety 1–early	Variety 2–medium
May	4 weekly irrigations; 60 mm per irrigation	1–2 irrigations, according to end of rainy season; 80 mm per irrigation
June	Every 20–30 days; 80–90 mm per irrigation	4 weekly irrigations; 60 mm per irrigation
July	As in June	2 irrigations at short intervals until harvest; 60 mm per irrigation
August	As in June	Every 20–30 days; 80–90 mm per irrigation

The total seasonal water application will be about 750 mm for the early peach variety, and more than 900 mm for the later variety. The irrigation program for plums will be similar to that for peaches, taking into account the date of ripening of the particular variety.

All the irrigation methods and schedules dicussed above were related to fully bearing trees. In the case of young fruit trees of all types and varieties, the most economical method of irrigation is to water tree individually. This can be done by preparing a small basin around each tree, or by using a static sprinkler for each tree. The standard practice is to irrigate every 18–20 days at the beginning and end of the season, and 12–14 days during the middle of the season. The total seasonal water application for young deciduous fruit trees will be approximately 600 mm, in approximately 12 irrigations.

Literature

ASSAF, R., LEVIN, I., BRAVDO, B., SHAPIRO, A.: Research report in apple response to water regime in Hulata (in Hebrew). Upper Galilee Growing Organization, Monograph **25**, pp. 1–50 (1968).

BRAVDO, B.: Photosynthesis, transpiration, leaf stomatal and mesophyll resistance measurements by the use of a ventilated diffusion porometer. Physiologia Plantarum (in press) (1972).

LEVIN, I., ASSAF, R., BRAVDO, B., SHAPIRO, A.: Research report in apple response to water regime in Hulata (in Hebrew). Upper Galilee Growing Organization, Monograph **40**, pp. 1–48 (1969).

SHOLANDER, P. F., HAMMEL, H. T., BRODSTREAT, D., HEMMINGSEN, E. A.: Sap pressure in vascular plants, Science **148**, 339–346 (1965).

Efficient Utilization of Water in Irrigation

E. SHMUELI

Owing to the rapid growth of world population and the resulting development of industry and agriculture, there is a constantly increasing demand for fresh water. At the same time, the quality of the available resources is being affected by salinity and pollution (MARSHALL, 1972). Consequently, the problem of water supply, which is directly related to the supply of food, is gradually becoming one of the crucial issues facing mankind.

FRAMJI and MAHAJAN (1969) have estimated that there are in the world about 500 million hectares of land suitable for irrigation from the standpoint of water availability; for social, political and economic reasons, however, less than one half of this potential is currently being exploited (Fig. 1). Moreover, the irrigable area could be increased considerably beyond the present potential if fresh water could be made available by desalination (CLAWSON and LANDSBERG, 1972). At present, the cost of desalinating sea water is still very high, which rules out the use of such water under field conditions. Future technological developments may permit the economic use of desalinated water in agriculture (CLAWSON and LANDSBERG, 1972).

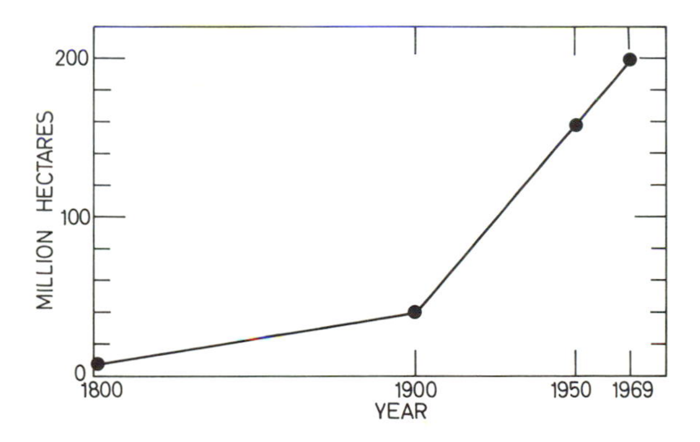

Fig. 1. Irrigated area of the world, from 1800 to 1969. Sources: 19th century – HAGAN *et al.*, 1967; 20th century – FRAMJI and MAHAJAN, 1969

If the future is promising, as indicated by the vast potential of irrigated agriculture in supplying food for the growing world population, the current situation is not very encouraging. Some countries with arid conditions are already seriously short of water and this problem will become more acute in the future. This leads us to the question whether successful agriculture indeed requires such large quantities of water as are applied today. The answer is that the present level of application of water in irrigation is, to some extent, wasteful (HAGAN *et al.*, 1967; JENSEN, 1972; VIETS, 1965).

Although the subject of rational use of water in irrigation is of worldwide interest, we shall restrict our discussion to the U.S.A. and Israel. In the seventeen western states of the U.S.A. the area under irrigation increased from 14.0 million acres in 1910 to 30.7 million acres in 1960. This was achieved by using mostly open irrigation methods and was made possible by the abundance of cheap water as well as much research on water-soil-plant-atmosphere relationships. This research has produced an extensive literature of much theoretical and practical interest (see "Irrigation of Agricultural Lands." Agronomy Monograph No. 11, 1967), but there has been almost no improvement in the efficiency of water application in U.S. agriculture over the past 30 years (JENSEN, 1972). In Israel, the amount of irrigation land has increased considerably over the last 25 years (Fig. 2). Because of water shortage, the development of Israel's irrigated agriculture

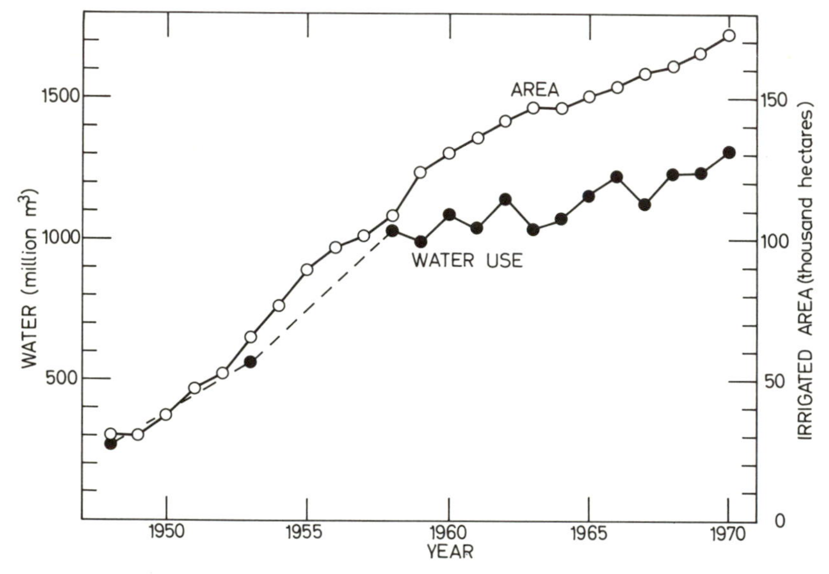

Fig. 2. Irrigated area and water use in Israel agriculture, 1948–1970. The yearly fluctuations in water use during the years 1958–1970 were affected mainly by three factors: (a) drought (in 1958, 1960, 1962, 1965 and 1969); (b) unfavorable rainfall distribution (1966 and 1968); and (c) changes in water use per hectare. The average water use was 87–100, 76–87, 70–83 and 71–78 cm in the periods 1948–1953, 1954–1958, 1959–1964 and 1965–1970, respectively. Sources: 1) Statistical Abstracts, Central Bureau of Statistics, Jerusalem, No. 15, 1964. 2) Water in Israel, Consumption and Extraction, 1962–1970. Ministry of Agriculture, Water Commission, Tel-Aviv, 1972

was accompanied by continuing attempts to minimize the amount of water applied per unit area, or per unit yield. As a result of research and Extension Service activities, sprinkling is widely used and less water is now applied, but yields are higher than they used to be.

Terminology and Some Basic Assumptions

In the last 100 years many terms have been introduced to describe various aspects of water use in irrigated agriculture. Some of these terms are mentioned in Chaps. 5.4 and 7.1. Here we shall restrict ourselves to terms referring to water supply and consumption by plants.

a) *Terms Relating to Quantity of Water in Irrigation and Water Consumption of Plants.* The terms *duty of water* and *water duty* were used at the end of the 19th century and beginning of the 20th. They referred to the amounts of water applied in irrigation, according to the prevailing agricultural practices. In the 20th century, other terms came into use: *water use, water requirements,* or *irrigation need,* defined as the quantity of water delivered to a crop (field), exclusive of precipitation. "Generally water requirement refers to the minimum amount that would be required under an assumed attainable irrigation efficiency, whereas water use refers to the amount actually being used, even though this may be twice as high as the water requirement or irrigation need."[1] Also widely used are the terms *water consumption, consumptive use* and *evapotranspiration.*

Consumptive use (C.U.) is defined as the water that evaporates from plant and soil surfaces plus the water used by plants in building their tissues. Under field conditions the amount of water retained within the plant tissue is generally less than 1% of the total amount that evaporates during the growing season (JENSEN *et al.,* 1967). Therefore the term *consumptive use* or *water consumption* actually includes all the water that evaporates from plant and soil surfaces (*evapotranspiration*).

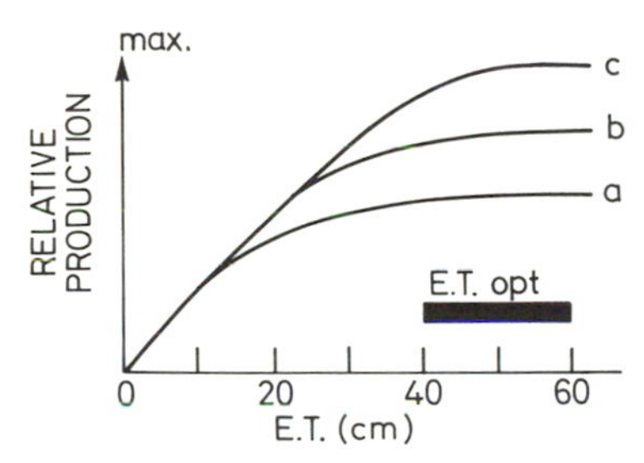

Fig. 3. Relation of evapotranspiration to 3 levels of plant production. The production levels and the shapes of the curves are hypothetical according to BERG *et al.,* 1967. $E.T._{opt}$—according to DOWNEY, 1971

b) *Terms Relating to Efficiency of Water Use.* The terms reflect two different approaches for expressing efficiency of water use in irrigation: 1) An engineering-planning approach with emphasis on the water factor and concerned primarily with evaluation of the technical performance of the irrigation system; and 2) an agronomic-economic approach, centered on yield as related to water input (Table 1).

The term *irrigation efficiency* (E_i) is widely used; some consider it synonymous with *water application efficiency* (E_a). These terms should be used with care, as they may be quite misleading to farmers and planners. E_{cu} was proposed in order "to provide more suitable methods of measuring and evaluating techniques of handling irrigation water" (HANSEN, 1960). *Field irrigation efficiency* has been defined by BLANEY and CRIDDLE (1962) as the percentage of irrigation water that is stored in the soil and available for consumptive use. This is synonymous with the term A_e (Table 1A) commonly used in Israel.

In forage crops D.W. (or Y) in Table 1B refers to the weight of the above-ground dry matter, while elsewhere it refers to the weight of marketable crops. Agronomists and water engineers usually express yields in tons per hectare per cm of water depth.

1 M. E. JENSEN, personal communication.

Table 1. Terms used to describe irrigation and water use efficiency

A. Engineering – planning approach

	Term	Symbol	Formula	Description	According to	Introduced by
1	Irrigation efficiency	E_i	$100\frac{W_{et}}{W_a}$	The ratio of the volume of irrigation water consumed by the crops of an irrigated area to the volume of water applied to this area.	HAGAN *et al.*, 1967	HALL, 1960
2	Water application efficiency	E_a	$100\frac{W_{et}+W_l}{W_a}$	As (1), including the volume of water applied for intentional leaching (W_l).	JENSEN *et al.*, 1967	ISRAELSEN, 1932
2_1	Water application efficiency	E_a	$100\frac{W_{et}+W_l-R_e}{W_a}$	As (2), including a correction for effective rainfall (R_e).	JENSEN *et al.*, 1967	JENSEN *et al.*, 1967
3	Consumptive use efficiency	E_{cu}	$100\frac{W_{cu}}{W_d}$	Ratio of the normal consumptive use of water to the net amount of water depleted from the root zone of the soil.	HAGAN *et al.*, 1967	HANSEN, 1960
4	Irrigation application efficiency	A_e	$100\frac{W_s}{W_a}$	Percentage of irrigation water that is stored in the soil root zone.	SHMUELI, 1973	BLANEY and CRIDDLE, 1962

All terms expressed as percent

W_{et} – Volume of water in a specified area (farm, field or plot) transpired by plants and evaporated from the soil.
W_a – Volume of water applied to the given area.
W_l – Volume of water necessary for leaching (salt control) in the given area.
R_e – Volume of effective rainfall in the given area.
W_s – Water stored in the root zone during irrigation.
W_{cu} – Normal consumptive use of water.
W_d – Net amount of water depleted from root zone soil.

B. Agronomic – economic approach

	Term	Symbol	Formula	Description	According to
5	Water-use efficiency	U_e	$\frac{D.W.}{E.T.}$	Ratio of dry weight of crop (ton/ha) to depth of evapotranspiration water (cm).	VIETS, 1965
6	Optimum irrigation efficiency	I_{opt}	$\max\left[\frac{Y}{W_a}\right]$	Maximum value of the ratio of yield to seasonal water applied.	SHMUELI, 1973

Economists often prefer other parameters, such as: *food value* (calorific value) of the usable crop per unit area per cm of water (CLAWSON and LANDSBERG, 1972), or *wheat equivalent* of the crop per cm of water (CLARK, 1970). Economists are also concerned with *production functions* (BERINGER, 1961; YARON, 1966) and *marginal return*, i.e. physical quantity of additional crop obtained from further watering, divided by the number of cubic meters of additional water used (CLARK, 1970).

Over the past fifty years great efforts have been made to determine the *water use efficiency* (U_e) for crops under different growing conditions (BERG *et al.*, 1967; BLACK, 1965; HAGAN *et al.*, 1967; SHMUELI, 1971; VIETS, 1965). By the Sixties it had become clear that each crop has a certain characteristic range of evapotranspiration ($E.T._{opt}$) which will ensure maximum yields (Fig. 3) under specific conditions of growth. In line with this, HAGAN *et al.*, (1967) have introduced the term *optimum water requirement* defined as "the seasonal depths of beneficial water use that result in maximum yields of different crops, where the depths include soil moisture supplied by precipitation as well as water delivered by irrigation." We propose the term *optimum irrigation efficiency*, I_{opt}, defined as the maximum value of the ratio of yield to the seasonal depth of *irrigation* water applied under a given set of crop and environmental conditions. Sometimes, even under arid conditions, a yield can be obtained without the application of any water. However, our definition for I_{opt} is restricted to irrigation conditions only.

c) Water-use Efficiency, Optimum Water Requirement and Optimum Irrigation Efficiency. In most general terms, *optimal* refers to the state of affairs which yields the best or most favorable degree of some desirable property. It is obvious that the expression for the efficiency of water use in agriculture should be based on economic considerations. Research on the economic aspects of water utilization for agriculture is only in its initial stages. A detailed discussion relating its assumptions and difficulties (BERINGER, 1961; CLARK, 1970; CLAWSON and LANDSBERG, 1972; YARON, 1966) to the interpretation of irrigation experiments and water use problems is not within the scope of this chapter. Following YARON (1966), we shall present only two essential points:

1) "In the economic analysis . . . irrigation water can and should be treated just like any other productive factor which takes part in the production process and contributes its share to the total output.

2) The basic variables of water quantity, irrigation frequency (timing) and irrigation depth are the ones which can be directly manipulated by the farmer; therefore, it seems desirable to perform the economic analysis[2] in terms of these variables . . . In nearly all the irrigation experiments analyzed, a high correlation was found between x_2 (total quantity of water applied) and x_3 (frequency of irrigation), and in effect, the function was usually reduced to $y = f(x_1, x_2)$ . . . Under conditions at one and the same location, the marginal yield of a crop is a function only of the quantity of water (other irrigation variables being equal)."

For the purpose of our discussion, concerning the terms in Table 1B and the term, optimum water requirement, it is not necessary to state accurately the maximal production

---

2 The following general formula was applied by YARON (1966) in analysis of 29 irrigation experiments:

$$y = f(x_1, x_2, x_3)$$

where: y = crop yield per land unit area; x_1 = depth of soil moistening; x_2 = total quantity of water applied; and x_3 = frequency of irrigation (or number of irrigations).

(or income) per unit water applied. It is enough to refer to ranges and not to exact points on a curve (Fig. 3). The terms water-use efficiency and optimum water requirement, in which crop yield is related to evapotranspiration express the *traditional approach* to the problem of irrigation efficiency. The term optimum irrigation efficiency, centered on the minimum seasonal water application, expresses the *optimization approach*. The former approach (JENSEN, 1972; VIETS, 1965) assumes that it is possible to increase the efficiency level by increasing the numerator of the efficiency equation (i.e. yield) without changing, or only slightly increasing, the denominator (E.T.). The optimization approach attempts to increase the irrigation efficiency by raising the crop yield and simultaneously reducing the amount of irrigation water applied to below the evapotranspiration level. This can be attained by the wetting of the root zone (see Fig. 4) with sprinkler irrigation.

d) *Comparison between Traditional and Optimizing Approaches in Irrigation Water Utilization*. Using the terminology presented in Table 1, we may now compare the two approaches in more detail.

The Traditional Approach. As interpreted here, the two basic assumptions of the traditional approach concerning water application are:

1) The volume of irrigation water should be closely related to evapotranspiration so as to replenish the water deficit throughout the root zone (ISRAELSEN, 1932; HANSEN, 1960).

2) The amount of irrigation water must include an additional volume of water (see Terms 2 and 2_1 in Table 1) for salt leaching (DONNAN and HOUSTON, 1967; HAGAN *et al.*, 1967; HALL, 1960; ISRAELSEN and HANSEN, 1967; JENSEN *et al.*, 1967).

In practice, the application of these assumptions under conditions of abundant water resources and open irrigation methods, which limit the possibilities of accurately controlling the quantity of water applied, led to a low irrigation efficiency. Under such conditions, "surface irrigation efficiencies are within the range of 60%, whereas in well-designed sprinkler irrigation systems, 75% efficiency is achieved" (HAGAN *et al.*, 1967). Water application efficiency of 80–85% is considered excellent for sprinkler irrigation (CHRISTIANSEN and DAVIS, 1967).

The traditional approach can to some extent justify the low irrigation efficiency, in terms of the overall water balance. "Excess water applied by either surface irrigation or sprinkling would return to the ground water supply and be available for reuse."[3] Therefore, "in the long run there may be very little change in the rate at which the available water resources are being depleted even though there is less water applied per unit area, because if excess water is applied presumably most of this returns to the ground water system".[3]

This may be true in regions where the water from deep percolation after irrigation (see Fig. 4) joins the ground water system and can be used again for irrigation. But even under these conditions one cannot disregard economic considerations (pumping costs, etc.) and problems of ground water pollution. These matters are beyond the scope of this chapter and will not be discussed here.

The Optimizing Approach. This approach was developed under conditions of an arid and semi-arid climate characterized by a rainy winter and a dry summer (Chap. I. 1). Under such conditions, it is possible to irrigate only part of the potentially irrigable land with water obtained close to the growing region. Much of the water must be brought

3 M. E. JENSEN, personal communication.

from relatively wet districts to those with little precipitation (in Israel, from the north to the south; see Fig. 5). Expanded use of conventional water sources or substantial additions of water from desalination processes will be exploited in semi-arid and arid regions, mainly to increase the irrigation areas in the drier regions which require water from outside sources. At the same time, it must be emphasized that in the dry regions there is generally no usable ground water. Thus, if water is applied in quantities exceeding those needed for leaching and plant use, deep percolation will be wasteful since the water will not recharge a ground water system which could serve as a source for irrigation water.

The optimizing approach reflects the need born of water resources dwindling down to the level of scarcity. The assumptions underlying this approach are:

1) Irrigation is complementary to the rain water stored in the soil. Together, these resources should cover the water requirement for maximum production or income.[4]

2) Partial wetting of the root zone by sprinkler irrigation does not cause much salt to accumulate in the course of one irrigation season, at least not to the extent of reducing the yields of most crops, provided that the irrigation water and soil conditions are normal (not saline, permeable soil, etc.). Therefore, it is possible to make a clear distinction between irrigation to meet plant needs and water application for salt leaching. In other words, *permanent* agriculture on a high production level can be maintained by supplying, during the growing season, a volume of water usually smaller than the evapotranspiration. The difference between the seasonal evapotranspiration and the amount of irrigation water applied is made up from the water stored in the root zone at the beginning of the irrigation season.

3) If the rainfall is insufficient to leach the salts which accumulate during the irrigation season, supplementary irrigation should be given, especially for this purpose, at the end of the rainy season (BRESLER and YARON, 1972).

4) Plant growth and production are directly related to integrated moisture stress (I.M.S.)[5] (WADLEIGH, 1946; TAYLOR, 1952). By wetting only part of the root zone, containing 70–80% of active roots, and proper timing, a very suitable I.M.S. for plant growth and production can be achieved (SHMUELI, 1971; SHMUELI *et al.*, 1972).

The differences between the traditional approach and the optimization approach are demonstrated in Fig. 4, which shows the moisture changes in the root zone of cotton following its first irrigation. It is clear that under arid and semiarid conditions it is possible to optimize the seasonal amount of irrigation water applied by wetting only

4 This principle is accepted, at least in theory, by those in favor of the traditional approach (HAGAN *et al.*, 1967; ISRAELSEN, 1932, 1967; JENSEN *et al.*, 1967); however, it has been neglected in practice, in view of the two basic assumptions of the traditional approach (see above).

5 According to TAYLOR (1952), I.M.S. represents the integrated soil moisture tension. It is obtained by integrating all the local and instantaneous moisture tension values over the soil profile of the root zone and the growth period. The defining formula is:

$$\text{I.M.S.} = \int_0^T \int_0^H S(h,t)\, dh\, dt$$

where: S(h,t) = soil moisture tension, expressed as a function of depth and time; H = depth of the soil profile; and T = duration of the growth period.

part of the root zone, thus avoiding water waste. However this applies only if the soil moisture in the potential root zone at the beginning of the irrigation season is near field capacity and there is no ground water table.

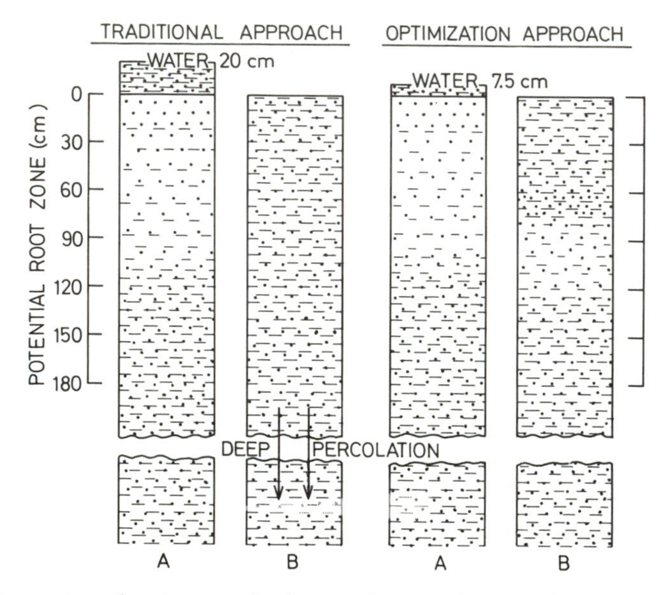

Fig. 4. The changes in soil moisture and salinity under irrigation according to the traditional and optimization approaches
Broken lines–soil moisture. Dots–soluble salts. A–when water applied. B–after water and salt redistribution.
Plant: Cotton, 60 days from the irrigation applied for germination or from the day of planting in wet soil; 70–80% of the roots in the 0–60 cm soil layer. **E. T.:** From the entire root zone – 10.0 cm; from the 0–60 layer–7.5 cm. W_t:–1.5 cm. W_s:–10.0 cm–traditional approach; 7.0 cm–optimization approach. E_a: 57.0%–traditional approach. E_i and A_e: 50% traditional approach. A_e: 93% optimization approach

Table 2. A: Seasonal water applications for cotton as recommended by the Israel Agricultural Planning Center; B: Irrigation requirements as determined by research; C: Seasonal evapotranspiration. All data in cm

No	Region	A[a]				B	C[b]
		1955	1960	1965	1970		
1	Hula Valley		70	50–60	45–60	35–50	50–65
2	Eastern Jezreel Valley	100	80	70(+ 15)[c]	70(+15)	70(+ 15)[d]	90
3	Western Jezreel Valley		55	45(+10)	40(+10)	40(+10)	60
4	Western Galilee						
5	Zevulun Valley	70	60	35(+10)	30(+10)	30(+10)	50–60
6	Coastal Plain						
7	North-western Negev	100	80	50(+15)	55(+15)	50(+15)	80

[a] In accepted use at the indicated years, also serving as basis for planning five years ahead.
[b] Calculated according to FUCHS and STANHILL (1963).
[c] Amounts in parentheses indicate preplanting irrigation in drought years.
[d] In the first two or three irrigations, water was applied by sprinkling and continued until the end of the season by furrows. In all other regions, only sprinkling was used.

The data for cotton in Table 2 show how the seasonal water application for summer field crops has been lowered by research conducted in Israel over the last 20 years. In the period 1958–1964, an annual saving of 100–250 million m^3 of water was achieved, out of a total consumption of about a billion m^3. It was thus possible to increase the irrigated area by about 25% without increasing water consumption (Fig. 2). This did not prevent the continued rise in yields, which occasionally reached world records. For example, in 1969, the average yield of cotton fiber for the country was 1.29 ton per hectare obtained with an average water application of 54 cm.

Maximum Yields (Y_{max}), Optimum Evapotranspiration (E. $T._{opt}$) Levels and Minimum Seasonal Depths of Irrigation Water

Studies of *water-use efficiency* (U_e) have led to a very important conclusion. E.$T._{opt}$ has a characteristic range for every crop and may be varied only slightly. With yields, the case is different. The yield potential obviously depends on the genetics of the crop, but under field conditions we are generally a long way from achieving the potential maximum productivity. Many factors other than water may limit production. The main avenue for increasing yields is through agronomic-physiologic research. Whereas the traditional approach was concerned only with maximum yield (BLACK, 1965; JENSEN, 1972; VIETS, 1965), in the optimization approach, persistent attempts are made to bring down the irrigation water applied per unit of yield to the lowest value compatible with optimum production, rather than to the optimum evapotranspiration level (E.$T._{opt}$).

The terms Y_{max} and E. $T._{opt}$ for corn are illustrated in Fig. 3. The values of E. T. are based on DOWNEY's (1971) critical review of about 30 studies. He concluded that: "a crop of maize will use 50 ± 10 cm of water from planting to maturity." In 19 experiments conducted in different regions of Israel, the E. $T._{opt}$ for corn was found to be in the range of 40–60 cm. The E. $T._{opt}$ range for corn is ± 20 % around average. The extent of the range in various crops is affected by the following factors:

a) *Climatic Differences between Different Crop Locations.* The evaporative demand (see Chaps. III.4 and V.4) has a direct influence on E. $T._{opt}$ obtained for different regions. Yet for most crops growing in their typical areas, the differences in evaporative demand are not large. In Israel, which is noted for the climatic variability of its agricultural regions, the range of differences in the values of evaporative demand for the main growing season is about 15% (Fig. 5).

Regarding E.$T._{opt}$ values of different crops: for cotton growing in Israel, the relative difference between the extreme values reached 100% (B and C in Table 2); for corn the relative difference between the extreme values was of the same order of magnitude as that of evaporative demand. In our opinion this difference reflects the botanical differences between two groups of field crops; corn with a limited growing period, and cotton with an unlimited growth period. In a hot and arid region like the Negev (see Fig. 5 and Table 2), daily evapotranspiration values of both corn and cotton exceed those in the Coastal Plain. But with corn the growing season in the hottest region will be shorter, with a concomitant reduction in the seasonal E.$T._{opt}$ level. On the other hand, the growing season of cotton is much longer in the Negev than in the Coastal Plain, further increasing the difference in the value E.$T._{opt}$ as compared with corn.

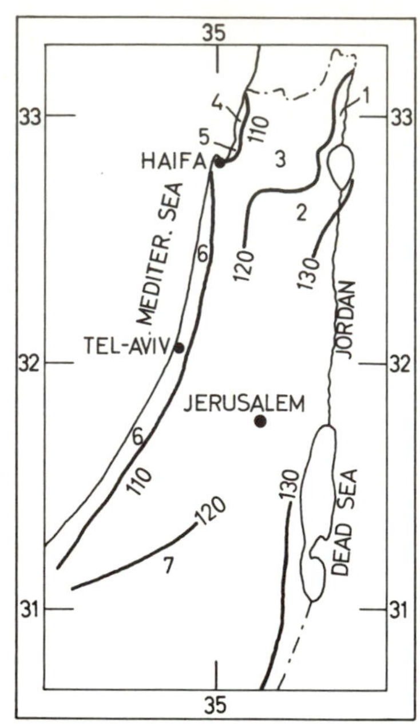

Fig. 5. Total summer (April 1–September 30) evaporation from an extensive open water surface, calculated by Penman's equation according to STANHILL (1963). The solid lines delineate areas of equal evaporation (cm). The numerals (1–7) represent the regions mentioned in Table 2

Table 3. Optimal seasonal water applications and yields from 10 citrus experiments during the period 1963–1971 compared with the water application recommended by the Israel Agricultural Planning Center in 1970

Exp. No.	Variety and rootstock	Region	Optimal treatment		Recommended seasonal water application (cm)
			Water application (cm)	Yield[a] (ton/ha)	
	Shamouti x sour orange				
1		Western Galilee and Coastal Plain	49	48.4	72
2			51	53.0	72
3			57	44.6[b]	83
4		Lod Plain	59	33.9	75
	Shamouti x sweet lime				
5		Coastal Plain	74	77.8[b]	83
6		Lod Plain	55	52.5	75
	Marsh grapefruit x sour orange				
7		Western Galilee	67	52.7	72
8		Western Jezreel Valley	70	60.1	80
9		Eastern Jezreel Valley	74	63.7	90
10		Northwestern Negev	74	57.1	90

[a] All yields higher than the country average during the period 1963–1971.
[b] Planted before 1953; all other experimental groves planted after that date.

b) *Methods of Measuring and Calculating E.T.* The inaccuracy in the measurement and calculation of E.T. is one of the main factors responsible for the width of the E.T.$_{opt}$ range for various crops (see also Chap. 5.4). Advances in the theory of soil physics and microclimatology, as well as in available instrumentation, make it possible to reduce errors to about $\pm$ 5% or even less. In Israel, we have actually approaced this value by a) partial wetting of the root zone, reducing the error in calculating E.T. due to deep percolation (COHEN and BRESLER, 1967), and b) the use of a neutron meter calibrated in the field (COHEN, 1964).

c) *Agrotechnical Factors* (varieties, fertilization, spacing, tillage, diseases, insects, etc.). It is generally accepted that the weight of these factors is relatively small (BLACK, 1965; JENSEN, 1972; VIETS, 1965). The influence of varieties is demonstrated for citrus in Table 3.

Possible Hazards Involved in Partial Wetting of the Root Zone

The success of partial wetting of the root zone may be jeopardized by a) a fall in yield and b) a rise in salinity.

a) *Partial Wetting and Yield Level.* It was pointed out by HANSEN (1960) and HALL (1960) in the United States that partial wetting of the root zone by sprinkling and the tendency to apply insufficient water gave a low financial return.

In Israel, too, when sprinkling came into general use, the amounts of water applied often were inadequate for good yields. The failure of partial wetting by sprinkling was found to occur mainly in regions there winter rainfall is insufficient to wet the entire potential root zone. Thus, use of the partial wetting method during the growing season must be restricted to crops planted in soil where the root zone moisture is close to field capacity.

The second reason for low yields may be incorrect timing of irrigation, as proper spacing of applications is no less important than the amount of water applied. In fact, these factors are interrelated (YARON, 1966). Optimal irrigation timing to meet plant needs will ensure maximum savings in water application and maximum yield (see p. 390 data for corn according to Shmueli and Leshem, 1967). This aspect received close attention, from the beginning of irrigation research but there is ample scope for further improvement. In practice, it is possible to determine the time of irrigation by means of physiological indicators (Chaps. V.5 and VIII.1), and the timing and depth of wetting by instruments which measure water status at various depths of the root zone (Chap. V. 2).

b) *Salinity.* If partial wetting of the root zone is practised and a distinction made between irrigation and leaching during the dry season, salinity may become more critical under certain conditions of water quality, soil and climate than with the traditional over-irrigation approach. But today there is no difficulty, theoretical or practical, in maintaining a suitable salt balance in the soil (Chaps. IV.2 and VII.1). However, the proper steps must be taken at the proper time. A modern agricultural production unit, based on the optimum utilization of irrigation water, must have current data on the amounts of water used and on the salt content of the soil and water available as part of routine farm operations. The salts can then be leached (if necessary) at the proper time and in the correct manner.

Future Possibilities

Optimization of irrigation involves a change from open gravity irrigation methods to the use of closed systems. Recently, different methods of supplying water from point sources have been developed (drip, trickle and subsurface irrigation methods). These methods may give good results under special conditions such as shallow soils, very steep slopes, gravelly soils, highly saline water, high susceptibility to leaf diseases etc., but it is unlikely that they will replace sprinkling under normal conditions. The use of sprinkling as a country-wide method is apparently an essential step in the development of irrigated agriculture under conditions of water scarcity, until desalinated water becomes cheap enough to be abundant.

From the point of view of agricultural development and exploitation of water resources, Israel can serve as a model of what can be achieved in arid and semi-arid countries. In the initial stage, water was exploited close to its source and applied by surface methods. The second stage in expanding the irrigated agriculture involved transporting water from the relatively heavy rainfall areas to drier districts and applying most of it by sprinkling. The problem of efficient water use arose during the first stage, but it became acute during the second stage, and especially towards the end of it, where Israel currently finds itself. The quantity of irrigation water in Israel is limited by the fact that water resources are being exploited to their near maximum. Consequently, any water saved in the process of producing a unit of agricultural product by withholding water per unit area without reducing yield, represents a tangible saving which permits the expansion of the irrigated area and hence total agricultural production (see Fig. 2).

It is evident that the amounts of water applied in traditional open irrigated agriculture can be decreased considerably. Reducing the irrigation applications by 25–50% will make this water available for use elsewhere, particularly in regions where the water situation is already critical, and will ease the pressure to develop new and expensive sources of water.

The necessary switch to closed irrigation methods to permit partial wetting and rationalization of water utilization and other production factors is not easy and involves large investments. The policy to be adopted in each situation will ultimately be determined by economic considerations.

The approaches developed to expand irrigated agriculture despite conditions of water scarcity and the steps taken to achieve optimum water utilization may serve as a guide to other arid and semi-arid countries in exploiting conventional water resources, and in due course, in utilizing desalinated water.

Literature

BERG, C., VAN DEN VISSER, W. C., KOVDA, V. A.: Water and salt balances. In: International Sourcebook on Irrigation and Drainage of Arid Lands in Relation to Salinity and Alkalinity. Draft edition, pp. 43–91. FAO/Unesco. Rome, Paris (1967).

BERINGER, C.: An economic model for determining the production function for water in agriculture. California Agricultural Experiment Station Research Report 240 (1961).

BLACK, C. A.: Crop yields in relation to water supply and soil fertility. *In* Plant Environment and Efficient Water Use. (W. H. PIERRE, D. KIRKHAM, J. PESEK, and R. SHAW, eds.) pp. 177–206. Amer. Soc. Agron. and Soil Sci. Soc. Amer., Madison, Wisconsin (1965).

BLANEY, B. F., CRIDDLE, W. D.: Determining consumptive use and irrigation water requirements. ARS, USDA Tech. Bull. 1275 (1962).

BRESLER, E., YARON, D.: Soil water regime in economic evaluation of salinity in irrigation. Water Resources Res. **8**(4), 791–800 (1972).

CLARK, C.: The Economics of Irrigation. 2nd ed. Oxford: Pergamon Press 1970.

CLAWSON, M., LANDSBERG, H. H.: Desalting Seawater: Achievements and Prospects. New York: Gordon and Breach 1972.

CHRISTIANSEN, J. E., DAVIS, J. R.: Sprinkler irrigation systems. in Irrigation of Agricultural Lands. Amer. Soc. Agron. Monograph **11**, 885–904 (1967).

COHEN, O. P.: A procedure for calibrating neutron moisture probes in the field. Israel J. agric. Res. **14**, 169–178 (1964).

COHEN, O. P., BRESLER, E.: The effect of non-uniform water application on soil moisture content, moisture depletion and irrigation efficiency. Soil Sci. Soc. Amer. Proc. **31**, 117–121 (1967).

DONNAN, W. W., HOUSTON, C. E.: Drainage related to irrigation management. *In* Irrigation of Agricultural Lands. Amer. Soc. of Agron. Monograph **11**, 974–987 (1967).

DOWNEY, L. A.: Water requirements of maize. J. Aust. Inst. Agric. Sci. (March, 1971), pp. 32–41. (1971)

FUCHS, M., STANHILL, G.: The use of Class A evaporation data to estimate the irrigation water requirements of the cotton crop. Israel J. agric. Res. **13**, 63–78 (1963).

FRAMJI, K. K., MAHAJAN, I. K.: Irrigation and drainage in the world. A global review. Int. Comm. on Irrig. and Drain. New Delhi (1969).

HAGAN, R. M., RIJOV, S. N., ASTON, M. M., BAVEL, C.M.H. van, RAHEJA, P. C.: Water plant growth and crop irrigation requirements. *in* International Sourcebook on Irrigation and Drainage of Arid Lands in Relation to Salinity and Alkalinity. Draft edition, pp. 282–342, FAO/Unesco, Rome, Paris. (1967).

HALL, W. A.: Performance parameters of irrigation systems. Amer. Soc. Agric. Engr., Trans. **3**, 75–76, 81 (1960).

HANSEN, V. E.: New concepts in irrigation efficiency. Amer. Soc. Agric. Engr. Trans. **3**, 55–57, 61, 64 (1960).

ISRAELSEN, O. W.: Irrigation Principles and Practices. lst. ed. New York: John Wiley 1932.

ISRAELSEN, O. W., HANSEN, V. E.: Irrigation Principles and Practices. Third printing. New York London-Sydney: John Wiley & Sons 1967.

JENSEN, M. E.: Programming irrigation for greater efficiency. *In* Optimizing the Soil Physical Environment Toward Greater Crop Yields. (D. HILLEL, ed.) pp. 133–161. New York: Academic Press 1972.

JENSEN, M. E., SWARNER, L. R., PHELAN, J. T.: Improving irrigation efficiencies. *in* Irrigation of Agricultural Lands. Amer. Soc. of Agron. Monograph **11**, 1120–1142 (1967).

MARSHALL, T. J.: Efficient management of water in agriculture. *in* Optimizing the Soil Physical Environment Toward Greater Crop Yields. (D. HILLEL, ed.) pp. 11–22. New York: Academic Press 1972.

REEVE, R. C., FIREMAN, M.: Salt problems in relation to irrigation. *In*: Irrigation of Agricultural Lands. Amer. Soc. of Agron. Monograph **11**, 988–1008 (1967).

SHMUELI, E.: The efficient use of water, and agricultural research. Haklauth B'Israel **12**(3), 20–27 (1966) (Hebrew with English summary)

SHMUELI, E.: The contribution of research to the efficient use of water in Israel agriculture. Z. f. Bewasserungswirtschaft **6**(1), 38–58 (1971).

SHMUELI, E., BIELORAI, H., HELLER, J., MANTELL, A.: Citrus water requirement experiments conducted in Israel during the 1960s. *In* Physical Aspects of Soil Water and Salts in Ecosystems. (A. HADAS *et al.*, eds.) Ecological Studies, Vol. 4, pp. 339–350 Berlin-Heidelber-New York: Springer 1973.

STANHILL, G.: Evaporation in Israel. Bull. Res. Counc. Israel. Vol. 11G. pp. 160–172 (1963).

TAYLOR, S. A.: Estimating the integrated soil moisture tension in the root zone of growing crops. Soil Sci. **73**, 331–340 (1952).

VIETS, F. G., Jr.: Increasing water use efficiency by soil management. *In*: Plant Environment and Efficient Water Use (W. H. PIERRE, D. KIRKHAM, J. PESEK, R. SHAW, eds.) pp. 259–274. Amer. Soc. of Agron. and Soil Sci. Soc. of Amer. Madison, Wisconsin (1965).

WADLEIGH, C. H.: The integrated moisture upon a root system in a large container of saline soil. Soil Sci. **61**, 225–238 (1946).

YARON, D.: Economic Criteria for Water Resource Development and Allocation. Monography. The Hebrew Univ. of Jerusalem, Part I. (mimeo) (1966).

Subject Index

Ecological Studies

Analysis and Synthesis

Edited by J. Jacobs, O. L. Lange, J. S. Olson, W. Wieser

Distribution rights for U. K., Commonwealth, and the Traditional British Market (excluding Canada): Chapman & Hall Ltd., London

Springer-Verlag Berlin Heidelberg New York

München Johannesburg London New Delhi Paris Rio de Janeiro Sydney Tokyo Wien

Volume 1

Analysis of Temperate Forest Ecosystems

Edited by D. E. Reichle
91 figs. XII, 304 pages. 1970
Cloth DM 52,–; US $ 21.40
ISBN 3-540-04793-X

The book answers the following basic questions about the ecological systems of the temperate forest zone of the earth: What are they? What do they do? What relations control them? What are the best uses of this zone, where so much of mankind dwells? Though suitable for general readers the book provides a thorough and detailed analysis of an ecosystem.

Volume 2

Integrated Experimental Ecology

Methods and Results of Ecosystem Research in the German Solling Project

Edited by H. Ellenberg
53 figs. XX. 214 pages. 1971
Cloth DM 58,–; US $ 23.80
ISBN 3-540-05074-4

This collective report by botany, zoology, agriculture climatology, and soil science experts in one of the IBP pilot projects provides a study of the functioning and productivity of forest and grassland ecosystems.

Volume 3

The Biology of the Indian Ocean

Edited by B. Zeitzschel in cooperation with S. A. Gerlach
286 figs. XIII, 549 pages. 1973
Cloth DM 123,–; US $ 50.50
ISBN 3-540-06004-9

The present volume contains much new information and some conclusions regarding the functioning and organization of the ecosystems of the Indian Ocean.

Volume 4

Physical Aspects of Soil Water and Salts in Ecosystems

Edited by A. Hadas, D. Swartzendruber, P. E. Rijtema, M. Fuchs, B. Yaron
221 figs. XVI, 460 pages. 1973
Cloth DM 94,–; US $ 38.60
ISBN 3-540-06 109-6

These collected research papers were read at a symposium in Rehovot, Israel. Theoretical and practical aspects are included and among the subjects covered are the physical aspects of the movement of water and ions in soil, the interactions of water with soil, evaporation from soil and plants, water requirements of crops, and the management of salinity.

Volume 7

Mediterranean Type Ecosystems

Origin and Structure

Edited by F. di Castri, H. A. Mooney
Approx. 88 figs.
Approx. 500 pages. 1973
Cloth DM 78,– US $ 32.00
ISBN 3-540-06106-1

This volume discusses the striking similarities between widely separated regions with mediterranean-type climate: the territories fringing the Mediterranean Sea, California, Central Chile and the southermost strips of South Africa and Australia. Similarities are not confined to climatic trends but are also reflected in the physiognomy of vegetation, in land use patterns, and frequently in the general appearance of the landscape.

Prices are subject to change without notice

Ecological Studies

In production:

Vol. 6: K. Stern and L. Roche: Genetics of Forest Ecosystems
Vol. 7: Mediterranean Type Ecosystems. Edited by F. di Castri and H. A. Mooney
Vol. 8: H. Lieth: Phenology and Seasonality Modeling

In preparation:

A. Baumgartner: Forest Energy Budgets
F. Golley: Tropical Ecology
A. Hasler: Land-Water Interactions
O. W. Heal and D. F. Perkins: The Ecology of British Moors and Mountain Grasslands
T. Kira et al.: Plants, Population and Logistic Growth
M. Kluge: Crassulacean Acid Metabolism
J. Kranz: Mathematical Analysis and Modeling of Epidemics
B. Osmond, O. Björkman and D. Andersen: Ecology and Ecophysiology of Atriplex
A. Poljakoff-Mayber: Plants in Saline Environments
B. Slavik: Methods of Studying Water Plant Relations
F. Wielgolaski: Scandinavian Tundra Biome